Microbial Biotechnology for Sustainable Pollution Control

Olubukola Oluranti Babalola
https://orcid.org/0000-0003-4344-1909
North-West University, South Africa

Olukayode Oladipo Amund
University of Lagos, Nigeria

Olayinka Olayemi Osuolale
https://orcid.org/0000-0001-6525-8502
Elizade University, Nigeria

IGI Global
Scientific Publishing
Publishing Tomorrow's Research Today

Vice President of Editorial Melissa Wagner
Director of Acquisitions Mikaela Felty
Director of Book Development Jocelynn Hessler
Production Manager Mike Brehm
Cover Design Jose Rosado

Published in the United States of America by
IGI Global Scientific Publishing
701 East Chocolate Avenue
Hershey, PA, 17033, USA
Tel: 717-533-8845 | Fax: 717-533-7115
Website: https://www.igi-global.com E-mail: cust@igi-global.com

Library of Congress Cataloging-in-Publication Data

LCCN: 2026032046 (CIP Data Pending)
ISBN13: 9798260029985
Isbn13Softcover: 9798260029992
EISBN13: 9798260030004

British Cataloguing in Publication Data
A Cataloguing in Publication record for this book is available from the British Library.

Table of Contents

Detailed Table of Contents

The rapid expansion of industrial, agricultural, and domestic activities has increased environmental contamination by persistent pollutants. Petroleum products, pharmaceuticals, petrochemicals, wood preservatives, and heavy metals pose serious ecological and public health risks due to their toxicity and persistence. Bioremediation has emerged as an eco-friendly and cost-effective alternative to conventional remediation technologies. Two key approaches, bioaugmentation and biostimulation, enhance pollutant degradation through the introduction of specialized microbial consortia or stimulation of indigenous microorganisms, respectively. Their effectiveness is influenced by factors such as pH, temperature, nutrient availability, oxygen levels, soil properties, and contaminant type. This chapter reviews the principles, applications, advantages, and limitations of both strategies, highlighting their importance in sustainable pollution management and in situ biotreatment.

Bioremediation is a sustainable and cost-effective approach for environmental cleanup, but single-strain systems often perform poorly under complex polluted conditions. Microbial consortia have therefore gained attention due to their synergistic interactions, metabolic diversity, and improved degradation efficiency. This review examines key mechanisms driving consortium performance, including metabolic cooperation, syntrophy, cross-feeding, and quorum sensing, which enhance stability and resilience. It also explores Artificially Designed Microbial Communities (ADMCs) as emerging tools for precision bioremediation using top-down and bottom-up approaches. Advanced technologies such as omics, synthetic biology, computational modeling, and bioreactor engineering are highlighted. Applications in hydrocarbon degradation, wastewater treatment, and heavy metal remediation demonstrate the practical value of microbial consortia. Several challenges and future prospects involving AI-guided engineering and integrated regulatory frameworks are extensively discussed.

The continued presence of hazardous contaminants remains a major global environmental challenge. Conventional remediation methods are often costly and may generate secondary environmental risks. In contrast, bioremediation provides a sustainable, eco-friendly, and cost-effective approach for pollutant removal. Recent advances in genetic engineering have enabled the development of genetically engineered microbes (GEMs) with enhanced capabilities for degrading environmental toxins. These engineered strains have shown significant potential in hydrocarbon degradation, heavy metal detoxification, and the breakdown of other xenobiotics. Coupled with modern genome-editing technologies, GEMs are increasingly used in environmental remediation, enabling innovative strategies for pollutant removal. Given the scale of global environmental problems, integrated technological approaches are essential. This review highlights recent advances, applications, and key challenges associated with the use of GEMs in environmental biotechnology.

Life on Earth solely depends on the state and quality of its environments. Technological advancements, unsustainable agricultural practices and industrialization are the contributing factors that influence the quality of life on earth. Industrialization and urbanization have immensely transformed the biosphere. However, there are certain drawbacks that we can highlight through these transformations. These include the diminution of natural resources, accumulation of pollutants and increased agro-industrial emissions such as chemicals, pesticides, chemical-based fertilizers and petroleum hydrocarbons. Recent advancements such as smart bioreactors, the use of genetically engineered microorganisms, and integrated treatment systems are at the forefront of modern bioremediation efforts and have expanded the versatility of microbial remediation. This chapter explores some of the advances in the use of bioreactors and engineered systems for microbial bioremediation in In-situ and Ex-situ applications for optimum removal of environmental pollutants.

The Anammox (anaerobic ammonia oxidation) wastewater treatment (WWT) process synchronously metabolizes ammonia nitrogen (NH3–N) and nitrite nitrogen (NO2–N) into dinitrogen gas (N2), is cost-effective and environmentally friendly, removing up to 90% of wastewater nitrogen. A Lab-scale expanded bed biofilm (biological mass attached to a solid surface) reactor (EBBR), equipped with ABDite® particles, was investigated for establishing the Anammox WWT process. Chemical and DNA analysis confirmed a successful establishment and that ammonia-oxidizing bacteria (AOB) Nitrosomonas spp. and Anammox bacteria - Brocadia, Kuenenia, and Jettenia spp. inhabited the EBBR-Anammox system; attaining a nitrogen loading rate (NLR), removal rate (NRR) and removal efficiency (NRE) of 9.0 kg N m^{-3} d^{-1}, 6.1 kg N m^{-3} d^{-1}, and 80 – 100% respectively. Establishment of the EBBR-Anammox process underscores its potential as a scalable technology for mitigating global eutrophic nitrogen water pollution, providing suitable treated WW for irrigation and reducing overall operational costs in the agricultural sector.

The textile industry drives economic growth and improves human welfare, yet it remains a significant source of industrial water pollution. Textile effluents contain dyes, salts, and auxiliary chemicals from dyeing, printing, and washing processes, many of which are toxic, carcinogenic, persistent, and resistant to degradation. These pollutants threaten human health and aquatic ecosystems. Conventional physicochemical treatments are often costly, energy-intensive, and generate secondary pollutants, limiting their sustainability. Biotechnological approaches are eco-friendly and cost-effective alternatives, but are limited by the complexity and toxicity of textile wastewater. This chapter reviews current biotechnological strategies for textile effluent management, emphasizing bacterial, fungal, and algal dye-decolourisation, their enzymatic degradation mechanisms, and factors influencing treatment performance. It also discusses biodegradation pathways, key operational

Preface

INTRODUCTION

Environmental pollution remains one of the most pressing global challenges of the twenty-first century. Rapid industrialization, urbanization, agricultural intensification, and expanding human activities have resulted in the widespread release of pollutants into terrestrial, aquatic, and atmospheric ecosystems. Persistent contaminants such as petroleum hydrocarbons, heavy metals, synthetic dyes, pharmaceuticals, pesticides, and other xenobiotic compounds continue to threaten biodiversity, ecosystem functionality, and human health. Conventional remediation technologies, while often effective, can be expensive, energy-intensive, and may generate secondary environmental impacts. Consequently, there is an increasing demand for sustainable, environmentally friendly, and economically viable alternatives for pollution management.

Microbial biotechnology has emerged as a transformative field offering innovative solutions for environmental restoration and pollution control. Microorganisms possess remarkable metabolic versatility, enabling them to degrade, transform, immobilize, or detoxify a wide range of contaminants. Advances in microbial ecology, molecular biology, systems biology, synthetic biology, and bioengineering have significantly expanded our understanding of microbial processes and their practical applications in environmental management. These developments have positioned microbial biotechnology at the forefront of sustainable remediation strategies worldwide.

Microbial Biotechnology for Sustainable Pollution Control brings together contributions from leading researchers and experts across multiple disciplines to provide a comprehensive overview of contemporary advances, emerging technologies, and future directions in microbial bioremediation. The volume explores fundamental principles, engineered approaches, industrial applications, technological innovations, policy considerations, and the integration of cutting-edge tools such as artificial intelligence and genome engineering. Collectively, the chapters demonstrate the immense potential of microbial systems to address environmental contamination while supporting sustainable development and resource recovery.

The book is organized into ten chapters that progressively examine the scientific foundations, technological advancements, practical applications, and governance dimensions of microbial pollution control.

CHAPTER OVERVIEW

Chapter 1 examines two fundamental bioremediation strategies that enhance pollutant degradation through either the stimulation of indigenous microbial populations or the introduction of specialized microbial communities. It discusses the principles, influencing factors, advantages, limitations, and practical applications of these approaches in sustainable pollution management.

Chapter 2 explores the role of microbial consortia in environmental remediation, emphasizing synergistic interactions, metabolic cooperation, syntrophy, cross-feeding mechanisms, and community engineering. The chapter also highlights advances in artificial microbial community design, omics technologies, synthetic biology, and computational approaches for improving bioremediation performance.

Chapter 3 focuses on genetically engineered microorganisms and their growing importance in environmental biotechnology. It reviews recent advances in microbial genetic engineering, applications in contaminant degradation, genome-editing technologies, and the challenges associated with biosafety, risk assessment, and regulatory oversight.

Chapter 4 discusses innovations in bioreactors and engineered remediation systems designed for both in situ and ex situ applications. Particular attention is given to smart bioreactors, integrated treatment technologies, and engineered microbial systems that improve remediation efficiency and scalability.

Chapter 5 presents the application of anaerobic ammonia oxidation technology within expanded bed biofilm reactor systems for nitrogen removal from wastewater. The chapter demonstrates the potential of this approach for addressing eutrophication while improving wastewater treatment sustainability and agricultural water reuse.

Chapter 6 reviews microbial and biotechnological approaches for treating textile wastewater. It examines bacterial, fungal, and algal processes involved in dye decolourisation and biodegradation, along with the operational factors and challenges influencing treatment effectiveness.

Chapter 7 investigates microbial remediation strategies for industrial wastes generated by mining, paint, and detergent industries. The chapter discusses various biological mechanisms for contaminant removal and highlights emerging technologies that enhance treatment efficiency and support resource recovery.

Chapter 8 examines the integration of microbial bioremediation within circular economy frameworks. It emphasizes waste-to-resource transformation strategies that combine environmental cleanup with resource recovery, promoting sustainability and reducing the environmental burden of waste generation.

Chapter 9 explores the future of microbial environmental biotechnology through the integration of artificial intelligence, CRISPR-based genome editing, and systems biology. The chapter discusses how these technologies can improve microbial discovery, engineering, predictive modeling, and the design of robust remediation systems while addressing associated ethical and regulatory considerations.

Chapter 10 addresses policy, governance, and regulatory dimensions of microbial bioremediation. It evaluates existing frameworks for risk assessment, biosafety, international standards, and stakeholder engagement, while proposing pathways toward harmonized global governance and responsible implementation of microbial technologies.

CONCLUSION

Together, these chapters provide a comprehensive perspective on the scientific, technological, and societal dimensions of microbial biotechnology for pollution control. By bridging fundamental research with practical applications and policy considerations, this volume serves as a valuable resource for researchers, environmental professionals, policymakers, industry practitioners, and postgraduate students interested in advancing sustainable solutions for environmental restoration.

The editors express their sincere appreciation to all contributing authors for sharing their expertise and insights. Their collective efforts have produced a volume that reflects the breadth, diversity, and rapid evolution of microbial biotechnology. It is our hope that this book will stimulate further research, foster interdisciplinary collaboration, and contribute meaningfully to the development of innovative and sustainable approaches for protecting environmental quality and promoting global ecological resilience.

Chapter 1
Biostimulation and Bioaugmentation Enhancing Native and Introduced Microbial Communities for Pollution Control

Kawther Hkeem Ibrahem
http://orcid.org/0009-0000-8788-2437
Department of Soil and Water Resources, College of Agriculture, University of Kirkuk, Kirkuk, Iraq

Swagata Ghosh
Applied Sciences Department, Symbiosis Institute of Technology, Symbiosis International (Deemed to be) University, India

Nidhi Bhardwaj
Faculty of Materials Science and Technology, VŠB–Technical University of Ostrava, Ostrava-Poruba, Czech Republic & Department of Chemistry, Faculty of Science, University of Hradec Kralove, Hradec Kralove, Czech Republic

Soumya Ghosh
http://orcid.org/0000-0002-4945-3516
Natural and Medical Sciences Research Center, University of Nizwa, Nizwa, Oman & University of the Free State, South Africa

DOI: 10.4018/979-8-2600-2998-5.ch001

ABSTRACT

The rapid expansion of industrial, agricultural, and domestic activities has increased environmental contamination by persistent pollutants. Petroleum products, pharmaceuticals, petrochemicals, wood preservatives, and heavy metals pose serious ecological and public health risks due to their toxicity and persistence. Bioremediation has emerged as an eco-friendly and cost-effective alternative to conventional remediation technologies. Two key approaches, bioaugmentation and biostimulation, enhance pollutant degradation through the introduction of specialized microbial consortia or stimulation of indigenous microorganisms, respectively. Their effectiveness is influenced by factors such as pH, temperature, nutrient availability, oxygen levels, soil properties, and contaminant type. This chapter reviews the principles, applications, advantages, and limitations of both strategies, highlighting their importance in sustainable pollution management and in situ biotreatment.

1 INTRODUCTION

Figure 1. Treatment of Contaminated Soil using Biostimulation and Bioaugmentation

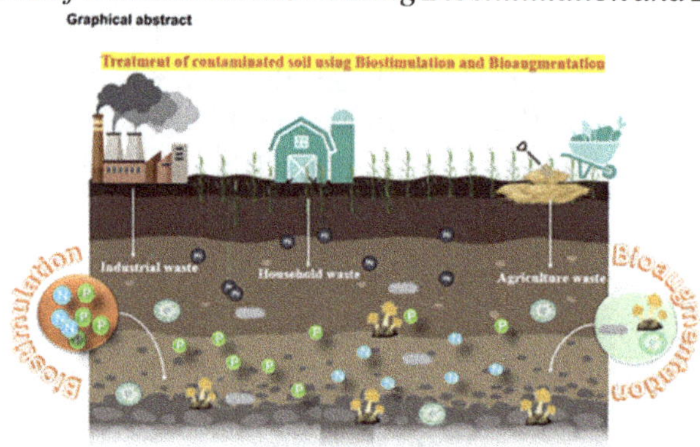

Environmental degradation has become a global concern, largely driven by industrial chemical production and improper waste disposal. Pollutants originating from oil refineries, gas stations, pharmaceuticals, petrochemical industries, and wood preservatives contribute significantly to soil and water contamination. Among these, aromatic compounds are particularly hazardous due to their mutagenic and carcinogenic properties (Mohd Noor, 2021, Nematollahi et al., 2022). Contaminated soils, especially those with fine-grained, clay-rich horizons, act as long-term pollutant reservoirs and present serious ecological and public health risks (Parajuli et al., 2018, Roslund et al., 2019, Khurshid et al., 2022).

Conventional remediation methods are often costly, energy-intensive, and associated with secondary environmental impacts, including the emergence of resistant microbes and the loss of beneficial microflora. Consequently, bioremediations, a biological approach including bioaugmentation, and biostimulation are gaining prominence due to their cost-effectiveness, sustainability, and selectivity (Puteri et al., 2025, Mrozik and Piotrowska-Seget, 2010). These techniques employ plants and microorganisms to degrade or transform pollutants into less toxic forms, with microbes generally preferred over plants due to their faster reproduction rates, greater adaptability, and superior ability to detoxify heavy metals and organic contaminants (Sonune, 2021, Hussain et al., 2022). Soil microbial communities, composed of diverse bacterial groups, play a central role in breaking down hydrocarbons such as PAHs into simpler molecules (Li et al., 2024, Patel et al., 2021). However, individual strains often lack the full enzymatic capacity for complete degradation, making microbial consortia more effective in ensuring thorough mineralization into CO_2, H_2O, or CH_4 (Ali et al., 2025).

Bioremediation technologies exploit these microbial processes to sustainably detoxify soil, water, and air without generating harmful by-products (Adams et al., 2015). Microorganisms, particularly bacteria, employ adaptive mechanisms, such as specialized membranes, extracellular polymeric substances (EPS), siderophores, and enzyme-based systems, to survive in toxic environments and transform contaminants (Joshi et al., 2023, Nnaji et al., 2024). When combined, bioaugmentation and biostimulation serve as complementary strategies in environmental biotechnology, providing efficient, economical, and eco-friendly alternatives to conventional remediation while restoring ecological balance (Sarkar et al., 2016, Mohamed and Samer, 2023). Bioaugmentation introduces specialized or engineered microbes with enhanced catabolic abilities to complement indigenous communities, while biostimulation enhances native microbial activity by optimizing nutrients and environmental conditions (Muter, 2023, Abbas et al., 2024).

2 BIOAUGMENTATION

Bioaugmentation has emerged as a cost-effective and promising bioremediation strategy, wherein efficient bacterial isolates or microbial consortia capable of degrading petroleum hydrocarbons are introduced into contaminated soils. This approach aims to supplement the native microbial population, thereby enhancing the overall efficiency of biodegradation (Zawierucha and Malina, 2006; Chettri et al., 2024). A key advantage of using microbial consortia over individual strains is their ability to overcome the inherent biological limitations of single organisms. Through complementary metabolic functions, consortia form a more resilient and diverse system, capable of adapting to a wider range of ecological and environmental conditions (Ram et al., 2022). Such diversity not only improves pollutant degradation but also stabilizes the microbial community under stress factors, including variations in pH, temperature, salinity, or nutrient availability. Moreover, synergistic interactions such as cross-feeding and cooperative enzymatic activity broaden the range of degradable substrates, making microbial consortia particularly effective for tackling complex and recalcitrant contaminants (Tabacchioni et al., 2021, Nunes et al., 2024). For successful bioaugmentation, microbial strains or consortia must exhibit rapid growth, ease of cultivation, tolerance to harsh environments (pH, temperature, salinity, nutrient availability), and the ability to withstand elevated pollutant levels (Table 1 and 2). Bioaugmentation is especially critical when indigenous microbes in soil or water display limited activity and slow pollutant removal rates (Figure 1). Although indigenous microflora are often multiplied for this purpose, the application of exogenous or genetically modified microbial cultures continues to raise concerns regarding ecological safety and sustainability (Zawierucha and Malina, 2014).

Figure 2. Bioaugmentation: a process for biological treatment of pollutants

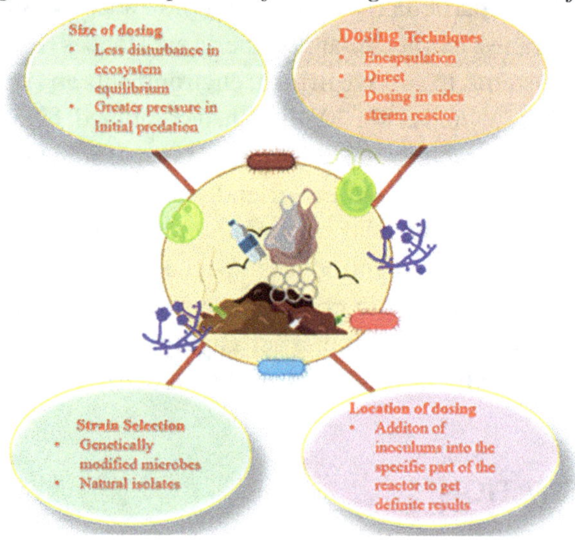

2.1 Micro-Organisms

Specialized microorganisms used in bioaugmentation for bioremediation can be naturally occurring, genetically modified strains, or microbial consortia specifically adapted to target pollutants. Their metabolic versatility is critical for enhancing bioremediation efficiency (El Fantroussi and Agathos, 2005). For instance, certain bacterial strains exhibit unique enzymatic pathways that enable the degradation of oil hydrocarbons into less toxic byproducts, thereby contributing significantly to oil spill cleanup and mitigation of industrial emissions. Fungi, on the other hand, are widely recognized for their exceptional ability to degrade complex organic pollutants through diverse extracellular enzymatic systems (Elshafei and Mansour, 2024). In particular, ligninolytic fungi, especially white-rot fungi, possess highly efficient oxidative enzymes such as laccases, lignin peroxidases (LiP), manganese peroxidases (MnP), and versatile peroxidases that enable the breakdown of recalcitrant organic compounds (Pointing, 2001, Janusz et al., 2017). These enzymes are non-specific in nature and can oxidize a broad range of pollutants, including polycyclic aromatic hydrocarbons (PAHs), synthetic dyes, pesticides, plastics, and petroleum hydrocarbons. White-rot fungi are especially effective due to their natural ability to decompose lignin, one of the most structurally complex aromatic polymers found in plant biomass (Arora and Sharma, 2010). During the biodegradation process, fungal hyphae penetrate contaminated substrates, increasing enzymatic contact with pollutants and enhancing mineralization efficiency. Furthermore, fungal-mediated degradation often results in lower environmental toxicity and improved bioremediation outcomes in contaminated soils and wastewater systems (Harms et al., 2011, Bilal et al., 2019). Therefore, selecting suitable microorganisms according to the chemical nature of pollutants and prevailing environmental conditions is essential for effective biomanipulation strategies (Rafeeq et al., 2023). In bioaugmentation, exogenous microorganisms are introduced to accelerate pollutant degradation by metabolizing chemicals directly. This approach is commonly applied in municipal wastewater treatment to activate sludge bioreactors. Various microorganisms, including *Bacillus licheniformis*, *P. polymyxa*, *B. stearothermophilus*, *Flavobacterium*, *B. thuringiensis*, *Penicillium* spp., *Aspergillus* spp., *Pseudomonas*, *Arthrobacter*, *Streptomyces*, and *Saccharomyces*, have been employed in bioaugmentation (Table 2). This method significantly increases both the efficiency and speed of pollutant removal from soil and water.

2.2 Mechanism of Bioaugmentation

The significance of bioaugmentation in bioremediation lies in its ability to enhance natural biodegradation processes, which are often too slow or ineffective at contaminated sites (Lebeau, 2011). Native microbial communities may lack the genetic traits required to efficiently degrade specific pollutants such as hydrocarbons, heavy metals, and other toxic compounds. Introducing specialized microorganisms with targeted metabolic pathways strengthens the community's capacity to degrade these pollutants more effectively (Mawang et al., 2021, Behera and Das, 2023). This approach not only reduces the concentration of hazardous substances in the environment but also accelerates the remediation timeline, underscoring the practical value of bioaugmentation (Figure 2).

Figure 3. Pathways of MICP (Microbially Induced Calcite Precipitation) bioaugmentation and biostimulation

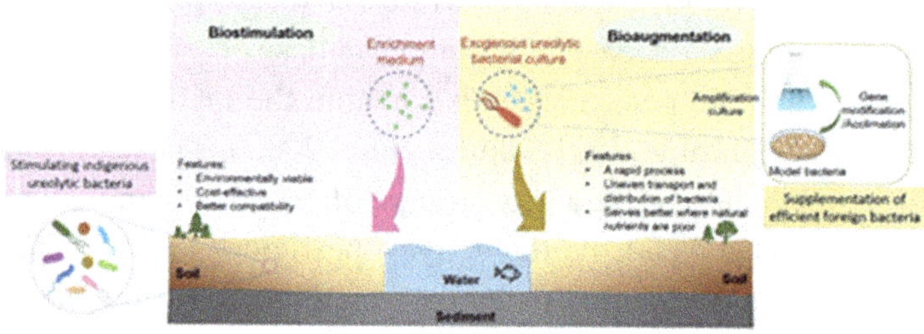

Beyond direct degradation, specialized microorganisms may also alter geochemical conditions to facilitate remediation. For example, anaerobic bacteria such as methanogens can reduce heavy metals like lead and chromium by mediating electron transfer, thereby converting them into less toxic forms (Aragaw, 2021, Chakraborty et al., 2025). The incorporation of *Pseudomonas putida* into oil-contaminated soil has been shown to accelerate hydrocarbon degradation (Aziz et al., 2024). Similarly, a previous study by Sharma et al. (Sharma et al., 2016) reported that in situ bioremediation using a microbial consortium of *Serratia marcescens*, *Streptomyces rochei*, and *Phanerochaete chrysosporium* achieved 60–70% degradation of polyaromatic hydrocarbons under controlled conditions. In the context of biostimulation under natural field conditions, soil amendments such as ammonium sulfate, paddy straw, and compost enhanced degradation rates, reaching 56–98% within seven days and up to 83.5–100% after 30 days. Another study demonstrated that *Rhodococcus* bioaugmentation was an efficient strategy for enhancing bacterial network interactions and promoting DBT degradation in petroleum-contaminated soils (Gentry et al., 2004, Chen et al., 2023).

Overall, the effectiveness of these strategies depends on multiple factors, including environmental conditions, the compatibility of introduced strains with indigenous microbial communities, and the chemical nature of the pollutants.

Table 1. Factors influencing bioaugmentation and biostimulation processes

Factor	Description	References
Loss of microbial viability during inoculation	Stress due to drastic changes in environmental conditions	(Tyagi et al., 2011).
Eventual cell death after inoculation	Depletion of nutrients or toxicity caused by contaminants	(Du and Liu, 2012).
Competition	Competition for nutrients by autochthonous microbes	(Thompson et al., 2005)
Predation	Bioaugmentation followed by disappearance of bacterial population and protozoa overgrowth	(Bouchez et al., 2000).
pH	Extremes in pH inhibit microbial degradation processes	(Chandra et al., 2013)
Temperature	Temperature influences microbial growth and degradation potential, and also the physical and chemical nature of oil	(Ribicic et al., 2018)
Moisture	Low moisture content limits microbial growth and metabolism; higher values reduce soil aeration	(Guo et al., 2012)

Table 2. Selected bacteria used in bioaugmentation studies

Microorganism used	Pollutants degraded	References
Pseudomonas fluorescens F113rifpcbrmBP1:gfpmut3	Biphenyl, polychlorinated biphenyl	(Chen et al., 2023)
Rhodococcus sp. Strain RHA1	4-chlorobenzoate	(Rodrigues et al., 2001)
Escherichia coli AtzA	Atrazine	(Strong et al., 2000)
Pseudomonas sp. / Pseudomonas putida B13ST1 (pPOB)	3-phenoxybenzoic acid	(Halden et al., 1999)
Pseudomonas fluorescens F113rifPCB	Biphenyl, polychlorinated	(Halden et al., 1999)
Pseudomonas fluorescens CS2	Biphenyl, Ethylbenzene	(Brazil et al., 1995)
Pseudomonas putida BCRc14349	Phenol, trichloroethane	(Chen et al., 2007)
Rhodococcus sp. F92	Various petroleum products	(Quek et al., 2006)
Arthrobacter, Burkholderia, Pseudomonas, Rhodococcus etc.	Petroleum hydrocarbons	(Salleh et al., 2003)

2.3 Advantages of Bioaugmentation

Bioaugmentation presents several advantages that contribute to its growing acceptance as a viable bio-restoration strategy. One of the main advantages is the potential for improving biodegradation rates. The use of specialized microorganisms suitable for growing in specific contaminated environments allows for more effective degradation of pollutants (Table 2). For example, studies indicate that bioaugmented systems can significantly increase the speed at which hydrocarbons are degraded in soils contaminated by oil, frequently reaching degradation efficiencies that are considerably higher than those observed in natural mitigation processes (Handore et al., 2023, Kuppan et al., 2024).

Targeted sanitation is another essential advantage of bioaugmentation. By selecting microorganisms with specific metabolic capacities, it becomes possible to treat particular contaminants that may be present on a site. This specificity permits a more personalized approach to sanitation compared to traditional methods, This acceleration of biodegradation is essential for timely sanitation, particularly in situations where rapid cleaning is necessary to mitigate environmental and health risks (Nzila et al., 2016).which can move towards a larger range of pollutants without the targeted efficiency that bioaugmentation can provide (Muter, 2023, Thompson et al., 2005)For example, the introduction of genetically modified microbes capable of degrading chloroaromatic compounds has demonstrated significant success in contaminated groundwater systems. The capacity to target particular compounds reduces global processing time and improves the efficiency of the bioaugmentation strategy (Angeles-De Paz et al., 2025, Rafeeq et al., 2023).

In addition, bioaugmentation can contribute to the restoration of ecological balance in contaminated ecosystems. By restoring or improving indigenous microbial populations, bioaugmentation can restore the dynamics of the essential natural microbial community to preserve ecosystem functions. This method not only facilitates the degradation of pollutants but also allows the recovery of beneficial microbial activities that support soil health and water quality (Kuppan et al., 2024). For example, the bioaugmentation strategies that focus on improving the populations of bacteria fixing nitrogen in contaminated soils have shown promising results in improving soil fertility along with the degradation of contaminants, thus promoting ecological recovery (Qattan, 2025).

Overall, the advantages of bioaugmentation, including improved biodegradation rates, targeted sanitation, ecological catering, and profitability, position it as a promising approach to meet the challenges posed by environmental contamination. While research continues to evolve and refine these methodologies, bioaugmentation can play an increasingly essential role in sustainable environmental management practices., Bioaugmentation, while an innovative approach to bioremediation, is not exempt from limitations and challenges (Cuadros-Pedraza et al., 2021). An important problem is the survival of the microorganisms introduced in the objective environment. It is well documented that environmental stressors, such as temperature fluctuations, pH levels, salinity, and nutrient availability, can negatively affect microbial viability. Many of the specialized microorganisms used in bioaugmentation are selected for their specific degradative abilities; However, their ability to prosper outside the controlled laboratory conditions remains uncertain (Kochhar et al., 2022). Hostile environments typical of contaminated sites, that is, landfills, oil spills, or heavy metal-loaded soils, often impose incompatible conditions with the survival of native species. In addition, indigenous microbial communities can overcome the introduced microorganisms, undermining their effectiveness and, in some cases, leading to a complete washing. The successful application of bioaugmentation requires not only the introduction of specialized microorganisms but also the optimization of environmental parameters, such as oxygen levels, nutrient availability, and moisture content. These factors are crucial to promote microbial growth and activity (Ahmed et al., 2024).

2.4 Factors Influencing Bioaugmentation

Microbial metabolism under aerobic or anaerobic conditions can mineralize many organic pollutants to carbon dioxide and water, or transform them into intermediate compounds with substantially lower toxicity (Ayilara and Babalola, 2023, Adetunji and Anani, 2021). The rhizosphere, the soil zone influenced by plant roots, supports dense and active microbial populations and supplies organic exudates that stimulate biodegradation processes, increase contaminant bioavailability, and enhance overall remediation efficiency and vegetation growth (Romantschuk et al., 2000).

The success of the bioaugmentation process largely depends on the ability of the introduced microbial consortia to survive and establish within the polluted site, while competing with the native microbial community and withstanding ecological pressures such as predation and interspecific competition. To provide effective physical support for the microbial biomass introduced into a polluted site, an appropriate carrier should be employed to ensure uniform distribution and accessibility, while also supplying adequate nutrients, moisture, and aeration to sustain microbial activity (Mrozik and Piotrowska-Seget, 2010, Nikolopoulou et al., 2013). To ensure the long-term survival of microbial cultures under stressful environmental conditions, bacterial or fungal strains are often encapsulated or immobilized. This approach not only maintains their viability over extended periods but also, when coupled with a continuous supply of nutrients, enhances their metabolic activity, resulting in more rapid and efficient pollutant degradation (Adams et al., 2015).

A variety of materials, including gelatin, agarose, alginate, agar, and polyurethane, have been employed to protect seeded microbial cultures from the inhibitory effects of toxic compounds, as well as from competition and predation. A previously reported study (Liu et al., 2009) made a comparison between immobilized cells of *Acinetobacter* sp. and *Sphingomonas* strains isolated from activated sludge-contaminated soil for the biodegradation of phenol. The results demonstrated that a mixed consortium of the two strains exhibited superior degradation efficiency compared to pure free-cell cultures alone (Goswami et al., 2018). Several parameters regulate the rate of bioaugmentation in soil, including temperature, pH, aeration, organic matter, nutrient availability, and soil type (Table 1). The absence or imbalance of any of these factors can significantly reduce the efficiency of the degradation process, leading to insufficient outcomes and prolonging the time required compared to well-controlled conditions. Soil pH, whether acidic, neutral, or alkaline, plays a critical role in regulating microbial population dynamics and metabolic activity in polluted environments. It not only influences microbial growth but also affects nutrient availability and the activity of key enzymes involved in contaminant degradation. Even slight variations in pH can significantly alter microbial metabolism, making soil pH a valuable predictor of microbial growth and bioremediation potential at a given site (Rajkumar and Kurinjimalar, 2021). One of the most critical factors influencing the efficiency of bioaugmentation is the presence of organic matter, which plays a key role in determining the bioavailability of pollutants. Organic materials not only affect the survival of the seeded microbial cultures but also enhance their ability to degrade contaminants. In addition, the presence of water in soil plays a crucial role in determining the rate of degradation . Water availability directly influences microbial activity by regulating the diffusion of substrates into cells and supporting normal metabolic processes. Insufficient moisture can lead to cellular dehydration of the seeded isolates, thereby reducing their survival rate and ultimately decreasing the efficiency of pollutant removal from the environment (Mashreghi and Prosser, 2006). Moisture in the environment directly influences the solubility and availability of heavy metals in water, as well as associated factors such as pH and osmotic pressure. Excessive moisture can reduce microbial degradation rates by creating anaerobic conditions, which hinder the survival and activity of aerobic microorganisms (Imron et al., 2020). Among environmental factors, two of the most critical determinants influencing the bioremediation of petroleum hydrocarbon–contaminated soils are temperature fluctuations and nutrient availability, particularly in high-latitude regions. Numerous microbial communities possess an inherent capacity to degrade hydrocarbon contaminants in soil ecosystems. Among these, bacterial genera such as *Pseudomonas*, *Acinetobacter*, *Bacillus*, *Rhodococcus*, and *Stenotrophomonas*, along with fungal taxa including *Candida*, *Torulopsis*, and other yeast species, have been extensively recognized for their hydrocarbon-degrading potential. These microorganisms contribute to hydrocarbon biodegradation primarily through the synthesis of biosurfactants, particularly glycolipid-type compounds, which may be produced both extracellularly and intracellularly. Biosurfactants enhance the bioavailability and emulsification of hydrophobic hydrocarbons, thereby facilitating their microbial uptake and subsequent metabolic degradation (Van Stempvoort and Biggar, 2008, Gomez and Sartaj, 2013) .Temperature plays a pivotal role in regulating microbial metabolism and activity, in addition to influencing the chemical and physical properties of hydrocarbons. At lower temperatures, microbial activity is significantly reduced, which in turn decreases the efficiency of pollutant degradation in soil (Coulon et al., 2005).Another challenge in the degradation of hydrocarbon pollutants in hot and arid soils is the effect of salinity, which strongly inhibits microbial growth and prolongs the adaptation period required for survival under such conditions. Salinity also decreases the efficiency of microbial degradation processes. The existence of salinity has an effect on the hydrocarbonoclastic activity of the halotolerant and halophilic microbes, and it also exposes the soil microorganisms to stress conditions (Kebede et al., 2021). To address these limitations, integrated bioremediation approaches involving both bioaugmentation and biostimulation are frequently employed to enhance contaminant degradation efficiency. Bioaugmentation involves the introduction of specialized pollutant-degrading microorganisms, whereas biostimulation enhances the activity of indigenous microbial communities through nutrient amendment and optimization of environmental conditions. The effectiveness of these bioremediation processes is commonly evaluated using biological indicators such as soil basal respiration, microbial abundance, and enzymatic activity, as reported in previous studies (Garousin et al., 2021, Li et al., 2021). Furthermore, contaminants adsorbed onto clay mineral surfaces may undergo desorption, subsequently migrating into groundwater systems or volatilizing into the atmosphere, thereby posing additional environmental risks . Both pathways can produce direct and indirect exposures that adversely affect human health. Therefore, understanding the physicochemical interactions between contaminants and clay minerals, including sorption/desorption equilibria, changes in redox conditions, and temperature-driven volatilization, is essential for accurate exposure assessment and the design of effective remediation strategies ((Romantschuk et al., 2023). Microorganisms, whether as native populations or introduced through processes such as bioaugmentation and biostimulation, play a vital role in bioattenuation. To achieve optimal results, it is essential to carefully plan and consider various factors, including the characteristics of the contaminated water or soil, the type and concentration of microbial populations, and the nature and levels of pollutants present. In the case of biodegradable organic compounds, such as petroleum hydrocarbons (PHCs), bacteria and archaea are particularly important in driving bioremediation also fungi, algae, protozoa, and viruses are the most widely used destroyed of hydrocarbon contaminants (Ubani et al., 2016, Towell et al., 2011). To enhance and accelerate this process, microbial activity and proliferation must be optimized by providing the most favorable environmental conditions (Cavazzoli et al., 2022, Cavazzoli et al., 2023).Genetic engineering plays an important role in modifying microorganisms already present in polluted sites, particularly when native strains lack the natural ability to degrade specific contaminants. Through genetic modifications, their degradative capacity can be enhanced, or entirely new metabolic pathways can be introduced. In some cases, non-indigenous genetically engineered strains are introduced to stimulate degradation. However, the effectiveness of this approach largely depends on the ability of the engineered microorganisms to survive, adapt, and compete with resident microbial communities in the contaminated environment (Fashola et al., 2016, Maglione et al., 2024).

Aerobic degradation is generally more effective than anaerobic degradation; therefore, stimulating the process by introducing oxygen, either directly into the soil or the aquatic environment, is crucial for accelerating the overall rate of contaminant breakdown (Kauppi et al., 2011). Numerous studies have investigated the relationship between soil enzymatic activity and the concentration of petroleum hydrocarbons, particularly in chronically contaminated soils. These studies highlight that effective degradation depends not only on the presence of hydrocarbon contaminants but also on their influence on the functional activity of soil microbial communities (Ossai et al., 2020, Mekonnen et al., 2024).

3 BIOSTIMULATION

Biostimulation as an eco-friendly and cost-effective remediation strategy that involves the addition of nutrients (such as nitrogen, carbon, phosphorus, and potassium), electron donors or acceptors, and in some cases enzymes or biosurfactants, to enhance the activity of indigenous microorganisms present in polluted environments. Compared to bioaugmentation, biostimulation is often considered more favorable, as it maintains the natural microbial diversity in the soil while promoting efficient pollutant degradation (Goswami et al., 2018, Sayed et al., 2021). A previous study by Perfumo et al. (Perfumo et al., 2007) has emphasized that the addition of nutrients, oxygen, or other electron donors and acceptors to contaminated environments can stimulate the activity of indigenous microbial populations, thereby enhancing bioremediation processes (Perfumo et al., 2007). Similarly, another study (Margesin and Schinner, 2001) described biostimulation as a natural remediation strategy that promotes the degradation of pollutants by optimizing environmental conditions such as aeration, nutrient availability, pH balance, and temperature control (Margesin and Schinner, 2001). This process refers to the stimulation of indigenous microorganisms within the local environment through the provision of appropriate nutrients and the optimization of physicochemical conditions, thereby enhancing their metabolic activity to accelerate the removal of hazardous substances and improve the efficiency of contaminant degradation. Biostimulation is more widely employed for the remediation of wastewater, soil, and groundwater, as it enhances the activity of indigenous microbial consortia. In contrast, bioaugmentation involves the introduction of exogenous microorganisms into the environment, which may cause greater disruption to the natural ecosystem (Selvasembian et al., 2023, Burns et al., 2013). Biostimulation focuses on sustaining indigenous microbial populations and enhancing the environmental conditions through various strategies, such as supplying adequate oxygen, maintaining a pH range suitable for local microorganisms, and providing essential nutrients to support their activity (Juárez-Maldonado et al., 2021). In some cases, the addition of sufficient nutrients, such as carbon and nitrogen sources, along with adequate oxygen supply to indigenous microorganisms, accelerates the degradation process, enabling pollutant removal in a significantly shorter time compared to natural attenuation, which may otherwise take decades to achieve (Toczyłowska-Mamińska et al., 2025). Furthermore, combining biostimulation with bioaugmentation has been shown to accelerate degradation processes and enhance the overall effectiveness of soil remediation (Varjani and Upasani, 2019). Elevated temperatures increase the diffusion rates of organic compounds and reduce their viscosity; however, higher temperatures also enhance the solubility and toxicity of hydrocarbons. These combined effects ultimately result in a reduction in remediation efficiency and prolong the time required for effective treatment (Zhang et al., 2021). Biostimulation showed an insignificant effect on the reduction of oil content in the soil. Microorganisms present in oil-contaminated soils, such as those polluted with diesel, are often unable to degrade hydrocarbons that are strongly adsorbed onto soil particles. Consequently, the bioavailability of hydrophobic compounds becomes a major limiting factor in the biodegradation process (Chaudhary et al., 2021). Microorganisms are capable of producing a variety of metabolic products, including organic acids, biosurfactants, and polymeric materials such as polysaccharides and polyphosphates, which play an important role in bioremediation. These metabolites contribute to the removal of pollutants from the environment. In particular, organic acids enhance the bioavailability, mobility, and solubility of metals, thereby facilitating their transformation or removal (Saha et al., 2021). Biosurfactants are structurally diverse surface-active biomolecules, including compounds such as polymyxins, glycolipids, lipopeptides, and glycoproteins, which significantly enhance the bioavailability of hydrophobic pollutants. By reducing surface and interfacial tension at the hydrocarbon–water interface, these biomolecules promote pollutant solubilization, emulsification, and dispersion, thereby improving microbial accessibility and facilitating subsequent biodegradation processes (Ojuederie and Babalola, 2017) . The major advantage of biostimulation lies in the utilization of indigenous microorganisms, which are environmentally compatible and widely distributed (Figure 3). This approach is cost-effective, sustainable, and applicable across diverse environments. Moreover, it promotes the in situ degradation of contaminants without causing ecological disturbance, thereby enhancing the overall efficiency of remediation processes (Goswami et al., 2018, Curiel-Alegre et al., 2022).

Figure 4. Effects of biodegradation, biotoxicity, and microbial community on biostimulation of sulfolane

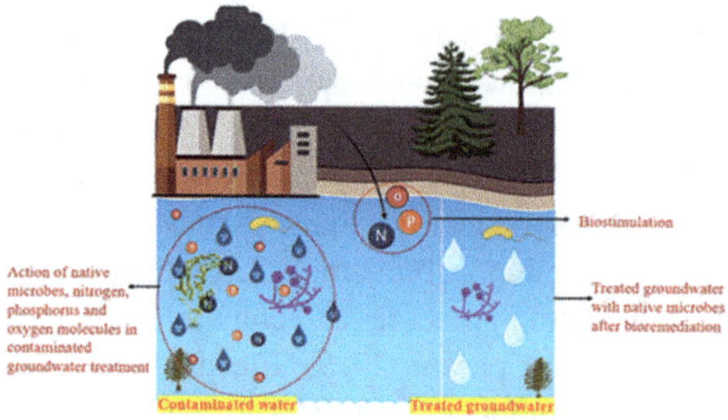

3.1 Factors Influencing Biostimulation

Biostimulation is considered an environmentally friendly, safe, and cost-effective approach compared to many other remediation techniques used for treating environmental contaminants. A key advantage of this strategy is its applicability across diverse environments, where it enhances the natural capacity of indigenous microorganisms to degrade harmful pollutants. Importantly, biostimulation promotes contaminant breakdown internally within ecosystems without causing additional environmental disturbance or introducing external hazards (Gaylarde and da Fonseca, 2025) . The effectiveness of biostimulation in remediation processes is strongly influenced by environmental factors such as temperature, moisture content, and pH, as well as the prevailing ecological conditions that shape microbial activity (Table 3) (Abdulsalam and Omale, 2009). These variables directly affect the efficiency of biodegradation, particularly in marine ecosystems, where remediation is more challenging than in soil environments. In aquatic systems, microorganisms often face difficulties in degrading complex polymers due to dilution effects and wave action, which reduce the concentration of pollutants available for microbial metabolism (Zawierucha and Malina, 2011). Moreover, the addition of nutrients to polluted soils is not always beneficial; in some cases, excessive nutrient supplementation can have adverse effects by reducing remediation efficiency (Cosgrove et al., 2010) For instance, studies by Nikolopoulou and Kalogerakis demonstrated that elevated concentrations of phosphorus and nitrogen can trigger eutrophication, leading to algal overgrowth and subsequent oxygen depletion, ultimately causing the death of aquatic organisms (Nikolopoulou and Kalogerakis, 2009). Therefore, optimizing biostimulation requires careful balancing of nutrient inputs, ensuring that the addition of limiting elements supports microbial degradation without causing ecological imbalances. The major pollutants that can be effectively addressed through biostimulation at contaminated sites include petroleum hydrocarbons, polyester-based polyurethanes, and sulfate-contaminated groundwater, all of which pose serious risks to both human health and ecosystems. Biostimulation has shown considerable potential in facilitating the degradation of these pollutants by enhancing the activity of indigenous microbial communities (Hamdi et al., 2007). In particular, synthetic polymers derived from various industrial and commercial products often contain intramolecular bonds, such as ester linkages, that are structurally similar to those present in natural macromolecules. This similarity enables microorganisms to recognize and enzymatically cleave these bonds, thereby accelerating the degradation process (Cosgrove et al., 2010).

4 COMBINED APPROACH: BIOSTIMULATION AND BIOAUGMENTATION

Bioremediation can be enhanced through two main approaches: bioaugmentation and biostimulation. The integrated application of biostimulation and bioaugmentation markedly enhances the degradation of environmental pollutants, particularly those that are recalcitrant and difficult to eliminate from soil and aquatic systems such as atrazine and pentachlorophenol (Table 3, and 4) (Dehnavi and Ebrahimipour, 2022). Both exogenous and indigenous microbial populations benefit from biostimulation, which involves the addition of electron acceptors or energy sources during the remediation process. A previous study by El Fantroussi and Agathos (El Fantroussi and Agathos, 2005) reported that combining bioaugmentation and biostimulation is particularly effective for the removal of pollutants from environments with high atrazine concentrations. Specifically, the use of citrate, succinate, and Pseudomonas sp. demonstrated promising results in enhancing degradation efficiency (El Fantroussi and Agathos, 2005). The addition of specific microorganisms along with essential nutrients significantly accelerates the degradation of pollutants at contaminated sites, achieving higher efficiency compared to the application of either approach alone, as shown in (Table 4) (Oro et al., 2024). Bioaugmentation demonstrated a marked enhancement in soil remediation when native hydrocarbon-degrading microorganisms were absent, showing greater effectiveness than biostimulation alone. A recent study (Liu et al., 2023) reported that the combined application of bioremediation strategies using *Achromobacter* and *Pseudomonas* for oil-contaminated soils significantly enhanced the remediation process, resulting in faster degradation and reduced treatment time compared to natural attenuation alone (Liu et al., 2023) (Figure 4). Similarly, another study (Chen et al., 2023) demonstrated that bioaugmentation with *Rhodococcus* was a simple and effective approach to improve the clustering of the bacterial network and enhance dibenzothiophene (DBT) degradation in petroleum-contaminated soil compared to natural attenuation (Chen et al., 2023). Effective bioremediation is typically achieved when environmental conditions at the contaminated site are well controlled, and this can be further enhanced through the use of specific bacterial or fungal strains, often genetically engineered and applied as pure cultures. The efficiency of the process also depends on employing suitable delivery vectors capable of transferring these microorganisms at high concentrations, with engineered strains carrying targeted genes to optimize pollutant degradation (Silva et al., 2004).

Figure 5. Flow chart of Biostimulation-bioaugmentation joint remediation process to deal with diesel-contaminated soil

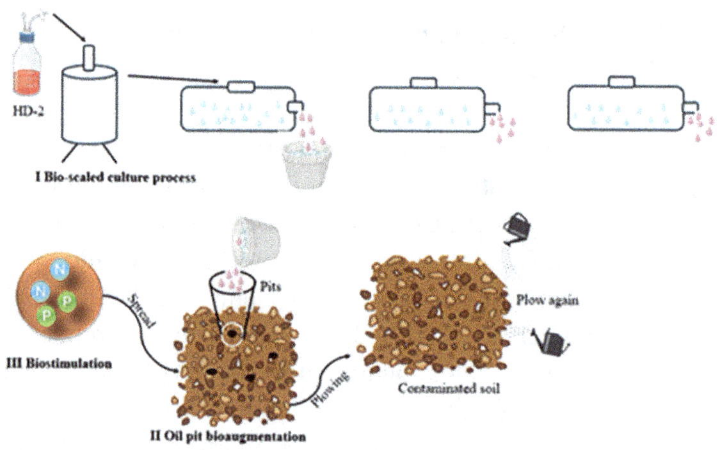

Previous studies have shown that understanding the degradative capacity of bacteria, their metabolic activity in culture, and their ability to persist for extended periods under site-specific conditions is crucial for achieving rapid and efficient pollutant degradation. Another study (Alisi et al., 2009) addressed this issue by demonstrating that the use of specific bacterial strains can successfully degrade toxic compounds in contaminated sites. Their findings showed that carefully selected microbial cultures achieved up to 75% degradation of diesel oil and phenanthrene within 42 days (Alisi et al., 2009).

Table 3. Comparison of biostimulation and bioaugmentation strategies

Bioremediation strategy	Principle	Major advantages	Limitations/ Challenges	Representative applications	References
Biostimulation	Enhancement of the activity of indigenous microbial communities through the addition of nutrients, electron acceptors, or optimization of environmental conditions	Cost-effective and environmentally sustainable; promotes the activity of native microorganisms already adapted to local environmental conditions; enhances natural biodegradation pathways	Effectiveness depends on the metabolic capability of indigenous microbes; may require careful optimization of pH, temperature, oxygen, and nutrient availability	Marine oil spill remediation; hydrocarbon-contaminated soil treatment using nutrient amendments	(Al-Mailem et al., 2017)
Bioaugmentation	Introduction of specialized or engineered microorganisms with high degradative capabilities into contaminated environments	Effective against recalcitrant and complex pollutants; accelerates biodegradation processes through the use of pollutant-specific microbial strains	Potential ecological imbalance due to introduction of non-native microorganisms; survival and adaptation of introduced strains may be limited; requires continuous monitoring and management	Heavy metal remediation in contaminated soils; wastewater treatment using specialized microbial consortia or engineered strains	(Pickering et al., 2024)

Table 4. Combined approach – Biostimulation and bioaugmentation

Aspect	Biostimulation	Bioaugmentation	Combined Biostimulation–Bioaugmentation Approach	Key References
Definition	Biostimulation involves the addition of nutrients, electron donors/ acceptors, surfactants, or other amendments to stimulate the metabolic activity of indigenous microbial communities already present at contaminated sites.	Bioaugmentation refers to the deliberate introduction of specialized, naturally occurring, or genetically engineered microorganisms into contaminated environments to enhance pollutant degradation.	The combined approach integrates nutrient amendment strategies with inoculation of exogenous microbial strains or consortia to achieve synergistic enhancement of biodegradation efficiency.	(Tyagi et al., 2011, Chettri et al., 2024, Shi et al., 2022)
Primary Objective	To enhance the growth and catabolic activity of native microbial populations capable of degrading contaminants.	To introduce microbial strains possessing specific degradative pathways for target pollutants lacking effective indigenous degraders.	To maximize microbial diversity, metabolic functionality, and biodegradation efficiency through cooperation between native and introduced microorganisms.	(Siles and Margesin, 2018, Huo et al., 2024, Shu et al., 2022)
Mechanism of Action	Stimulates existing microbial communities by supplying essential nutrients (C, N, P), oxygen, surfactants, moisture, or growth-promoting factors, thereby accelerating enzymatic degradation processes.	Enhances biodegradation by increasing the abundance of pollutant-degrading microorganisms and expanding the catabolic gene pool through inoculation of selected strains or microbial consortia.	Provides optimal nutrients and environmental conditions to support the survival, establishment, and cooperative interaction of both indigenous and introduced microbial populations.	(Tyagi et al., 2011, Nzila, 2018, Nunes et al., 2024)
Microbial Source	Indigenous/native microorganisms naturally present in contaminated environments.	Exogenous microbial strains isolated from contaminated environments or developed through laboratory selection/engineering.	Combination of indigenous microbial communities and introduced specialized degraders.	(OJ et al., 2023, Nzila, 2018, Kumar and Agrawal, 2025)
Key Advantages	Economical and environmentally sustainable; Minimal ecological disturbance; Utilizes microbes already adapted to local conditions; Suitable for large-scale in situ applications	Highly effective for recalcitrant and toxic pollutants, Enables targeted degradation pathways, Enhances degradation beyond native microbial capacity	Combines ecological adaptability with pollutant specificity, Improves degradation efficiency and resilience, Enhances microbial cooperation and long-term remediation performance	(Saeed et al., 2024, Huo et al., 2024, Megharaj et al., 2011)

continued on following page

Table 4. Continued

Aspect	Biostimulation	Bioaugmentation	Combined Biostimulation–Bioaugmentation Approach	Key References
Major Limitations	Dependent on indigenous microbial diversity and metabolic capability, Less effective for persistent or highly toxic pollutants, Excess nutrient addition may cause eutrophication	Introduced microorganisms may fail to survive or compete with native microbiota; Potential ecological imbalance, Requires biosafety monitoring and management	More operationally complex. Requires optimization of nutrient concentrations and inoculum density, Increased monitoring and management requirements	(OJ et al., 2023, Nzila, 2018, Kumar and Agrawal, 2025)
Environmental Factors Affecting Efficiency	Nutrient availability, oxygen concentration, moisture, pH, salinity, temperature, and contaminant bioavailability.	Adaptability and survival of introduced strains, microbial competition, environmental stress, pH, temperature, and nutrient status.	Interaction between indigenous and introduced microorganisms, nutrient balance, environmental compatibility, and contaminant complexity.	(Huo et al., 2024; Nunes et al., 2024)
Representative Applications	Hydrocarbon degradation, nutrient-poor contaminated soils, landfill bioremediation, wastewater treatment, oil spill remediation.	Remediation of PCBs, PAHs, pesticides, chlorinated compounds, heavy metals, and xenobiotic contaminants.	Treatment of mixed contaminants (organic pollutants + heavy metals), long-term contaminated soils, industrial wastewater, and complex aquatic systems.	(OJ et al., 2023, Nzila, 2018, Kumar and Agrawal, 2025)
Monitoring Parameters	Nutrient concentrations, microbial biomass, respiration rate, enzymatic activity, and pollutant reduction.	Survival and colonization of introduced strains, catabolic gene abundance, PCR/qPCR analysis, and degradation kinetics.	Nutrient dynamics, microbial community structure, functional gene expression, enzymatic activity, and pollutant degradation efficiency.	(Saeed et al., 2024, Huo et al., 2024, Megharaj et al., 2011)
Typical Outcomes	Enhanced biodegradation rates but limited by native microbial potential and contaminant complexity.	High pollutant specificity and accelerated degradation, although effectiveness may vary depending on strain survival.	Synergistic biodegradation, improved stability, enhanced pollutant removal efficiency, and better long-term remediation performance.	(OJ et al., 2023, Nzila, 2018, Tyagi et al., 2011)

5. CONCLUSION

At the global level, the world is increasingly facing severe risks of environmental contamination as a result of industrialization and human activities. These threats are further exacerbated by rising global temperatures and the growing incidence of acid rain. Additionally, the loss of microbial diversity and functionality, combined with the continuous release of toxic and hazardous wastes into soil and water, poses a serious challenge to ecosystem health and sustainability. Among the available approaches, bioremediation is considered the most effective, eco-friendly, and cost-efficient method. This study focuses on two major strategies-bioaugmentation and biostimulation-as highly effective techniques for pollutant removal, with particular emphasis on their advantages and wide-ranging applications. Further studies and research are needed to refine the methodologies of bioremediation, particularly in the areas of bioaugmentation and biostimulation, to ensure the development of reliable and environmentally sustainable strategies. Advancing these approaches is essential for addressing global pollution challenges and preventing the environmental disasters that may arise from them.

REFERENCES

Abbas, M. M., Husien, K. S., & Zainal, I. G. (2024). Assessment of Lead and Cadmium in Some Industrial Wastewater in Kirkuk City. *Journal of Pioneering Medical Sciences, 13*(1), 13. DOI: 10.61091/jpms202413101

Abdulsalam, S., & Omale, A. B. (2009). Comparison of biostimulation and bioaugmentation techniques for the remediation of used motor oil contaminated soil. *Brazilian Archives of Biology and Technology, 52*(3), 747–754. DOI: 10.1590/S1516-89132009000300027

Adams, G. O., Fufeyin, P. T., Okoro, S. E., & Ehinomen, I. (2015). Bioremediation, biostimulation and bioaugmention: A review. *International Journal of Environmental Bioremediation & Biodegradation, 3*(1), 28–39. DOI: 10.12691/ijebb-3-1-5

Adetunji, C. O., & Anani, O. A. (2021). *Bioaugmentation: a powerful biotechnological techniques for sustainable Ecorestoration of soil and groundwater contaminants. Microbial Rejuvenation of Polluted Environment* (Vol. 1). Springer.

Ahmed, R., Obaidi, A. & Bahaa, M. 2024. Influence of Environmental Conditions on Bacterial Remediation of Oil-Contaminated Soil: A Case Study at Baiji Refinery. *Al-Rafidain Journal of Engineering Sciences*, 383-391.

Al-Mailem, D. M., Al-Deieg, M., Eliyas, M., & Radwan, S. S. (2017). Biostimulation of indigenous microorganisms for bioremediation of oily hypersaline microcosms from the Arabian Gulf Kuwaiti coasts. *Journal of Environmental Management, 193*, 576–583. DOI: 10.1016/j.jenvman.2017.02.054 PMID: 28262419

Ali, S. S., Al-Tohamy, R., & Sun, J. (2025). *Unmasking microplastics in anaerobic digestion: Hidden threats, synergistic pollutants, and biodegradation Frontiers—A comprehensive hotspot review.* Environmental Chemistry and Ecotoxicology.

Alisi, C., Musella, R., Tasso, F., Ubaldi, C., Manzo, S., Cremisini, C., & Sprocati, A. R. (2009). Bioremediation of diesel oil in a co-contaminated soil by bioaugmentation with a microbial formula tailored with native strains selected for heavy metals resistance. *The Science of the Total Environment, 407*(8), 3024–3032. DOI: 10.1016/j.scitotenv.2009.01.011 PMID: 19201450

Angeles-De Paz, G., Cubero-Cardoso, J., Pozo, C., Calvo, C., Aranda, E., & Robledo-Mahón, T. (2025). Optimizing bioaugmentation for pharmaceutical stabilization of sewage sludge: A study on short-term composting under real conditions. *Journal of Fungi (Basel, Switzerland), 11*(1), 67. DOI: 10.3390/jof11010067 PMID: 39852486

Aragaw, T. (2021). Functions of various bacteria for specific pollutants degradation and their application in wastewater treatment: A review. *International Journal of Environmental Science and Technology*, *18*(7), 2063–2076. DOI: 10.1007/s13762-020-03022-2

Ayilara, M. S., & Babalola, O. O. (2023). Bioremediation of environmental wastes: The role of microorganisms. *Frontiers in Agronomy*, *5*, 1183691. DOI: 10.3389/fagro.2023.1183691

Aziz, Z. S., Jazza, S. H., Dageem, H. N., Banoon, S. R., Balboul, B. A., & Abdelzaher, M. (2024). Bacterial biodegradation of oil-contaminated soil for pollutant abatement contributing to achieve sustainable development goals: A comprehensive review. *Results in Engineering*, *22*, 102083. DOI: 10.1016/j.rineng.2024.102083

Behera, S., & Das, S. (2023). Potential and prospects of Actinobacteria in the bioremediation of environmental pollutants: Cellular mechanisms and genetic regulations. *Microbiological Research*, *273*, 127399. DOI: 10.1016/j.micres.2023.127399 PMID: 37150049

Bilal, M., Ashraf, S. S., Barceló, D., & Iqbal, H. M. (2019). Biocatalytic degradation/redefining "removal" fate of pharmaceutically active compounds and antibiotics in the aquatic environment. *The Science of the Total Environment*, *691*, 1190–1211. DOI: 10.1016/j.scitotenv.2019.07.224 PMID: 31466201

BOUCHEZ, PATUREAU, DABERT, JURETSCHKO, DORÉ, DELGENÈS, MOLETTA & WAGNER (2000). Ecological study of a bioaugmentation failure. *Environmental microbiology,* 2, 179-190.

Brazil, G., Kenefick, L., Callanan, M., Haro, A., De Lorenzo, V., Dowling, D., & O'Gara, F. (1995). Construction of a rhizosphere pseudomonad with potential to degrade polychlorinated biphenyls and detection of bph gene expression in the rhizosphere. *Applied and Environmental Microbiology*, *61*(5), 1946–1952. DOI: 10.1128/aem.61.5.1946-1952.1995 PMID: 7646029

Burns, R. G., Deforest, J. L., Marxsen, J., Sinsabaugh, R. L., Stromberger, M. E., Wallenstein, M. D., Weintraub, M. N., & Zoppini, A. (2013). Soil enzymes in a changing environment: Current knowledge and future directions. *Soil Biology & Biochemistry*, *58*, 216–234. DOI: 10.1016/j.soilbio.2012.11.009

Cavazzoli, S., Selonen, V., Rantalainen, A.-L., Sinkkonen, A., Romantschuk, M., & Squartini, A. (2022). Natural additives contribute to hydrocarbon and heavy metal co-contaminated soil remediation. *Environmental Pollution*, *307*, 119569. DOI: 10.1016/j.envpol.2022.119569 PMID: 35680061

Cavazzoli, S., Squartini, A., Sinkkonen, A., Romantschuk, M., Rantalainen, A.-L., Selonen, V., & Roslund, M. I. (2023). Nutritional additives dominance in driving the bacterial communities succession and bioremediation of hydrocarbon and heavy metal contaminated soil microcosms. *Microbiological Research, 270*, 127343. DOI: 10.1016/j.micres.2023.127343 PMID: 36841130

Chakraborty, S., Talukdar, A., Dey, S., & Bhattacharya, S. (2025). Role of fungi, bacteria and microalgae in bioremediation of emerging pollutants with special reference to pesticides, heavy metals and pharmaceuticals. *Discover Environment, 3*(1), 91. DOI: 10.1007/s44274-025-00217-7

Chandra, S., Sharma, R., Singh, K., & Sharma, A. (2013). Application of bioremediation technology in the environment contaminated with petroleum hydrocarbon. *Annals of Microbiology, 63*(2), 417–431. DOI: 10.1007/s13213-012-0543-3

Chaudhary, D. K., Bajagain, R., Jeong, S.-W., & Kim, J. (2021). Effect of consortium bioaugmentation and biostimulation on remediation efficiency and bacterial diversity of diesel-contaminated aged soil. *World Journal of Microbiology & Biotechnology, 37*(3), 46. DOI: 10.1007/s11274-021-02999-3 PMID: 33554294

Chen, W., Zuo, Y., Hou, Z., Wang, B., Xiong, S., Ding, X., Peng, B., Zhou, K., Li, J., Liu, R., Ding, G., Wei, Y., & Xu, T. (2023). Effect of Rhodococcus bioaugmentation and biostimulation on dibenzothiophene biodegradation and bacterial community interaction in petroleum-contaminated soils. *Frontiers in Environmental Science, 11*, 1270599. DOI: 10.3389/fenvs.2023.1270599

Chen, Y.-M., Lin, T.-F., Huang, C., Lin, J.-C., & Hsieh, F.-M. (2007). Degradation of phenol and TCE using suspended and chitosan-bead immobilized Pseudomonas putida. *Journal of Hazardous Materials, 148*(3), 660–670. DOI: 10.1016/j.jhazmat.2007.03.030 PMID: 17434262

Chettri, D., Verma, A. K., & Verma, A. K. (2024). Bioaugmentation: An approach to biological treatment of pollutants. *Biodegradation, 35*(2), 117–135. DOI: 10.1007/s10532-023-10050-5 PMID: 37684525

Cosgrove, L., Mcgeechan, P., Handley, P., & Robson, G. (2010). Effect of biostimulation and bioaugmentation on degradation of polyurethane buried in soil. *Applied and Environmental Microbiology, 76*(3), 810–819. DOI: 10.1128/AEM.00534-09 PMID: 19948849

Coulon, F., Pelletier, E., Gourhant, L., & Delille, D. (2005). Effects of nutrient and temperature on degradation of petroleum hydrocarbons in contaminated sub-Antarctic soil. *Chemosphere, 58*(10), 1439–1448. DOI: 10.1016/j.chemosphere.2004.10.007 PMID: 15686763

Cuadros-Pedraza, A., Blanco-Zuñiga, C. R., & Rojas-Arias, N. (2021). Efficiency of biostimulation and bioaugmentation processes in the treatment of soils contaminated with hydrocarbons. *Journal of Applied Research and Technology*, *19*(5), 540–547. DOI: 10.22201/icat.24486736e.2021.19.5.1589

Curiel-Alegre, S., Velasco-Arroyo, B., Rumbo, C., Khan, A. H. A., Tamayo-Ramos, J. A., Rad, C., Gallego, J. L. R., & Barros, R. (2022). Evaluation of biostimulation, bioaugmentation, and organic amendments application on the bioremediation of recalcitrant hydrocarbons of soil. *Chemosphere*, *307*, 135638. DOI: 10.1016/j.chemosphere.2022.135638 PMID: 35817192

Dehnavi, S., & Ebrahimipour, G. (2022). Comparative remediation rate of biostimulation, bioaugmentation, and phytoremediation in hydrocarbon contaminants. *International Journal of Environmental Science and Technology*, *19*(11), 11561–11586. DOI: 10.1007/s13762-022-04343-0

Du, L., & Liu, W. (2012). Occurrence, fate, and ecotoxicity of antibiotics in agro-ecosystems. A review. *Agronomy for Sustainable Development*, *32*(2), 309–327. DOI: 10.1007/s13593-011-0062-9

El Fantroussi, S., & Agathos, S. N. (2005). Is bioaugmentation a feasible strategy for pollutant removal and site remediation? *Current Opinion in Microbiology*, *8*(3), 268–275. DOI: 10.1016/j.mib.2005.04.011 PMID: 15939349

Elshafei, A. M., & Mansour, R. (2024). Microbial bioremediation of soils contaminated with petroleum hydrocarbons. *Discover Soil*, *1*(1), 9. DOI: 10.1007/s44378-024-00004-5

Fashola, M. O., Ngole-Jeme, V. M., & Babalola, O. O. (2016). Heavy metal pollution from gold mines: Environmental effects and bacterial strategies for resistance. *International Journal of Environmental Research and Public Health*, *13*(11), 1047. DOI: 10.3390/ijerph13111047 PMID: 27792205

Garousin, H., Pourbabaee, A. A., Alikhani, H. A., & Yazdanfar, N. (2021). A combinational strategy mitigated old-aged petroleum contaminants: Ineffectiveness of biostimulation as a bioremediation technique. *Frontiers in Microbiology*, *12*, 642215. DOI: 10.3389/fmicb.2021.642215 PMID: 33717040

Gaylarde, C. C. & Da Fonseca, E. M. Biofertilization and Bioremediation—How Can Microbiological Technology Assist the Ecological Crisis in Developing Countries? Micro, 2025. MDPI, 18.

Gentry, T., Rensing, C., & Pepper, I. (2004). New approaches for bioaugmentation as a remediation technology. *Critical Reviews in Environmental Science and Technology, 34*(5), 447–494. DOI: 10.1080/10643380490452362

Gomez, F., & Sartaj, M. (2013). Field scale ex-situ bioremediation of petroleum contaminated soil under cold climate conditions. *International Biodeterioration & Biodegradation, 85*, 375–382. DOI: 10.1016/j.ibiod.2013.08.003

Goswami, M., Chakraborty, P., Mukherjee, K., Mitra, G., Bhattacharyya, P., Dey, S., & Tribedi, P. (2018). Bioaugmentation and biostimulation: A potential strategy for environmental remediation. *Journal of Microbiology & Experimentation, 6*, 223–231.

Guo, R., Li, G., Jiang, T., Schuchardt, F., Chen, T., Zhao, Y., & Shen, Y. (2012). Effect of aeration rate, C/N ratio and moisture content on the stability and maturity of compost. *Bioresource Technology, 112*, 171–178. DOI: 10.1016/j.biortech.2012.02.099 PMID: 22437050

Halden, R. U., Tepp, S. M., Halden, B. G., & Dwyer, D. F. (1999). Degradation of 3-phenoxybenzoic acid in soil by Pseudomonas pseudoalcaligenes POB310 (pPOB) and two modified Pseudomonas strains. *Applied and Environmental Microbiology, 65*(8), 3354–3359. DOI: 10.1128/AEM.65.8.3354-3359.1999 PMID: 10427019

Hamdi, H., Benzarti, S., Manusadžianas, L., Aoyama, I., & Jedidi, N. (2007). Bioaugmentation and biostimulation effects on PAH dissipation and soil ecotoxicity under controlled conditions. *Soil Biology & Biochemistry, 39*(8), 1926–1935. DOI: 10.1016/j.soilbio.2007.02.008

Handore, A. V., Jagtap, V. S., Ghangale, S. S., Khandelwal, S. R., Bholay, A. D., Karmakar, R., & Handore, D. V. (2023). *Bioaugmentation: an emerging ecofriendly technology for wastewater treatment. Integrative Strategies for Bioremediation of Environmental Contaminants, Volume Two*. Elsevier. DOI: 10.1016/B978-0-443-14013-6.00004-4

Harms, H., Schlosser, D., & Wick, L. Y. (2011). Untapped potential: Exploiting fungi in bioremediation of hazardous chemicals. *Nature Reviews. Microbiology, 9*(3), 177–192. DOI: 10.1038/nrmicro2519 PMID: 21297669

Huo, J., Song, B., Lin, X., Riaz, M., Zhao, X., Liu, S., & She, Q. (2024). Ecological characteristics of sugar beet plant and rhizosphere soil in response to high boron stress: A study of the remediation potential. *Journal of Environmental Management, 356*, 120655. DOI: 10.1016/j.jenvman.2024.120655 PMID: 38513589

Hussain, A., Rehman, F., Rafeeq, H., Waqas, M., Asghar, A., Afsheen, N., Rahdar, A., Bilal, M., & Iqbal, H. M. (2022). In-situ, Ex-situ, and nano-remediation strategies to treat polluted soil, water, and air–A review. *Chemosphere*, *289*, 133252. DOI: 10.1016/j.chemosphere.2021.133252 PMID: 34902385

Imron, M. F., Kurniawan, S. B., Ismail, N. I., & Abdullah, S. R. S. (2020). Future challenges in diesel biodegradation by bacteria isolates: A review. *Journal of Cleaner Production*, *251*, 119716. DOI: 10.1016/j.jclepro.2019.119716

Janusz, G., Pawlik, A., Sulej, J., Świderska-Burek, U., Jarosz-Wilkołazka, A., & Paszczyński, A. (2017). Lignin degradation: Microorganisms, enzymes involved, genomes analysis and evolution. *FEMS Microbiology Reviews*, *41*(6), 941–962. DOI: 10.1093/femsre/fux049 PMID: 29088355

Joshi, H. K., Vishwakarma, M. C., Kumar, R., Sharma, H., Bhandari, N. S., & Joshi, S. K. (2023). The biosorption of Zn2+ by various biomasses from wastewater: A review. *Journal of Water Process Engineering*, *56*, 104389. DOI: 10.1016/j.jwpe.2023.104389

Juárez-Maldonado, A., Tortella, G., Rubilar, O., Fincheira, P., & Benavides-Mendoza, A. (2021). Biostimulation and toxicity: The magnitude of the impact of nanomaterials in microorganisms and plants. *Journal of Advanced Research*, *31*, 113–126. DOI: 10.1016/j.jare.2020.12.011 PMID: 34194836

Kauppi, S., Sinkkonen, A., & Romantschuk, M. (2011). Enhancing bioremediation of diesel-fuel-contaminated soil in a boreal climate: Comparison of biostimulation and bioaugmentation. *International Biodeterioration & Biodegradation*, *65*(2), 359–368. DOI: 10.1016/j.ibiod.2010.10.011

Kebede, G., Tafese, T., Abda, E. M., Kamaraj, M., & Assefa, F. (2021). Factors influencing the bacterial bioremediation of hydrocarbon contaminants in the soil: Mechanisms and impacts. *Journal of Chemistry*, *2021*, 9823362. DOI: 10.1155/2021/9823362

Khurshid, C. A., Mahdi, K., Ahmed, O. I., Osman, R., Rahman, M., & Ritsema, C. (2022). Assessment of potentially toxic elements in the urban soil and plants of Kirkuk City in Iraq. *Sustainability (Basel)*, *14*(9), 5655. DOI: 10.3390/su14095655

Kochhar, N., Shrivastava, S., Ghosh, A., Rawat, V. S., Sodhi, K. K., & Kumar, M. (2022). Perspectives on the microorganism of extreme environments and their applications. *Current Research in Microbial Sciences*, *3*, 100134. DOI: 10.1016/j.crmicr.2022.100134 PMID: 35909612

Kumar, P., & Agrawal, K. (2025). *Ecological Restorations of Wetlands by Geomicrobes. Mineral Transformation and Bioremediation by Geo-Microbes*. Springer.

Kuppan, N., Padman, M., Mahadeva, M., Srinivasan, S., & Devarajan, R. (2024). A comprehensive review of sustainable bioremediation techniques: Eco friendly solutions for waste and pollution management. *Waste Management Bulletin, 2*(3), 154–171. DOI: 10.1016/j.wmb.2024.07.005

Lebeau, T. (2011). *Bioaugmentation for in situ soil remediation: how to ensure the success of such a process. Bioaugmentation, biostimulation and biocontrol*. Springer.

Li, X., Shen, X., Jiang, W., Xi, Y., & Li, S. (2024). Comprehensive review of emerging contaminants: Detection technologies, environmental impact, and management strategies. *Ecotoxicology and Environmental Safety, 278*, 116420. DOI: 10.1016/j.ecoenv.2024.116420 PMID: 38701654

Li, X., Wang, A., Wan, W., Luo, X., Zheng, L., He, G., Huang, D., Chen, W., & Huang, Q. (2021). High salinity inhibits soil bacterial community mediating nitrogen cycling. *Applied and Environmental Microbiology, 87*(21), e01366–e21. DOI: 10.1128/AEM.01366-21 PMID: 34406835

Liu, Q., Sun, S., Chen, S., Su, Y., Wang, Y., Tang, F., Zhao, C., & Li, L. (2023). A novel dehydrocoenzyme activator combined with a composite microbial agent TY for enhanced bioremediation of crude oil-contaminated soil. *Journal of Environmental Management, 331*, 117246. DOI: 10.1016/j.jenvman.2023.117246 PMID: 36642048

Liu, Y. J., Nikolausz, M., & Wang, X. C. (2009). Biodegradation and detoxication of phenol by using free and immobilized cells of Acinetobacter sp. XA05 and Sphingomonas sp. FG03. *Journal of Environmental Science and Health. Part A, Toxic/Hazardous Substances & Environmental Engineering, 44*(2), 130–136. DOI: 10.1080/10934520802539673 PMID: 19123092

Maglione, G., Zinno, P., Tropea, A., Mussagy, C. U., Dufossé, L., Giuffrida, D., & Mondello, A. (2024). Microbes' role in environmental pollution and remediation: A bioeconomy focus approach. *AIMS Microbiology, 10*(3), 723–755. DOI: 10.3934/microbiol.2024033 PMID: 39219757

Margesin, R., & Schinner, F. (2001). Bioremediation (natural attenuation and biostimulation) of diesel-oil-contaminated soil in an alpine glacier skiing area. *Applied and Environmental Microbiology, 67*(7), 3127–3133. DOI: 10.1128/AEM.67.7.3127-3133.2001 PMID: 11425732

Mashreghi, M. & Prosser, J. I. 2006. Survival and activity of lux-marked phenanthrene-degrading Pseudomonas stutzeri P16 under different conditions.

Mawang, C.-I., Azman, A.-S., Fuad, A.-S. M., & Ahamad, M. (2021). Actinobacteria: An eco-friendly and promising technology for the bioaugmentation of contaminants. *Biotechnology Reports (Amsterdam, Netherlands)*, *32*, e00679. DOI: 10.1016/j.btre.2021.e00679 PMID: 34660214

Megharaj, M., Ramakrishnan, B., Venkateswarlu, K., Sethunathan, N., & Naidu, R. (2011). Bioremediation approaches for organic pollutants: A critical perspective. *Environment International*, *37*(8), 1362–1375. DOI: 10.1016/j.envint.2011.06.003 PMID: 21722961

Mekonnen, B. A., Aragaw, T. A., & Genet, M. B. (2024). Bioremediation of petroleum hydrocarbon contaminated soil: A review on principles, degradation mechanisms, and advancements. *Frontiers in Environmental Science*, *12*, 1354422. DOI: 10.3389/fenvs.2024.1354422

Mohamed, B. A., & Samer, M. (2023). *Biobutanol production from agricultural wastes. Valorization of Wastes for Sustainable Development*. Elsevier.

Mohd Noor, M. (2021). Environmental impacts of oil industry: An overview of the impacts and source management EEEE Environmental Impacts of Oil Industry: An overview of the impacts and source management. *Environmental Management*, *4*, 1–8.

Mrozik, A., & Piotrowska-Seget, Z. (2010). Bioaugmentation as a strategy for cleaning up of soils contaminated with aromatic compounds. *Microbiological Research*, *165*(5), 363–375. DOI: 10.1016/j.micres.2009.08.001 PMID: 19735995

Muter, O. (2023). Current trends in bioaugmentation tools for bioremediation: A critical review of advances and knowledge gaps. *Microorganisms*, *11*(3), 710. DOI: 10.3390/microorganisms11030710 PMID: 36985282

Nematollahi, M. J., Rastegari Mehr, M., Shakeri, A., Amjadian, K., Ebrahimi, P., & Pirouei, M. (2022). Polycyclic aromatic hydrocarbons (PAHs) in soils around oil refineries; Case study: Erbil and Kirkuk refineries, Iraq-A modelling approach. *Journal of Environmental Health Science & Engineering*, *20*(2), 699–716. DOI: 10.1007/s40201-022-00809-y PMID: 36406609

Nikolopoulou, M., & Kalogerakis, N. (2009). Biostimulation strategies for fresh and chronically polluted marine environments with petroleum hydrocarbons. *Journal of Chemical Technology & Biotechnology: International Research in Process. Environmental & Clean Technology*, *84*, 802–807.

Nikolopoulou, M., Pasadakis, N., & Kalogerakis, N. (2013). Evaluation of autochthonous bioaugmentation and biostimulation during microcosm-simulated oil spills. *Marine Pollution Bulletin, 72*(1), 165–173. DOI: 10.1016/j.marpolbul.2013.04.007 PMID: 23660443

Nnaji, N. D., Anyanwu, C. U., Miri, T., & Onyeaka, H. (2024). Mechanisms of heavy metal tolerance in bacteria: A review. *Sustainability (Basel), 16*(24), 11124. DOI: 10.3390/su162411124

Nunes, P. S., Lacerda-Junior, G. V., Mascarin, G. M., Guimarães, R. A., Medeiros, F. H., Arthurs, S., & Bettiol, W. (2024). Microbial consortia of biological products: Do they have a future? *Biological Control, 188*, 105439. DOI: 10.1016/j.biocontrol.2024.105439

Nzila, A. (2018). Biodegradation of high-molecular-weight polycyclic aromatic hydrocarbons under anaerobic conditions: Overview of studies, proposed pathways and future perspectives. *Environmental Pollution, 239*, 788–802. DOI: 10.1016/j.envpol.2018.04.074 PMID: 29751337

Nzila, A., Razzak, S. A., & Zhu, J. (2016). Bioaugmentation: An emerging strategy of industrial wastewater treatment for reuse and discharge. *International Journal of Environmental Research and Public Health, 13*(9), 846. DOI: 10.3390/ijerph13090846 PMID: 27571089

Oj, A.-S., Meiczinger, M., Al-Juboori, R., Grmasha, R., Andredaki, M., Somogyi, V., Idowu, I., Stenger-Kovács, C., Jakab, M. & Lengyel, E. 2023. The synergistic effect of degradation and adsorption for emerging pollutants removal from water and real wastewater by using immobilized activated carbon from pomegranate peels.

Ojuederie, O. B., & Babalola, O. O. (2017). Microbial and plant-assisted bioremediation of heavy metal polluted environments: A review. *International Journal of Environmental Research and Public Health, 14*(12), 1504. DOI: 10.3390/ijerph14121504 PMID: 29207531

Oro, C. E. D., Saorin Puton, B. M., Venquiaruto, L. D., Dallago, R. M., & Tres, M. V. (2024). Effective microbial strategies to remediate contaminated agricultural soils and conserve functions. *Agronomy (Basel), 14*(11), 2637. DOI: 10.3390/agronomy14112637

Ossai, I. C., Ahmed, A., Hassan, A., & Hamid, F. S. (2020). Remediation of soil and water contaminated with petroleum hydrocarbon: A review. *Environmental Technology & Innovation, 17*, 100526. DOI: 10.1016/j.eti.2019.100526

Parajuli, A., Grönroos, M., Siter, N., Puhakka, R., Vari, H. K., Roslund, M. I., Jumpponen, A., Nurminen, N., Laitinen, O. H., Hyöty, H., Rajaniemi, J., & Sinkkonen, A. (2018). Urbanization reduces transfer of diverse environmental microbiota indoors. *Frontiers in Microbiology*, *9*, 84. DOI: 10.3389/fmicb.2018.00084 PMID: 29467728

Patel, A. B., Manvar, T., Jain, K. R., Desai, C., & Madamwar, D. (2021). Metagenomic insights into bacterial communities' structures in polycyclic aromatic hydrocarbons degrading consortia. *Journal of Environmental Chemical Engineering*, *9*(6), 106578. DOI: 10.1016/j.jece.2021.106578

Perfumo, A., Banat, I. M., Marchant, R., & Vezzulli, L. (2007). Thermally enhanced approaches for bioremediation of hydrocarbon-contaminated soils. *Chemosphere*, *66*(1), 179–184. DOI: 10.1016/j.chemosphere.2006.05.006 PMID: 16782171

Pickering, L., Castro-Gutierrez, V., Holden, B., Haley, J., Jarvis, P., Campo, P., & Hassard, F. (2024). How bioaugmentation for pesticide removal influences the microbial community in biologically active sand filters. *Chemosphere*, *363*, 142956. DOI: 10.1016/j.chemosphere.2024.142956 PMID: 39074664

Pointing, S. (2001). Feasibility of bioremediation by white-rot fungi. *Applied Microbiology and Biotechnology*, *57*(1-2), 20–33. DOI: 10.1007/s002530100745 PMID: 11693920

Puteri, M. N., Gew, L. T., Ong, H. C., & Ming, L. C. (2025). Technologies to eliminate microplastic from water: Current approaches and future prospects. *Environment International*, *199*, 109397. DOI: 10.1016/j.envint.2025.109397 PMID: 40279687

Qattan, S. Y. (2025). Harnessing bacterial consortia for effective bioremediation: Targeted removal of heavy metals, hydrocarbons, and persistent pollutants. *Environmental Sciences Europe*, *37*(1), 85. DOI: 10.1186/s12302-025-01103-y

Quek, E., Ting, Y.-P., & Tan, H. M. (2006). Rhodococcus sp. F92 immobilized on polyurethane foam shows ability to degrade various petroleum products. *Bioresource Technology*, *97*(1), 32–38. DOI: 10.1016/j.biortech.2005.02.031 PMID: 16154500

Rafeeq, H., Afsheen, N., Rafique, S., Arshad, A., Intisar, M., Hussain, A., Bilal, M., & Iqbal, H. M. (2023). Genetically engineered microorganisms for environmental remediation. *Chemosphere*, *310*, 136751. DOI: 10.1016/j.chemosphere.2022.136751 PMID: 36209847

Rajkumar, R., & Kurinjimalar, C. (2021). *Microbes and plant mineral nutrition. Microbiological activity for soil and plant health management*. Springer.

Ram, R. M., Debnath, A., Negi, S. & Singh, H. 2022. Use of microbial consortia for broad spectrum protection of plant pathogens: Regulatory hurdles, present status and future prospects. *Biopesticides*, 319-335.

Ribicic, D., Mcfarlin, K. M., Netzer, R., Brakstad, O. G., Winkler, A., Throne-Holst, M., & Størseth, T. R. (2018). Oil type and temperature dependent biodegradation dynamics-Combining chemical and microbial community data through multivariate analysis. *BMC Microbiology, 18*(1), 83. DOI: 10.1186/s12866-018-1221-9 PMID: 30086723

Rodrigues, J. L., Maltseva, O. V., Tsoi, T. V., Helton, R. R., Quensen, J. F., Fukuda, M., & Tiedje, J. M. (2001). Development of a Rhodococcus recombinant strain for degradation of products from anaerobic dechlorination of PCBs. *Environmental Science & Technology, 35*(4), 663–668. DOI: 10.1021/es001308t PMID: 11349275

Romantschuk, M., Lahti-Leikas, K., Kontro, M., Galitskaya, P., Talvenmäki, H., Simpanen, S., Allen, J. A., & Sinkkonen, A. (2023). Bioremediation of contaminated soil and groundwater by in situ biostimulation. *Frontiers in Microbiology, 14*, 1258148. DOI: 10.3389/fmicb.2023.1258148 PMID: 38029190

Romantschuk, M., Sarand, I., Petänen, T., Peltola, R., Jonsson-Vihanne, M., Koivula, T., Yrjälä, K., & Haahtela, K. (2000). Means to improve the effect of in situ bioremediation of contaminated soil: An overview of novel approaches. *Environmental Pollution, 107*(2), 179–185. DOI: 10.1016/S0269-7491(99)00136-0 PMID: 15092994

Roslund, M. I., Rantala, S., Oikarinen, S., Puhakka, R., Hui, N., Parajuli, A., Laitinen, O. H., Hyöty, H., Rantalainen, A.-L., Sinkkonen, A., Grönroos, M., Hui, N., Kalvo, R., Nurminen, N., Oikarinen, S., Parajuli, A., Puhakka, R., Rantala, S., Roslund, M. I., & Sinkkonen, A. (2019). Endocrine disruption and commensal bacteria alteration associated with gaseous and soil PAH contamination among daycare children. *Environment International, 130*, 104894. DOI: 10.1016/j.envint.2019.06.004 PMID: 31220749

Saeed, H., Padmesh, S., Singh, A., Nandy, A., Singh, S. P., & Deshwal, R. K. (2024). Impact of veterinary pharmaceuticals on environment and their mitigation through microbial bioremediation. *Frontiers in Microbiology, 15*, 1396116. DOI: 10.3389/fmicb.2024.1396116 PMID: 39040911

Saha, L., Tiwari, J., Bauddh, K., & Ma, Y. (2021). Recent developments in microbe–plant-based bioremediation for tackling heavy metal-polluted soils. *Frontiers in Microbiology, 12*, 731723. DOI: 10.3389/fmicb.2021.731723 PMID: 35002995

Salleh, A. B., Ghazali, F. M., Rahman, R., & Basri, M. (2003). Bioremediation of petroleum hydrocarbon pollution. *Indian Journal of Biotechnology, 2*, 411–425.

Sarkar, J., Kazy, S. K., Gupta, A., Dutta, A., Mohapatra, B., Roy, A., Bera, P., Mitra, A., & Sar, P. (2016). Biostimulation of indigenous microbial community for bioremediation of petroleum refinery sludge. *Frontiers in Microbiology, 7*, 1407. DOI: 10.3389/fmicb.2016.01407 PMID: 27708623

Sayed, K., Baloo, L., & Sharma, N. K. (2021). Bioremediation of total petroleum hydrocarbons (TPH) by bioaugmentation and biostimulation in water with floating oil spill containment booms as bioreactor basin. *International Journal of Environmental Research and Public Health, 18*(5), 2226. DOI: 10.3390/ijerph18052226 PMID: 33668225

Selvasembian, R., Azelee, N. I. B. W., Saravanan, R., Ponnusami, V., & Mishra, A. K. (2023). *Valorization of Wastes for Sustainable Development: Waste to Wealth.* Elsevier.

Sharma, A., Singh, S. B., Sharma, R., Chaudhary, P., Pandey, A. K., Ansari, R., Vasudevan, V., Arora, A., Singh, S., Saha, S., & Nain, L. (2016). Enhanced biodegradation of PAHs by microbial consortium with different amendment and their fate in in-situ condition. *Journal of Environmental Management, 181*, 728–736. DOI: 10.1016/j.jenvman.2016.08.024 PMID: 27558829

Shi, L., Liu, Z., Yang, L., & Fan, W. (2022). Effect of plant waste addition as exogenous nutrients on microbial remediation of petroleum-contaminated soil. *Annals of Microbiology, 72*(1), 22. DOI: 10.1186/s13213-022-01679-3

Shu, X., He, J., Zhou, Z., Xia, L., Hu, Y., Zhang, Y., Zhang, Y., Luo, Y., Chu, H., Liu, W., Yuan, S., Gao, X., & Wang, C. (2022). Organic amendments enhance soil microbial diversity, microbial functionality and crop yields: A meta-analysis. *The Science of the Total Environment, 829*, 154627. DOI: 10.1016/j.scitotenv.2022.154627 PMID: 35306065

Siles, J. A., & Margesin, R. (2018). Insights into microbial communities mediating the bioremediation of hydrocarbon-contaminated soil from an Alpine former military site. *Applied Microbiology and Biotechnology, 102*(10), 4409–4421. DOI: 10.1007/s00253-018-8932-6 PMID: 29594357

Silva, E., Fialho, A. M., Sá-Correia, I., Burns, R. G., & Shaw, L. J. (2004). Combined bioaugmentation and biostimulation to cleanup soil contaminated with high concentrations of atrazine. *Environmental Science & Technology, 38*(2), 632–637. DOI: 10.1021/es0300822 PMID: 14750741

Singh Arora, D., & Kumar Sharma, R. (2010). Ligninolytic fungal laccases and their biotechnological applications. *Applied Biochemistry and Biotechnology, 160*(6), 1760–1788. DOI: 10.1007/s12010-009-8676-y PMID: 19513857

Sonune, N. (2021). *Microbes: a potential tool for bioremediation. Rhizobiont in bioremediation of hazardous waste*. Springer.

Strong, L. C., Mctavish, H., Sadowsky, M. J., & Wackett, L. P. (2000). Field-scale remediation of atrazine-contaminated soil using recombinant Escherichia coli expressing atrazine chlorohydrolase. *Environmental Microbiology*, *2*(1), 91–98. DOI: 10.1046/j.1462-2920.2000.00079.x PMID: 11243266

Tabacchioni, S., Passato, S., Ambrosino, P., Huang, L., Caldara, M., Cantale, C., Hett, J., Del Fiore, A., Fiore, A., Schlüter, A., Sczyrba, A., Maestri, E., Marmiroli, N., Neuhoff, D., Nesme, J., Sørensen, S. J., Aprea, G., Nobili, C., Presenti, O., & Bevivino, A. (2021). Identification of beneficial microbial consortia and bioactive compounds with potential as plant biostimulants for a sustainable agriculture. *Microorganisms*, *9*(2), 426. DOI: 10.3390/microorganisms9020426 PMID: 33669534

Thompson, I. P., Van Der Gast, C. J., Ciric, L., & Singer, A. C. (2005). Bioaugmentation for bioremediation: The challenge of strain selection. *Environmental Microbiology*, *7*(7), 909–915. DOI: 10.1111/j.1462-2920.2005.00804.x PMID: 15946288

Toczyłowska-Mamińska, R., Mamiński, M. Ł., & Kwasowski, W. (2025). Microbial Fuel Cell Technology as a New Strategy for Sustainable Management of Soil-Based Ecosystems. *Energies*, *18*(4), 970. DOI: 10.3390/en18040970

Towell, M. G., Paton, G. I., & Semple, K. T. (2011). The biodegradation of cable oil components: Impact of oil concentration, nutrient addition and bioaugmentation. *Environmental Pollution*, *159*(12), 3777–3783. DOI: 10.1016/j.envpol.2011.06.043 PMID: 21872976

Tyagi, M., Da Fonseca, M. M. R., & De Carvalho, C. C. (2011). Bioaugmentation and biostimulation strategies to improve the effectiveness of bioremediation processes. *Biodegradation*, *22*(2), 231–241. DOI: 10.1007/s10532-010-9394-4 PMID: 20680666

Ubani, O., Atagana, H. I., Thantsha, M. S. & Adeleke, R. 2016. Characterisation of oil degrading bacteria from tailored compost containing crude oil sludge.

Van Stempvoort, D., & Biggar, K. (2008). Potential for bioremediation of petroleum hydrocarbons in groundwater under cold climate conditions: A review. *Cold Regions Science and Technology*, *53*(1), 16–41. DOI: 10.1016/j.coldregions.2007.06.009

Varjani, S., & Upasani, V. N. (2019). Influence of abiotic factors, natural attenuation, bioaugmentation and nutrient supplementation on bioremediation of petroleum crude contaminated agricultural soil. *Journal of Environmental Management*, *245*, 358–366. DOI: 10.1016/j.jenvman.2019.05.070 PMID: 31158688

Zawierucha, I., & Malina, G. (2006). Bioaugmentation as a method of biodegradation enhancement in oil hydrocarbons contaminated soil. *Ecohydrology & Hydrobiology*, *6*(1-4), 163–169. DOI: 10.1016/S1642-3593(06)70138-0

Zawierucha, I., & Malina, G. (2011). *Bioremediation of contaminated soils: effects of bioaugmentation and biostimulation on enhancing biodegradation of oil hydrocarbons. Bioaugmentation, biostimulation and biocontrol*. Springer.

Zawierucha, I., & Malina, G. (2014). Application of Zeolite to Trace Metal Removal from Groundwater in Permeable Reactive Barrier Systems (PRB). *Ochrona Środowiska*, *36*, 39–44.

Zhang, K., Wang, S., Guo, P., & Guo, S. (2021). Characteristics of organic carbon metabolism and bioremediation of petroleum-contaminated soil by a mesophilic aerobic biopile system. *Chemosphere*, *264*, 128521. DOI: 10.1016/j.chemosphere.2020.128521 PMID: 33039688

Chapter 2
Harnessing Microbial Consortia for Bioremediation:
Synergistic Interactions and Community Engineering

Chioma Blaise Chikere
http://orcid.org/0000-0003-3004-9780
University of Port Harcourt, Nigeria

Chioma Bertha Ehis-Eriakha
http://orcid.org/0000-0002-3284-2021
Faculty of Science, Admiralty University of Nigeria, Ibusa, Nigeria

Rosemary Agbeko
http://orcid.org/0009-0004-9813-3676
World Bank Africa Centre of Excellence, Centre for Oilfield Chemicals Research (ACE-CEFOR), University of Port Harcourt, Nigeria

Peace Omoikhudu Oleghe
Faculty of Science, Edo State University, Iyamho, Nigeria

Chidinma Peace Okafor
http://orcid.org/0000-0001-7381-9605
Faculty of Biological Sciences, Imo State University, Owerri, Nigeria

Uzoma Chukwuka Okorafor
http://orcid.org/0009-0009-3102-9475
World Bank Africa Centre of Excellence, Centre for Oilfield Chemicals Research (ACE-CEFOR), University of Port Harcourt, Nigeria

Memory Tekere
http://orcid.org/0000-0002-0606-7826
College of Agriculture and Environmental Sciences (CAES), University of South Africa, South Africa

ABSTRACT

Bioremediation is a sustainable and cost-effective approach for environmental cleanup, but single-strain systems often perform poorly under complex polluted con-

DOI: 10.4018/979-8-2600-2998-5.ch002

ditions. Microbial consortia have therefore gained attention due to their synergistic interactions, metabolic diversity, and improved degradation efficiency. This review examines key mechanisms driving consortium performance, including metabolic cooperation, syntrophy, cross-feeding, and quorum sensing, which enhance stability and resilience. It also explores Artificially Designed Microbial Communities (ADMCs) as emerging tools for precision bioremediation using top-down and bottom-up approaches. Advanced technologies such as omics, synthetic biology, computational modeling, and bioreactor engineering are highlighted. Applications in hydrocarbon degradation, wastewater treatment, and heavy metal remediation demonstrate the practical value of microbial consortia. Several challenges and future prospects involving AI-guided engineering and integrated regulatory frameworks are extensively discussed.

1 INTRODUCTION

Microorganisms possess remarkable metabolic abilities that allow them to transform or degrade a wide array of environmental contaminants into less harmful forms. Their presence in polluted habitats demonstrates their capacity to adapt to stress and utilize unusual substrates for growth (Bhandari et al., 2021; Okoye et al., 2024). Traditional bioremediation methods rely on single microbial strains known to target specific contaminants. Although such isolates can be effective under controlled conditions, they typically underperform in complex natural environments where multiple pollutants coexist, nutrient availability fluctuates, and environmental parameters such as pH, moisture, and temperature vary. Furthermore, many pollutants undergo multi-step degradation processes that exceed the metabolic range of individual microorganisms (Okafor et al., 2021; Massot et al., 2022; Kumar et al., 2025).

This limitation has driven growing interest in microbial consortia, which are structured communities of diverse microorganisms such as bacteria, archaea, fungi, and occasionally algae that interact and cooperate to carry out biochemical transformations that individual species often cannot achieve alone (McCarty et al., 2019; Qattan, 2025; Mishra et al., 2025). Compared with monocultures, microbial consortia bring together a broader metabolic repertoire, more robust physiological resilience, and a higher capacity to adapt to dynamic and stressful environmental conditions. This coordinated metabolic networking enhances degradation efficiency and minimizes the buildup of toxic by-products that could inhibit microbial activity. Their combined activities enable more efficient breakdown of contaminants, making them powerful agents in bioremediation (Ibo et al., 2024; Santos et al., 2025; Qattan, 2025).

The effectiveness of microbial consortia is largely rooted in ecological principles such as mutualism, syntrophy, metabolic complementarity, and resource partitioning (Yamagishi et al., 2020). Consortia often exhibit synergistic behavior, where the cooperative actions of community members enhance pollutant degradation beyond what each species could achieve individually. One species may transform a complex compound into intermediate products that other species subsequently utilize or degrade, forming a chain of metabolic reactions. This phenomenon is particularly evident in the degradation of petroleum hydrocarbons, where diverse bacteria participate sequentially in breaking down long-chain and aromatic compounds (Hawley et al., 2014; Chen et al., 2024). In metal-polluted environments, microbial consortia may collectively detoxify contaminants through oxidation, reduction, sequestration, and biosorption, with each organism contributing a different mechanism (Gadd, 2010; Chen et al., 2024). These coordinated interactions allow consortia to exploit multiple substrates and thrive under chemically stressful conditions.

Recent scientific advances have revealed that microbial communities are sophisticated networks governed by communication signals, spatial organization, and metabolic exchanges. Quorum sensing is one of the processes that supports microbes to regulate group behaviors, enhance biofilm development, and synchronize community-wide metabolic responses. Biofilms create protective niches that shelter microorganisms from environmental stress while fostering close physical contact that promotes nutrient sharing and metabolic cooperation. Understanding these dynamics has expanded opportunities to manipulate and optimize microbial consortia for targeted bioremediation applications (Scott & Hasty, 2016; Kylilis et al., 2018).

Building on this knowledge, the field of microbial community engineering has emerged as a promising strategy for creating high-performance consortia. Through genome-guided selection, metabolic modeling, and synthetic biology tools, researchers can design communities with complementary capabilities tailored to specific pollutants. Engineered consortia can be structured to enhance cooperation, reduce antagonism, and stabilize community composition under field conditions (Rezaei & Moghimi, 2024). These engineered systems may become essential for degrading persistent organic pollutants, microplastics, and other emerging contaminants that conventional methods struggle to remove (Qattan, 2025).

Technological progress in high-throughput omics, including metagenomics, transcriptomics, proteomics, and metabolomics, has revolutionized how microbial communities are studied and characterized. These tools offer detailed insight into community structure, metabolic pathways, gene expression, and stress responses. Integrated with computational modeling and machine-learning approaches, they allow scientists to predict microbial interactions, assess community stability, and identify key functional species. Techniques such as microscopy, spectroscopy, and microfluidic platforms enable real-time observation of microbial activity and pollutant degradation, bridging the gap between laboratory studies and environmental applications (Mahajan & Prakash, 2025).

This chapter provides a comprehensive examination of the ecological principles and biochemical mechanisms underlying microbial consortia for bioremediation. It also highlights microbial community engineering and recent advances in synthetic biology, demonstrating how rational design approaches can enhance the performance of consortia. The chapter also presents real-world applications and emerging innovations as powerful and sustainable solutions for restoring contaminated ecosystems. Finally, it discusses the challenges and future directions in the development and deployment of microbial consortia.

2. MICROBIAL CONSORTIA IN BIOREMEDIATION: CONCEPTS AND MECHANISTIC FOUNDATIONS

2.1 Ecological Principles and Functional Architecture of Microbial Consortia

The ecological and mechanistic bases that confer superior performance to consortia, emerge from interactions that focus on key interspecies processes such as extracellular enzyme release, metabolic cooperation leading to metabolite exchange, syntrophic electron-sharing partnerships, spatial clustering, and quorum-sensing horizontal gene transfer mechanisms **(Figure 1),** all of which gradually expand or reshape the metabolic potential of individual members, thereby promoting community stability, resilience, and degradative efficiency (Kumar et al., 2025; PatSnap, 2025). Other mechanistic foundations and concepts are presented in **Table 1.**

Figure 1. Key Interspecies Processes Driving Enhanced Performance in Microbial Consortia

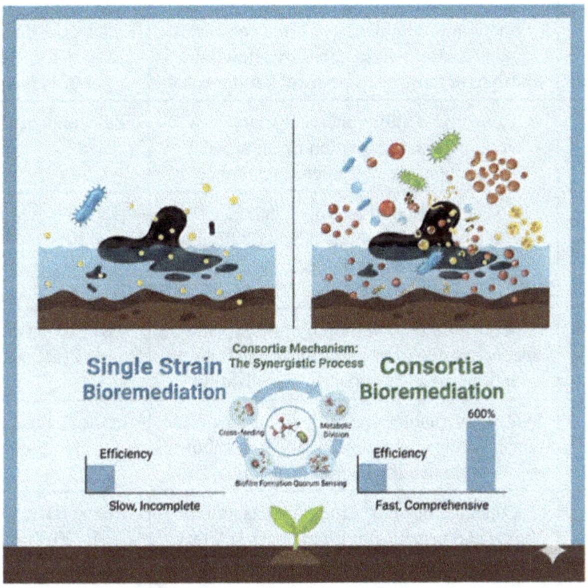

Table 1. Concepts and Mechanistic Foundations of Microbial Consortia

Mechanism	Concept	Reference
Metabolic Complexity and Multi-Step Degradation Pathways	Consortia enable sequential and parallel degradation of complex pollutants via multi-step transformations, preventing toxic intermediate accumulation and enhancing complete mineralization.	Zuo et al., 2022; Zhili et al., 2023; Schroven et al., 2021; Stams & Plugge, 2021; Satpathy et al., 2021; Kumari et al., 2023
Division of Metabolic Labour	Microbial communities divide metabolic tasks among members, reducing energy burden, increasing efficiency, and enabling simultaneous degradation of diverse pollutants.	Lawson et al., 2021; Zelezniak et al., 2020; Delgado-Baquerizo et al., 2020; Zuo et al., 2022
Cross-Feeding and Metabolic Coupling	Metabolite exchange between microbes enhances energy efficiency and stabilizes degradation pathways, especially under anaerobic conditions.	Zelezniak et al., 2020; Schroven et al., 2021; D'Souza et al., 2018; Stams & Plugge, 2021; Zhili et al., 2023
Role of Rare and Keystone Taxa	Rare and keystone taxa, though low in abundance, strongly influence degradation efficiency and community stability; their loss destabilizes consortia.	Kumari et al., 2023; Chao et al., 2024; Lawson et al., 2021; Satpathy et al., 2021

continued on following page

Table 1. Continued

Mechanism	Concept	Reference
Ecological Stability and Functional Redundancy	Functional redundancy ensures continued degradation under stress by allowing multiple organisms to perform similar roles.	Delgado-Baquerizo et al., 2020; Zuo et al., 2022; Satpathy et al., 2021; Kumari et al., 2023
Emergent Properties and Bioremediation Efficiency	Consortia exhibit emergent traits such as improved degradation rates, broader substrate range, and higher resilience.	Zelezniak et al., 2020; Zhili et al., 2023
Synergistic Interactions in Microbial Communities	Synergy among microbes enhances degradation performance beyond individual capabilities.	Lawson et al., 2021; Satpathy et al., 2021
Metabolic Cooperation and Cooperative Catabolism	Sequential degradation by primary and secondary degraders prevents toxic buildup and improves pollutant mineralization.	Kumari et al., 2023; Zuo et al., 2022; Zhili et al., 2023
Syntrophy and Thermodynamic Interdependence	Syntrophy enables energetically unfavorable reactions through interdependent metabolite exchange in anaerobic systems.	Stams & Plugge, 2021; Schroven et al., 2021
Cross-Feeding Networks and Resource Recycling	Exchange of metabolites such as amino acids and organic acids enhances stability and reduces metabolic cost.	Zelezniak et al., 2015; D'Souza et al., 2018; Kumari et al., 2023
Quorum Sensing and Cell-to-Cell Communication	Microbial communication coordinates enzyme production and biofilm formation, optimizing degradation efficiency.	Papenfort & Bassler, 2016; Whiteley et al., 2017; Satpathy et al., 2021
Biofilm Formation and Spatial Organization	Biofilms enhance metabolite exchange, protect against stress, and enable simultaneous aerobic-anaerobic degradation.	Flemming et al., 2023; Li et al., 2021; Rather et al., 2021; Michaelis & Grohmann, 2023
Functional Redundancy and Community Resilience	Redundant metabolic roles ensure stability and sustained pollutant degradation under environmental stress.	Delgado-Baquerizo et al., 2020; Lawson et al., 2021
Emergent Properties of Synergistic Communities	Integration of microbial interactions leads to improved degradation, resilience, and stability in consortia.	Zelezniak et al., 2020; Satpathy et al., 2021

3. SYNERGISTIC INTERACTIONS SHAPING MICROBIAL COMMUNITY PERFORMANCE

3.1 Synergy in Mixed Microbial Communities

Synergy in mixed microbial systems arises from multiple interacting mechanisms rather than a single dominant process. A key mechanism is metabolic handoff, where one organism partially transforms a substrate and another rapidly consumes the intermediate, preventing toxic accumulation and improving overall energy efficiency.

This is particularly important in PAH degradation, where sequential transformation by different taxa enhances breakdown efficiency, as seen in interactions between *Sphingobium* and *Burkholderia* that improve phenanthrene degradation (Mataigne et al. 2021; Macchi et al., 2021; Nieto et al., 2023).

In anoxic environments, cooperation is strongly driven by electron transfer processes. Indirect interspecies electron transfer (IIET) occurs through carriers like hydrogen or formate, helping maintain favourable thermodynamics by keeping product levels low and supporting upstream metabolism (Shen et al., 2016). In contrast, direct interspecies electron transfer (DIET) bypasses soluble carriers entirely, using conductive structures such as pili, cytochromes, magnetite, or biochar to enable direct electron flow between cells, which enhances processes like anaerobic digestion and methane production (Rotaru et al., 2014; Xu et al., 2019).

Another important layer involves microbial "public goods" such as extracellular enzymes, biosurfactants, and redox cofactors that benefit the entire community regardless of which organisms produce them. In hydrocarbon-contaminated environments, mixed communities of *Pseudomonas*, *Thalassospira*, and *Alcanivorax* often outperform single strains because shared biosurfactants increase substrate accessibility, although this cooperation can be destabilized by cheating or imbalanced community structure (Kaczorek et al., 2018; Shahidi-Rizi et al., 2017; Bayatian et al., 2025).

In addition, physical structure such as biofilms further enhances these interactions by creating spatial organiation that supports metabolic coupling, electron transfer, and sequential reactions across oxygen gradients. The extracellular matrix stabilizes enzymes and intermediates, protects against stress, and promotes horizontal gene transfer, which can reshape community function over time (Rather et al., 2021; Flemming et al., 2023b; Michaelis & Grohmann, 2023). Biochar-supported bacterial–fungal consortia illustrate these combined effects by enabling biofilm formation, detoxification, and even electron transfer in redox-driven processes (Rotaru et al., 2014; Upadhyay et al., 2023).

3.2 Ecological Constraints, Emergent Properties, and Performance Trade-Offs

Although synergy often improves degradation efficiency, it is not guaranteed. Many environmentally relevant reactions sit close to the thermodynamic edge, and even small shifts, like oxygen intrusion, pH swings, or sudden temperature drops, can disrupt syntrophic balances (Dang & Chen, 2017). Sometimes, most microbes selected for their unique attributes based on compatibility fails to replicate the rate of their growth response creating a complete disparity in substrate utilization and unintentional rivalry. In some cases, DIET-capable partners may lose their advantage when alternative electron acceptors appear, short-circuiting the intended electron flow (Lovley, 2017). These performance swings are part of why community assembly still feels, at times, more like ecological negotiation than engineering.

Where cooperation succeeds, it often does so because environmental constraints and community traits line up in just the right way. Detoxifying members remove metals or reactive intermediates before they inhibit downstream degraders, a pattern commonly observed in mixed-contamination soils where organic breakdown would otherwise stall. Structured systems, such as biofilms or granular aggregates, often show higher resilience because spatial organization reduces exposure to toxins and preserves local metabolic loops. But these same structures can also trap inhibitory end-products, slow down substrate penetration, or detach under hydrodynamic stress, introducing new vulnerabilities.

Taken together, these interactions- metabolic, electrochemical, structural, and ecological-shape the emergent performance of microbial communities. Synergy is less a single property than an evolving balance: one that appears when thermodynamics, microenvironments, and community composition align and fades when they drift out of sync. Understanding these moving parts is what makes the design and application of microbial consortia both powerful and, at times, unexpectedly fragile.

4. PRINCIPLES, TYPES, AND DESIGN FRAMEWORKS FOR ARTIFICIALLY DESIGNED MICROBIAL COMMUNITIES

The continuous influx of anthropogenic pollutants such as hydrocarbons, heavy metals, pesticides, and industrial wastes poses critical threats to ecosystems and human health. Conventional single-strain bioremediation strategies often fail to address the complexity and heterogeneity of polluted sites because they lack the metabolic diversity and interspecies cooperation needed for complete degradation. In contrast, microbial communities—natural or engineered—display rich metabolic networks capable of degrading complex pollutant mixtures. Artificially Designed Microbial Communities (ADMCs) represent a rational and next-generation paradigm for bioremediation, leveraging ecological and systems biology insights to construct microbial consortia with desired functional traits and stability (Chen et al., 2025).

Environmental pollutants often occur as complex mixtures and under heterogeneous physicochemical conditions, making single-strain remediation strategies ineffective. Artificially Designed Microbial Communities (ADMCs) overcome this limitation by exploiting metabolic complementarity, syntrophy, and functional redundancy, thereby enhancing degradation efficiency, robustness, and resilience in contaminated environments. Designing artificial microbial communities or synthetic microbial communities leverages principles from ecology, evolution, engineering, and computation to build engineered mini-ecosystems, a community of microorganisms with well-known phenotypic and functional traits, to solve real-life issues in various areas like bioremediation, agriculture, bioenergy, and medicine. The foundational principles for ADMCs emphasize stability, functional robustness, and predictability in programming multispecies behaviour (Dexter & Lu, 2022; Liang et al., 2022).

Engineering microbial consortia offers an effective strategy for optimizing interactions among microorganisms and between microbes and their environment, thereby enhancing long-term community stability. Within microbial communities, individual species compete for limited resources while simultaneously utilizing metabolic by-products released by other organisms to gain competitive and functional advantages (Pacheco et al., 2019; Li et al., 2021).

4.1 Artificially Designed Microbial Communities (ADMCs)

Artificially Designed Microbial Communities (ADMCs) in bioremediation are designed based on ecological and systems biology principles that emphasize microbial interactions within functional networks rather than isolated activity (Lü et al., 2024). Key concepts include division of labour, metabolic complementarity, syntrophy, and community stability (Liang et al., 2022). In these systems, different microbes are selected to perform specific steps in pollutant degradation, reducing metabolic load and improving efficiency. This allows sequential breakdown of contaminants, where some strains initiate degradation while others process intermediates. Synthetic consortia have shown enhanced degradation of polycyclic aromatic hydrocarbons through such cooperative interactions, achieving more complete removal than monocultures (Li et al., 2021).

4.2 Metabolite Exchange and Cross-Feeding in ADMCs

Metabolite exchange is a common feature of microbial ecosystems and provides functional advantages through metabolic complementarity and syntrophy. In engineered consortia, cross-feeding enables one organism's metabolic by-products to serve as substrates for another, enhancing pollutant degradation and reducing inhibitory compound buildup (Lu et al., 2024). Understanding these metabolite pathways is essential for designing efficient synthetic communities with improved bioremediation performance.

Cross-feeding in artificially designed microbial consortia involves both contact-independent (diffusion, transport, vesicles) and contact-dependent mechanisms (nanotubes, flagella-like structures, and direct contact) (Lü et al., 2024). Its efficiency is influenced by community composition and environmental factors such as pH, temperature, and nutrients (Xu & Yu, 2021; Luo et al., 2025). Controlled nutrient supplementation can further optimize interspecies interactions, supporting the design of stable and efficient engineered microbial systems (Liao et al., 2020).

Losoi et al. (2019) developed a synthetic consortium of Escherichia coli *and* Acinetobacter baylyi *ADP1 with engineered metabolic interdependence through reciprocal gluconate and acetate exchange, enabling controlled growth without external inducers. The system showed stable performance across varying conditions and demonstrated reliable population balance using fluorescent reporters. This study highlights cross-feeding as an effective strategy for controlling microbial community stability, supported by broader evidence from ADMC research (Ziesack et al., 2019; Culp and Goodman, 2023).*

4.3 Construction of ADMCs: Engineering and Ecological Strategies

Since the composition and spatial organization of microbial communities directly influence the efficiency and pathways of pollutant degradation, the deliberate design of microbial consortia plays a pivotal role in enhancing bioremediation outcomes. Numerous studies have demonstrated that artificially designed microbial communities (ADMCs) can markedly improve degradation efficiency and environmental adaptability by exploiting metabolic interactions and cooperative behaviour among constituent strains (Chen et al., 2025). Accordingly, the construction of ADMCs using top-down and bottom-up approaches has been extensively reviewed in the context of synthetic microbial community design (Lyu et al., 2024; Peng et al., 2025).

Growing concerns beyond bioremediation such as climate change and global environmental degradation have further accelerated interest in engineered microbial ecosystems. ADMCs are increasingly being explored as sustainable biocatalysts for diverse applications, including the replacement of petrochemical-based processes with bio-based chemical production, the development of biofertilizers as alternatives to synthetic fertilizers, and other environmentally friendly biotechnological solutions. These emerging applications underscore the transformative potential of microbial communities in supporting a circular and low-carbon bioeconomy.

However, the inherent biological complexity of microbial communities poses significant challenges to their rational design and functional optimization. Achieving reliable performance in biotechnological applications requires a deeper understanding of microbial interactions, community dynamics, and emergent properties, which can only be addressed through systems-level approaches. While many reviews have focused on either rational bottom-up engineering or evolutionary top-down strategies (Dahabieh et al., 2020), recent studies increasingly emphasize the advantages of integrating both paradigms. This hybrid top-down/bottom-up framework leverages the strengths of each approach, offering enhanced control, robustness, and functional performance in ADMCs (León and Nogales, 2022).

4.4 Bottom-up Approach

The bottom-up approach to engineering microbial systems involves assembling simpler biological components into increasingly complex structures, such that individual parts and organisms function as subsystems of an emergent community (Leon and Nogales, 2022). This strategy relies on detailed knowledge of the physicochemical and kinetic properties of community components, with complexity increasing along the biological hierarchy from DNA parts and metabolic pathways to whole organisms and ecosystems. Because bottom-up engineering depends on a thorough understanding of fundamental biological mechanisms, it is guided by rational design principles and requires extensive expertise to support informed engineering decisions. While computational tools for designing low-complexity components such as genetic parts, pathways, or individual organisms are well established, tools specifically aimed at the rational design of whole microbial communities are only beginning to emerge (Jones et al., 2022).

4.5 Top Down Approach

Top-down engineering reduces microbial community complexity by using ecological and evolutionary selection pressures to steer natural or semi-natural communities toward desired functions without requiring detailed prior knowledge of individual members' roles. It relies on adaptive processes and community assembly principles to develop stable, cooperative microbial systems that can outperform monocultures in tasks such as organic matter degradation or synthesis. Functional insights are later derived using genome-wide and meta-omics analyses, which link molecular interactions to observed outcomes. Although less controlled than bottom-up approaches, evolutionary engineering has successfully revealed novel functional modules, and emerging computational tools are improving its rational design potential. However, challenges remain, particularly in cultivating diverse microorganisms and maintaining long-term stability, as competitive dynamics may lead to dominance by a few taxa (Díaz-García et al., 2021; Chang et al., 2021a; Xie et al., 2021).

An alternative framing describes engineered microbial communities (EngComs) and synthetic communities (SynComs) as corresponding to top-down and bottom-up approaches, respectively. EngComs are developed through ecological selection processes such as enrichment, artificial selection, or directed evolution, where environmental pressures shape existing microbial communities toward enhanced biodegradation. In contrast, SynComs are built bottom-up through the deliberate assembly of isolated or engineered strains designed to perform specific biodegradation functions (Chen et al., 2025).

The distinction between these terms reflects differences in design philosophy rather than fundamentally different engineering strategies. "Top-down" and "bottom-up" emphasize the direction of construction, while "EngComs" and "SynComs" highlight the level of intentional control and ecological complexity involved. Together, they describe a continuum ranging from ecological selection-driven systems to highly rationally assembled microbial communities used in bioremediation.

4.6 Hybrid Approach

Bottom-up and top-down approaches each have clear strengths but also important weaknesses. Bottom-up design offers precise control through rational assembly, but it can limit metabolic diversity and lead to suboptimal performance. Top-down strategies explore a wider metabolic and ecological space through selection and evolution, but they are constrained by limited mechanistic understanding, which reduces design precision. To address these challenges, researchers are increasingly adopting integrated "middle-out" or hybrid approaches that combine rational community construction with ecological regulation using environmental factors **(Fig. 9.2)** (Liang et al., 2022; Leon and Nogales, 2022).

Figure 2. Approaches for Artificially Designed Microbial Consortia

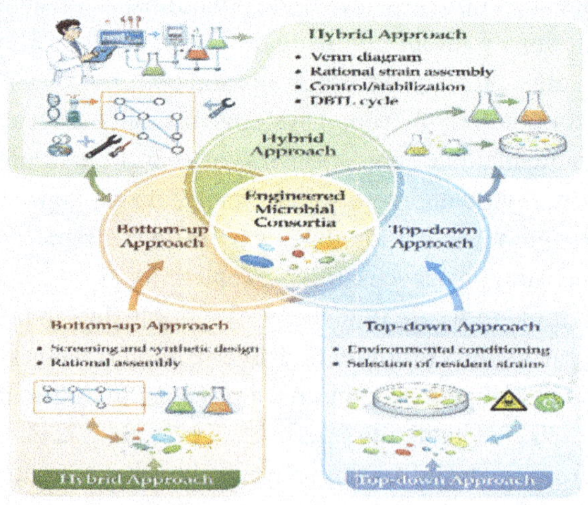

In these hybrid systems, consortia are first assembled by adjusting inoculation ratios and then optimized through environmental conditions such as temperature, nutrients, and cultivation settings. For example, temperature cycling has been shown to improve the stability and composition of a synthetic *Escherichia coli* and *Pseudomonas putida* consortium (Krieger et al., 2021). This integrated strategy improves robustness and performance, as microbial communities often benefit from functional exchange between bottom-up and top-down design elements.

4.7 Design Frameworks for ADMCs

The design framework aims to assemble a stable microbial community that is resilient to environmental perturbations. Hence, it provides step-by-step guidance on turning principles into functional consortia, integrating theory with iterative experimentation. A good design framework starts with a clear definition of the problem, develops a conceptual ecosystem model, creates quantitative models, identifies the microbial community processes of interest, and develops and evaluates candidate design strategies.

The Design-Build-Test-Learn (DBTL) cycle proposed by Lawson and team (Lawson et al., 2019) serves as the blueprint for ADMCs. The Design phase employs computational models to predict interactions; Build assembles strains through high-throughput methods such as *in situ* metagenomic engineering (delivery of engineered mobile genetic elements to resident microorganisms) and CRISPR-Cas gene editing techniques; Test evaluates using multi-omics and function. For instance, high-throughput phenotypic testing using droplet microfluidics or fully automated microbioreactor platforms that combine liquid handling and advanced sensing with microlitre plates or scaled-down bioreactors; and Learn refines models with data, accelerating optimization.

Ecological frameworks emphasize interaction engineering (e.g., balancing mutualism with antagonism), hierarchical orchestration (keystone governance), and metabolic division of labour (SDOL/MDOL for static/dynamic partitioning). These address stability through niche allocation and anti-cheater mechanisms, often hybridizing top-down refinement with bottom-up assembly.

Computational frameworks like constraint-based modelling with defined boundaries (e.g., environmental limits), simulation of multi-scale dynamics, flux balance analysis (FBA) for exploring combinations of chemical transformations, optimization of metabolism, agent-based models that track spatial competition, and AI/ML that integrates omics for strain prediction.

Evolutionary frameworks apply artificial selection via top-down culturing, bottom-up CRISPR editing, or model-driven hybrids to resolve trade-offs, enabling responsive ADMCs under environmental flux.

Integrated paradigms, such as SuperCC platforms, combine these for predictive engineering. An example is the herbicide-degrading consortia formed through ecological filtering and cross-feeding simulation. Efforts to standardize these frameworks, including metadata protocols and modular ecosystems, will further scale frameworks toward reproducible ecotechnologies.

5. APPLICATIONS OF MICROBIAL CONSORTIA AND ADMCS IN BIOREMEDIATION

Microbial consortia and Artificially Designed Microbial Communities (ADMCs) have significantly increased the potential and ecological span of bioremediation technologies more than any other technology in the last twenty years. Natural microbial consortia also exhibit metabolic plasticity, synergies, and functional complementarities that are not possible in isolated organisms (Brenner et al., 2008; Shade et al., 2018). These characteristics play a key role in the dismantling of persistent organic pollutants (POPs), heavy metals, hydrocarbons, dye contaminants, pesticides, microplastics, and newer pollutants such as pharmaceuticals and polyfluoroalkyl substances (PFAS). The principles of synthetic biology, metabolic engineering, genome editing, and microbial ecology are used to expand these natural advantages, enabling ADMCs to design predictable, modular communities that can control the breakdown of pollutants with greater efficacy and stability (Johns et al., 2022). Microbial consortia can undergo a sequence of steps needed to fully mineralize contaminants under the influence of synergistic interaction vested in cross-feeding, cross-feeding regulation, niche partitioning, and cooperative stress tolerance (Li et al., 2023).

5.1 Hydrocarbons and Petroleum Waste Degradation

The chemically diverse quantities of compounds found in crude oil and petroleum products include linear and branched alkanes, aromatic hydrocarbons, and polycyclic aromatic hydrocarbons (PAHs), which are broken down by a variety of enzymes and redox reactions (Varjani, 2017). In natural microbial consortia, the division of labour means that taxa exhibit complementary metabolic competencies; that is, various taxa have specific metabolic activities. For example, alkane-degrading taxa such as *Alcanivorax*, *Marinobacter*, and *Rhodococcus* use monooxygenase and dioxygenase systems for cleavage. In contrast, aromatic hydrocarbon-mineralising genera, such as *Sphingomonas*, *Mycobacterium*, and *Cycloclasticus*, use ring cleavage (Prince et al., 2018). Biodegradation is significantly boosted by synergistic interactions in these communities, especially in the marine environment, where biosurfactant-producing bacteria also augment bioavailability of the hydrocarbons (Dell'Anno et al., 2021).

Hydrocarbon bioremediation can be enhanced through ADMCs with engineered catabolic pathways or improved biosurfactant production capabilities. Recombinant *alkB* (for degradation of alkanes) or "ndo" (naphthalene degradation operon) gene clusters have been incorporated into synthetic consortia to enable degradation of high-molecular-weight molecules or tolerance of the toxic intermediate (Santala et al., 2018). Overengineered consortia have been observed to perform better in harsh conditions, such as Arctic waters and deep subsea reservoirs, where individual strains exhibit slowed metabolic activity (Kubota et al., 2020).

5.2 Detoxification of Heavy Metals and Metalloids

Heavy metals are not biodegradable as organic pollutants; hence, bioremediation aims to convert them to less toxic or non-toxic forms. Microbial consortia are useful in the detoxification of metals through different mechanisms such as biosorption, biomineralization, redox transformation, methylation and demethylation, and metal precipitation (Valls & De Lorenzo, 2002). The H_2S gas produced by sulphate-reducing bacteria (SRB) reacts with dissolved heavy metal cations such as lead, cadmium, and zinc to form insoluble metal sulphides. Methanogens and iron-reducing bacteria (IRB) found in consortia stabilize the flow of electrons, providing the best redox environment for metal transformation (Wu et al., 2019).

ADMCs have been designed to increase metal transformation through microbial redoxing. Demonstrated improvements in the reduction of chromium (Cr(VI)) and arsenic (As) in groundwater and industrial wastewater include modified strains of *Shewanella oneidensis* and *Geobacter sulfurreducans* expressing enhanced cytochromes or metal-binding peptides (Poulos et al., 2017). Multi-species engineered to sustain redox stratification allow metals with multiple oxidation states to react sequentially.

5.3 Degradation of Pesticides and Agrochemicals

Pesticides are prone to degradation owing to various factors that influence the eventual release of their constituents in the soil. Pesticides, such as organophosphates, neonicotinoids, carbamates, and chlorinated aromatic compounds, pose a grave environmental concern because they are stable and toxic. By taking advantage of sequential hydrolysis, dechlorination, side-chain cleavage, and ultimate mineralization, microbial consortia offer an expanded enzymatic repertoire compared to their individual counterparts (Singh & Walker, 2006). *Paracoccus*, *Bacillus*, and *Sphingobium* host natural consortia that consume organophosphates by acting synergistically with phosphatases and hydrolases (Cycoń *et al.,* 2019). These types of consortia restore microbial balance and plant-microbe relationships destroyed by prolonged pesticide use. Engineered hydrolase genes, e.g., Organophosphorus Hydrolase (OPH) and Methyl Parathion Hydrolase (MPH) encoding genes introduced into ADMCs can enhance degradation levels and increase specificity (Li et al., 2022).

5.4 Treatment of Wastewater and Industrial Effluents

The use of microbial consortia in wastewater treatment is based on trophic networks, which are complex and allow the elimination of organic, nitrogen, phosphorus, and xenobiotics. There are thousands of microbial taxa in activated sludge that participate in interactions mediating nitrification, denitrification, ammonification, fermentation, and methanogenesis (Wagner & Loy, 2002). In anaerobic digesters, polysaccharides are digested in fermenters, where they are converted to volatile fatty acids, metabolized to acetate by syntrophic bacteria, and finally converted to methane by methanogens. In the absence of such interdependent metabolism, wastewater treatment performance would break down.

Wastewater system-orientated ADMCs include modified ammonia-oxidizing microorganisms, modified denitrifiers, and refractory compound degradation strains, e.g., chlorinated solvents and pharmaceuticals (Dinh *et al.,* 2020). Artificial quorum-sensing pathways enable ADMCs to self-regulate metabolism in response to pollutant load, enhancing shock load resilience and response to changes in influent composition.

5.5 Degradation of Plastics

The environmental problem of plastic pollution is one of the most enduring because the rubber materials, including PET (polyethylene terephthalate), PE (polyethylene), PU (polyurethane), and PVC (polyvinyl chloride) are polymeric in their structure and have chemical uncrystallization.

Natural microbial consortia drive degradation through oxidative changes and both bacterial and fungal hydrolytic and mineralization processes (Urbanek *et al.,* 2018). The oxidation of surfaces using enzymes such as laccases and peroxidases, catalyzed by fungal species such as *Aspergillus* and *Penicillium*, increases polymer hydrophilicity. Oligomers and monomers are then hydrolyzed by bacteria using enzymes called PETase and MHETase (Yoshida et al., 2016).

6. CASE STUDIES OF MICROBIAL CONSORTIA ADMCS IN BIOREMEDIATION

The case studies provided below are a heterogeneous sample of contaminants, including hydrocarbons, halogenated solvents, heavy metals, plastics, pesticides, pharmaceutical residues, and PFAS (Per- and polyfluoroalkyl substances), each demonstrating the peculiarities of microbial collaboration, their persistence, and their conversion of pollution.

Case Study 1: Hydrocarbon-Degrading Marine Consortia Following the Deepwater Horizon Spill

The Deepwater Horizon oil spill demonstrated the critical role of natural microbial consortia in marine hydrocarbon biodegradation. The spill stimulated the enrichment of specialized hydrocarbon-degrading microorganisms, including *Alcanivorax, Cycloclasticus, Colwellia,* and *Marinobacter*, which participated in alkane, PAH, methane, and gaseous hydrocarbon degradation through specialized metabolic pathways (Hazen et al., 2010). Biosurfactant-producing bacteria further enhanced hydrocarbon bioavailability and accelerated biodegradation processes (Dell'Anno et al., 2021). Overall, the incident highlighted the adaptive capacity of marine microbial communities and showed how environmental disturbances can naturally select microorganisms with strong catabolic potential for ecosystem recovery.

Case Study 2: Dehalococcoides mccartyi ADMC for Chlorinated Solvent Remediation

The KB-1 microbial consortium has proven effective in the bioremediation of chlorinated solvents such as trichloroethylene (TCE) and perchloroethylene (PCE), which commonly contaminate groundwater due to industrial activities. The consortium is driven by *Dehalococcoides*, a bacterium capable of converting these toxic compounds into harmless ethene under anaerobic conditions (Maymó-Gatell et al., 1997). Its efficiency depends on cooperative interactions among microorganisms, where fermentative bacteria supply hydrogen needed for dechlorination, while methanogens help maintain system stability (Hendrickson et al., 2016). Field applications across several continents have demonstrated over 95% reduction in chlorinated solvents, highlighting the practical value of microbial consortia in sustainable groundwater remediation.

Case Study 3: Sulphate-Reducing Consortia for Acid Mine Drainage (AMD) Remediation

Acid mine drainage is a serious environmental challenge characterized by high acidity and elevated levels of dissolved metals such as Fe, Mn, Pb, and As. Sulphate-reducing bacterial (SRB) consortia offer an effective bioremediation strategy by producing hydrogen sulphide, which precipitates metals as insoluble sulphides, reducing both toxicity and acidity. These systems function through synergistic interactions with fermentative microorganisms and methanogens that maintain favourable redox and pH conditions. Field applications at Iron Mountain Mine (USA) and Dexing Copper Mine (China) achieved up to 90–95% iron removal efficiency and substantial heavy metal reduction (Wu et al., 2019).

Case Study 4: Plastic Biodegradation Through *Ideonella sakaiensis*–Fungal Consortia

The increasing accumulation of polyethylene terephthalate (PET) plastics has driven research into microbial degradation strategies. Consortia combining bacteria and fungi have shown enhanced PET breakdown through complementary enzymatic actions. *Ideonella sakaiensis* produces PETase and MHETase for hydrolysis, while ligninolytic fungi such as *Aspergillus* and *Penicillium* initiate surface oxidation using laccase and peroxidase, improving polymer accessibility (Yoshida et al., 2016; Urbanek et al., 2018). Such synergistic systems have achieved 50–60% PET mass reduction in pilot studies. This highlights the effectiveness of engineered microbial consortia and ADMC principles for modular, multi-step plastic degradation processes.

Case Study 5: Soil Remediation of Organophosphate Pesticides Using Multi-Strain Consortia

The increasing accumulation of polyethylene terephthalate (PET) plastics has driven research into microbial degradation strategies. Consortia combining bacteria and fungi have shown enhanced PET breakdown through complementary enzymatic actions. *Ideonella sakaiensis* produces PETase and MHETase for hydrolysis, while ligninolytic fungi such as *Aspergillus* and *Penicillium* initiate surface oxidation using laccase and peroxidase, improving polymer accessibility (Yoshida et al., 2016; Urbanek et al., 2018). Such synergistic systems have achieved 50–60% PET mass reduction in pilot studies. This highlights the effectiveness of engineered microbial consortia and ADMC principles for modular, multi-step plastic degradation processes.

7. CHALLENGES AND FUTURE DIRECTIONS

Despite their promising advances for sustainable pollution control, significant scientific and practical barriers and regulatory challenges still limit their widespread application. Understanding these constraints, while identifying potential technological and ecological solutions, is essential for advancing the next generation of bioremediation strategies (Kumar et al., 2025).

A primary barrier to the effective application of microbial consortia is the difficulty associated with scaling up from controlled experimental systems to real-world contaminated sites. Laboratory studies are typically conducted under optimized and stable conditions that do not accurately reflect the spatial and temporal heterogeneity of natural environments. In the field, microorganisms are exposed to fluctuating temperatures, variable pH, limited nutrient availability, and inconsistent contaminant distribution. These factors can significantly influence microbial metabolism and interaction patterns, often resulting in reduced remediation efficiency compared to laboratory predictions (Garbisu et al., 2017; Wang et al., 2024). Moreover, achieving uniform distribution and sustained activity of introduced consortia across large or heterogeneous sites remains a major logistical challenge.

Another significant concern is community stability. The effectiveness of microbial consortia depends on maintaining balanced interactions such as metabolic cooperation, cross-feeding, and syntrophy among member species. However, these interactions are inherently sensitive to environmental disturbances. Changes in pollutant concentrations, nutrient inputs, or redox conditions can shift competitive dynamics within the consortium, allowing certain species to dominate while others decline or disappear. This loss of functional diversity may disrupt key degradation pathways, leading to incomplete remediation or the accumulation of toxic intermediates. In addition, competition with indigenous microbial populations can further destabilize introduced consortia, making long-term performance unpredictable (Potts et al., 2022).

Closely related to stability is the challenge of predictability and control. Microbial communities exhibit complex, nonlinear behaviour that are difficult to forecast, particularly in open environmental systems. Interactions among community members are often context-dependent, meaning that a consortium optimized for one site may perform poorly in another. Fujita et al. 2023 reported that lack of predictability to consortia response to environmental stressors or management hinders reproducibility and complicates the design of universally applicable remediation strategies.

Furthermore, another challenge lies in the complexity of pollutants commonly encountered at polluted sites. Most contaminated areas contain a mixture of hydrocarbons, heavy metals, pesticides, and synthetic recalcitrant compounds, each requiring distinct metabolic pathways and microbial capabilities to degrade effectively.

The multifunctionality required from microbial consortia to simultaneously act on diverse contaminants necessitates diverse and complementary metabolic activities among the consortium members. Designing such multifunctional consortia involves a deep understanding of microbial ecology, physiology, and interspecies interaction networks that are still being elucidated. Additionally, the degradation of pollutants can generate toxic intermediate metabolites, which can inhibit microbial growth or affect the stability of the community (Zhang & Zhang, 2022).

Beyond scientific and technical constraints, regulatory limitations pose substantial hurdles to the deployment of microbial consortia. The challenge of regulatory frameworks inability to address the utilization of complex microbial communities, especially those that are engineered or introduced into new environments. Concerns regarding ecological imbalance, unintended persistence, horizontal gene transfer, and potential risks to human and environmental health have resulted in cautious regulatory approaches. The lack of standardized guidelines for evaluating the safety, efficacy, and long-term impacts of microbial consortia further slows approval processes and discourages commercial investment (Chemla et al., 2025; Nunes et al., 2024).

7.1 Emerging Opportunities: Omics, Artificial Intelligence, and Biotechnological Innovation

Despite these challenges, rapid advancements in analytical and computational tools offer new avenues for improving the design and application of microbial consortia. Omics technologies, including metagenomics, metatranscriptomics, metaproteomics, and metabolomics, have revolutionized the study of microbial communities by enabling comprehensive profiling of their genetic potential, functional activity, and metabolic outputs. These omics technologies provide insights into the roles of individual species, the identification of keystone organisms, and the mechanisms underlying cooperative and competitive interactions. Such knowledge can guide the selection or enrichment of microbial communities with enhanced pollutant-degrading capabilities (Ceballos Rodriguez-Conde et al., 2025).

The integration of artificial intelligence (AI) and machine learning with omics data represents a powerful strategy for overcoming the complexity of microbial systems. It is common knowledge that AI-based models have the capacity of processing large data in other to unlock patterns that are unknown and to propose solutions to future occurrences. This predictive capacity can support the rational assembly of microbial consortia, optimization of environmental parameters, and anticipation of stability issues before field implementation (Chang et al., 2021b; Toussaint et al., 2024). Over time, AI-driven frameworks may enable adaptive bioremediation approaches, where microbial communities are dynamically managed based on real-time monitoring and performance feedback.

Progress in biotechnology and synthetic biology further enhances the potential for precise community engineering. Genetic modification and metabolic pathway optimization can be used to improve degradation efficiency, stress tolerance, or interspecies communication. Synthetic consortia composed of well-characterized strains offer greater reproducibility and controllability than naturally assembled communities (Johns et al., 2016). Additionally, engineering signaling networks such as quorum sensing systems may allow coordinated regulation of metabolic activities across consortium members, strengthening synergistic interactions and reducing competitive imbalances (Guo et al., 2025).

Complementary technologies also play an important role in advancing consortia-based remediation. The incorporation of microbial communities into immobilization matrices, biofilms, or bioelectrochemical systems can protect microorganisms from environmental stress while enhancing electron transfer and substrate accessibility. Such hybrid systems can improve process efficiency and operational stability, particularly in challenging environments (Lovley & Nevin, 2011; Armanu et al., 2025).

7.2 Strategic Recommendations for Future Research and Application

To fully realize the potential of microbial consortia, future research must move beyond short-term laboratory experiments and prioritize long-duration field trials. These studies are essential for evaluating community persistence, functional resilience, and real-world effectiveness under fluctuating environmental conditions. Greater emphasis should also be placed on applying ecological theory, including concepts such as succession, redundancy, and resilience, to guide consortium design and management (Allison & Martiny, 2008; Grandel et al., 2025).

The development of standardized protocols for consortium construction, monitoring, performance evaluation, and risk assessment is equally critical. Standardization would enhance reproducibility, facilitate regulatory review, and promote wider adoption of consortia-based technologies. Achieving this goal will require interdisciplinary collaboration among microbiologists, environmental engineers, computational scientists, and regulatory authorities (Novak et al., 2025).

From an applied perspective, microbial consortia should be integrated into multifunctional remediation frameworks. Combining microbial processes with phytoremediation, constructed wetlands, or resource recovery systems can enhance pollutant removal while contributing to ecosystem restoration and circular economy objectives. Such integrated approaches can deliver environmental, economic, and social benefits simultaneously (Ayub et al., 2025).

Finally, addressing regulatory and public concerns must be a central component of future efforts. Transparent risk assessment, long-term environmental monitoring, and proactive stakeholder engagement will be essential for building trust and acceptance. Clear and adaptive regulatory pathways that balance innovation with environmental protection will accelerate the responsible deployment of microbial consortia (Chemla et al., 2025).

8. CONCLUSION

Microbial consortia represent a transformative approach to bioremediation by harnessing cooperative interactions and collective metabolic potential. However, challenges related to scale-up, stability, predictability, and regulation currently limit their widespread application. Advances in omics technologies, artificial intelligence, and biotechnology provide a robust foundation for overcoming these constraints. By integrating ecological understanding with innovative engineering and supportive policy frameworks, microbial consortia can be effectively developed into reliable and sustainable solutions for mitigating complex environmental pollutions.

REFERENCES

Allison, S. D., & Martiny, J. B. H. (2008). Resistance, resilience, and redundancy in microbial communities. *Proceedings of the National Academy of Sciences of the United States of America, 105*(Suppl. 1), 11512–11519. DOI: 10.1073/pnas.0801925105 PMID: 18695234

Armanu, E. G., Bertoldi, S., Chrzanowski, Ł., Volf, I., Heipieper, H. J., & Eberlein, C. (2025). Benefits of Immobilized Bacteria in Bioremediation of Sites Contaminated with Toxic Organic Compounds. *Microorganisms, 13*(1), 155. DOI: 10.3390/microorganisms13010155 PMID: 39858923

Ayub, A., Wani, A. K., Malik, S. M., Ayub, M., Chopra, C., Singh, R., & Malik, T. (2025). Harnessing microbes and plants for bioremediation of heavy metal contaminants: Current paradigms and future perspectives. *Environmental Challenges, 20*, 101220. DOI: 10.1016/j.envc.2025.101220

Bayatian, M., Pourbabaee, A. A., & Amoozegar, M. A. (2025). Revealing the composition of bacterial communities in various oil-contaminated soils and investigating their intrinsic traits in hydrocarbon degradation. *Scientific Reports, 15*(1), 22016. Advance online publication. DOI: 10.1038/s41598-025-05519-7 PMID: 40596159

Bhandari, S., Poudel, D. K., Marahatha, R., Dawadi, S., Khadayat, K., Phuyal, S., & Parajuli, N. (2021). Microbial enzymes used in bioremediation. *Journal of Chemistry, 2021*(1), 8849512.

Brenner, K., You, L., & Arnold, F. H. (2008). Engineering microbial consortia: A new frontier in synthetic biology. *Trends in Biotechnology, 26*(9), 483–489. DOI: 10.1016/j.tibtech.2008.05.004 PMID: 18675483

Ceballos Rodriguez-Conde, F., Zhu, S., & Dikicioglu, D. (2025). Harnessing microbial division of labor for biomanufacturing: A review of laboratory and formal modeling approaches. *Critical Reviews in Biotechnology, 45*(6), 1–19. DOI: 10.1080/07388551.2025.2455607 PMID: 39972973

Chang, C. Y., Vila, J. C., Bender, M., Li, R., Mankowski, M. C., Bassette, M., Borden, J., Golfier, S., Sanchez, P. G. L., Waymack, R., Zhu, X., Diaz-Colunga, J., Estrela, S., Rebolleda-Gomez, M., & Sanchez, A. (2021a). Engineering complex communities by directed evolution. *Nature Ecology & Evolution, 5*(7), 1011–1023. DOI: 10.1038/s41559-021-01457-5 PMID: 33986540

Chang, Y., Guo, X., Chen, Y., & Wang, J. (2021b). Machine learning approaches for microbial community analysis in environmental biotechnology. *Biotechnology Advances, 47*, 107696.

Chao, H., Cai, A., Heimbuger, B., Wu, Y., Zhao, D., Sun, M., & Hu, F. (2024). Keystone taxa enhance the stability of soil bacterial communities and multifunctionality under steelworks disturbance. *Journal of Environmental Management*, *356*, 120664. DOI: 10.1016/j.jenvman.2024.120664 PMID: 38508006

Chemla, Y., Sweeney, C. J., Wozniak, C. A., & Voigt, C. A. (2025). Design and regulation of engineered bacteria for environmental release. *Nature Microbiology*, *10*(2), 281–300. DOI: 10.1038/s41564-024-01918-0 PMID: 39905169

Chen, C., Li, K., Lin, X., Li, W., & Xu, X. (2025). Engineering natural microbial communities: Harnessing synthetic communities for bioremediation. *Current Opinion in Microbiology*, *87*, 102652. DOI: 10.1016/j.mib.2025.102652 PMID: 40768900

Chen, W., Sun, J., Ji, R., Min, J., Wang, L., Zhang, J., Qiao, H., & Cheng, S. (2024). Crude Oil Biodegradation by a Biosurfactant-Producing Bacterial Consortium in High-Salinity Soil. *Journal of Marine Science and Engineering*, *12*(11), 2033. DOI: 10.3390/jmse12112033

Culp, E. J., & Goodman, A. L. (2023). Cross-feeding in the gut microbiome: Ecology and mechanisms. [External Link]. *Cell Host & Microbe*, *31*(4), 485–499. DOI: 10.1016/j.chom.2023.03.016 PMID: 37054671

Cycoń, M., Mrozik, A., & Piotrowska-Seget, Z. (2019). Antibiotics in the soil environment—Degradation and their impact on microbial activity and diversity. *Frontiers in Microbiology*, *10*, 338. DOI: 10.3389/fmicb.2019.00338 PMID: 30906284

D'Souza, G., Shitut, S., Preussger, D., Yousif, G., Waschina, S., & Kost, C. (2018). Ecology and evolution of metabolic cross-feeding interactions in bacteria. *Natural Product Reports*, *35*(5), 455–488. DOI: 10.1039/C8NP00009C PMID: 29799048

Dahabieh, M. S., Thevelein, J. M., & Gibson, B. (2020). Multimodal microorganism development: Integrating top-down biological engineering with bottom-up rational design. *Trends in Biotechnology*, *38*(3), 241–253. DOI: 10.1016/j.tibtech.2019.09.006 PMID: 31653446

Dang, H., & Chen, C. A. (2017). Ecological Energetic Perspectives on Responses of Nitrogen-Transforming Chemolithoautotrophic Microbiota to Changes in the Marine Environment. *Frontiers in Microbiology*, *8*, 1246. DOI: 10.3389/fmicb.2017.01246 PMID: 28769878

Delgado-Baquerizo, M., Reich, P. B., Trivedi, C., Eldridge, D. J., Abades, S., Alfaro, F. D., Bastida, F., Berhe, A. A., Cutler, N. A., Gallardo, A., García-Velázquez, L., Hart, S. C., Hayes, P. E., He, J.-Z., Hseu, Z.-Y., Hu, H.-W., Kirchmair, M., Neuhauser, S., Pérez, C. A., & Singh, B. K. (2020). Multiple elements of soil biodiversity drive ecosystem functions across biomes. *Nature Ecology & Evolution*, *4*(2), 210–220. DOI: 10.1038/s41559-019-1084-y PMID: 32015427

Dell'Anno, F., Rastelli, E., Sansone, C., Brunet, C., Ianora, A., & Dell'Anno, A. (2021). Bacteria, fungi and microalgae for the bioremediation of marine sediments contaminated by petroleum hydrocarbons in the omics era. *Microorganisms*, *9*(8), 1695. DOI: 10.3390/microorganisms9081695 PMID: 34442774

Dexter, H. S., & Lu, T. (2022). Engineering microbial consortia with rationally designed cellular interactions. *Current Opinion in Biotechnology*, *76*, 102730. DOI: 10.1016/j.copbio.2022.102730 PMID: 35609504

Díaz-García, L., Huang, S., Spröer, C., Sierra-Ramírez, R., Bunk, B., Overmann, J. & Overmann D. J. (2021). Dilution-to-stimulation/extinction method: a combination enrichment strategy to develop a minimal and versatile lignocellulolytic bacterial consortium. *Applied Environmental Microbiology,* (2) 2021, 87:e02427-20, https://doi.org/10. 1128/AEM.02427-20

Dinh, C. T., Tran, N. H., Nguyen, M. N., Hong, C. S., Bao, W., Zomaya, A. Y., & Gramoli, V. (2020). Federated learning over wireless networks: Convergence analysis and resource allocation. *IEEE/ACM Transactions on Networking*, *29*(1), 398–409. DOI: 10.1109/TNET.2020.3035770

Flemming, H. C., van Hullebusch, E. D., Neu, T. R., Nielsen, P. H., Seviour, T., Stoodley, P., Wingender, J., & Wuertz, S. (2023a). The biofilm matrix: Multitasking in a shared space. *Nature Reviews. Microbiology*, *21*(2), 70–86. DOI: 10.1038/s41579-022-00791-0 PMID: 36127518

Flemming, H.-C., Wuertz, S., Wingender, J., & Szewzyk, U. (2023b). Biofilms: An emergent form of bacterial life. *Nature Reviews. Microbiology*, *21*(2), 70–85. DOI: 10.1038/s41579-022-00791-0 PMID: 36127518

Fujita, H., Ushio, M., Suzuki, K., Abe, M. S., Yamamichi, M., Iwayama, K., Canarini, A., Hayashi, I., Fukushima, K., Fukuda, S., Kiers, E. T., & Toju, H. (2023). Alternative stable states, nonlinear behavior, and predictability of microbiome dynamics. *Microbiome*, *11*(1), 63. DOI: 10.1186/s40168-023-01474-5 PMID: 36978146

Gadd, G. M. (2010). Metals, minerals and microbes: Geomicrobiology and bioremediation. *Microbiology (Reading, England)*, *156*(3), 609–643. DOI: 10.1099/mic.0.037143-0 PMID: 20019082

Garbisu, C., Garaiyurrebaso, O., Epelde, L., Grohmann, E., & Alkorta, I. (2017). Plasmid-mediated bioaugmentation for the bioremediation of contaminated soils. *Frontiers in Microbiology*, *8*, 1966. DOI: 10.3389/fmicb.2017.01966 PMID: 29062312

Grandel, N. E., Alexander, A. M., Peng, X., Palamountain, C., Alnahhas, R. N., Hirning, A. J., Josić, K., & Bennett, M. R. (2025). Long-term homeostasis in microbial consortia via auxotrophic cross-feeding. *Nature Communications*, *16*(1), 8573. DOI: 10.1038/s41467-025-63575-z PMID: 41022726

Guo, Y., Gao, M., Jiang, L., Huang, H., Kang, G., & Yu, H. (2025). Synthetic microbial consortia based on quorum-sensing for disease therapy. *Bioresources and Bioprocessing*, *12*(1), 125. DOI: 10.1186/s40643-025-00962-w PMID: 41182556

Hawley, E. R., Piao, H., Scott, N. M., Malfatti, S., Pagani, I., Huntemann, M., Chen, A., Glavina del Rio, T., Foster, B., Copeland, A., Jansson, J., Pati, A., Tringe, S., Gilbert, J. A., Lorenson, T. D., & Hess, M. (2014). Metagenomic analysis of microbial consortium from natural crude oil that seeps into the marine ecosystem offshore Southern California. *Standards in Genomic Sciences*, *9*(3), 1259–1274. DOI: 10.4056/sigs.5029016 PMID: 25197496

Hazen, T. C., Dubinsky, E. A., DeSantis, T. Z., Andersen, G. L., Piceno, Y. M., Singh, N., Jansson, J. K., Probst, A., Borglin, S. E., Fortney, J. L., Stringfellow, W. T., Bill, M., Conrad, M. E., Tom, L. M., Chavarria, K. L., Alusi, T. R., Lamendella, R., Joyner, D. C., Spier, C., & Mason, O. U. (2010). Deep-sea oil plume enriches indigenous oil-degrading bacteria. *Science*, *330*(6001), 204–208. DOI: 10.1126/science.1195979 PMID: 20736401

Hendrickson, D., Kelley, D. R., Tenen, D., Bernstein, B., & Rinn, J. L. (2016). Widespread RNA binding by chromatin-associated proteins. *Genome Biology*, *17*(1), 28. DOI: 10.1186/s13059-016-0878-3 PMID: 26883116

Ibo, E. M., Adeogun, A. O., Orji, M. U., & Umeh, O. R. (2024). Harnessing microbial synergy: A comprehensive evaluation of consortia-mediated bioremediation strategies for petroleum refinery wastewater treatment. *Cleaner Water*, *2*, 100055. DOI: 10.1016/j.clwat.2024.100055

Johns, G., Samuel, V., Freemantle, L., Lewis, J., & Waddington, L. (2022). The global prevalence of depression and anxiety among doctors during the covid-19 pandemic: Systematic review and meta-analysis. *Journal of Affective Disorders*, *298*, 431–441. DOI: 10.1016/j.jad.2021.11.026 PMID: 34785264

Johns, N. I., Blazejewski, T., Gomes, A. L. C., & Wang, H. H. (2016). Principles for designing synthetic microbial communities. *Current Opinion in Microbiology, 31*, 146–153. DOI: 10.1016/j.mib.2016.03.010 PMID: 27084981

Jones, T. S., Oliveira, S. M., Myers, C. J., Voigt, C. A., & Densmore, D. (2022). Genetic circuit design automation with Cello 2.0. *Nature Protocols, 17*(4), 1097–1113. DOI: 10.1038/s41596-021-00675-2 PMID: 35197606

Kaczorek, E., Pacholak, A., Zdarta, A., & Smułek, W. (2018). The Impact of Biosurfactants on Microbial Cell Properties Leading to Hydrocarbon Bioavailability Increase. *Colloids and Interfaces, 2*(3), 35. DOI: 10.3390/colloids2030035

Krieger, A. G., Zhang, J., & Lin, X. N. (2021). Temperature regulation as a tool to program synthetic microbial community composition. *Biotechnology and Bioengineering, 118*(3), 1381–1392. DOI: 10.1002/bit.27662 PMID: 33399224

Kubota, Y., Shiono, T., Kusumoto, B., & Fujinuma, J. (2020). Multiple drivers of the COVID-19 spread: The roles of climate, international mobility, and region-specific conditions. *PLoS One, 15*(9), e0239385. DOI: 10.1371/journal.pone.0239385 PMID: 32966315

Kumar, P., Chitara, D., Sengupta, S., Banerjee, P., & Rai, S. N. (2025). Microbial consortia in biotechnology: Applications and challenges in industrial processes. *3 Biotech, 15*(386). DOI: 10.1007/s13205-025-04558-1

Kumari, R., Singha, L. P., & Shukla, P. (2023). Biotechnological potential of microbial bio-surfactants, their significance, and diverse applications. *FEMS Microbes, 4*, xtad015. Advance online publication. DOI: 10.1093/femsmc/xtad015 PMID: 37614639

Kylilis, N., Tuza, Z. A., Stan, G. B., & Polizzi, K. M. (2018). Tools for engineering coordinated system behaviour in synthetic microbial consortia. *Nature Communications, 9*(1), 2677. DOI: 10.1038/s41467-018-05046-2 PMID: 29992956

Lawson, C. E., Harcombe, W. R., Hatzenpichler, R., Lindemann, S. R., Löffler, F. E., O'Malley, M. A., García Martín, H., Pfleger, B. F., Raskin, L., Venturelli, O. S., Weissbrodt, D. G., Noguera, D. R., & McMahon, K. D. (2019). Common principles and best practices for engineering microbiomes. *Nature Reviews. Microbiology, 17*(12), 725–741. DOI: 10.1038/s41579-019-0255-9 PMID: 31548653

Lawson, C. E., Harcombe, W. R., Hatzenpichler, R., Lindemann, S. R., Löffler, F. E., O'Malley, M. A., García Martín, H., Pfleger, B. F., Raskin, L., Venturelli, O. S., Weissbrodt, D. G., Noguera, D. R., & McMahon, K. D. (2021). Common principles and best practices for engineering microbial communities. *Nature Reviews. Microbiology, 19*(11), 725–741. DOI: 10.1038/s41579-019-0255-9 PMID: 31548653

León, D. S., & Nogales, J. (2022). Toward merging bottom–up and top–down model-based designing of synthetic microbial communities. *Current Opinion in Microbiology, 69*, 102169. DOI: 10.1016/j.mib.2022.102169 PMID: 35763963

Li, B., Deng, A., Li, K., Hu, Y., Li, Z., Shi, Y., & Lu, J. (2022). Viral infection and transmission in a large, well-traced outbreak caused by the SARS-CoV-2 Delta variant. *Nature Communications, 13*(1), 460. DOI: 10.1038/s41467-022-28089-y PMID: 35075154

Li, D., Ji, S., Guo, Y., & Sang, N. (2021). Ambient NO_2 exposure sex-specifically impairs myelin and contributes to anxiety and depression-like behaviors of C57BL/6J mice. *Journal of Hazardous Materials, 416*, 125836. DOI: 10.1016/j.jhazmat.2021.125836 PMID: 34492793

Li, J. F., Wang, J., Liu, Z., Liu, Q., Guo, C., Liao, C., Li, W., Guo, C., Zhang, T., & Liu, J. (2023). Comparative study on reproductive biology of two co-invasion tilapia species in the Shanmei reservoir. *Shui Sheng Sheng Wu Hsueh Bao, 47*, 657–665.

Liang, Y., Ma, A., & Zhuang, G. (2022). Construction of environmental synthetic microbial consortia: Based on engineering and ecological principles. *Frontiers in Microbiology, 13*, 829717. DOI: 10.3389/fmicb.2022.829717 PMID: 35283862

Liao, C., Wang, T., Maslov, S., & Xavier, J. B. (2020). Modeling microbial cross-feeding at intermediate scale portrays community dynamics and species coexistence. *PLoS Computational Biology, 16*(8), e1008135. DOI: 10.1371/journal.pcbi.1008135 PMID: 32810127

Losoi, P. S., Santala, V. P., & Santala, S. M. (2019). Enhanced population control in a synthetic bacterial consortium by interconnected carbon cross-feeding. *ACS Synthetic Biology, 8*(12), 2642–2650. DOI: 10.1021/acssynbio.9b00316 PMID: 31751122

Lovley, D. R. (2017). Syntrophy Goes Electric: Direct Interspecies Electron Transfer. *Annual Review of Microbiology, 71*(1), 643–664. DOI: 10.1146/annurev-micro-030117-020420 PMID: 28697668

Lovley, D. R., & Nevin, K. P. (2011). A shift in the current paradigm for electron exchange within microbial communities. *Energy & Environmental Science, 4*(12), 4896–4906. DOI: 10.1039/c1ee02229f

Lü, H., Wei, J. L., Tang, G. X., Chen, Y. S., Huang, Y. H., Hu, R., & Li, Q. X. (2024). Microbial consortium degrading of organic pollutants: Source, degradation efficiency, pathway, mechanism and application. *Journal of Cleaner Production, 451*, 141913. DOI: 10.1016/j.jclepro.2024.141913

Luo, L., Chen, X., Liu, B., Nie, Y., & Wu, X. L. (2025). Strengthen or Weaken: Evolutionary Directions of Cross-Feeding After Formation. *Environmental Microbiology Reports, 17*(4), e70175. DOI: 10.1111/1758-2229.70175 PMID: 40784676

Lyu, X., Nuhu, M., Candry, P., Wolfanger, J., Betenbaugh, M., Saldivar, A., & Shrestha, S. (2024). Top-down and bottom-up microbiome engineering approaches to enable biomanufacturing from waste biomass. *Journal of Industrial Microbiology & Biotechnology, 51*, kuae025. Advance online publication. DOI: 10.1093/jimb/kuae025 PMID: 39003244

Macchi, M., Festa, S., Nieto, E., Irazoqui, J. M., Vega-Vela, N. E., Junca, H., Valacco, M. P., Amadio, A. F., Morelli, I. S., & Coppotelli, B. M. (2021). Design and evaluation of synthetic bacterial consortia for optimized phenanthrene degradation through the integration of genomics and shotgun proteomics. *Biotechnology Reports (Amsterdam, Netherlands), 29*, e00588. DOI: 10.1016/j.btre.2021.e00588 PMID: 33489789

Mahajan, M., & Prakash, A. (2025). Bacterial consortia as potential bioremediation agents for wastewater treatment: A comprehensive review. *International Journal of Advancement in Life Sciences Research, 8*(1), 16–33. DOI: 10.31632/ijalsr.2025.v08i01.002

Massot, F., Bernard, N., Alvarez, L. M. M., Martorell, M. M., Mac Cormack, W. P., & Ruberto, L. A. (2022). Microbial associations for bioremediation. What does "microbial consortia" mean? *Applied Microbiology and Biotechnology, 106*(7), 2283–2297. DOI: 10.1007/s00253-022-11864-8 PMID: 35294589

Mataigne, V., Vannier, N., Vandenkoornhuyse, P., & Hacquard, S. (2021). Microbial Systems Ecology to Understand Cross-Feeding in Microbiomes. *Frontiers in Microbiology, 12*, 780469. DOI: 10.3389/fmicb.2021.780469 PMID: 34987488

Maymo-Gatell, X., Chien, Y. T., Gossett, J. M., & Zinder, S. H. (1997). Isolation of a bacterium that reductively dechlorinates tetrachloroethene to ethene. *Science, 276*(5318), 1568–1571. DOI: 10.1126/science.276.5318.1568 PMID: 9171062

McCarty, N. S., & Ledesma-Amaro, R. (2019). Synthetic biology tools to engineer microbial communities for biotechnology. *Trends in Biotechnology, 37*(2), 181–197. DOI: 10.1016/j.tibtech.2018.11.002 PMID: 30497870

Michaelis, C., & Grohmann, E. (2023). Horizontal Gene Transfer of Antibiotic Resistance Genes in Biofilms. *Antibiotics (Basel, Switzerland), 12*(2), 328. DOI: 10.3390/antibiotics12020328 PMID: 36830238

Mishra, T., Tiwari, P. B., Kanchan, S., & Kesheri, M. (2025). Advances in Microbial Bioremediation for Effective Wastewater Treatment. *Water (Basel), 17*(22), 3196. DOI: 10.3390/w17223196

Nieto, E. E., Macchi, M., Valacco, M. P., Festa, S., Morelli, I. S., & Coppotelli, B. M. (2023). Metaproteomic and gene expression analysis of interspecies interactions in a PAH-degrading synthetic microbial consortium constructed with the key microbes of a natural consortium. *Biodegradation, 34*(2), 181–197. DOI: 10.1007/s10532-022-10012-3 PMID: 36596914

Novak, V., Andeer, P. F., King, E., Calabria, J., Fitzpatrick, C., Kelm, J. M., & Northen, T. R. (2025). Breaking the reproducibility barrier with standardized protocols for plant–microbiome research. *PLoS Biology, 23*(9), e3003358. DOI: 10.1371/journal.pbio.3003358 PMID: 40920748

Nunes, P. S., Lacerda-Junior, G. V., Mascarin, G. M., Guimarães, R. A., Medeiros, F. H., Arthurs, S., & Bettiol, W. (2024). Microbial consortia of biological products: Do they have a future? *Biological Control, 188*, 105439. DOI: 10.1016/j.biocontrol.2024.105439

Okafor, C. P., Udemang, N. L., Chikere, C. B., Akaranta, O., & Ntushelo, K. (2021). Indigenous microbial strains as bioresource for remediation of chronically polluted Niger Delta soils. *Scientific African, 11*, e00682. DOI: 10.1016/j.sciaf.2020.e00682

Okoye, A. U., Selvarajan, R., Chikere, C. B., Okpokwasili, G. C., & Mearns, K. (2024). Characterization and identification of long-chain hydrocarbon-degrading bacterial communities in long-term chronically polluted soil in Ogoniland: An integrated approach using culture-dependent and independent methods. *Environmental Science and Pollution Research International, 31*(21), 30867–30885. DOI: 10.1007/s11356-024-33326-6 PMID: 38622422

Pacheco, A. R., Moel, M., & Segrè, D. (2019). Costless Metabolic Secretions as Drivers of Interspecies Interactions in Microbial Ecosystems. *Nature Communications, 10*(1), 1–12. DOI: 10.1038/s41467-018-07946-9 PMID: 30626871

Papenfort, K., & Bassler, B. L. (2016). Quorum sensing signal-response system in Gram-negative bacteria. *Nature Reviews. Microbiology, 14*(9), 576–588. DOI: 10.1038/nrmicro.2016.89 PMID: 27510864

PatSnap. (2025, April 29). *What is a microbial consortium and how does it work in synthetic biology?* Synapse. https://synapse.patsnap.com/article/what-is-a-microbial-consortium-and-how-does-it-work-in-synthetic-biology

Peng, Q., Quan, L., Zheng, H., Li, J., & Xie, G. (2025). Analyzing the contribution of top-down and bottom-up methods to the construction of synthetic microbial communities in Jiuyao. *Food Microbiology*, *129*, 104759. DOI: 10.1016/j.fm.2025.104759 PMID: 40086988

Potts, L. D., Douglas, A., Perez Calderon, L. J., Anderson, J. A., Witte, U., Prosser, J. I., & Gubry-Rangin, C. (2022). Chronic environmental perturbation influences microbial community assembly patterns. *Environmental Science & Technology*, *56*(4), 2300–2311. DOI: 10.1021/acs.est.1c05106 PMID: 35103467

Poulos, C. J., Bayer, A., Beaupre, L., Clare, L., Poulos, R. G., Wang, R. H., & McGilton, K. S. (2017). A comprehensive approach to reablement in dementia. *Alzheimer's & Dementia: Translational Research & Clinical Interventions*, *3*(3), 450–458. DOI: 10.1016/j.trci.2017.06.005 PMID: 29067351

Prince, J. F., van den Bosch, M. A., Nijsen, J. F., Smits, M. L., van den Hoven, A. F., Nikolakopoulos, S., Wessels, F. J., Bruijnen, R. C. G., Braat, M. N. G. J. A., Zonnenberg, B. A., & Lam, M. G. (2018). Efficacy of radioembolization with 166Ho-microspheres in salvage patients with liver metastases: A phase 2 study. *Journal of Nuclear Medicine*, *59*(4), 582–588. DOI: 10.2967/jnumed.117.197194 PMID: 28916623

Qattan, S. Y. (2025). Harnessing bacterial consortia for effective bioremediation: Targeted removal of heavy metals, hydrocarbons, and persistent pollutants. *Environmental Sciences Europe*, *37*(1), 85. DOI: 10.1186/s12302-025-01103-y

Rather, M. A., Gupta, K., & Mandal, M. (2021). Microbial biofilm: formation, architecture, antibiotic resistance, and control strategies. *Brazilian journal of microbiology: [publication of the Brazilian Society for Microbiology]*, *52*(4), 1701–1718. https://doi.org/DOI: 10.1007/s42770-021-00624-x

Rezaei, Z., & Moghimi, H. (2024). Fungal-bacterial consortia: A promising strategy for the removal of petroleum hydrocarbons. *Ecotoxicology and Environmental Safety*, *280*, 116543. DOI: 10.1016/j.ecoenv.2024.116543 PMID: 38833981

Rotaru, A. E., Shrestha, P. M., Liu, F., Markovaite, B., Chen, S., Nevin, K. P., & Lovley, D. R. (2014). Direct interspecies electron transfer between *Geobacter metallireducens* and *Methanosarcina barkeri*. *Applied and Environmental Microbiology*, *80*(15), 4599–4605. DOI: 10.1128/AEM.00895-14 PMID: 24837373

Santala, S., Efimova, E., & Santala, V. (2018). Dynamic decoupling of biomass and wax ester biosynthesis in Acinetobacter baylyi by an autonomously regulated switch. *Metabolic Engineering Communications, 7*, e00078. DOI: 10.1016/j.mec.2018.e00078 PMID: 30271720

Santos, B., Araújo, J., Carvalho, B., Cotrim, C., Bernardino, R., Freitas, F., & Encarnação, T. (2025). Bioremediation of Synthetic Wastewater with Contaminants of Emerging Concern by Nannochloropsis sp. and Lipid Production: A Circular Approach. *Bioengineering (Basel, Switzerland), 12*(3), 246. DOI: 10.3390/bioengineering12030246 PMID: 40150710

Satpathy, R. (2021). Application of Dehalogenase Enzymes in Bioremediation of Halogenated Pollutants: A Short Review. In Malik, J. (Ed.), *Handbook of Research on Microbial Remediation and Microbial Biotechnology for Sustainable Soil* (pp. 373–385)., DOI: 10.4018/978-1-7998-7062-3.ch014

Schroven, K., Aertsen, A. & Lavigne, R. (2021). Bacteriophages as drivers of bacterial virulence and their potential for biotechnological exploitation, *FEMS Microbiology Reviews*, 45(1), 041, https://doi.org/DOI: 10.1093/femsre/fuaa041

Scott, S. R., & Hasty, J. (2016). Quorum sensing communication modules for microbial consortia. *ACS Synthetic Biology, 5*(9), 969–977. DOI: 10.1021/acssynbio.5b00286 PMID: 27172092

Shade, A., Dunn, R. R., Blowes, S. A., Keil, P., Bohannan, B. J., Herrmann, M., Küsel, K., Lennon, J. T., Sanders, N. J., Storch, D., & Chase, J. (2018). Macroecology to unite all life, large and small. *Trends in Ecology & Evolution, 33*(10), 731–744. DOI: 10.1016/j.tree.2018.08.005 PMID: 30209011

Shahidi-Rizi, M. S. S., Emtiazi, G., Sepahi, A. A., Maal, K. B., & Hosseini, F. (2017). Synergistic effect of nanoparticles and *Thalassospira* sp. isolated from Bahregan area in biodegradation of oil sludge. *Petroleum Science and Technology, 35*(9), 831–838. DOI: 10.1080/10916466.2016.1278386

Shen, L., Zhao, Q., Wu, X., Li, X., Li, Q., & Wang, Y. (2016). Interspecies electron transfer in syntrophic methanogenic consortia: From cultures to bioreactors. *Renewable & Sustainable Energy Reviews, 54*, 1358–1367. DOI: 10.1016/j.rser.2015.10.102

Singh, B. K., & Walker, A. (2006). Microbial degradation of organophosphorus compounds. *FEMS Microbiology Reviews, 30*(3), 428–471. DOI: 10.1111/j.1574-6976.2006.00018.x PMID: 16594965

Stams, A. J. M., & Plugge, C. M. (2021). Electron transfer in syntrophic communities. *Nature Reviews. Microbiology, 19*(7), 457–469. DOI: 10.1038/nrmicro2166 PMID: 19609258

Toussaint, P. A., Leiser, F., Thiebes, S., Schlesner, M., Brors, B., & Sunyaev, A. (2024). Explainable artificial intelligence for omics data: A systematic mapping study. *Briefings in Bioinformatics, 25*(1), bbad453. DOI: 10.1093/bib/bbad453 PMID: 38113073

Upadhyay, S. K., Rani, N., Kumar, V., Mythili, R., & Jain, D. (2023). A review on simultaneous heavy metal removal and organo-contaminants degradation by potential microbes: Current findings and future outlook. *Microbiological Research, 273*, 127419. DOI: 10.1016/j.micres.2023.127419 PMID: 37276759

Urbanek, A. K., Rymowicz, W., & Mirończuk, A. M. (2018). Degradation of plastics and plastic-degrading bacteria in cold marine habitats. *Applied Microbiology and Biotechnology, 102*(18), 7669–7678. DOI: 10.1007/s00253-018-9195-y PMID: 29992436

Valls, M., & De Lorenzo, V. (2002). Exploiting the genetic and biochemical capacities of bacteria for the remediation of heavy metal pollution. *FEMS Microbiology Reviews, 26*(4), 327–338. DOI: 10.1016/S0168-6445(02)00114-6 PMID: 12413663

Varjani, S. J. (2017). Microbial degradation of petroleum hydrocarbons. *Bioresource Technology, 223*, 277–286. DOI: 10.1016/j.biortech.2016.10.037 PMID: 27789112

Wagner, M., & Loy, A. (2002). Bacterial community composition and function in sewage treatment systems. *Current Opinion in Biotechnology, 13*(3), 218–227. DOI: 10.1016/S0958-1669(02)00315-4 PMID: 12180096

Wang, S., Zhan, Y., Jiang, X., & Lai, Y. (2024). Engineering microbial consortia as living materials: Advances and prospectives. *ACS Synthetic Biology, 13*(9), 2653–2666. DOI: 10.1021/acssynbio.4c00313 PMID: 39174016

Whiteley, M., Diggle, S. P., & Greenberg, E. P. (2017). Progress in and promise of bacterial quorum sensing research. *Nature, 551*(7680), 313–320. DOI: 10.1038/nature24624 PMID: 29144467

Wu, K., Su, D., Liu, J., Saha, R., & Wang, J. P. (2019). Magnetic nanoparticles in nanomedicine: A review of recent advances. *Nanotechnology, 30*(50), 502003. DOI: 10.1088/1361-6528/ab4241 PMID: 31491782

Xie, B. B., Li, M., Anantharaman, K., & Ravin, N. V. (2021). Editorial: the uncultured microorganisms: novel technologies and applications. *Frontiers in Microbiology, 12*, 756287. DOI: 10.3389/fmicb.2021.756287 PMID: 34899641

Xu, C., & Yu, H. (2021). Insights into constructing a stable and efficient microbial consortium. *Chinese Journal of Chemical Engineering*, *30*, 112–120. 112-120.doi .org/10.1016/j.cjche.2020.12.012. DOI: 10.1016/j.cjche.2020.12.012

Xu, H., Chang, J., Wang, H., Liu, Y., Zhang, X., Liang, P., & Huang, X. (2019). Enhancing direct interspecies electron transfer in syntrophic-methanogenic associations with (semi) conductive iron oxides: Effects and mechanisms. *The Science of the Total Environment*, *695*, 133876. DOI: 10.1016/j.scitotenv.2019.133876 PMID: 31756846

Yamagishi, J. F., Saito, N., & Kaneko, K. (2020). Advantage of leakage of essential metabolites for cells. *Physical Review Letters*, *124*(4), 048101. DOI: 10.1103/PhysRevLett.124.048101 PMID: 32058757

Yoshida, S., Hiraga, K., Takehana, T., Taniguchi, I., Yamaji, H., Maeda, Y., Toyohara, K., Miyamoto, K., Kimura, Y., & Oda, K. (2016). A bacterium that degrades and assimilates poly (ethylene terephthalate). *Science*, *351*(6278), 1196–1199. DOI: 10.1126/science.aad6359 PMID: 26965627

Zelezniak, A., Andrejev, S., Ponomarova, O., Mende, D. R., Bork, P., & Patil, K. R. (2015). Metabolic dependencies drive species co-occurrence in diverse microbial communities. *Proceedings of the National Academy of Sciences of the United States of America*, *112*(20), 6449–6454. DOI: 10.1073/pnas.1421834112 PMID: 25941371

Zhang, T., & Zhang, H. (2022). Microbial consortia are needed to degrade soil pollutants. *Microorganisms*, *10*(2), 261. DOI: 10.3390/microorganisms10020261 PMID: 35208716

Zhili, N., Lilin, Z., Ziyang, L., Bin, K., Gefu, Z., Jianbo, J., & Tao, W. (2023). Iron-modified biochar boosts anaerobic digestion of sulfamethoxazole pharmaceutical wastewater: Performance and microbial mechanism. *Journal of Hazardous Materials*, *452*, 131314. DOI: 10.1016/j.jhazmat.2023.131314 PMID: 37030222

Ziesack, M., Gibson, T., Oliver, J. K., Shumaker, A. M., Hsu, B. B., Riglar, D. T., Giessen, T. W., DiBenedetto, N. V., Bry, L., Way, J. C., Silver, P. A., & Gerber, G. K. (2019). Engineered interspecies amino acid cross-feeding increases population evenness in a synthetic bacterial consortium. *mSystems*, *4*(4), 10–1128. DOI: 10.1128/mSystems.00352-19 PMID: 31409662

Zuo, D., Han, Y., Gao, X., Ma, G., Xu, Z., Bi, Y., Abbaspour, K. C., & Yang, H. (2022). Identification of priority management areas for non-point source pollution based on critical source areas in an agricultural watershed of Northeast China. *Environmental Research*, *214*(2), 113892. DOI: 10.1016/j.envres.2022.113892 PMID: 35863450

Chapter 3
Genetically Engineered Microbes in Environmental Biotechnology:
Advances, Challenges, and Risk Assessment

Ayansina Segun Ayangbenro
http://orcid.org/0000-0002-3220-1873

Department of Microbiology, Federal University of Medicine and Medical Sciences, Abeokuta, Nigeria & Food Security and Safety Focus Area, Faculty of Natural and Agricultural Sciences, North-West University, South Africa

Funso R. Kutu
http://orcid.org/0000-0002-8162-0329

School of Agricultural Sciences, Faculty of Agriculture and Natural Sciences, University of Mpumalanga, Mbombela, South Africa

ABSTRACT

The continued presence of hazardous contaminants remains a major global environmental challenge. Conventional remediation methods are often costly and may generate secondary environmental risks. In contrast, bioremediation provides a sustainable, eco-friendly, and cost-effective approach for pollutant removal. Recent advances in genetic engineering have enabled the development of genetically engineered microbes (GEMs) with enhanced capabilities for degrading environmental toxins. These engineered strains have shown significant potential in hydrocarbon degradation, heavy metal detoxification, and the breakdown of other xenobiotics.

DOI: 10.4018/979-8-2600-2998-5.ch003

Coupled with modern genome-editing technologies, GEMs are increasingly used in environmental remediation, enabling innovative strategies for pollutant removal. Given the scale of global environmental problems, integrated technological approaches are essential. This review highlights recent advances, applications, and key challenges associated with the use of GEMs in environmental biotechnology.

1. INTRODUCTION

Contamination from both organic and inorganic pollutants is a major global environmental issue, endangering human health and the provision of essential ecosystem goods and services. According to a recent World Health Organization (WHO) assessment, environmental risk factors, including pollution, radiation, unsafe working conditions, and climate change, contribute to approximately 24% of annual human mortality worldwide (WHO, 2024). A major contributor to environmental pollution, heavy metal ions and metalloids, including As, Cd, Cr, Hg, Pb, and Zn, are increased when poorly managed and untreated wastewaters are discharged. Despite being noncovalent, some heavy metal ions can form large supramolecular complexes and exert their toxicity through microRNAs, thereby contributing to diseases such as cancer (Kumar et al., 2025). Endocrine-disrupting compounds, polypropylene copolymers, microplastics, antimicrobial resistance genes, and polyfluoroalkyl substances (PFASs) are among the other contaminants that pose significant threats to human health through food contamination and toxicity, causing diseases such as hormone disruption and cancer (Kumar et al., 2025). All these compounds, released from various industrial activities, pose substantial environmental hazards when regulatory guidelines and proper waste-management practices are not followed. Such pollutants can exert genotoxic effects on diverse biological systems (Kugarajah et al., 2023).

Conventional pollution-control strategies rely predominantly on chemical and physical remediation methods. In parallel with these approaches, bioremediation has gained widespread recognition as an effective and environmentally sustainable alternative (Ayangbenro & Babalola, 2018). Bioremediation supports ecosystem recovery by facilitating the natural breakdown of contaminants and helping prevent further environmental degradation. Its application has grown rapidly in recent years, addressing an increasingly broad spectrum of pollution challenges.

Microbes are among the biological agents with the most environmental flexibility and dietary versatility. Microorganisms significantly influence the various processes that occur in natural environments (Poursat et al., 2019). Microorganisms are ubiquitous and play roles, directly or indirectly, in nearly all processes occurring on Earth (Kugarajah et al., 2023). They serve as key drivers of biogeochemical cycles, and due to their ecological importance, can be harnessed for numerous applications, including environmental remediation. Many microbial species naturally possess the ability to degrade, detoxify, transform, accumulate, or metabolize a wide array of organic and inorganic pollutants (Dangi et al., 2019; Kuppan et al., 2024; Sharma & Shukla, 2022). The breakdown of hazardous compounds is primarily mediated through microbial metabolic pathways, particularly those involving oxidation–reduction reactions facilitated by microbial enzymes (Kuppan et al., 2024).

Numerous technologies, including engineering biology, are needed to address environmental pollution through targeted interventions that integrate ecological principles and functional enhancement. Engineering biology can be used to detect and break down pollutants and to convert waste streams, particularly those that are resistant and non-biodegradable, into value-added products, thereby replacing production from fossil fuels with biological alternatives (Lea-Smith et al., 2025). A variety of technologies are employed in engineering biology, including synthetic biology techniques that can be automated to facilitate accurate chromosomal modification and rapid plasmid assembly (Lea-Smith et al., 2025). For example, a CRISPR-edited *Pseudomonas aeruginosa* strain was recently engineered to enhance metabolite and hydrocarbon-emulsification traits, thereby improving hydrocarbon handling and making oil more susceptible to degradation (Salazar-García et al., 2025). Similarly, *Pseudomonas putida* KT2440 was engineered to simultaneously degrade three classes of pesticides (organophosphates, pyrethroids, and carbamates) through the co-expression of multiple degradation pathways (Gong et al., 2018). Furthermore, *Bacillus* laccases and other extracellular enzymes have been used to detoxify industrial effluents, decolorize textile dyes, and support site-directed mutagenesis for solid-waste bioconversion (Bu et al., 2020).

Although modified microbes hold potential for remediating pollutants, biosafety issues persist, especially regarding horizontal gene transfer (HGT) of catabolic functions. Environmental setting significantly influences both plasmid transmission and pollutant breakdown activity, according to research on degradative plasmids. In phenol-contaminated farmlands, for example, *P. fluorescens* TP13 carrying a catechol-2,3-dioxygenase plasmid enhanced phenol breakdown and plant biomass concurrently with plasmid transfer to several native rhizosphere bacteria in the field (Wang et al., 2014). Socio-regulatory issues, in addition to biosafety and HGT hazards, hinder the deployment of engineered microbial remediation. Different countries have varying regulatory frameworks governing the introduction of genetically modified organisms and engineered microbes, which often necessitate distinct risk assessments, institutional coordination, and biosafety oversight (Brewer & George, 2025; Lea-Smith et al., 2025).

Recent developments in GEMs for environmental biotechnology, aimed at enhancing the effectiveness of bioremediation in contaminated media, are summarized in this review. It evaluates ecological, regulatory, and biosafety hurdles to adoption while highlighting the use of CRISPR-based gene editing and AI-driven pathway design to customize metabolic networks for targeted degradation of pollutants.

2. GENOME EDITING TECHNOLOGIES AND SYNTHETIC BIOLOGY TOOLS

Synthetic biology is a multidisciplinary area of science that aims to design and construct artificial biological systems that either reproduce specific cellular functions and processes or display entirely new capabilities absent in natural organisms (Benner & Sismour, 2005; Hirschi et al., 2022). The term synthetic biology first appeared in the scientific literature in 1980, where it was used in the context of genetically modified bacteria (Hobom, 1980). Initially, the concept was narrowly associated with genetic manipulation and was often treated as synonymous with bioengineering. However, as scientific knowledge and technological tools advanced, biology and chemistry increasingly converged, each contributing distinct engineering principles (Hirschi et al., 2022).

Recent progress in synthetic biology has greatly enhanced efforts in addressing environmental problems. Genetically engineered microorganisms are now widely employed in environmental biotechnology to remediate hazardous substances, including xenobiotics and pesticides (Rafeeq et al., 2023). However, designing effective synthetic microbial communities requires a deep understanding of natural microbial ecosystems, where identifying the specific species contributing to bioremediation remains a major challenge. This section surveys the toolkits used to design, build, test, and model GEMs for environmental biotechnology.

2.1 CRISPR as the Genome-Editing Technology

Clustered Regularly Interspaced Short Palindromic Repeats (CRISPR) technology has quickly become an essential tool in modern biological research. The development of CRISPR-based genome editing, together with advances in computational biology and imaging, has ushered in a transformative era in which human diseases can be diagnosed, genetic susceptibilities can be predicted, and targeted interventions can be designed with unprecedented precision (Wang & Doudna, 2023). Similarly, the ability to identify and rapidly modify genes governing plant traits is revolutionizing agricultural research and accelerating crop improvement (Verdezoto-Prado et al., 2025). Beyond medicine and agriculture, these tools are also being applied to environmental biotechnology, enabling the engineering of microorganisms for pollution control and the bioremediation of hazardous compounds (Fig. 1).

Figure 1. Applications of the CRISPR-Cas system in bioremediation of pollutants and hazardous materials.

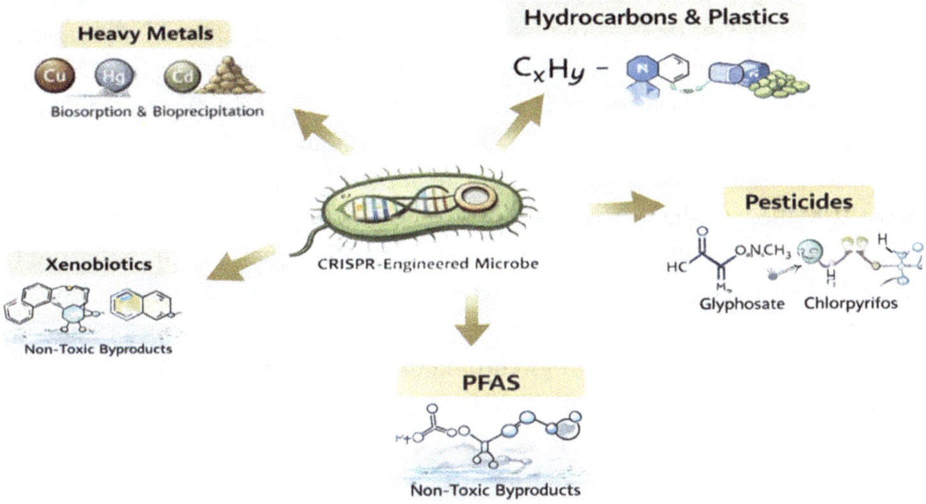

Originally identified as a bacterial adaptive immune system that protects against viral invasion, the programmable activity of the Cas9 enzyme has since revolutionized a wide range of fields, including medicine, biotechnology, and agriculture (Adli, 2018). CRISPR-Cas systems can be precisely engineered to generate sequence-specific DNA breaks at defined genomic loci, enabling precise genetic alteration through host-dependent DNA repair pathways (Selle & Barrangou, 2015). Curiosity-driven studies further demonstrated that these systems use RNA molecules transcribed from repeat-spacer arrays to direct Cas proteins to complementary sequences in invading viral DNA or RNA, facilitating targeted cleavage (Barrangou et al., 2007; Brouns et al., 2008). Subsequent studies demonstrated that this RNA-directed cleavage mechanism could be harnessed to introduce targeted modifications in the genomes of virtually any cell type with remarkable precision and simplicity (Gasiunas et al., 2012; Villiger et al., 2024; Wang & Doudna, 2023). The application of CRISPR-Cas technology is providing critical insights that could pave the way for broader adoption in microbial environmental applications. Beyond its original role in gene editing, catalytically inactive Cas9 variants have expanded CRISPR applications to encompass gene regulation, epigenetic modification, chromatin engineering, and live-cell imaging, thereby surpassing the conventional genome-editing capacity of wild-type Cas9 (Adli, 2018; Ansori et al., 2023).

The CRISPR–Cas9 system remains the most widely adopted genome-editing tool, consisting of the Cas9 nuclease paired with a sequence-specific guide RNA

(Wang & Doudna, 2023). Its exceptional flexibility stems from an RNA-directed mechanism that cleaves DNA at user-defined genomic sites through simple base complementarity. Because target recognition depends on guide RNA-DNA pairing, the Cas9 enzyme can be readily reprogrammed to target diverse DNA sequences simply by altering the guide RNA. In eukaryotic cells, the resulting double-strand breaks are processed by intrinsic DNA repair pathways, enabling precise and programmable alterations to the genome (Wang & Doudna, 2023).

Mutations introduced into the catalytic residues of Cas9 have generated variants capable of either single-strand DNA nicking or merely binding to DNA without cleavage. These catalytically inactive Cas9 derivatives have been engineered to modulate gene expression through targeted transcriptional repression or activation (Qi et al., 2013). Subsequent innovations have yielded Cas9 fusion proteins that facilitate site-specific base editing, chromatin remodeling, or sequence insertion (Anzalone et al., 2019; Komor et al., 2016; Liu et al., 2022). In addition, other Cas effectors, including RNA-targeting variants, have expanded the CRISPR toolbox, supported by extensive biochemical and structural studies (Wang & Doudna, 2023).

The CRISPR-Cas system has been successfully used to improve the removal of various environmental pollutants (Table 1). The application includes the successful targeted gene-editing in members of the *Cupriavidus* genus to elucidate the molecular mechanisms underlying organophosphorus pesticide degradation (Zhang et al., 2023) (Table 1). Furthermore, CRISPR-Cas systems have been used to construct enzymatic cascade assays to investigate the breakdown pathways of organophosphorus pollutants, including paraoxon, demeton, and dichlorvos (Fu et al., 2022). Beyond single-gene modifications, this advanced approach enables simultaneous editing of multiple genomic loci, thereby enabling precise control and reprogramming of microbial metabolic networks that mediate the biodegradation of environmental pollutants (Hassan & Ganai, 2023). CRISPR technology has catalyzed an explosion of research across molecular and cellular biology, including agriculture and synthetic biology.

Table 1. Application of the CRISPR–Cas systems in environmental biotechnology.

Pollutant class	Organism	CRISPR-Cas application	Reference
Copper	*Acidomyces richmondensis*	CRISPR-Cas9 system for efficient targeted gene disruption for copper tolerance.	Li et al. (2023)
Biomining of nickel and reduction of toxic by-products (Fe^{2+}, H_2SO_4)	*Acidithiobacillus ferridurans*	The CRISPR-dCas9 system was used to knock down genes, reducing toxic iron ions and sulfuric acid in leachate and improving Ni recovery.	Chen et al. (2023)
Organophosphorus pesticides	*Cupriavidus nantongensis* X1	CRISPR-Cas9 deletion of the *opdB* organophosphate degradation gene was used to confirm its role and enable targeted metabolic manipulation of organophosphorus insecticide breakdown.	Zhang et al. (2023)
PFAS (Per- and Polyfluoroalkyl Substances)	*Pseudomonas pohangensis* or *E. coli*	Engineering microbes to express optimized dehalogenases/oxygenases for partial defluorination and breakdown of PFAS; potential use of CRISPR and synthetic biology to enhance PFAS degradation pathways in *Pseudomonas* or *E. coli*	Xue et al. (2020)
Hydrocarbon	*Methylorubrum extorquens*	Using the plasmid-treated CRISPR-Cas9 system, the expression of the *MxaF* gene (encoding methanol dehydrogenase enzyme) was increased. This facilitates increased remediation of n-alkane and polycyclic aromatic hydrocarbons.	Makia et al. (2024)
Gasoline	*Pseudomonas aeruginosa*	Using a CRISPR-Cas9–APOBEC1–UGI base-editing system, a targeted cytosine-to-thymine substitution was introduced into the rpoS gene, generating a premature stop codon. The resulting mutant showed increased pyocyanin and rhamnolipid production, leading to enhanced gasoline emulsification in cell-free supernatants and demonstrating effective modulation of a key global regulator.	Salazar-García et al. (2025)

2.2 Horizontal Gene Transfer and Plasmid-Mediated Degradation

Horizontal gene transfer (HGT) is increasingly recognized as a fundamental process shaping bacterial evolution, diversification, and ecological adaptation across diverse environments (Moura de Sousa et al., 2023). It occurs through multiple mechanisms, including natural transformation, transduction, and conjugation. Among these, conjugation is often regarded as particularly important for enabling rapid microbial responses to environmental fluctuations (Bobate et al., 2023; Carvalho et al., 2020). The dominance of specific HGT pathways most likely reflects the interplay between bacterial community composition, environmental conditions, and the inherent properties of the mobile genetic elements (MGEs) involved (Heuer & Smalla, 2012).

The genes responsible for degrading naturally occurring and xenobiotic organic compounds are frequently carried on MGEs, including plasmids, integrons, and transposons (Garbisu et al., 2017; Rios Miguel et al., 2020). Through HGT, these elements can be mobilized across bacterial populations, enabling recipient strains to acquire novel catabolic functions and thereby degrade a wide range of organic contaminants. HGT enables bacteria to exchange genetic material, even between distantly related taxa, allowing rapid adaptation to changing environmental conditions (Garbisu et al., 2017). Although spontaneous mutations also contribute to bacterial evolution, their rates are typically low within natural populations. Moreover, an elevated mutation rate is generally associated with a higher likelihood of deleterious effects and cell death, limiting its evolutionary advantage (Martinez et al., 2009).

Among the major HGT mechanisms, transformation, transduction, and conjugation, conjugative transfer is considered the most effective method of delivering plasmid-encoded genetic information through direct cell-to-cell contact between the donor and recipient cells (Furuya & Lowy, 2006; Tao et al., 2022). Conjugation is pivotal in the spread of antibiotic- and metal-resistance genes, as well as genes involved in the breakdown of organic contaminants. Despite its recognized importance, the extent to which conjugation contributes to HGT among soil bacteria, and the environmental and biological factors influencing the transfer and persistence of plasmid-bearing populations in soil ecosystems, remains incompletely understood (Garbisu et al., 2017).

Bacterial adaptation over evolutionary timescales has been driven in part by the remarkable plasticity of bacterial genomes, which enables genomic rearrangements and the exchange of DNA fragments that facilitate the acquisition of advantageous traits (Sørensen et al., 2005). The deletion, recombination, or horizontal uptake of functional gene modules can profoundly influence the rate and extent of bacterial evolutionary change (Bertels & Rainey, 2011). Mobile genetic elements are key mediators of these processes, promoting both intracellular genome reorganization and intercellular gene exchange (Garbisu et al., 2017).

Most research on lateral gene transfer among bacteria has focused on the spread of antibiotic and metal resistance. In comparison, the horizontal spread of catabolic genes responsible for degrading organic pollutants in natural settings such as soil has received less attention. Nonetheless, several catabolic plasmids, including those involved in the biodegradation of 2,4-dichlorophenoxyacetic acid (2,4-D), have been extensively characterized, providing valuable insights into the ecological roles and molecular mechanisms underlying environmental gene transfer (Dejonghe et al., 2000; Newby & Pepper, 2002).

The TOL plasmid pWW0 from *Pseudomonas putida*, which encodes the upper and meta-cleavage pathways for toluene/xylene degradation, can be transferred into other hosts (Greated et al., 2002). In addition, plasmids carrying *nah* operons (NAH plasmids) that encode enzymes for naphthalene degradation have been characterized in *Pseudomonas* spp., allowing recipients to degrade naphthalene via the salicylate pathway (Mohapatra & Phale, 2021). These catabolic plasmids are frequently associated with transposons and integrative elements, facilitating HGT (Silva-Rocha et al., 2011). Through HGT, microbial communities under pollutant stress can rapidly disseminate degradative traits, enabling non-catabolic organisms to acquire new functions and expand their ecological roles (Varner et al., 2022). This dynamic exchange accelerates adaptation and is a key driver of biodegradation in contaminated ecosystems, thereby amplifying *in situ* remediation capacity.

Attention has also turned toward the application of GEMs for bioaugmentation in contaminated environments. However, the intentional release of GMOs into natural ecosystems remains subject to stringent regulatory oversight and biosafety concerns (Garbisu et al., 2017). Horizontal exchange of catabolic genes between GEMs and native soil bacteria may accelerate the emergence and spread of novel degradative traits within indigenous microbial communities. In this context, Massa et al. (2009) engineered *Pseudomonas putida* PaW340/pDH5 by introducing dehalogenase genes from *Arthrobacter* sp. FG1, creating a recombinant strain capable of degrading 4-chlorobenzoic acid (CBA) in soil slurry. When this engineered strain was inoculated into slurry systems, CBA removal was markedly enhanced relative to slurries supplemented with pre-adapted *Arthrobacter* sp. FG1, underscoring the capacity of engineered bacteria to boost biodegradation efficiency under controlled environmental conditions. Thus, plasmid-mediated systems epitomize both the promise and the risks of GEM-based remediation: they enable adaptive, flexible pollutant degradation but demand careful design to prevent unintended ecological impacts.

2.3 Metabolic Engineering and Pathway Design

Metabolic engineering refers to the deliberate alteration of cellular metabolic pathways by introducing or modifying genes using tools such as recombinant DNA technology, PCR, and high-throughput sequencing (Kim et al., 2023; Oldiges et al., 2014). This approach optimizes the genetic and regulatory networks within living cells, thereby redirecting metabolic fluxes to increase the production of target compounds (Sharma & Shukla, 2022). It has been widely used to reconstruct and optimize biosynthetic pathways in microorganisms (Dangi et al., 2019). It is now a standard strategy for producing recombinant proteins, often by expressing heterologous genes in hosts such as *Escherichia coli* (Kim et al., 2023). Microbial systems have found diverse applications across the food, agriculture, and environmental biotechnology sectors, including bioremediation. These systems have been used to produce numerous industrially relevant products, such as amino acids, bioactive secondary metabolites, biofuels, human insulin and growth hormone, polymers, and recombinant proteins (Zhang & Wang, 2016).

Subsequent research efforts focused on engineering metabolic networks to develop strains that overproduce specific chemicals. Interest in biologically derived fuels and chemicals intensified during the 1970s energy crisis, driving intensive studies of cellular metabolism and its regulatory mechanisms through both experimental and computational approaches, including metabolic control analysis (Kacser, 1973) and metabolic pathway synthesis (Seressiotis & Bailey, 1986). Parallel to these advances, continuous efforts were made to optimize fermentation strategies for enhanced product yields. The concurrent advances in recombinant DNA technology enabled the rational reconstruction of metabolic pathways to achieve desired phenotypic traits, thereby laying the foundation for modern metabolic engineering. A common strategy involves establishing new biosynthetic routes in microbial hosts by introducing heterologous genes that encode the relevant operons or gene clusters responsible for product formation (Kim et al., 2023). A representative example is the metabolic engineering of *P. putida* strain KT2440 for the aerobic degradation of 1,2,3-trichloropropane (TCP). A synthetic catabolic pathway was reconstructed in *P. putida* KT2440 through the heterologous expression of three key enzymes, epoxide hydrolase, haloalkane dehalogenase, and haloalcohol dehalogenase, sourced from different microorganisms. These enzymes collectively mediated the conversion of TCP into glycerol, thereby enabling the engineered strain to grow using TCP as its sole carbon source (Gong et al., 2016).

Microbial strains used in bioremediation often encounter various environmental stresses, including high concentrations of toxic pollutants and solvents, extreme pH, temperature, or ionic strength (Ibrahim et al., 2025; Pandey et al., 2009; Pushkar et al., 2021). Consequently, stress-tolerant bacterial strains are considered ideal hosts for constructing GEMs designed to degrade pollutants (Wang et al., 2012). Metabolic engineering has also been applied to complete and incomplete catabolic pathways and to design strains that harbor multiple degradation routes (Samin et al., 2014). This approach facilitates the complete mineralization of pollutants while preventing the accumulation of toxic intermediates arising from misrouted degradation pathways.

2.4 Synthetic Microbial Communities

Synthetic microbial consortia are multi-population systems composed of diverse microbial species, either native or metabolically engineered, that collectively perform complex metabolic functions, such as the degradation of toxic pollutants (Lee et al., 2018). These defined assemblies are highly adaptable, exhibiting enhanced tolerance to environmental fluctuations and improved efficiency in performing intricate biotransformation compared to single-species systems (Sharma & Shukla, 2022). The stability of such synthetic communities depends on a network of microbial interactions, encompassing metabolite exchange, quorum-sensing-mediated communication, antibiosis, and other cooperative or competitive cellular processes (Nozari et al., 2018). Recent advances have focused on engineering strategies to precisely regulate intercellular communication, spatial organization, functional stability, and biocontainment, thereby harnessing the full potential of microbial consortia (Deter & Lu, 2022; Johns et al., 2016). In bioremediation, the application of synthetic communities is gaining momentum as a promising approach, given that the degradation of complex and recalcitrant pollutants often requires multiple complementary catabolic pathways originating from distinct microbial taxa (Contreras-Salgado et al., 2024).

The application of metabolic engineering has also been instrumental in improving the performance of microbial consortia used for pollutant degradation. Achieving this, however, requires detailed insight into microbial physiology, population-level dynamics, and the complex interactions that occur among species in natural ecosystems (Sharma & Shukla, 2022). In their native habitats, microorganisms engage in a variety of ecological relationships, such as mutualism, commensalism, parasitism, and predation, which collectively influence community stability and metabolic function. Consequently, the rational design of a synthetic microbial community requires deliberate manipulation of both intracellular processes and interspecies interactions to ensure coordinated, efficient functional activity (Aparicio et al., 2018). Research has increasingly focused on engineering microbial consortia to modulate population behavior and optimize them for specific biotechnological applications (Table 2). For instance, a synthetic consortium comprising *Advenella* sp., *Arthrobacter sulfonivorans*, and *Variovorax soli* demonstrated efficient biodegradation of the herbicide diuron (Villaverde et al., 2017). While individual strains exhibited limited degradation capacity, co-cultivation of all three strains resulted in 98.8% mineralization of diuron in soil and water systems. This highlights the potential of combining metabolic engineering with microbial community design to enhance pollutant removal through synergistic metabolic interactions.

Table 2. Synthetic microbial community applications in pollutant removal

Application / Function	Description	Reference
Bioremediation of pesticide (bromoxynil octanoate and 3,5-dibromo-4-hydroxybenzoate) using synthetic microbiomes based on 18 keystone species identified from natural microbiomes.	Construction of synthetic microbiomes with enhanced pollutant-degrading capability through combined top-down and bottom-up engineering of native soil microbiomes to improve herbicide degradation; includes metabolic interaction modeling and simplification to functional consortia for testing in soil systems.	Ruan et al. (2024)
Stable synthetic microbial consortium, comprising *Rhodococcus jostii* strain PET and *Acinetobacter baylyi* ADP1, for plastic upcycling	The microbial consortium synergistically metabolizes mixed plastic waste depolymerization products and funnels them into valuable compounds, illustrating division of labor and functional collaboration within engineered consortia.	Diao et al. (2025)
Enhanced pyrene degradation (three-fold higher rate than single strains) through synergistic metabolic interactions and biosurfactant production.	The artificially reconstructed consortia *Mycobacterium* spp. PO1 & PO2, *Novosphingobium pentaromativorans* PY1, *Ochrobactrum* sp. PW1, and *Bacillus* sp. FW1. The enhanced pyrene biodegradation was due to cooperative interactions within the bacterial mixture.	Wanapaisan et al. (2018)
A synthetic microbial consortium composed of two genetically engineered microbes. *S. elongatus* fixes CO_2 through photosynthetic metabolism and secretes sufficient carbohydrates to support the growth and active metabolism of *P. putida*, which has been engineered to consume sucrose and to degrade the environmental pollutant 2,4-dinitrotoluene (2,4-DNT)	*Synechococcus elongatus* PCC 7942 and *Pseudomonas putida* EM173. *S. elongatus* engineered to secrete sucrose from CO_2 fixation; P. putida engineered for sucrose utilization and 2,4-DNT degradation.	Fedeson et al. (2020)
Arthrobacter sulfonivorans, Variovorax soli, and Advenella sp. consortium were used for the removal of diuron (a phenylurea herbicide).	Each strain contributes distinct catabolic steps in diuron degradation; individual strains or strain pairs showed limited activity, whereas the full three-member consortium enabled complete mineralization, indicating strong metabolic interdependence.	Villaverde et al. (2017)

2.5 Computational Approaches

Systems biology and computational modeling have emerged as indispensable tools for understanding microbial interactions with environmental contaminants and for uncovering the molecular processes that govern biodegradation (Hassan & Ganai, 2023). Advances in bioinformatics and scientific computing now enable the integration of multi-omics (genomics, metagenomics, metabolomics, proteomics, and transcriptomics) to design and optimize bioremediation strategies (De Sousa et al., 2018). In-silico approaches have proven particularly valuable for developing enzyme-based remediation processes, as they enable the prediction, design, and engineering of catalytic activities with enhanced substrate specificity and stability (Malla et al., 2018).

Computational biology facilitates the analysis of genes, proteins, and cellular processes, offering insights into the metabolic networks that drive microbial degradation of xenobiotics (Liu et al., 2018; Purohit et al., 2018). Using approaches such as flux balance analysis, pathway modeling, and metabolic flux analysis, researchers can mine large metabolic datasets, simulate system-level responses, and identify rate-limiting steps in degradation pathways (Mandpe et al., 2022; Singh et al., 2021). These models help elucidate the kinetics of biodegradation reactions and can guide the modification of microbial strains to enhance pesticide breakdown (Cuperlovic-Culf, 2018; Tudi et al., 2021).

Quantitative structure–activity relationship (QSAR) and three-dimensional QSAR (3D-QSAR) models are powerful computational methods that predict the interaction of pesticide molecules with microbial enzymes at the molecular level (Hassan & Ganai, 2023). These approaches have been successfully applied to assess pesticide toxicity, bioaccumulation potential, and degradability across different environmental matrices (Dreher et al., 2018). Furthermore, constraint-based modeling frameworks such as OptKnock enable the prediction of gene knockouts or additions that can enhance the bioremediation potential of microbial hosts (Mandeep et al., 2021). Similarly, OptStrain enables the mining and incorporation of catabolic genes from diverse microbial genomes into engineered chassis for optimal degradation performance (Ibrahim et al., 2021), while OptReg provides in silico control of regulatory and metabolic fluxes to fine-tune enzyme expression and pathway efficiency (Dash & Osborne, 2023). These computational and systems-level tools facilitate a deeper understanding of genome-scale interactions and metabolic behavior in microbes, guiding the design of next-generation biocatalysts and GEMs for sustainable bioremediation of pollutants.

2.6 Machine Learning Applications in Prediction and Optimization

At present, no universal machine learning (ML) model exists that can accurately predict all types of bioremediation outcomes (Hao et al., 2025). Each ML algorithm is optimized for specific datasets, prediction goals, or problem categories. Consequently, the design and development of an ML framework must be tailored to the specific objectives of each study. The modeling process typically involves several key stages: data acquisition and preprocessing, model selection and construction, and model validation and interpretation (Hao et al., 2025).

Building on predictive outcomes and analyses of critical influencing factors, ML is increasingly pivotal in optimizing and refining bioremediation protocols. Its principal goal is to reduce experimental duration and costs while enhancing pollutant removal efficiency and degradation rates (Hao et al., 2025). By analyzing complex datasets generated across numerous remediation trials, ML algorithms can identify the most effective combinations of experimental parameters. Recent advances have markedly accelerated progress in microbial bioremediation, enabling rapid identification of degradative microorganisms and the rational engineering of high-performance enzymes, thereby reshaping modern environmental biotechnology (Alavian & Khodabakhshi, 2025).

Machine learning and deep learning (DL) models have been successfully applied to predict enzyme function, optimize metabolic pathways, and enhance the early detection of environmental pollutants (Meng et al., 2023). These computational frameworks enable the exploration of vast genomic and proteomic datasets, allowing the discovery of novel biocatalysts with customized catalytic or stability profiles tailored for specific remediation applications.

One prominent example is the AI-assisted engineering of enzymes for the degradation of polyethylene terephthalate (PET). Deep learning has guided the redesign of PETase, leading to the development of FAST-PETase, a thermostable variant capable of degrading post-consumer PET in a few days under mild conditions (Lu et al., 2022). Computational analyses of protein dynamics have also uncovered key stabilizing mutations that enhance catalytic robustness (Stockinger et al., 2025). Furthermore, the integration of a 3D convolutional neural network (3DCNN) platform, MutCompute, enabled mutational optimization, resulting in a 29-fold improvement in PET degradation efficiency at 40°C (Jayasekara et al., 2023).

Beyond plastic degradation, AI has expanded capabilities in detecting, monitoring, and characterizing diverse environmental contaminants (Alavian & Khodabakhshi, 2025). Image-based AI and machine vision systems now allow real-time, non-destructive detection of microplastics in soil and aquatic environments (Sarker et al., 2024). For heavy metal pollution, ML-driven predictive models integrate physicochemical and geospatial datasets to estimate contamination levels in soils and sediments, improving ecological risk assessments (Yaseen et al., 2024).

However, a growing body of evidence underscores this optimization capability. Naizabayeva et al. (2024) demonstrated how ML-assisted predictive analysis of pollutant remediation cycles provided valuable time-dependent insights for soil cleanup, significantly reducing process uncertainty. Similarly, Chakraborty et al. (2024) employed ML models to forecast changes in soil microbial communities and identify key environmental stressors that most strongly influence microbial health.

This approach mitigates the labor-intensive and time-consuming nature of traditional monitoring while enabling real-time evaluation of soil quality. In another example, Yu et al. (2024) validated an ML-based model for optimizing phenol biodegradation in wastewater using *P. aeruginosa* PAO1. The predictive framework minimized the need for extensive experimental trials, thereby saving time and resources while ensuring efficient strain performance across different conditions. These studies highlight the dual potential of machine learning as both a predictive and prescriptive tool, offering data-driven guidance to streamline bioremediation processes, improve system performance, and achieve high efficiency at reduced operational costs.

3. CHALLENGES AND LIMITATIONS OF GENETICALLY ENGINEERED MICROBES IN ENVIRONMENTAL BIOTECHNOLOGY

The deployment of GEMs in environmental biotechnology continues to face scientific, technical, ecological, and socio-regulatory barriers, despite the remarkable progress in genome editing, synthetic biology, and computational approaches. These constraints have curbed early optimism and limited large-scale applications. Key scientific and technical hurdles include maintaining the genetic stability and functionality of GEMs in complex, fluctuating environments, as well as ensuring effective competition with native microbial communities. Ecological uncertainties, such as potential horizontal gene transfer, unintended ecosystem impacts, and loss of engineered traits, further complicate field deployment. Beyond these biological constraints, stringent biosafety regulations, ethical considerations, and public perception issues continue to influence the acceptance and development of policies. Addressing these challenges requires an integrated approach that combines robust biocontainment strategies, multi-omics-guided strain optimization, and adaptive regulatory frameworks, balancing innovation with environmental safety. Such coordinated efforts are essential to realizing the full potential of engineered strains as reliable tools for sustainable environmental remediation and resource recovery.

3.1 Genetic Stability and Evolutionary Loss

Engineered traits encoded on plasmids or synthetic circuits are susceptible to loss due to mutation or recombination, particularly under the selective pressures and fluctuating conditions prevalent in natural environments (Renda et al., 2014). Plasmid loss and instability reduce the long-term effectiveness, particularly in complex and open ecosystems where engineered populations are exposed to fluctuating environmental conditions, resource limitations, and competition from native species (Johns et al., 2016). The loss and instability are often driven by a single mutational event that results in the complete loss of the intended function. In such cases, stochastic mutations can progressively erode the genetic information encoded within carefully designed circuits, while natural selection amplifies these failures by favoring mutants that eliminate costly synthetic burdens (Renda et al., 2014). Consequently, a central challenge is to halt or at least decelerate the evolutionary processes that compromise system stability and performance over time. Studies have demonstrated that plasmid maintenance requires either strong selective pressure or maintenance systems, which themselves have drawbacks, such as fitness costs and the potential for the selection of escape mutants (Garbisu et al., 2017; Kroll et al., 2010). To mitigate this evolutionary instability, various strategies have been proposed, including reducing host mutation rates, minimizing the use of mutation-prone genetic architectures, and designing more robust regulatory networks (Renda et al., 2014; Son et al., 2021). Nonetheless, ensuring the long-term robustness and functional stability of microbial communities remains a major challenge.

By incorporating the dynamics of mutation and selection into predictive frameworks, recent developments have started to quantitatively model and mitigate the evolutionary loss of engineered. For example, computational models that link DNA design parameters with host population dynamics have been developed to predict how engineered traits evolve over time and under varying selective pressures (Ingram & Stan, 2023). These models assign explicit parameters for mutation rates and fitness costs, allowing the simulation of mutation spread and functional decay across generations, thereby providing a theoretical basis for forecasting evolutionary failure and guiding design choices that maximize genetic "shelf life." Such frameworks highlight that the interplay between genetic stability, cellular burden, and host growth rate critically determines when and how engineered functions are lost, and suggest that design strategies that minimize expression burden or mitigate mutation hotspots can extend stability (Ingram & Stan, 2023).

Recent research has also explored how plasmid properties themselves shape evolutionary outcomes. In a study by Pisera and Liu (2025), controlling plasmid copy number revealed that low copy numbers can accelerate the fixation of beneficial alleles, whereas high copy numbers help maintain allelic diversity, which in turn can delay loss-of-function by preserving multiple sequence variants within a population. However, high copy number also slows the purging of deleterious mutations and increases the likelihood that non-functional alleles will hitchhike with beneficial ones (Pisera & Liu, 2025). These findings underscore that plasmid biology, particularly copy number and mutation rate, must be considered alongside host-circuit interactions when seeking evolutionary stability, as plasmid design decisions can fundamentally alter how engineered sequences fare over many generations.

3.2. Biosafety and Risk Management

Biosafety refers to the principles, practices, and regulatory frameworks designed to prevent unintended exposure to, release of, or harm from biological agents, including genetically engineered microorganisms (Gao et al., 2025; Wang & Zhang, 2019). In the context of genetic engineering and synthetic biology, biosafety encompasses not only laboratory containment but also environmental, ecological, and evolutionary risk assessment across the full life cycle of engineered organisms. Closely linked to biosafety is biosecurity, which addresses the prevention of misuse, diversion, or unintended proliferation of engineered biological systems with dual-use potential(Wang & Zhang, 2019). Together, biosafety and biosecurity form the foundation for the responsible development, deployment, and governance of GEMs.

Ensuring the safe deployment of GEMs demands robust biological containment and a deep understanding of microbe-environment interactions. Key biosafety risks include unintended impacts on non-target species and the propagation of antibiotic resistance determinants (Zhao et al., 2025). To minimize the effects of these challenges, engineered microbes are often equipped with self-limiting mechanisms, such as programmed cell death modules or regulatory circuits that confine survival to highly specific environmental conditions (Oliveira et al., 2022). One well-established containment system is the *hok/sok* toxin–antitoxin module, which enhances plasmid stability by eliminating plasmid-free descendants and thereby reducing reliance on antibiotic selection (Khodak et al., 2023).

A further critical issue involves the potential for HGT from engineered strains to native microbial populations, which could lead to unintended ecological consequences (Zhao et al., 2025). Monitoring in field settings is technically demanding. A recent review highlights environmental stressors that stimulate plasmid mobility and promote the transfer of antibiotic resistance genes, underscoring the dual-use risk posed by mobile constructs (Li & Zhang, 2023). Detecting low-frequency HGT events, quantifying ecological impacts on non-target taxa, and attributing observed changes to the engineered intervention requires deep sampling, high-resolution metagenomics, and longitudinal studies (Garbisu et al., 2017). Without rigorous monitoring, subtle or delayed adverse effects may go unnoticed until after release.

Advances in precision genome editing, such as CRISPR-based systems, help reduce off-target integration and limit the spread of mobile genetic elements, thereby lowering HGT risk (Wang & Doudna, 2023). Insights from microbial ecology further inform containment strategies by elucidating GEM dynamics and interactions within complex soil ecosystems. Looking forward, integrating *in situ* biosensing systems offers a proactive surveillance mechanism, enabling prompt detection and control of GEMs proliferation or escape (Zhao et al., 2025). Equally important is cultivating regulatory transparency and stakeholder engagement, which are essential to building public trust in biotechnological interventions. The combination of robust containment, real-time monitoring, and inclusive governance provides a realistic path toward the responsible and effective employment of GEMs in environmental remediation (Zhao et al., 2025).

Beyond technical containment systems, effective governance must integrate biosafety considerations, such as biocontainment strategies, ecological risk assessment, and post-release surveillance, with social and political dimensions of public trust and accountability (Lea-Smith et al., 2025). This perspective underscores that biosafety risk assessment cannot be isolated from decision-making processes, including transparent reporting, stakeholder engagement, and adaptive governance capable of responding to evolving technologies and environmental feedback. Mitigating misuse or unintended proliferation must be embedded in regulatory structures, particularly as synthetic biology tools become more accessible and powerful (Lea-Smith et al., 2025).

Biosafety and biosecurity are not static compliance requirements but dynamic processes that must evolve alongside advances in synthetic biology. Responsible deployment of GEMs, therefore, depends on integrating robust biological containment, predictive ecological risk assessment, continuous monitoring, and coherent regulatory oversight. Equally important is sustained public and stakeholder engagement, which is essential for maintaining trust and ensuring that environmental applications of synthetic biology align with societal values and environmental protection goals (Zhao et al., 2025).

3.3 Poor Translation From Lab to Field

Bridging the gap between laboratory findings and successful field-scale deployment of environmental remediation technologies, particularly microbial bioremediation, remains a formidable challenge. The primary concerns are scalability and economic viability, both of which are crucial to ensuring that remediation strategies can be effectively implemented and maintained under real-world conditions (Zhao et al., 2025). One of the most persistent barriers to scaling laboratory successes lies in reproducing controlled experimental conditions within the inherently variable and complex environments found in nature (Fernández et al., 2018). Laboratory experiments are typically performed under carefully controlled conditions that maximize microbial growth and degradative activity; yet such optimized settings rarely exist *in situ*. In natural ecosystems, fluctuating temperatures, variable moisture levels, pH changes, and intense microbial competition can significantly influence the effectiveness of bioremediation processes (Zhao et al., 2025).

To address these barriers and improve the practical effectiveness of bioremediation, future research should prioritize strategies that improve microbial adaptability and efficiency. This includes developing microbial strains capable of sustaining activity across diverse environmental conditions and designing microbial consortia that can dynamically respond to environmental fluctuations (Scheuerl et al., 2020; Shu & Huang, 2022). Integrating microbial bioremediation with complementary approaches, such as chemical oxidation, phytoremediation, or physical removal techniques, can further enhance treatment efficiency, especially in complex contamination scenarios. Such combined approaches not only enhance overall remediation efficiency but also provide economic benefits by reducing long-term operational costs and promoting sustainable, publicly acceptable remediation practices.

3.4 Balancing Costs and Improving Efficiency

Cost-effectiveness remains a central challenge in implementing GEM-based bioremediation, particularly in challenging environments, where constraints on mass transport and microbial mobility magnify technical difficulties (Zhao et al., 2025). High expenditures associated with microbial cultivation, site conditioning, and extended monitoring greatly constrain the scalability of these technologies. Producing robust GEMs typically requires expensive growth media and strict environmental controls, making large-scale deployment financially demanding (Compernolle et al., 2013).

To mitigate these barriers, emerging strategies aim to optimize process efficiency while curbing costs. One promising strategy is to combine microbial degradation with electrokinetic remediation, which has been shown to reduce energy use and operational expenses while improving pollutant removal in challenging matrices (Yuan et al., 2016). Another cost-reducing measure involves substituting specialized culture media with low-cost substrates or industrial byproducts for microbial growth, thereby improving economic feasibility (Kadier et al., 2021).

3.5 Regulatory, Ethical, and Public-Acceptance Barriers

The environmental release of GEMs is subject to stringent regulatory oversight to prevent unintended ecological consequences. Regulatory agencies differ by jurisdiction, but they universally demand rigorous risk assessments before permitting the deployment of GEMs (Hanlon & Sewalt, 2021). Key considerations include evaluating the likelihood of HGT to native microbial populations, the survival and spread of GEMs beyond targeted areas, and the potential for unanticipated deleterious ecological effects (Zhao et al., 2025). In the United States, GEMs fall under the EPA's purview under the Toxic Substances Control Act (TSCA), which mandates detailed reviews of genetic constructs, intended applications, and post-release monitoring protocols (Murphy & Krimsky, 2003). Within the European Union, the directive on the deliberate release of genetically modified organisms implements analogous safeguards to ensure environmental safety (Wesseler et al., 2023). The ethical imperative to avoid irreversible ecosystem changes often leads regulators to require staged, highly contained trials and extensive environmental risk assessment.

Compliance with these regulations entails addressing critical biosafety concerns, including confining GEMs within designated treatment systems and preventing their persistence or HGT beyond controlled environments. To this end, developers must demonstrate robust containment strategies, such as multilayered physical, genetic, and operational barriers, alongside continuous environmental monitoring, rather than relying solely on conventional engineering safeguards, such as explosion-proof infrastructure or toxic gas-handling systems (Palladino et al., 2024). Moreover, public perception and acceptance of GEMs play a pivotal role in shaping regulatory outcomes, as societal apprehensions regarding safety and ethics can delay or restrict field approvals (Olawade et al., 2025). Consequently, these challenges underscore the need for comprehensive risk–benefit evaluations and the establishment of transparent, science-based regulatory frameworks that balance innovation with environmental protection (Olawade et al., 2025).

To support responsible use, regulatory frameworks should increasingly incorporate real-time monitoring data and bioinformatics-driven insights into their decision-making processes (Zhao et al., 2025). The advent of environmental "omics" tools and high-resolution biosensors enables precise *in situ* tracking of GEM behavior, facilitating adaptive regulation that responds to evolving field data (Zhao et al., 2025). By incorporating scientific evidence and stakeholder input into regulatory processes, agencies can more effectively balance ecological integrity with the promise of engineered bioremediation.

4. RISK ASSESSMENT, GOVERNANCE, AND REGULATION

The design and application of GEMs for environmental biotechnology must prioritize safety, reliability, and ecological integrity. The deliberate release of GEMs into natural ecosystems requires comprehensive ecological risk assessments to mitigate potential adverse impacts on biodiversity, soil, and human health (Lea-Smith et al., 2025). While strategies that enhance microbial efficiency offer substantial biotechnological advantages, they must be carefully balanced with biosafety measures to prevent unintended ecological disturbances, including the disruption of native microbial assemblages or cascading effects on higher trophic levels (Lea-Smith et al., 2025).

Regulatory frameworks for GEM release vary internationally. In the United Kingdom, the deliberate release of GEMs is governed by stringent legislation, with approvals granted only in exceptional cases. In the United States, regulatory oversight is shared between the Department of Agriculture (USDA) and the Environmental Protection Agency (EPA). Within the European Union, the regulatory landscape for GEMs is evolving to reflect advances in new genomic techniques. The European Commission is currently reviewing frameworks to distinguish between conventionally modified organisms and those generated through precision gene-editing technologies, a move that could lead to proportionate, risk-based regulations and potentially facilitate the deployment of engineered microbes in environmental biotechnology (Lea-Smith et al., 2025). Regulatory oversight for GEMs combines international agreements, regional statutes, and agency guidance that mandate case-by-case environmental risk assessment (ERA) before any intentional environmental release. At the international level, the Cartagena Protocol on Biosafety establishes Annex III "Risk Assessment" as the methodological backbone for assessing living modified organisms (LMOs), requiring parties to undertake scientifically sound ERAs that consider both biodiversity and human health impacts (Diversity, 2000).

The release of microbes engineered with antibiotic resistance genes is highly restricted, as such genes pose a risk of horizontal transfer to pathogenic species. However, advances in precise gene-editing technologies have enabled the generation of microbial strains devoid of foreign DNA biotechnology (Lea-Smith et al., 2025). The use of counter-selectable markers, such as *sacB*, facilitates the removal of antibiotic resistance genes after selection, ensuring cleaner genetic constructs. Moreover, the expanding application of CRISPR-Cas systems across diverse microbial taxa, in combination with directed accelerated revision technology, presents both novel opportunities and biosafety challenges for environmental synthetic biology and biosecurity (Jeong et al., 2023; Yang et al., 2019).

A further challenge lies in achieving global regulatory consistency. Variations in national regulations and standards hinders collaboration and the deployment of innovations across borders. A more harmonized and integrated global framework for biosafety and biosecurity, one that promotes interoperability between jurisdictions, will be critical for the safe and equitable advancement of environmental biotechnology (Lea-Smith et al., 2025). Finally, as the field becomes increasingly data-driven, data privacy, biosecurity, and intellectual property (IP) considerations take on renewed importance. Integrating biotechnology with artificial intelligence and automation generates vast biological datasets, raising challenges in safeguarding sensitive information and ensuring fair IP protection (Lea-Smith et al., 2025).

Monitoring, long-term stewardship, and governance capacity are often limiting factors in the responsible deployment of resources. Detecting low-frequency HGT events or attributing ecological changes to a GEMs release requires intensive, long-term sampling and high-resolution molecular analyses that are seldom budgeted in early trials. International reviews, therefore, emphasize the need for transparent stakeholder engagement, independent oversight, and adaptive management mechanisms as preconditions for open-environment applications, especially for high-consequence technologies such as gene drives (Diversity, 2000; Lassoued et al., 2020).

To address these challenges, a dynamic and adaptable governance approach is required, one centered on proactive sequence screening for sequences of concern, complemented by biocontainment measures to prevent harm and traceability systems for identification, liability, and IP management. Developing coherent global standards, measurable metrics, and interoperable informatics infrastructure will be crucial to ensuring biosafety, enhancing transparency, and strengthening societal confidence in the responsible use of engineered microbes for environmental applications (Lea-Smith et al., 2025).

5. CONCLUSIONS

Genetically engineered microbes present a transformative approach to addressing persistent environmental pollution. Their ability to degrade various xenobiotics through precisely engineered enzymatic pathways represents a significant improvement over conventional remediation methods, which often fail to achieve complete mineralization and can pose considerable process-safety risks. Recent advances in genetic engineering, particularly CRISPR-based editing, metabolic pathway optimization, synthetic biology, and AI-driven pathway design, have enabled the enhancement of microbial capabilities for environmental degradation, making efficient, sustainable biodegradation increasingly feasible under safer operational conditions. Despite these promising developments, challenges remain in translating laboratory-scale successes to field applications, including potential horizontal gene transfer, unintended ecosystem impacts, the management of degradation intermediates, and the loss of engineered traits, which further complicate field deployment. Beyond these constraints, stringent biosafety regulations, ethical considerations, and public perception issues continue to influence the acceptance and development of policies. Addressing these challenges requires an integrated framework that combines advanced biocontainment measures, multi-omics-driven strain optimization, and adaptive regulatory oversight, balancing scientific innovation with environmental protection. Such coordinated efforts are crucial to unlocking the full potential of GEMs as dependable and sustainable tools for environmental remediation and resource recovery.

Ultimately, sustained research, innovation, and responsible governance are essential to realize the potential of GEMs in environmental biotechnology. With thoughtful design, layered containment strategies, and continuous monitoring, GEM-based biotechnologies can significantly mitigate the environmental and health impacts of xenobiotics, providing a safer and more sustainable pathway toward ecosystem restoration and pollution mitigation.

REFERENCES

Adli, M. (2018). The CRISPR tool kit for genome editing and beyond. *Nature Communications*, *9*(1), 1911. DOI: 10.1038/s41467-018-04252-2 PMID: 29765029

Alavian, F., & Khodabakhshi, F. (2025). Integrating artificial intelligence with microbial biotechnology for sustainable environmental remediation. *Environmental Monitoring and Assessment*, *197*(11), 1183. DOI: 10.1007/s10661-025-14666-3 PMID: 41055779

Ansori, A. N., Antonius, Y., Susilo, R. J., Hayaza, S., Kharisma, V. D., Parikesit, A. A., Zainul, R., Jakhmola, V., Saklani, T., Rebezov, M., Ullah, M. E., Maksimiuk, N., Derkho, M., & Burkov, P. (2023). Application of CRISPR-Cas9 genome editing technology in various fields: A review. *Narra J*, *3*(2), e184. DOI: 10.52225/narra.v3i2.184 PMID: 38450259

Anzalone, A. V., Randolph, P. B., Davis, J. R., Sousa, A. A., Koblan, L. W., Levy, J. M., Chen, P. J., Wilson, C., Newby, G. A., Raguram, A., & Liu, D. R. (2019). Search-and-replace genome editing without double-strand breaks or donor DNA. *Nature*, *576*(7785), 149–157. DOI: 10.1038/s41586-019-1711-4 PMID: 31634902

Aparicio, J. D., Raimondo, E. E., Gil, R. A., Benimeli, C. S., & Polti, M. A. (2018). Actinobacteria consortium as an efficient biotechnological tool for mixed polluted soil reclamation: Experimental factorial design for bioremediation process optimization. *Journal of Hazardous Materials*, *342*, 408–417. DOI: 10.1016/j.jhazmat.2017.08.041 PMID: 28854393

Ayangbenro, A. S., & Babalola, O. O. (2018). Metal(loid) bioremediation: Strategies employed by microbial polymers. *Sustainability (Basel)*, *10*(9), 3028. DOI: 10.3390/su10093028

Barrangou, R., Fremaux, C., Deveau, H., Richards, M., Boyaval, P., Moineau, S., Romero, D. A., & Horvath, P. (2007). CRISPR provides acquired resistance against viruses in prokaryotes. *Science*, *315*(5819), 1709–1712. DOI: 10.1126/science.1138140 PMID: 17379808

Benner, S. A., & Sismour, A. M. (2005). Synthetic biology. *Nature Reviews. Genetics*, *6*(7), 533–543. DOI: 10.1038/nrg1637 PMID: 15995697

Bertels, F., & Rainey, P. B. (2011). Within-genome evolution of REPINs: A new family of miniature mobile DNA in bacteria. *PLOS Genetics*, *7*(6), e1002132. DOI: 10.1371/journal.pgen.1002132 PMID: 21698139

Bobate, S., Mahalle, S., Dafale, N. A., & Bajaj, A. (2023). Emergence of environmental antibiotic resistance: Mechanism, monitoring and management. *Environmental Advances, 13*, 100409. DOI: 10.1016/j.envadv.2023.100409

Brewer, A. M., & George, D. R. (2025). Meeting the moment: Opportunities and limitations for genetically engineered microbes for bioremediation. *Journal of Science Policy & Governance, 26*(2), 1–13. DOI: 10.38126/JSPG260201

Brouns, S. J., Jore, M. M., Lundgren, M., Westra, E. R., Slijkhuis, R. J., Snijders, A. P., Dickman, M. J., Makarova, K. S., Koonin, E. V., & Van Der Oost, J. (2008). Small CRISPR RNAs guide antiviral defense in prokaryotes. *Science, 321*(5891), 960–964. DOI: 10.1126/science.1159689 PMID: 18703739

Bu, T., Yang, R., Zhang, Y., Cai, Y., Tang, Z., Li, C., Wu, Q., & Chen, H. (2020). Improving decolorization of dyes by laccase from Bacillus licheniformis by random and site-directed mutagenesis. *PeerJ, 8*, e10267. DOI: 10.7717/peerj.10267 PMID: 33240620

Carvalho, G., Fouchet, D., Danesh, G., Godeux, A.-S., Laaberki, M.-H., Pontier, D., Charpentier, X., & Venner, S. (2020). Bacterial transformation buffers environmental fluctuations through the reversible integration of mobile genetic elements. *mBio, 11*(2), e02443-19. DOI: 10.1128/mBio.02443-19 PMID: 32127449

Chakraborty, S., Ghosh, S., Banerjee, S., Kumar, S., & Bhattacharyya, P. (2024). Elucidating the synergistic effect of acidity and metalloid poisoning on the microbiome through metagenomics and machine learning approaches. *Environmental Research, 243*, 117885. DOI: 10.1016/j.envres.2023.117885 PMID: 38072100

Chen, J., Liu, Y., & Mahadevan, R. (2023). Genetic engineering of *Acidithiobacillus ferridurans* using CRISPR systems to mitigate toxic release in biomining. *Environmental Science & Technology, 57*(33), 12315–12324. DOI: 10.1021/acs.est.3c02492 PMID: 37556825

Compernolle, T., Van Passel, S., & Lebbe, L. (2013). Bioremediation: How to deal with removal efficiency uncertainty? An economic application. *Journal of Environmental Management, 127*, 77–85. DOI: 10.1016/j.jenvman.2013.04.016 PMID: 23685120

Contreras-Salgado, E. A., Sánchez-Morán, A. G., Rodríguez-Preciado, S. Y., Sifuentes-Franco, S., Rodríguez-Rodríguez, R., Macías-Barragán, J., & Díaz-Zaragoza, M. (2024). Multifaceted applications of synthetic microbial communities: Advances in biomedicine, bioremediation, and industry. *Microbiology Research, 15*(3), 1709–1727. DOI: 10.3390/microbiolres15030113

Cuperlovic-Culf, M. (2018). Machine learning methods for analysis of metabolic data and metabolic pathway modeling. *Metabolites*, *8*(1), 4. DOI: 10.3390/metabo8010004 PMID: 29324649

Dangi, A. K., Sharma, B., Hill, R. T., & Shukla, P. (2019). Bioremediation through microbes: Systems biology and metabolic engineering approach. *Critical Reviews in Biotechnology*, *39*(1), 79–98. DOI: 10.1080/07388551.2018.1500997 PMID: 30198342

Dash, D. M., & Osborne, W. J. (2023). A systematic review on the implementation of advanced and evolutionary biotechnological tools for efficient bioremediation of organophosphorus pesticides. *Chemosphere*, *313*, 137506. DOI: 10.1016/j.chemosphere.2022.137506 PMID: 36526134

De Sousa, C. S., Hassan, S. S., Pinto, A. C., Silva, W. M., De Almeida, S. S., Soares, S. D. C., Azevedo, M. S., Rocha, C. S., Barh, D., & Azevedo, V. (2018). Microbial omics: applications in biotechnology. In Debmalya, B., & Vasco, A. (Eds.), *Omics technologies and bio-engineering* (Vol. 2, pp. 3–20). Elsevier. DOI: 10.1016/B978-0-12-815870-8.00001-2

Dejonghe, W., Goris, J., El Fantroussi, S., Höfte, M., De Vos, P., Verstraete, W., & Top, E. M. (2000). Effect of dissemination of 2, 4-dichlorophenoxyacetic acid (2, 4-D) degradation plasmids on 2, 4-D degradation and on bacterial community structure in two different soil horizons. *Applied and Environmental Microbiology*, *66*(8), 3297–3304. DOI: 10.1128/AEM.66.8.3297-3304.2000 PMID: 10919784

Deter, H. S., & Lu, T. (2022). Engineering microbial consortia with rationally designed cellular interactions. *Current Opinion in Biotechnology*, *76*, 102730. DOI: 10.1016/j.copbio.2022.102730 PMID: 35609504

Diao, J., Tian, Y., Park, S., Cho, S.-M., & Moon, T. S. (2025). Engineering microbial consortia for mixed plastic upcycling. *Nature Communications*, *17*(1), 637. DOI: 10.1038/s41467-025-67409-w PMID: 41381587

Diversity, C. P. B. C. B. (2000). *Cartagena Protocol on Biosafety to the Convention on Biological Diversity: text and annexes.* Secretariat of the Convention of Biological Diversity.

Dreher, J., Scheiber, J., Stiefl, N., & Baumann, K. (2018). xMaP An interpretable alignment-free four-dimensional quantitative structure–activity relationship technique based on molecular surface properties and conformer ensembles. *Journal of Chemical Information and Modeling*, *58*(1), 165–181.

Fedeson, D. T., Saake, P., Calero, P., Nikel, P. I., & Ducat, D. C. (2020). Biotransformation of 2,4-dinitrotoluene in a phototrophic co-culture of engineered *Synechococcus elongatus* and *Pseudomonas putida. Microbial Biotechnology, 13*(4), 997–1011. DOI: 10.1111/1751-7915.13544 PMID: 32064751

Fernández, P. M., Vinarta, S. C., Bernal, A. R., Cruz, E. L., & Figueroa, L. I. (2018). Bioremediation strategies for chromium removal: Current research, scale-up approach and future perspectives. *Chemosphere, 208*, 139–148. DOI: 10.1016/j.chemosphere.2018.05.166 PMID: 29864705

Fu, R., Wang, Y., Liu, Y., Liu, H., Zhao, Q., Zhang, Y., Wang, C., Li, Z., Jiao, B., & He, Y. (2022). CRISPR-Cas12a based fluorescence assay for organophosphorus pesticides in agricultural products. *Food Chemistry, 387*, 132919. DOI: 10.1016/j.foodchem.2022.132919 PMID: 35421656

Furuya, E. Y., & Lowy, F. D. (2006). Antimicrobial-resistant bacteria in the community setting. *Nature Reviews. Microbiology, 4*(1), 36–45. DOI: 10.1038/nrmicro1325 PMID: 16357859

Gao, W., Wu, Z., Zuo, K., Xiang, Q., Chen, X., Zhang, L., & Liu, H. (2025). Biosafety concept: Origins, evolution, and prospects. *Biosafety and Health, 7*(4), 209–217. DOI: 10.1016/j.bsheal.2025.07.003 PMID: 40918204

Garbisu, C., Garaiyurrebaso, O., Epelde, L., Grohmann, E., & Alkorta, I. (2017). Plasmid-mediated bioaugmentation for the bioremediation of contaminated soils. *Frontiers in Microbiology, 8*, 1966. DOI: 10.3389/fmicb.2017.01966 PMID: 29062312

Gasiunas, G., Barrangou, R., Horvath, P., & Siksnys, V. (2012). Cas9–crRNA ribonucleoprotein complex mediates specific DNA cleavage for adaptive immunity in bacteria. *Proceedings of the National Academy of Sciences of the United States of America, 109*(39), E2579–E2586. DOI: 10.1073/pnas.1208507109 PMID: 22949671

Gong, T., Liu, R., Zuo, Z., Che, Y., Yu, H., Song, C., & Yang, C. (2016). Metabolic engineering of *Pseudomonas putida* KT2440 for complete mineralization of methyl parathion and γ-hexachlorocyclohexane. *ACS Synthetic Biology, 5*(5), 434–442. DOI: 10.1021/acssynbio.6b00025 PMID: 26854500

Gong, T., Xu, X., Dang, Y., Kong, A., Wu, Y., Liang, P., Wang, S., Yu, H., Xu, P., & Yang, C. (2018). An engineered *Pseudomonas putida* can simultaneously degrade organophosphates, pyrethroids and carbamates. *The Science of the Total Environment, 628-629*, 1258–1265. DOI: 10.1016/j.scitotenv.2018.02.143 PMID: 30045547

Greated, A., Lambertsen, L., Williams, P. A., & Thomas, C. M. (2002). Complete sequence of the IncP-9 TOL plasmid pWW0 from *Pseudomonas putida. Environmental Microbiology*, *4*(12), 856–871. DOI: 10.1046/j.1462-2920.2002.00305.x PMID: 12534468

Hanlon, P., & Sewalt, V. (2021). GEMs: Genetically engineered microorganisms and the regulatory oversight of their uses in modern food production. *Critical Reviews in Food Science and Nutrition*, *61*(6), 959–970. DOI: 10.1080/10408398.2020.1749026 PMID: 32274948

Hao, L., Zhang, D., Zhou, H., Wang, Z., Zhang, J., Zhao, Z., & Li, M. (2025). Machine learning applications in prediction and optimization of heavy metal bioremediation: A review. *Journal of Environmental Chemical Engineering*, *13*(5), 118845. DOI: 10.1016/j.jece.2025.118845

Hassan, S., & Ganai, B. A. (2023). Deciphering the recent trends in pesticide bioremediation using genome editing and multi-omics approaches: A review. *World Journal of Microbiology and Biotechnology*, *39*(6), 151. DOI: 10.1007/s11274-023-03603-6 PMID: 37029313

Heuer, H., & Smalla, K. (2012). Plasmids foster diversification and adaptation of bacterial populations in soil. *FEMS Microbiology Reviews*, *36*(6), 1083–1104. DOI: 10.1111/j.1574-6976.2012.00337.x PMID: 22393901

Hirschi, S., Ward, T. R., Meier, W. P., Müller, D. J., & Fotiadis, D. (2022). Synthetic biology: Bottom-up assembly of molecular systems. *Chemical Reviews*, *122*(21), 16294–16328. DOI: 10.1021/acs.chemrev.2c00339 PMID: 36179355

Hobom, B. (1980). Gene surgery: On the threshold of synthetic biology. *Medizinische Klinik*, *75*(24), 834–841. PMID: 6160378

Ibrahim, A., Oginga, B., Zhang, Y., Ling, W., Tang, L., Elatafi, E., Abady, M., & Gao, Y. (2025). Bioremediation of soils with emerging organic contaminants using immobilized microorganisms. *Environmental Technology and Innovation*, *40*, 104345. DOI: 10.1016/j.eti.2025.104345

Ibrahim, M., Raajaraam, L., & Raman, K. (2021). Modelling microbial communities: Harnessing consortia for biotechnological applications. *Computational and Structural Biotechnology Journal*, *19*, 3892–3907. DOI: 10.1016/j.csbj.2021.06.048 PMID: 34584635

Ingram, D., & Stan, G.-B. (2023). Modelling genetic stability in engineered cell populations. *Nature Communications*, *14*(1), 3471. DOI: 10.1038/s41467-023-38850-6 PMID: 37308512

Jayasekara, S. K., Joni, H. D., Jayantha, B., Dissanayake, L., Mandrell, C., Sinharage, M. M. S., Molitor, R., Jayasekara, T., Sivakumar, P., & Jayakody, L. N. (2023). Trends in in-silico guided engineering of efficient polyethylene terephthalate (PET) hydrolyzing enzymes to enable bio-recycling and upcycling of PET. *Computational and Structural Biotechnology Journal, 21*, 3513–3521. DOI: 10.1016/j.csbj.2023.06.004 PMID: 37484494

Jeong, S. H., Lee, H. J., & Lee, S. J. (2023). Recent advances in CRISPR-Cas technologies for synthetic biology. *The Journal of Microbiology, 61*(1), 13–36. DOI: 10.1007/s12275-022-00005-5 PMID: 36723794

Johns, N. I., Blazejewski, T., Gomes, A. L., & Wang, H. H. (2016). Principles for designing synthetic microbial communities. *Current Opinion in Microbiology, 31*, 146–153. DOI: 10.1016/j.mib.2016.03.010 PMID: 27084981

Kacser, H. (1973). The control of flux. Symposia of the Society for Experimental Biology, Kadier, A., Ilyas, R., Huzaifah, M., Harihastuti, N., Sapuan, S., Harussani, M., Azlin, M., Yuliasni, R., Ibrahim, R., & Atikah, M. (2021). Use of industrial wastes as sustainable nutrient sources for bacterial cellulose (BC) production: Mechanism, advances, and future perspectives. *Polymers, 13*(19), 3365.

Khodak, Y. A., Shaifutdinov, R. R., Khasanov, D. S., Orlova, N. A., & Vorobiev, I. I. (2023). Location and orientation of the genetic toxin–antitoxin element hok/sok in the plasmid affect expression of pharmaceutically significant proteins in bacterial cells. *Biochemistry. Biokhimiia, 88*(9), 1326–1337. DOI: 10.1134/S0006297923090122 PMID: 37770399

Kim, G. B., Choi, S. Y., Cho, I. J., Ahn, D.-H., & Lee, S. Y. (2023). Metabolic engineering for sustainability and health. *Trends in Biotechnology, 41*(3), 425–451. DOI: 10.1016/j.tibtech.2022.12.014 PMID: 36635195

Komor, A. C., Kim, Y. B., Packer, M. S., Zuris, J. A., & Liu, D. R. (2016). Programmable editing of a target base in genomic DNA without double-stranded DNA cleavage. *Nature, 533*(7603), 420–424. DOI: 10.1038/nature17946 PMID: 27096365

Kroll, J., Klinter, S., Schneider, C., Voß, I., & Steinbüchel, A. (2010). Plasmid addiction systems: Perspectives and applications in biotechnology. *Microbial Biotechnology, 3*(6), 634–657. DOI: 10.1111/j.1751-7915.2010.00170.x PMID: 21255361

Kugarajah, V., Nisha, K. N., Jayakumar, R., Sahabudeen, S., Ramakrishnan, P., & Mohamed, S. (2023). Significance of microbial genome in environmental remediation. *Microbiological Research, 271*, 127360. DOI: 10.1016/j.micres.2023.127360 PMID: 36931127

Kumar, M., Rohilla, A., Singh, D., Panwar, A., & Manna, S. (2025). Microbial enzyme-based water treatment in smart cities: A synoptic review. *Current Opinion in Environmental Science and Health, 46*, 100640. DOI: 10.1016/j.coesh.2025.100640

Kuppan, N., Padman, M., Mahadeva, M., Srinivasan, S., & Devarajan, R. (2024). A comprehensive review of sustainable bioremediation techniques: Eco-friendly solutions for waste and pollution management. *Waste Management Bulletin, 2*(3), 154–171. DOI: 10.1016/j.wmb.2024.07.005

Lassoued, R., Macall, D. M., Smyth, S. J., Phillips, P. W., & Hesseln, H. (2020). How should we regulate products of new breeding techniques? Opinion of surveyed experts in plant biotechnology. *Biotechnology Reports (Amsterdam, Netherlands), 26*, e00460. DOI: 10.1016/j.btre.2020.e00460 PMID: 32617264

Lea-Smith, D. J., Hassard, F., Coulon, F., Partridge, N., Horsfall, L., Parker, K. D. J., Smith, R. D. J., McCarthy, R. R., McKew, B., Gutierrez, T., Kumar, V., Dotro, G., Yang, Z., Curtis, T. P., Golyshin, P., Heaven, S., Jefferson, B., Jeffrey, P., Jones, D. L., & Krasnogor, N. (2025). Engineering biology applications for environmental solutions: Potential and challenges. *Nature Communications, 16*(1), 3538. DOI: 10.1038/s41467-025-58492-0 PMID: 40229265

Lee, Y., Jeong, S. E., Hur, M., Ko, S., & Jeon, C. O. (2018). Construction and evaluation of a Korean native microbial consortium for the bioremediation of diesel fuel-contaminated soil in Korea. *Frontiers in Microbiology, 9*, 2594. DOI: 10.3389/fmicb.2018.02594 PMID: 30425703

Li, J., Wang, X., Zou, J., Yang, K., Wang, X., Wang, Y., Zhang, H., Huang, H., Su, X., Yao, B., Luo, H., & Qin, X. (2023). Identification and characterization of the determinants of copper resistance in the acidophilic fungus *Acidomyces richmondensis* MEY-1 using the CRISPR/Cas9 System. *Applied and Environmental Microbiology, 89*(3), e02107–e02122. DOI: 10.1128/aem.02107-22 PMID: 36912653

Li, L.-G., & Zhang, T. (2023). Plasmid-mediated antibiotic resistance gene transfer under environmental stresses: Insights from laboratory-based studies. *The Science of the Total Environment, 887*, 163870. DOI: 10.1016/j.scitotenv.2023.163870 PMID: 37149187

Liu, B., Dong, X., Cheng, H., Zheng, C., Chen, Z., Rodríguez, T. C., Liang, S.-Q., Xue, W., & Sontheimer, E. J. (2022). A split prime editor with untethered reverse transcriptase and circular RNA template. *Nature Biotechnology, 40*(9), 1388–1393. DOI: 10.1038/s41587-022-01255-9 PMID: 35379962

Liu, Z., Liu, Y., Zeng, G., Shao, B., Chen, M., Li, Z., Jiang, Y., Liu, Y., Zhang, Y., & Zhong, H. (2018). Application of molecular docking for the degradation of organic pollutants in the environmental remediation: A review. *Chemosphere, 203*, 139–150. DOI: 10.1016/j.chemosphere.2018.03.179 PMID: 29614407

Lu, H., Diaz, D. J., Czarnecki, N. J., Zhu, C., Kim, W., Shroff, R., Acosta, D. J., Alexander, B. R., Cole, H. O., Zhang, Y., Lynd, N. A., Ellington, A. D., & Alper, H. S. (2022). Machine learning-aided engineering of hydrolases for PET depolymerization. *Nature, 604*(7907), 662–667. DOI: 10.1038/s41586-022-04599-z PMID: 35478237

Makia, A. A., Al-Taeea, A. M., & Atwanb, Z. W. (2024). Enhancing the performance of *Methylorubrum extorquens* AAZ-using CRISPR/Cas9 for the degradation of hydrocarbon compounds. *Asia-Pacific Journal of Molecular Biology and Biotechnology, 32*, 39–48. DOI: 10.35118/apjmbb.2024.032.3.05

Malla, M. A., Dubey, A., Yadav, S., Kumar, A., Hashem, A., & Abd Allah, E. F. (2018). Understanding and designing the strategies for the microbe-mediated remediation of environmental contaminants using omics approaches. *Frontiers in Microbiology, 9*, 1132. DOI: 10.3389/fmicb.2018.01132 PMID: 29915565

Mandeep, L., Liu, H., & Shukla, P. (2021). Synthetic biology and biocomputational approaches for improving microbial endoglucanases toward their innovative applications. *ACS Omega, 6*(9), 6055–6063. DOI: 10.1021/acsomega.0c05744 PMID: 33718696

Mandpe, A., Bombaywala, S., Paliya, S., & Kumar, S. (2022). Efficacy of microbes in removal of pesticides from watershed system. In Kapoor, R. T., & Shah, M. P. (Eds.), *Synergistic Approaches for Bioremediation of Environmental Pollutants: Recent Advances and Challenges* (pp. 27–51). Academic Press. DOI: 10.1016/B978-0-323-91860-2.00010-5

Martinez, J. L., Sánchez, M. B., Martínez-Solano, L., Hernandez, A., Garmendia, L., Fajardo, A., & Alvarez-Ortega, C. (2009). Functional role of bacterial multidrug efflux pumps in microbial natural ecosystems. *FEMS Microbiology Reviews, 33*(2), 430–449. DOI: 10.1111/j.1574-6976.2008.00157.x PMID: 19207745

Massa, V., Infantino, A., Radice, F., Orlandi, V., Tavecchio, F., Giudici, R., Conti, F., Urbini, G., Di Guardo, A., & Barbieri, P. (2009). Efficiency of natural and engineered bacterial strains in the degradation of 4-chlorobenzoic acid in soil slurry. *International Biodeterioration and Biodegradation, 63*(1), 112–115. DOI: 10.1016/j.ibiod.2008.07.006

Meng, S., Li, Z., Zhang, P., Contreras, F., Ji, Y., & Schwaneberg, U. (2023). Deep learning guided enzyme engineering of Thermobifida fusca cutinase for increased PET depolymerization. *Chinese Journal of Catalysis*, *50*, 229–238. DOI: 10.1016/S1872-2067(23)64470-5

Mohapatra, B., & Phale, P. S. (2021). Microbial degradation of naphthalene and substituted naphthalenes: Metabolic diversity and genomic insight for bioremediation. *Frontiers in Bioengineering and Biotechnology*, *9*, 602445. DOI: 10.3389/fbioe.2021.602445 PMID: 33791281

Moura de Sousa, J., Lourenço, M., & Gordo, I. (2023). Horizontal gene transfer among host-associated microbes. *Cell Host & Microbe*, *31*(4), 513–527. DOI: 10.1016/j.chom.2023.03.017 PMID: 37054673

Murphy, N., & Krimsky, S. (2003). Implicit precaution, scientific inference, and indirect evidence: The basis for the US Environmental Protection Agency's regulation of genetically modified crops. *New Genetics and Society*, *22*(2), 159–175. DOI: 10.1080/1463677032000102602

Naizabayeva, L., Nurzhanov, C. A., Satymbekov, M. N., & Elle, V. Z. (2024). Information system for remediation and cleanup of contaminated soil with machine learning. *Procedia Computer Science*, *231*, 145–150. DOI: 10.1016/j.procs.2023.12.186

Newby, D., & Pepper, I. (2002). Dispersal of plasmid pJP4 in unsaturated and saturated 2, 4-dichlorophenoxyacetic acid contaminated soil. *FEMS Microbiology Ecology*, *39*(2), 157–164. DOI: 10.1111/j.1574-6941.2002.tb00917.x PMID: 19709195

Nozari, M., Samaei, M. R., Dehghani, M., & Ebrahimi, A. A. (2018). Bioremediation of alkane hydrocarbons using bacterial consortium from soil. *Health Scope*, *7*(In Press), 12524. DOI: 10.5812/jhealthscope.12524

Olawade, D. B., Oladipo, P., Ajisafe, O., Egbon, E., Fapohunda, O., & Kade, A. (2025). Genetically engineered microorganisms: A promising frontier for PFAS bioremediation. [REMOVED HYPERLINK FIELD]. *Process Safety and Environmental Protection*, *202*, 107699. DOI: 10.1016/j.psep.2025.107699

Oldiges, M., Eikmanns, B. J., & Blombach, B. (2014). Application of metabolic engineering for the biotechnological production of L-valine. *Applied Microbiology and Biotechnology*, *98*(13), 5859–5870. DOI: 10.1007/s00253-014-5782-8 PMID: 24816722

Oliveira, N. M., Wheeler, J. H., Deroy, C., Booth, S. C., Walsh, E. J., Durham, W. M., & Foster, K. R. (2022). Suicidal chemotaxis in bacteria. *Nature Communications*, *13*(1), 7608. DOI: 10.1038/s41467-022-35311-4 PMID: 36494355

Palladino, F., Marcelino, P. R. F., Schlogl, A. E., José, Á. H. M., Rodrigues, R. C. L. B., Fabrino, D. L., Santos, I. J. B., & Rosa, C. A. (2024). Bioreactors: Applications and innovations for a sustainable and healthy future—a critical review. *Applied Sciences (Basel, Switzerland), 14*(20), 9346. DOI: 10.3390/app14209346

Pandey, J., Chauhan, A., & Jain, R. K. (2009). Integrative approaches for assessing the ecological sustainability of *in situ* bioremediation. *FEMS Microbiology Reviews, 33*(2), 324–375. DOI: 10.1111/j.1574-6976.2008.00133.x PMID: 19178567

Pisera, A. O., & Liu, C. C. (2025). The role of plasmid copy number and mutation rate in evolutionary outcomes. *Nature Ecology and Evolution, 9*(9), 1694–1704. DOI: 10.1038/s41559-025-02792-7 PMID: 40659873

Poursat, B. A., van Spanning, R. J., de Voogt, P., & Parsons, J. R. (2019). Implications of microbial adaptation for the assessment of environmental persistence of chemicals. *Critical Reviews in Environmental Science and Technology, 49*(23), 2220–2255. DOI: 10.1080/10643389.2019.1607687

Purohit, H. J., Tikariha, H., & Kalia, V. C. (2018). Current scenario on application of computational tools in biological systems. In Purohit, H., Kalia, V., & More, R. (Eds.), *Soft Computing for Biological Systems* (pp. 1–12). Springer. DOI: 10.1007/978-981-10-7455-4_1

Pushkar, B., Sevak, P., Parab, S., & Nilkanth, N. (2021). Chromium pollution and its bioremediation mechanisms in bacteria: A review. *Journal of Environmental Management, 287*, 112279. DOI: 10.1016/j.jenvman.2021.112279 PMID: 33706095

Qi, L. S., Larson, M. H., Gilbert, L. A., Doudna, J. A., Weissman, J. S., Arkin, A. P., & Lim, W. A. (2013). Repurposing CRISPR as an RNA-guided platform for sequence-specific control of gene expression. *Cell, 152*(5), 1173–1183. DOI: 10.1016/j.cell.2013.02.022 PMID: 23452860

Rafeeq, H., Afsheen, N., Rafique, S., Arshad, A., Intisar, M., Hussain, A., Bilal, M., & Iqbal, H. M. N. (2023). Genetically engineered microorganisms for environmental remediation. *Chemosphere, 310*, 136751. DOI: 10.1016/j.chemosphere.2022.136751 PMID: 36209847

Renda, B. A., Hammerling, M. J., & Barrick, J. E. (2014). Engineering reduced evolutionary potential for synthetic biology. *Molecular BioSystems, 10*(7), 1668–1678. DOI: 10.1039/C3MB70606K PMID: 24556867

Rios Miguel, A. B., Jetten, M. S. M., & Welte, C. U. (2020). The role of mobile genetic elements in organic micropollutant degradation during biological wastewater treatment. *Water Research X*, *9*, 100065. DOI: 10.1016/j.wroa.2020.100065 PMID: 32984801

Ruan, Z., Chen, K., Cao, W., Meng, L., Yang, B., Xu, M., Xing, Y., Li, P., Freilich, S., Chen, C., Gao, Y., Jiang, J., & Xu, X. (2024). Engineering natural microbiomes toward enhanced bioremediation by microbiome modeling. *Nature Communications*, *15*(1), 4694. DOI: 10.1038/s41467-024-49098-z PMID: 38824157

Salazar-García, L. M., Damas-Ramos, L. C., Trejo-Alarcón, L. M., Rago, D., Ahonen, L., Cruz-Morales, P., Ponce-Noyola, P., & Licona-Cassani, C. (2025). CRISPR-driven enhanced hydrocarbon emulsification in an environmental *Pseudomonas aeruginosa* strain. *Microbial Cell Factories*, *24*(1), 151. DOI: 10.1186/s12934-025-02769-y PMID: 40605043

Samin, G., Pavlova, M., Arif, M. I., Postema, C. P., Damborsky, J., & Janssen, D. B. (2014). A *Pseudomonas putida* strain genetically engineered for 1, 2, 3-trichloropropane bioremediation. *Applied and Environmental Microbiology*, *80*(17), 5467–5476. DOI: 10.1128/AEM.01620-14 PMID: 24973068

Sarker, M. A. B., Imtiaz, M. H., Holsen, T. M., & Baki, A. B. (2024). Real-time detection of microplastics using an ai camera. *Sensors (Basel)*, *24*(13), 4394. DOI: 10.3390/s24134394 PMID: 39001173

Scheuerl, T., Hopkins, M., Nowell, R. W., Rivett, D. W., Barraclough, T. G., & Bell, T. (2020). Bacterial adaptation is constrained in complex communities. *Nature Communications*, *11*(1), 754. DOI: 10.1038/s41467-020-14570-z PMID: 32029713

Selle, K., & Barrangou, R. (2015). Harnessing CRISPR–Cas systems for bacterial genome editing. *Trends in Microbiology*, *23*(4), 225–232. DOI: 10.1016/j.tim.2015.01.008 PMID: 25698413

Seressiotis, A., & Bailey, J. E. (1986). MPS: An algorithm and data base for metabolic pathway synthesis. *Biotechnology Letters*, *8*(12), 837–842. DOI: 10.1007/BF01078641

Sharma, B., & Shukla, P. (2022). Futuristic avenues of metabolic engineering techniques in bioremediation. *Biotechnology and Applied Biochemistry*, *69*(1), 51–60. DOI: 10.1002/bab.2080 PMID: 33242354

Shu, W.-S., & Huang, L.-N. (2022). Microbial diversity in extreme environments. *Nature Reviews. Microbiology*, *20*(4), 219–235. DOI: 10.1038/s41579-021-00648-y PMID: 34754082

Silva-Rocha, R., de Jong, H., Tamames, J., & de Lorenzo, V. (2011). The logic layout of the TOL network of *Pseudomonas putida* pWW0 plasmid stems from a metabolic amplifier motif (MAM) that optimizes biodegradation of m-xylene. *BMC Systems Biology*, *5*(1), 191. DOI: 10.1186/1752-0509-5-191 PMID: 22078029

Singh, A. K., Bilal, M., Iqbal, H. M., & Raj, A. (2021). Trends in predictive biodegradation for sustainable mitigation of environmental pollutants: Recent progress and future outlook. *The Science of the Total Environment*, *770*, 144561. DOI: 10.1016/j.scitotenv.2020.144561 PMID: 33736422

Son, H.-I., Weiss, A., & You, L. (2021). Design patterns for engineering genetic stability. *Current Opinion in Biomedical Engineering*, *19*, 100297. DOI: 10.1016/j.cobme.2021.100297 PMID: 34308010

Sørensen, S. J., Bailey, M., Hansen, L. H., Kroer, N., & Wuertz, S. (2005). Studying plasmid horizontal transfer in situ: A critical review. *Nature Reviews. Microbiology*, *3*(9), 700–710. DOI: 10.1038/nrmicro1232 PMID: 16138098

Stockinger, P., Niederhauser, C., Farnaud, S., & Buller, R. (2025). Computational analysis reveals temperature-induced stabilization of FAST-PETase. *Computational and Structural Biotechnology Journal*, *27*, 969–977. DOI: 10.1016/j.csbj.2025.03.006 PMID: 40151525

Tao, S., Chen, H., Li, N., Wang, T., & Liang, W. (2022). The spread of antibiotic resistance genes *in vivo* model. *The Canadian Journal of Infectious Diseases & Medical Microbiology*, *2022*(1), 3348695. DOI: 10.1155/2022/3348695 PMID: 35898691

Tudi, M., Daniel Ruan, H., Wang, L., Lyu, J., Sadler, R., Connell, D., Chu, C., & Phung, D. T. (2021). Agriculture development, pesticide application and its impact on the environment. *International Journal of Environmental Research and Public Health*, *18*(3), 1112. DOI: 10.3390/ijerph18031112 PMID: 33513796

Varner, P. M., Allemann, M. N., Michener, J. K., & Gunsch, C. K. (2022). The effect of bacterial growth strategies on plasmid transfer and naphthalene degradation for bioremediation. *Environmental Technology & Innovation*, *28*, 102910. DOI: 10.1016/j.eti.2022.102910 PMID: 37091576

Verdezoto-Prado, J., Chicaiza-Ortiz, C., Mejía-Pérez, A. B., Freire-Torres, C., Viteri-Yánez, M., Deng, L., Barba-Ostria, C., & Guamán, L. P. (2025). Advances in environmental biotechnology with CRISPR/Cas9: Bibliometric review and cutting-edge applications. *Discover Applied Sciences*, *7*(3), 167. DOI: 10.1007/s42452-025-06609-x

Villaverde, J., Rubio-Bellido, M., Merchán, F., & Morillo, E. (2017). Bioremediation of diuron contaminated soils by a novel degrading microbial consortium. *Journal of Environmental Management, 188*, 379–386. DOI: 10.1016/j.jenvman.2016.12.020 PMID: 28011373

Villiger, L., Joung, J., Koblan, L., Weissman, J., Abudayyeh, O. O., & Gootenberg, J. S. (2024). CRISPR technologies for genome, epigenome and transcriptome editing. *Nature Reviews. Molecular Cell Biology, 25*(6), 464–487. DOI: 10.1038/s41580-023-00697-6 PMID: 38308006

Wanapaisan, P., Laothamteep, N., Vejarano, F., Chakraborty, J., Shintani, M., Muangchinda, C., Morita, T., Suzuki-Minakuchi, C., Inoue, K., Nojiri, H., & Pinyakong, O. (2018). Synergistic degradation of pyrene by five culturable bacteria in a mangrove sediment-derived bacterial consortium. *Journal of Hazardous Materials, 342*, 561–570. DOI: 10.1016/j.jhazmat.2017.08.062 PMID: 28886568

Wang, F., & Zhang, W. (2019). Synthetic biology: Recent progress, biosafety and biosecurity concerns, and possible solutions. *Journal of Biosafety and Biosecurity, 1*(1), 22–30. DOI: 10.1016/j.jobb.2018.12.003

Wang, H., Liu, X., Liu, S., Yu, Y., Lin, J., Lin, J., Pang, X., & Zhao, J. (2012). Development of a markerless gene replacement system for *Acidithiobacillus ferrooxidans* and construction of a pfkB mutant. *Applied and Environmental Microbiology, 78*(6), 1826–1835. DOI: 10.1128/AEM.07230-11 PMID: 22210219

Wang, J. Y., & Doudna, J. A. (2023). CRISPR technology: A decade of genome editing is only the beginning. *science, 379*(6629), eadd8643.

Wang, Y., Jiang, Q., Zhou, C., Chen, B., Zhao, W., Song, J., Fang, R., Chen, J., & Xiao, M. (2014). In-situ remediation of contaminated farmland by horizontal transfer of degradative plasmids among rhizosphere bacteria. *Soil Use and Management, 30*(2), 303–309. DOI: 10.1111/sum.12105

Wesseler, J., Kleter, G., Meulenbroek, M., & Purnhagen, K. P. (2023). EU regulation of genetically modified microorganisms in light of new policy developments: Possible implications for EU bioeconomy investments. *Applied Economic Perspectives and Policy, 45*(2), 839–859. DOI: 10.1002/aepp.13259

WHO. (2024). *Compendium of WHO and other UN guidance on health and environment: version with International Classification of Health Intervention (ICHI) codes*. World Health Organization.

Xue, F., Yu, X., Shang, Y., Peng, C., Zhang, L., Xu, Q., & Li, A. (2020). Heterologous overexpression of a novel halohydrin dehalogenase from *Pseudomonas pohangensis* and modification of its enantioselectivity by semi-rational protein engineering. *International Journal of Biological Macromolecules, 146*, 80–88. DOI: 10.1016/j.ijbiomac.2019.12.203 PMID: 31891700

Yang, J., Kim, B., Kim, G. Y., Jung, G. Y., & Seo, S. W. (2019). Synthetic biology for evolutionary engineering: From perturbation of genotype to acquisition of desired phenotype. *Biotechnology for Biofuels, 12*(1), 113. DOI: 10.1186/s13068-019-1460-5 PMID: 31086565

Yaseen, Z. M., Melini Wan Mohtar, W. H., Homod, R. Z., Alawi, O. A., Abba, S. I., Oudah, A. Y., Togun, H., Goliatt, L., Ul Hassan Kazmi, S. S., & Tao, H. (2024). Heavy metals prediction in coastal marine sediments using hybridized machine learning models with metaheuristic optimization algorithm. *Chemosphere, 352*, 141329. DOI: 10.1016/j.chemosphere.2024.141329 PMID: 38296204

Yu, L., Gong, S., & Liang, J. (2024). Assistance of aminated straw on enhanced biodegradation of phenol through bacteria and potential prediction of machine learning for bioremediation. *Water Science and Technology, 90*(7), 2174–2185. DOI: 10.2166/wst.2024.323

Yuan, Y., Guo, S., Li, F., Wu, B., Yang, X., & Li, X. (2016). Coupling electrokinetics with microbial biodegradation enhances the removal of cycloparaffinic hydrocarbons in soils. *Journal of Hazardous Materials, 320*, 591–601. DOI: 10.1016/j.jhazmat.2016.07.043 PMID: 27501882

Zhang, H., & Wang, X. (2016). Modular co-culture engineering, a new approach for metabolic engineering. *Metabolic Engineering, 37*, 114–121. DOI: 10.1016/j.ymben.2016.05.007 PMID: 27242132

Zhang, Y., Geng, Y., Li, S., Shi, T., Ma, X., Hua, R., & Fang, L. (2023). Efficient knocking out of the organophosphorus insecticides degradation gene opdB in *Cupriavidus nantongensis* X1T via CRISPR/Cas9 with red system. *International Journal of Molecular Sciences, 24*(6), 6003. DOI: 10.3390/ijms24066003 PMID: 36983076

Zhao, S., Su, X., Xu, C., Gao, X., & Lu, S. (2025). Microbial adaptation and genetic modifications for enhanced remediation in low-permeability soils. *The Science of the Total Environment, 958*, 177916. DOI: 10.1016/j.scitotenv.2024.177916 PMID: 39647202

Chapter 4
Bioreactors and Engineered Systems for Microbial Remediation:
Innovations in In-Situ and Ex-Situ Applications

Luyanda Msolo
http://orcid.org/0000-0002-7272-437X
University of Fort Hare, South Africa

Anthony Ifeanyi Okoh
University of Fort Hare, South Africa

Etinosa Ogbomoede Igbinosa
University of Benin, Nigeria

ABSTRACT

Life on Earth solely depends on the state and quality of its environments. Technological advancements, unsustainable agricultural practices and industrialization are the contributing factors that influence the quality of life on earth. Industrialization and urbanization have immensely transformed the biosphere. However, there are certain drawbacks that we can highlight through these transformations. These include the diminution of natural resources, accumulation of pollutants and increased agro-industrial emissions such as chemicals, pesticides, chemical-based fertilizers and petroleum hydrocarbons. Recent advancements such as smart bioreactors, the use of genetically engineered microorganisms, and integrated treatment systems are at the forefront of modern bioremediation efforts and have expanded the versatility

DOI: 10.4018/979-8-2600-2998-5.ch004

of microbial remediation. This chapter explores some of the advances in the use of bioreactors and engineered systems for microbial bioremediation in In-situ and Ex-situ applications for optimum removal of environmental pollutants.

1 INTRODUCTION

Life on Earth solely depends on the state and quality of its environments. Technological advancements, developments, and industrialization are the most contributing factors that influence the quality of life on earth. Industrialization and urbanization have immensely transformed the biosphere. However, there are certain drawbacks that we can highlight through these transformations. These include the diminution of natural resources, accumulation of pollutants, increased carbon emissions and other processes associated with human health (Malla et al., 2018). These industrial revolutions continue to present adverse health effects towards the earth's inhabitants including humans and amplify environmental degradation.

Microorganisms attenuate pollutants in the environment through enzymatic processes and convert them to safe products. This process is very slow and requires optimum environmental parameters that will allow microbes to proliferate, for bioremediation to take full effect (Karigar and Rao, 2011). There are several microbial species that are recognized as effective pollutant degraders. These include Bacterial species, Fungi species and Algae.

Human exposure to environmental pollutants (such as air pollutants) has been associated with numerous health defects which include but are not limited to the following: respiratory complications, developmental disorders, cancers, cardiovascular diseases and other health complications (Ayilara and Babalola, 2023). Aquatic pollutants have also been associated with various human health issues which include cancer, neurological disorders, hypertension, kidney defects and reproductive issues. These pollutants include but are not limited to pesticides, heavy metals, and organic compounds which can be transmitted to humans through direct ingestion or indirectly through the consumption of aquatic fauna (Ayilara and Babalola, 2023).

The contamination of aquatic and terrestrial environments through agro-industrial emissions such as chemicals, pesticides, chemical-based fertilizers and petroleum hydrocarbons remain the most bothersome public and environmental health concern globally (Karigar and Rao, 2011). Recent research highlights the use of bioremediation technology as a complimentary safe and cost-effective tool for removing these environmental contaminants (Ahmed et al., 2021).

The ubiquitous nature of microbes allows them to thrive almost everywhere in unimaginable matrices, where they exploit a wide range of environmental substrates, thus making them remarkable bioremediation agents (Kour et al., 2022; Ayilara and

Babalola, 2023). Furthermore, microbes have additional properties which further justify them as potent pollutant removers. These properties include reduced genome sizes, cellular simplicity, rapid production cycles and prolific adaptive ability to new ecological niches (Bhatt et al., 2021).

This chapter explores some of the recent advances in the use of bioreactors and engineered systems for microbial bioremediation in In-situ and Ex-situ applications for optimum removal of environmental pollutants.

2 MICROBIAL PROCESSES FOR BIOREMEDIATION

Microbes such as bacteria, fungi, and archaea have evolved to metabolize or bioremediate a variety of natural and manmade substances. This process uses metabolic properties to breakdown, convert, or immobilize contaminants. Microbial bioremediation is based on microorganisms' metabolic capabilities to detoxify or alter toxic molecules, which is determined by accessibility and bioavailability (Odelade and Babalola, 2019; Kulikova and Perminova, 2021). Conversely, there are numerous disputes in literature about "what defines the bioavailable fraction" and how to assess it (Chowdhury et al., 2003; Choppala et al., 2014; Sinduja et al., 2023).

Pollutants quickly bond to mineral and organic materials (solid phases) after entering the soil environment through a variety of physical and chemical mechanisms. Pollutant-soil interaction consists of sorption, complexation, and precipitation. Soil's propensity to release (desorb) contaminants dictates their sensitivity to microbial breakdown, which influences the efficacy of bioremediation procedure. Bioavailability in soil aggregates the smallest composite units in the complex soil environment which is limited by the delivery of pollutant molecules to a microbial cell. Sorption, which determines a contaminant's bioavailability, is an important but incomprehensible mechanism in bioremediation. There are two schools of thought regarding bioavailability and the subsequent biodegradation of organic pollutants: (i) the pre-requisite release of the contaminant from the sorbed phase to the aqueous phase for degradation by microbes and (ii) biodegradation of the pollutant in the sorbed phase by enzymes without being desorbed (Singh, 2008; Megharaj et al., 2011; Hazra et al., 2012; Zhang and Wang, 2020). Bioremediation techniques are typically characterized as either in-situ or ex-situ. In-situ bioremediation involves treating the polluted material on-site, whereas ex-situ bioremediation requires removing the polluted material to be treated somewhere else (Kuppan et al., 2024). In situ bioremediation is the biological degradation of organic pollutants under natural settings into carbon dioxide and water or an attenuated transformation product. It is an affordable, low-maintenance, environmentally benign, and long-lasting method of cleaning up polluted environments. Ex-situ bioremediation methods can be more costly compared to in-situ approaches since they require the excavation of contaminated samples for treatment. Certain environmental factors play crucial roles in bioremediation. For instance, different microbes require different pH (acidic or alkaline), optimum temperature (30°C to 40°C), moisture content, oxygen availability, nutrient availability, and salinity for optimal microbial activity. (Kebede et al., 2021; Alori et al., 2022) Figure 10.1 summarizes the notable environmental factors affecting the bioremediation techniques.

Figure 1. Notable environmental factors affecting natural bioremediation processes.

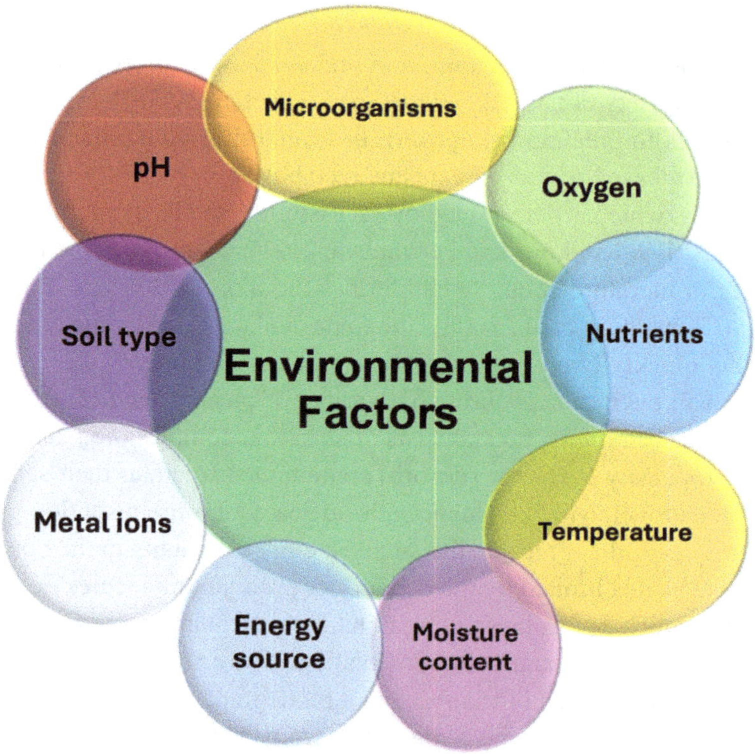

3 VARIOUS MICROORGANISMS UTILIZED AS BIOREMEDIATION AGENTS

Microorganisms that biodegrade in a variety of settings have been discovered as active members of microbial consortia. *Acinethobacter, Alcaligenes, Actinobacter, Arthrobacter, Bacillins, Berijerinckia, Flavobacterium, Methylosinus, Mycrobacterium,*

Mycococcus, Nitrosomonas, Nocardia, Penicillium, Phanerochaete, Pseudomonas, Rhizoctomia, Serratio, Trametes, and *Xanthofacter* are among the bacteria. However, Bacteria, Fungi and Algae are the most common bioremediation agents with an elevated capacity to remove pollutants such as heavy metal from the environment. These microorganisms have been utilized to clean environments that are polluted with heavy metals through the application of metal-resistant strains, the uptake of these heavy metals by metal-resistant strains occurs through bioaccumulation or adsorption (Medfu et al., 2020).

According to previous reports, using bioremediation agents reduces heavy metal stress, increases phytoremediation capability when combined with plant growth-promoting bacteria, and improves plant growth characteristics (Hassan et al., 2017).

3.1 Bacteria

The bioremediation of environmental contaminants, including polyaromatic, aromatic, and aliphatic hydrocarbons, involves a wide range of bacterial strains and their derivatives (Ayilara et al., 2023). Bacterial-mediated methods have been employed to reduce heavy metal toxicity. Endophytic bacteria are better able to resist and remediate heavy metal stress. Bacterial species have demonstrated the aptitude to remove heavy metals from the rhizosphere of several plant species and boost their growth (Hassan et al., 2017). Notably, these bacterial strains require ideal conditions for optimal biodegradability. A temperature of 30°C to 40°C, a pH of 5 to 8, and an oxygen level ranging from 10% to 40%, as well as a carbon and nitrogen source. These conditions promote optimal bioremediation (Kebede et al., 2021). Table 1 highlights some bacterial species and enzymes utilized in various processes of microbial bioremediation.

Table 1. Different Bacterial species and enzymes used in microbial bioremediation

Bacterial Species	Pollutants degraded	Enzymes used	Mechanisms of remediation	References
Bacillus* sp., *B. pumuilus, B. cereus, B. subtilis, B. coagulans	Nickel, Zinc, Lead, Oil, Palm oil, Copper, casein, and feather	Lipase, Protease, Amylase	Biosorption, Bioaugmentation, Bioimmobilization	Saranya et al., 2019; Malakar et al., 2020; Bhandari et al., 2021
Lysinibacillus sphaericus, Oudemansiella canarii	Azo dyes, Congo red dye	Laccase, Lignin peroxidase	Enzymatic reductase	Iark et al., 2019
E. coli	Synthetic dyes, steroids, Polyvinyl alcohol	Dehydrogenase, Lignin peroxidase	Enzymatic reduction	Ye et al., 2019
***Rhodococcus* sp.**	Aluminium, Lead, Copper	Lipase, Laccase	Bioaugmentation	Nanda et al., 2019
***Burkholderia* sp., *Ralstonia* sp. and *Achromobacter* sp.**	Hydrocarbons	Hydrolase	Bioprecipitation, Biosorption	Gricajeva et al., 2022
***Pseudomonas spp., P. aeruginosa, P. cepacian, Aeromonas* sp.**	Chromium, Lead, Uranium, Nickel, and Copper	Oxygenase, Laccase, Hydrolase	Bioattenuation, Biosorption	Malakar et al., 2020; Bhandari et al., 2021; Dave and Das, 2021
S. rhizophila	Polyvinyl alcohol	Dehydrogenase	Enzymatic reduction	Wei et al., 2018
Bacillus cereus	Wastewater pollutants	Amylase	Enzymatic reduction	Sonune and Garode, 2018

3.2 Fungi

Mycoremediation is regarded as a less-explored gold mine. The process involves the use of fungi or its derivatives to degrade environmental pollutants and has shown to be a comparatively cost-effective and environmentally friendly remediation process (Akhtar and Mannan, 2020). There are multiple reports on fungal species as effective agents of environmental bioremediation, which involves the remediation of diverse products including organic, inorganic, and emerging pollutants such as antibiotics, pharmaceuticals, polycyclic aromatic hydrocarbons, dyes, pesticides, detergents, heavy metals, herbicides, cyanotoxins, Phthalate and Pharma drugs (Akpasi et al., 2023; Ayilara et al., 2023).

Arbuscular mycorrhizal fungi reduce heavy metal toxicity by preventing their absorption and transport in plant tissues. Furthermore, fungi generate a wide range of morphological and physiological alterations. Macro and microfungal species have demonstrated the aptitude to reduce heavy metal toxicity, primarily in marine systems. Figure 10.2 illustrates the application of fungal species to bioremediation of various pollutants.

Figure 2. Notable applications of fungal species in bioremediation of various pollutants.

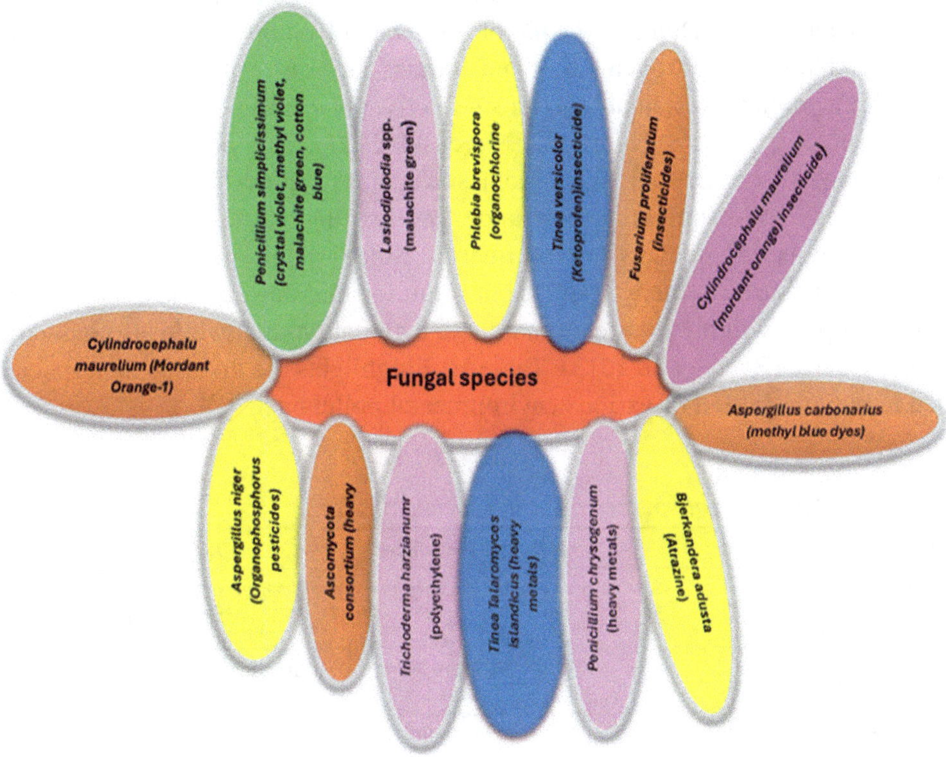

3.3 Algae

In recent times algae have been used to remove dissolved nutrients from aquaculture effluents, a process known as algal bioremediation or phycoremediation. Algal bioremediation involves the adsorption of contaminants to the surface of the algae independent of cell metabolism (Ayilara et al., 2023). It is widely accepted as an efficient and cost-effective method in wastewater treatment as this is a sustainable means to get rid of food and water-based pollutants especially heavy metals (Maroušek and Maroušková 2021).

Microbial species are also being explored for the potential of producing bioplastics instead of terrestrial crops. To create plastics in a sustainable manner while keeping an uninterrupted supply of food, researchers investigated the aquatic world and discovered microalgae as a potential and viable option. Seaweeds are rich in polysaccharides such as carrageenan and alginate, which are the primary components for creating bioplastics (Sudhakar et al., 2021).

Algae breeding can be done on industrial effluents and sewage without depleting available water sources for biomass production. Algae have a high ability to reduce the Chemical Oxygen Demand (COD) of wastewater by consuming and absorbing organic and inorganic nitrogen (ammonia, nitrates, etc.) and carbon for growth. Additionally, it promotes sustainable and cost-effective phosphorus recovery (Maroušek and Maroušková 2021; Stávková and Maroušek 2021). This provides microalgae an advantage over other sources for bioplastic synthesis. It may now follow a zero-waste strategy while simultaneously upcycling existing wastes and residual algal biomass into useful products in biorefineries. Since water covers over 3/4 of the overall earth's surface, there is sufficient room for microalgae. Thus, using them for this purpose will turn waste biomass into a global product while also reducing climate change (Balaji et al. 2013). Table 10.2 highlights the diverse Algal species used in bioremediation of various environmental pollutants.

Table 2. Different Algal species used in bioremediation

Algal species	Pollutant remediated	References
Spirulina platensis	Crude oil, Copper	El-Sheek and Hamouda (2014); Ayilara et al., 2023
Nostoc punctiforme	Heavy metals, Petroleum hydrocarbons	
Chlorella vulgaris		Fu and Secundo (2016); Kalhor et al., 2017; Ayilara et al., 2023
Cystoseira barbata	Chromium	Yalçın and Özyürek (2018)
Cystoseira crinite	Zinc, Cadmium and Nickel	
Sargassum filipendula		Moino et al., 2017; Nishikawa et al., 2018; Costa et al., 2020
Ulva lactuca	Zinc, Mercury	Senthikumar et al., 2019; Fabre et al., 2020
Caulerpa scalpelliformis	Cadmium	Jayakumar et al., 2021
Sargassum polycystum		Jayakumar et al., 2022
***Gracilariacorticata* sp.**	Nickel	Raju et al., 2021
***Synechocystis* sp.**	Pyrene	Patel et al., 2015
Turbinararia ornata	Lead	Al-Dhabi and Arasu 2022
Phormidium mucicola, Oscillatoria pranceps, Anabaena variabilia	Hydrocarbons	Al-Hussieny et al., 2020

4. ADVANCES IN MICROBIAL BIOREMEDIATION

The employment of microorganisms as bioremediation agents has recently resulted in numerous advancements. Recent advances in microbial bioremediation involve the use of microbial glycoconjugates, microbial biofilms, bioelectrochemical systems, and nanotechnology. Microbial glycoconjugates aid to lower surface tension, boost bioavailability, and form a solvent interface for organic contaminants. This improves the removal of contaminants in the environment (Visvanathan et al., 2024). Table 10.3 highlights recent technological developments in bioremediation processes with the aid of existing and emerging microorganisms.

Table 3. Recent technological developments in bioremediation

S/N	Emerging technology	Methods/Microbes used	Bioremediation process
1.	Use of microbial glycoconjugates	*Scedosporium* sp., *Acinetobacter* sp.	Biodegration of petroleum hydrocarbons
2.	Microbial biofilms	*Exiguobacterium profundum*	Bioremediation of organic pollutants
3.	Bioelectrochemical system	Biological and electrochemical methods	Remediate petroleum hydrocarbons
4.	Nanotechnology	Nanomaterials	Remediation of environmental pollutants

5 PHYTOREMEDIATION

Phytoremediation uses plants and their microbial partners such as mycorrhizal fungi to clean soil polluted with heavy metals, pesticides, and hydrocarbons (Raklami et al., 2022). Plant root exudates attract microbes that improve pollutant breakdown, detoxify metals, regulate hormones and protect against pathogens (Raklami et al., 2022). Microbes enhance pollutant breakdown, metal accumulation, plant resilience, further improved by inoculants and genetic engineering (Ma et al., 2022).

These microorganisms are beneficial to plants because they promote bioavailability of pollutants, detoxify heavy metals, and promote growth through bio-resources which allow plants to assimilate more resources on hormone regulation and hence growth and pathogen protection. This microbial enhancement produces better breakdown of contaminants, more metal accumulation, and plant resilience and all the processes improve phytoremediation for account of the plant to be used as a biofilter while also reducing ecological dose for the microbe. This allows application of microbial inoculants with desirable traits to a contaminated soil to enhance these capabilities, and the further aim of using genetic engineering strategies to improve both microbial and plant capabilities.

6 BIOFILTRATION AND MICROBIAL POLLUTION CONTROL

Biofiltration uses bacteria and fungi on filter materials to degrade harmful air and water pollutants (Sheoran et al., 2022). Biofilms formed by these microbes transform contaminants into harmless byproducts like CO_2 while protozoa and invertebrates can also aid removal (Sheoran et al., 2022). In wastewater treatment, biofilters perform nitrification and denitrification, and advanced types like bio-trickling filters enhance removal of volatile organic compounds (VOC's), odors, and industrial contaminants (Hassan et al., 2021).

In wastewater treatment processes, biofilters are used to degrade organic matter and nitrogen compounds through nitrification and denitrification. For example, ammonium is converted to first nitrate and finally, nitrogen gas that is lost in the process. The biofilm forms a layer over the biofilter and absorbs the contaminants while transforming them into harmless byproducts such as CO_2, water, and biomass. Sub-categories of bio-filters, such as bio-trickling filters, allow for continuous nutrient irrigation and microbial biofilm activity, thereby, enhancing pollutant removal efficiency. They can remove very high portions of VOCs, odors, methane, and other contaminants from industrial emissions and wastewater (Hassan et al., 2021).

In oil-contaminated salt-waters, the bacteria *Alcanivorax borkumensis* thrives in breaking down hydrocarbons and converting them into less harmful compounds (Baltaci et al., 2025). The lignin-degrading enzymes produced by white-rot fungi, *Phanerochaete chrysosporium*, degrade persistent organic pollutants, such as dyes and pesticides (Kumar et al., 2025). This can clean industrial waste, mining sites, and oil spills.

7 BIOAUGMENTATION

Bioaugmentation is a site-specific method that enhances the biological breakdown of pollutants such as petroleum hydrocarbons, pesticides, and heavy metals by breaking them down with specific microbes or specialized enzymes. This in situ technique is frequently used to clean contaminated soil (soil bioremediation), purify industrial wastewater, and remove airborne contaminants (air biofiltration). In essence, bioaugmentation enhances the natural ability of microorganisms to degrade toxicants by increasing microbial activity in the defined area (Muter, 2023). In natural wastewater treatment processes, bioaugmentation is a highly effective bioremediation technique for treating complex wastewater by introducing modified microbial strains with targeted functions into the treatment system, thereby increasing microbial community diversity and improving the degradation activity of key functional microbes. This modern technique has been successfully applied in various aspects such as optimized degradation of harmful substances, in optimizing the removal efficacy of nutrients, adjusting to unfavorable environmental conditions and thus, improving the wastewater treatment systems (Ma et al., 2022).

8 BIOSTIMULATION AND BIOVENTING

Biostimulation and Bioventing enhance aerobic biodegradation by injecting air or oxygen into the unsaturated zone. The bioventing technology uses regulated airflow through the supply of oxygen to the unsaturated zone, thereby boosting the activity of native bacteria for bioremediation. During the bioventing process, nutrients and moisture are added to aid in the bioremediation process. As a result, contaminants are converted by microbes into innocuous compounds. In recent years, other in-situ bioremediation techniques have converged on this one. In order to improve the ability of the different bacteria to biodegrade and encourage the purification of the heavy metal contaminants by precipitation, bioventing is a technique that helps to stimulate the native microflora through sufficient aeration (Bala et al., 2022). In order to increase the activity of the resident microorganisms and encourage the breakdown of contaminants, nutrients or other materials are added to the soil. Nitrogen, phosphorus, potassium, metabolites, enzymes, electron acceptors, and biosurfactants are a few examples of these nutrients. Biostimulation depends on the activity of native or resident microbes in the soil, as opposed to bioaugmentation. Because native microbes are frequently more competitive and more suited to the local environment, this approach is often used. Biostimulation can boost these microbes' metabolic activity by giving them the resources they need, which will speed up and improve the breakdown of pollutants (Visvanathan et al., 2024).

9 BIOSPARGING

This process involves the injection of air or gases directly into saturated zones to stimulate aerobic degradation. Similar to bioventing, air is inoculated into the core of the soil to promote microbial activity, which removes pollutants from contaminated areas. By infusing air into a saturated zone, bioventing encourages the flow of combustible organic compounds upward to an unsaturated zone nearby, as an alternative to traditional biodegradation procedures. Both soil porosity and pollutant biodegradability are necessary for biosparging to be successful. Regarding bioventing and soil vapour extraction (SVE), biosparging promotes microbial degradation, whereas in-situ air sparging (IAS) utilizes high air-flow rates to volatilize contaminants (Atlas and Philp, 2005; Tripathi and Singh, 2022; Bala et al., 2022). Biosparging is frequently used to extract fuel from water sources. Advanced bioremediation involves supplying microorganisms with oxygen to accelerate the biodegradation processes. Various technologies such as adsorption, microbial degradation, biosparging, PRBs, and the use of modified or synthetic zeolites, can be used to remove organic impurities (Maitra, 2018; Ahmadnezhad et al., 2021).

10 BIOSLURPING

This method combines the techniques of soil vapor extraction and vacuum-enhanced pumping to produce bioremediation. Products from capillary, light non-aqueous phase liquids (LNAPL's) in saturated and unsaturated zones are frequently recovered using this technology. This method can eliminate volatile and semi-volatile organic substances. The pumping mechanism moves LNAPLs to the surface by moving upward. Here, the LNAPL separates from the water and air, which reduces microbial activity. Although this method is less effective for clean-up in low-permeable soil, it is less expensive to operate than some other in-situ methods (De Pourcq et al., 2015; Sharma, 2020; Alori et al., 2022).

11 PERMEABLE REACTIVE BARRIERS (PRB's)

Passive systems filled with reactive media (including microbes) installed in the flow path of contaminants. For treating groundwater contamination, the PRB system provides a passive, in-situ, economical and sustainable alternative to the more traditional methods. Placing reactive media in the direction of a pollutant flow is the basic idea behind how a PRB system operates. Contaminated water is removed from groundwater by immobilizing or converting the target contaminant species into harmless compounds as it flows through this reactive material under a natural hydraulic gradient. The PRB has historically been used to clean up chlorinated solvents in contaminated groundwater. In order to improve the application of PRBs in the remediation of various contaminants, additional materials are now being used, including zeolites, organic matter, limestone, alkaline substances, sulfate, and organic carbon-based waste products (Zhang et al., 2022; Budania and Dangayach, 2023).

12 BIOREACTORS

Bioreactors convert raw materials into desired products after a sequence of biological reactions. Bioreactors offer the perfect environment for bioremediation to flourish. A bioreactor holds the cleanup samples. When treating polluted soil, a bioreactor offers several benefits over ex-situ bioremediation techniques. The time required for bioremediation is significantly decreased by an effective bioremediation method based on bioreactors that can accurately regulate pH, agitation, temperature, aeration, substrate concentration and inoculum concentration. Controlling and manipulating the environmental or abiotic factors within the bioreactor allows optimal biological responses to occur. Because of their versatility, bioreactor designs can minimize abiotic losses while optimizing microbial decomposition (Gomes et al., 2013; Davoodi et al., 2020; Bala et al., 2022).

13 NATURAL ATTENUATION

Molecular approaches such as DNA fingerprinting, quantitative PCR, and mRNA expression have been used to improve natural attenuation bioremediation. These techniques have been used to detect, measure, and characterize environmentally beneficial microorganisms, as well as to monitor changes within microbial communities (Scow and Hicks, 2005). Modern techniques such as Remediation by Enhanced Natural Attenuation (RENA) have been implemented in recent times. This approach entails turning the soil to improve aeration of the substrate as to enhance microbial proliferation, nutrient availability, moisture and subsequently, the breakdown of pollutants. RENA is a modern microbial-assisted remediation technique that degrades soil pollutants through a combination of remediation mechanisms such as bioaugmentation, biodegradation and bioaccumulation. This technique improves the function and the ability of the microbes in bioremediation and offers a more affordable alternative compared to other remediation processes (Ayilara et al., 2023).

Figure 3. Notable in-situ and ex-situ bioremediation techniques

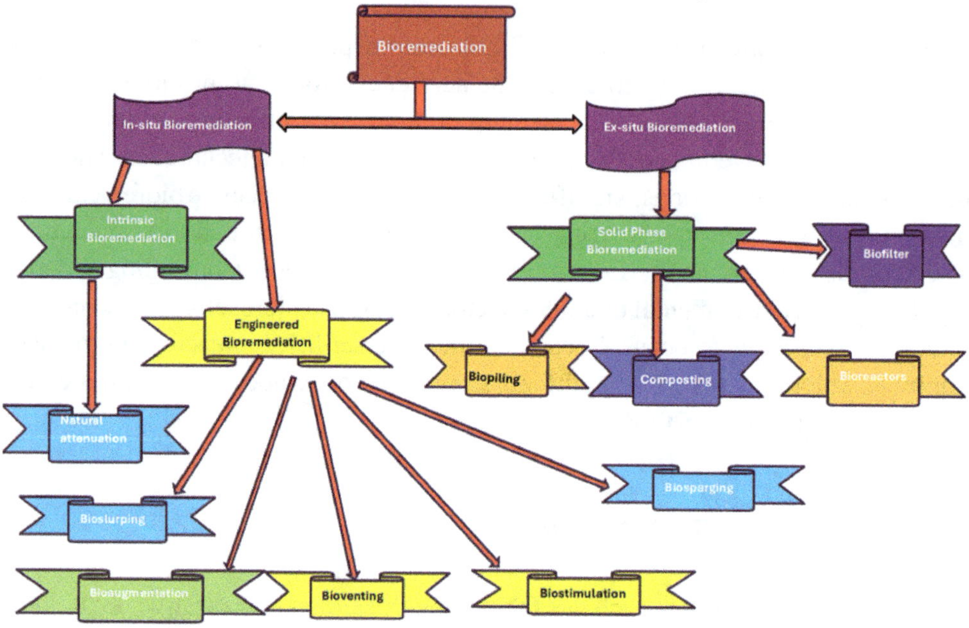

14 CONCLUSION AND FUTURE PERSPECTIVES

Several studies have explored the use of microbial enzymes for waste bioremediation. However, further improvements are necessary to maintain a safer and sustainable environment. More research is needed to identify microorganisms capable of quickly and efficiently removing toxins from industrial sources. Novel microorganisms and enzymes may be more effective in bioremediating pollutants than current methods. More studies are required to develop quick detection tools that can track progress and certify entire biodegradation of contaminants in the environment. Microbes employed in bioremediation can be genetically engineered to create more enzymes, improving their biodegradation potential. Combining many microbial consortia, rather than one, is a more effective way for bioremediation. This allows the presence of diverse species that use different substrates, boosting the rate of microbial breakdown. Microbes often break down organic substrates while removing inorganic contaminants. Therefore, efforts should be increased to identify microorganisms capable of digesting manmade contaminants. Nuclear waste from research, hospitals, fuel processing companies, and reactors remains a major source of pollution worldwide. It is worth noting to consider employing microorganisms and microbial enzymes for nuclear waste bioremediation.

Microbes can occasionally create pollution rather than eliminate it. Furthermore, microbial biostimulation can cause an algal bloom. To ensure a sustainable ecology, it is critical to establish prevention strategies.

In nature, different chemicals degrade at varying temperatures, and microbial survival is likewise temperature dependent, thus, further research is needed to determine the optimal temperature for natural chemical breakdown. Maintaining temperature equilibrium is crucial for microbial survival. This can prevent bioremediation failure when new bacteria are introduced into an ecosystem. Continuous research is necessary to understand the effects of positive and effective microbes on the environment after bioremediation. Introduced organisms may cause pollution through mutations or other means. Regulatory agencies should therefore monitor the risks associated with microorganisms in specific ecosystems. Enzymes and microorganisms may lose their efficacy if applied directly to soil. Combining microbes with other technologies, such as nanoparticles, may improve their activity. Raising awareness of microbial deterioration can be used by both policymakers and the public. Many individuals are unaware of this strategy; therefore, they may opt for traditional procedures that are less safe or effective than microbial bioremediation.

CONFLICT OF INTEREST

The authors declare no conflict of interest.

ACKNOWLEDGEMENTS

The authors would like to thank the South African Medical Research Council and the National Research Foundation for the financial support.

REFERENCES

Ahmadnezhad, Z., Vaezihir, A., Schüth, C., & Zarrini, G. (2021). Combination of zeolite barrier and bio sparging techniques to enhance efficiency of organic hydrocarbon remediation in a model of shallow groundwater. *Chemosphere, 273*, 128555. DOI: 10.1016/j.chemosphere.2020.128555 PMID: 33087257

Ahmed, S. F., Mofijur, M., Nuzhat, S., Chowdhury, A. T., Rafa, N., Uddin, M. A., Inayat, A., Mahlia, T. M. I., Ong, H. C., Chia, W. Y., & Show, P. L. (2021). Recent developments in physical, biological, chemical, and hybrid treatment techniques for removing emerging contaminants from wastewater. *Journal of Hazardous Materials, 416*, 125912. DOI: 10.1016/j.jhazmat.2021.125912 PMID: 34492846

Akhtar, N., & Mannan, M. A. U. (2020). Mycoremediation: Expunging environmental pollutants. *Biotechnology Reports (Amsterdam, Netherlands), 26*, e00452. DOI: 10.1016/j.btre.2020.e00452 PMID: 32617263

Akpasi, S. O., Anekwe, I. M. S., Tetteh, E. K., Amune, U. O., Shoyiga, H. O., Mahlangu, T. P., & Kiambi, S. L. (2023). Mycoremediation as a potentially promising technology: current status and prospects: a review. *Applied Sciences (Basel, Switzerland), 13*(8), 4978. DOI: 10.3390/app13084978

Al-Dhabi, N. A., & Arasu, M. V. (2022). Biosorption of hazardous waste from the municipal wastewater by marine algal biomass. *Environmental Research, 204*, 112115. DOI: 10.1016/j.envres.2021.112115 PMID: 34563525

Al-Hussieny, A. A., Imran, S. G., & Jabur, Z. A. (2020). The use of local blue-green algae in the bioremediation of hydrocarbon pollutants in wastewater from oil refineries. *Plant Archives, 20*, 797–802.

Alori, E. T., Gabasawa, A. I., Elenwo, C. E., & Agbeyegbe, O. O. (2022). Bioremediation techniques as affected by limiting factors in soil environment. *Frontiers in Soil Science, 2*, 937186. DOI: 10.3389/fsoil.2022.937186

Ayilara, M. S., Adeleke, B. S., Adebajo, M. T., Akinola, S. A., Fayose, C. A., Adeyemi, U. T., Gbadegesin, L. A., Omole, R. K., Johnson, R. M., Edhemuino, M., Ogundolie, F. A., & Babalola, O. O. (2023). Remediation by enhanced natural attenuation; an environment-friendly remediation approach. *Frontiers in Environmental Science, 11*, 1182586. DOI: 10.3389/fenvs.2023.1182586

Ayilara, M. S., & Babalola, O. O. (2023). Bioremediation of environmental wastes: The role of microorganisms. *Frontiers in Agronomy, 5*, 1183691. DOI: 10.3389/fagro.2023.1183691

Bala, S., Garg, D., Thirumalesh, B. V., Sharma, M., Sridhar, K., Inbaraj, B. S., & Tripathi, M. (2022). Recent strategies for bioremediation of emerging pollutants: A review for a green and sustainable environment. *Toxics, 10*(8), 484. DOI: 10.3390/toxics10080484 PMID: 36006163

Balaji, S., Gopi, K., & Muthuvelan, B. (2013). A review on production of poly β hydroxybutyrates from cyanobacteria for the production of bio plastics. *Algal Research, 2*(3), 278–285. DOI: 10.1016/j.algal.2013.03.002

Bhandari, S., Poudel, D. K., Marahatha, R., Dawadi, S., Khadayat, K., Phuyal, S., & Parajuli, N. (2021). Microbial enzymes used in bioremediation. *Journal of Chemistry, 2021*(1), 8849512.

Bhatt, P., Gangola, S., Bhandari, G., Zhang, W., Maithani, D., Mishra, S., & Chen, S. (2021). New insights into the degradation of synthetic pollutants in contaminated environments. *Chemosphere, 268*, 128827. DOI: 10.1016/j.chemosphere.2020.128827 PMID: 33162154

Budania, R., & Dangayach, S. (2023). A comprehensive review on permeable reactive barrier for the remediation of groundwater contamination. *Journal of Environmental Management, 332*, 117343. DOI: 10.1016/j.jenvman.2023.117343 PMID: 36758361

Choppala, G., Saifullah, , Bolan, N., Bibi, S., Iqbal, M., Rengel, Z., Kunhikrishnan, A., Ashwath, N., & Ok, Y. S. (2014). Cellular mechanisms in higher plants governing tolerance to cadmium toxicity. *Critical Reviews in Plant Sciences, 33*(5), 374–391. DOI: 10.1080/07352689.2014.903747

Chowdhury, U. K., Rahman, M. M., Samanta, G., Biswas, B. K., Basu, G. K., Chanda, C. R., & Chakraborti, D. (2003). In Naidu, R., Gupta, V. V. S. R., Rogers, S., Kookana, R. S., Bolan, N. S., & Adriano, D. (Eds.), *Groundwater arsenic contamination in West Bengal-India and Bangladesh: case study on bioavailibility of geogenic arsenic. Bioavailibility, Toxicity and Risk Relationships in Ecosystems* (pp. 291–329). Science Publishers.

Costa, C. S. D., Bertagnolli, C., Boos, A., da Silva, M. G. C., & Vieira, M. G. A. (2020). Application of a dealginated seaweed derivative for the simultaneous metal ions removal from real and synthetic effluents. *Journal of Water Process Engineering*, *37*, 101546. DOI: 10.1016/j.jwpe.2020.101546

Dave, S., & Das, J. (2021). Role of microbial enzymes for biodegradation and bioremediation of environmental pollutants: challenges and future prospects. Bioremediation for environmental sustainability, 325-346.

Davoodi, S. M., Miri, S., Taheran, M., Brar, S. K., Galvez-Cloutier, R., & Martel, R. (2020). Bioremediation of unconventional oil contaminated ecosystems under natural and assisted conditions: A review. *Environmental Science & Technology*, *54*(4), 2054–2067. DOI: 10.1021/acs.est.9b00906 PMID: 31904944

De Pourcq, K., Ayora, C., García-Gutiérrez, M., Missana, T., & Carrera, J. (2015). A clay permeable reactive barrier to remove Cs-137 from groundwater: Column experiments. *Journal of Environmental Radioactivity*, *149*, 36–42. DOI: 10.1016/j.jenvrad.2015.06.029 PMID: 26197347

Fabre, E., Dias, M., Costa, M., Henriques, B., Vale, C., Lopes, C. B., & Pereira, E. (2020). Negligible effect of potentially toxic elements and rare earth elements on mercury removal from contaminated waters by green, brown and red living marine macroalgae. *The Science of the Total Environment*, *724*, 138133. DOI: 10.1016/j.scitotenv.2020.138133 PMID: 32268287

Fu, P., & Secundo, F. (2016). Algae and their bacterial consortia for soil bioremediation. *Chemical Engineering Transactions*, *49*, 427–432.

Gomes, H. I., Dias-Ferreira, C., & Ribeiro, A. B. (2013). Overview of in-situ and ex-situ remediation technologies for PCB-contaminated soils and sediments and obstacles for full-scale application. *The Science of the Total Environment*, *445*, 237–260. DOI: 10.1016/j.scitotenv.2012.11.098 PMID: 23334318

Gricajeva, A., Nadda, A. K., & Gudiukaite, R. (2022). Insights into polyester plastic biodegradation by carboxyl ester hydrolases. *Journal of Chemical Technology and Biotechnology*, *97*(2), 359–380. DOI: 10.1002/jctb.6745

Hassan, M., Zhu, G., Yang, Z., & Lu, Y. (2021). Simultaneous removal of sulfamethoxazole and enhanced denitrification process from simulated municipal wastewater by a novel 3D-BER system. *Journal of Environmental Health Science & Engineering*, *19*(1), 23–38. DOI: 10.1007/s40201-020-00562-0 PMID: 34150216

Hassan, Z. U., Ali, S., Rizwan, M., Ibrahim, M., Nafees, M., & Waseem, M. 2017. Role of bioremediation agents (bacteria, fungi, and algae) in alleviating heavy metal toxicity. Probiotics in agroecosystem, pp.517-537.

Hazra, C., Kundu, D., & Chaudhari, A. (2012). Biosurfactant-assisted bioaugmentation in bioremediation. Microorganisms in Environmental Management, 631-664.

Iark, D., dos Reis Buzzo, A. J., Garcia, J. A. A., Côrrea, V. G., Helm, C. V., Corrêa, R. C. G., & Peralta, R. M. (2019). Enzymatic degradation and detoxification of azo dye Congo red by a new laccase from Oudemansiella canarii. *Bioresource Technology*, *289*, 121655. DOI: 10.1016/j.biortech.2019.121655 PMID: 31247524

Jayakumar, V., Govindaradjane, S., Rajamohan, N., & Rajasimman, M. (2022). Biosorption potential of brown algae, Sargassum polycystum, for the removal of toxic metals, cadmium and zinc. *Environmental Science and Pollution Research International*, *29*(28), 41909–41922. DOI: 10.1007/s11356-021-15185-7 PMID: 34275071

Jayakumar, V., Govindaradjane, S., & Rajasimman, M. (2021). Efficient adsorptive removal of zinc by Green marine macro alga Caulerpa scalpelliformis–characterization, optimization, modeling, isotherm, kinetic, thermodynamic, desorption and regeneration studies. *Surfaces and Interfaces*, *22*, 100798. DOI: 10.1016/j.surfin.2020.100798

Kalhor, A. X., Movafeghi, A., Mohammadi-Nassab, A. D., Abedi, E., & Bahrami, A. (2017). Potential of the green alga Chlorella vulgaris for biodegradation of crude oil hydrocarbons. *Marine Pollution Bulletin*, *123*(1-2), 286–290. DOI: 10.1016/j.marpolbul.2017.08.045 PMID: 28844453

Karigar, C. S., & Rao, S. S. (2011). Role of microbial enzymes in the bioremediation of pollutants: A review. *Enzyme Research*, *2011*, 2011. DOI: 10.4061/2011/805187 PMID: 21912739

Kebede, G., Tafese, T., Abda, E. M., Kamaraj, M., & Assefa, F. (2021). Factors influencing the bacterial bioremediation of hydrocarbon contaminants in the soil: Mechanisms and impacts. *Journal of Chemistry*, *2021*(1), 9823362. DOI: 10.1155/2021/9823362

Kour, D., Khan, S. S., Kour, H., Kaur, T., Devi, R., Rai, P. K., Judy, C., McQuestion, C., Bianchi, A., Spells, S., & Mohan, R. (2022). Microbe-mediated bioremediation: Current research and future challenges. *Journal of Applied Biology and Biotechnology*, *10*(2), 6–24. DOI: 10.7324/JABB.2022.10s202

Kulikova, N. A., & Perminova, I. V. (2021). Interactions between humic substances and microorganisms and their implications for nature-like bioremediation technologies. *Molecules (Basel, Switzerland), 26*(9), 2706. DOI: 10.3390/molecules26092706 PMID: 34063010

Kumar, R., Farda, B., Mignini, A., Djebaili, R., Koolman, L., Paul, A., Mondal, S., Joel, J. M., Pandit, A., Panneerselvam, P., Pellegrini, M., & Mitra, D. (2025). Microbial Solutions in Agriculture: Enhancing Soil Health and Resilience through Bio-Inoculants and Bioremediation. *Bacteria, 4*(3), 28. DOI: 10.3390/bacteria4030028

Kuppan, N., Padman, M., Mahadeva, M., Srinivasan, S., & Devarajan, R. (2024). A comprehensive review of sustainable bioremediation techniques: Eco friendly solutions for waste and pollution management. *Waste Management Bulletin, 2*(3), 154–171. DOI: 10.1016/j.wmb.2024.07.005

Ma, H., Zhao, Y., Yang, K., Wang, Y., Zhang, C., & Ji, M. (2022). Application oriented bioaugmentation processes: Mechanism, performance improvement and scale-up. *Bioresource Technology, 344*, 126192. DOI: 10.1016/j.biortech.2021.126192 PMID: 34710609

Ma, W., Tang, S., Dengzeng, Z., Zhang, D., Zhang, T., & Ma, X. (2022). Root exudates contribute to belowground ecosystem hotspots: A review. *Frontiers in Microbiology, 13*, 937–940. DOI: 10.3389/fmicb.2022.937940 PMID: 36274740

Maitra, S. (2018). *In situ bioremediation: An overview*. Responsibility of Life Science Informatics Publications.

Malakar, N., Mitra, S., Toppo, P., & Mathur, P. (2020). Understanding the functional attributes of different microbial enzymes in bioremediation.

Malla, M. A., Dubey, A., Yadav, S., Kumar, A., Hashem, A., & Abd Allah, E. F. (2018). Understanding and designing the strategies for the microbe-mediated remediation of environmental contaminants using omics approaches. *Frontiers in Microbiology, 9*, 1132. DOI: 10.3389/fmicb.2018.01132 PMID: 29915565

Maroušek, J., & Maroušková, A. (2021). Economic considerations on nutrient utilization in wastewater management. *Energies, 14*(12), 3468. DOI: 10.3390/en14123468

Medfu Tarekegn, M., Zewdu Salilih, F., & Ishetu, A. I. (2020). Microbes used as a tool for bioremediation of heavy metal from the environment. *Cogent Food & Agriculture, 6*(1), 1783174. DOI: 10.1080/23311932.2020.1783174

Megharaj, M., Ramakrishnan, B., Venkateswarlu, K., Sethunathan, N., & Naidu, R. (2011). Bioremediation approaches for organic pollutants: A critical perspective. *Environment International, 37*(8), 1362–1375. DOI: 10.1016/j.envint.2011.06.003 PMID: 21722961

Moino, B. P., Costa, C. S., da Silva, M. G., & Vieira, M. G. (2017). Removal of nickel ions on residue of alginate extraction from Sargassum filipendula seaweed in packed bed. *Canadian Journal of Chemical Engineering, 95*(11), 2120–2128. DOI: 10.1002/cjce.22859

Muter, O. (2023). Current trends in bioaugmentation tools for bioremediation: A critical review of advances and knowledge gaps. *Microorganisms, 11*(3), 710. DOI: 10.3390/microorganisms11030710 PMID: 36985282

Nanda, M., Kumar, V., & Sharma, D. K. (2019). Multimetal tolerance mechanisms in bacteria: The resistance strategies acquired by bacteria that can be exploited to 'clean-up'heavy metal contaminants from water. *Aquatic Toxicology, 212*, 1–10. DOI: 10.1016/j.aquatox.2019.04.011 PMID: 31022608

Nishikawa, E., da Silva, M. G. C., & Vieira, M. G. A. (2018). Cadmium biosorption by alginate extraction waste and process overview in life cycle assessment context. *Journal of Cleaner Production, 178*, 166–175. DOI: 10.1016/j.jclepro.2018.01.025

Odelade, K. A., & Babalola, O. O. (2019). Bacteria, fungi and archaea domains in rhizospheric soil and their effects in enhancing agricultural productivity. *International Journal of Environmental Research and Public Health, 16*(20), 3873. DOI: 10.3390/ijerph16203873 PMID: 31614851

Patel, J. G., Nirmal Kumar, J. I., Kumar, R. N., & Khan, S. R. (2015). Enhancement of pyrene degradation efficacy of Synechocystis sp., by construction of an artificial microalgal-bacterial consortium. *Cogent Chemistry, 1*(1), 1064193. DOI: 10.1080/23312009.2015.1064193

Raju, C. A., Anitha, J., Kalyani, R. M., Satyanandam, K., & Jagadeesh, P. (2021). Sorption of cobalt using marine macro seaweed graciliariacorticatared algae powder. *Materials Today: Proceedings, 44*, 1816–1827. DOI: 10.1016/j.matpr.2020.12.009

Raklami, A., Meddich, A., Oufdou, K., & Baslam, M. (2022). Plants Microorganisms-based bioremediation for heavy metal cleanup: Recent developments, phytoremediation techniques, regulation mechanisms, and molecular responses. *International Journal of Molecular Sciences, 23*(9), 5031. DOI: 10.3390/ijms23095031 PMID: 35563429

Saranya, K., Sundaramanickam, A., Shekhar, S., & Swaminathan, S. (2019). Biosorption of mercury by Bacillus thuringiensis (CASKS3) isolated from mangrove sediments of southeast coast India. *Indian Journal of Geo-Marine Sciences, 48*, 143–150.

Scow, K. M., & Hicks, K. A. (2005). Natural attenuation and enhanced bioremediation of organic contaminants in groundwater. *Current Opinion in Biotechnology, 16*(3), 246–253. DOI: 10.1016/j.copbio.2005.03.009 PMID: 15961025

Senthilkumar, R., Prasad, D. R., Govindarajan, L., Saravanakumar, K., & Prasad, B. N. (2019). Green alga-mediated treatment process for removal of zinc from synthetic solution and industrial effluent. *Environmental Technology, 40*(10), 1262–1270. DOI: 10.1080/09593330.2017.1420696 PMID: 29284361

Sharma, I. (2020). Bioremediation techniques for polluted environment: concept, advantages, limitations, and prospects. In *Trace metals in the environment-new approaches and recent advances*. IntechOpen.

Sheoran, K., Siwal, S. S., Kapoor, D., Singh, N., Saini, A. K., Alsanie, W. F., & Thakur, V. K. (2022). Air pollutants removal using biofiltration technique: A challenge at the frontiers of sustainable environment. *ACS Engineering Au, 2*(5), 378–396. DOI: 10.1021/acsengineeringau.2c00020 PMID: 36281334

Sinduja, M., Sathya, V., Maheswari, M., Kalpana, P., Dhevagi, P., Dinesh, G. K., & Chitdeshwari, T. (2023). Chemical transformation and bioavailability of chromium in the contaminated soil amended with bioamendments. *Bioremediation Journal, 27*(3), 229–250. DOI: 10.1080/10889868.2022.2049677

Singh, D. K. (2008). Biodegradation and bioremediation of pesticide in soil: Concept, method and recent developments. *Indian Journal of Microbiology, 48*(1), 35–40. DOI: 10.1007/s12088-008-0004-7 PMID: 23100698

Sonune, N. A., & Garode, A. M. (2018). Isolation and Characterization of Efficient Bacterial Isolates for Treatment of Municipal Wastewater. Int. J. Curr. Microbiol. App. Sci, 4(7), 577-584.

Stávková, J., & Maroušek, J. (2021). Novel sorbent shows promising financial results on P recovery from sludge water. *Chemosphere, 276*, 130097. DOI: 10.1016/j.chemosphere.2021.130097 PMID: 33711796

Sudhakar, P., Santosh, R., Asthalakshmi, B., Kumaresan, G., & Velraj, R. J. R. E. (2021). Performance augmentation of solar photovoltaic panel through PCM integrated natural water circulation cooling technique. *Renewable Energy, 172*, 1433–1448. DOI: 10.1016/j.renene.2020.11.138

Tripathi, M., & Singh, D. N. (Eds.). (2022). *Bioremediation: challenges and advancements*. Bentham Science Publishers. DOI: 10.2174/97898150360391220101

Tyagi, B., & Kumar, N. (2021). Bioremediation: Principles and applications in environmental management. In Bioremediation for Environmental Sustainability (pp. 3-28). Elsevier.

Visvanathan, P., Kirubakaran, D., Selvam, K., Manimegalai, P., Rajkumar, M., & Vasantharaj, K. (2024). Recent strategies for natural bioremediation of emerging pollutants: Development of a green and sustainable environment. *ACS Symposium Series*, *2*, 1–19. DOI: 10.1021/bk-2024-1476.ch001

Wei, Y., Fu, J., Wu, J., Jia, X., Zhou, Y., Li, C., & Chen, F. (2018). Bioinformatics analysis and characterization of highly efficient polyvinyl alcohol (PVA)-degrading enzymes from the novel PVA degrader Stenotrophomonas rhizophila QL-P4. *Applied and Environmental Microbiology*, *84*(1), e01898–e17. DOI: 10.1128/AEM.01898-17 PMID: 29079625

Yalçın, S., & Özyürek, M. (2018). Biosorption potential of two brown seaweeds in the removal of chromium. *Water Science and Technology*, *78*(12), 2564–2576. DOI: 10.2166/wst.2019.007 PMID: 30767921

Ye, X., Peng, T., Feng, J., Yang, Q., Pratush, A., Xiong, G., & Hu, Z. (2019). A novel dehydrogenase 17β-HSDx from Rhodococcus sp. P14 with potential application in bioremediation of steroids contaminated environment. *Journal of Hazardous Materials*, *362*, 170–177. DOI: 10.1016/j.jhazmat.2018.09.023 PMID: 30236938

Zhang, M., Tao, S., & Wang, X. (2020). Interactions between organic pollutants and carbon nanomaterials and the associated impact on microbial availability and degradation in soil: A review. *Environmental Science. Nano*, *7*(9), 2486–2508. DOI: 10.1039/D0EN00515K

Zhang, Y., Cao, B., Yin, H., Meng, L., Jin, W., Wang, F., & Al-Tabbaa, A. (2022). Application of zeolites in permeable reactive barriers (PRBs) for in-situ groundwater remediation: A critical review. *Chemosphere*, *308*, 136290. DOI: 10.1016/j.chemosphere.2022.136290 PMID: 36058373

Chapter 5
Application of Anammox Process in an Expanded Bed Biofilm Reactor for Eutrophic Nitrogen Removal

Comfort Chinyere Anyanwu
http://orcid.org/0009-0006-5256-4264
Manchester Metropolitan University, UK

Devine Akhidime
Manchester Metropolitan University, UK

Louise Melling
http://orcid.org/0009-0008-3349-5261
Manchester Metropolitan University, UK

ABSTRACT

The Anammox (anaerobic ammonia oxidation) wastewater treatment (WWT) process synchronously metabolizes ammonia nitrogen (NH3–N) and nitrite nitrogen (NO2–N) into dinitrogen gas (N2), is cost-effective and environmentally friendly, removing up to 90% of wastewater nitrogen. A Lab-scale expanded bed biofilm (biological mass attached to a solid surface) reactor (EBBR), equipped with ABDite® particles, was investigated for establishing the Anammox WWT process. Chemical and DNA analysis confirmed a successful establishment and that ammonia-oxidizing bacteria (AOB) Nitrosomonas spp. and Anammox bacteria - Brocadia, Kuenenia, and Jettenia spp. inhabited the EBBR-Anammox system; attaining a nitrogen loading

DOI: 10.4018/979-8-2600-2998-5.ch005

rate (NLR), removal rate (NRR) and removal efficiency (NRE) of 9.0 kg N m^{-3} d^{-1}, 6.1 kg N m^{-3} d^{-1}, and 80 – 100% respectively. Establishment of the EBBR-Anammox process underscores its potential as a scalable technology for mitigating global eutrophic nitrogen water pollution, providing suitable treated WW for irrigation and reducing overall operational costs in the agricultural sector.

1 INTRODUCTION

Eutrophication is the overloading of natural water bodies with nutrients, mainly nitrogen and phosphorus. Eutrophic nitrogen is the accumulation of high concentrations of nitrite nitrogen (NO_2-N), nitrate nitrogen (NO_3-N), and ammonia nitrogen (NH_3-N) within a natural water body (Boyd, 2019). High levels of eutrophic contaminants pose a significant global environmental challenge, threatening public health and aquatic ecosystems. Large quantities of nitrogen pollutants enter rivers and lakes from human-related sources, including agricultural, industrial effluents, and domestic wastewater. (Delkash et al., 2018; Glibert, 2017; Ngatia et al., 2019). The discharge of approximately 80% of untreated sewage into water bodies has reduced water availability for drinking, recreation, and fisheries, resulting in economic losses, particularly in developing countries (World Water Development Report 2018).

1.1 The Biological Nitrogen Recycling Processes

Processes of the biological nitrogen cycle that supply nutrients to all life forms on Earth can cause eutrophication and simultaneously remove eutrophic nitrogen from wastewater (Kuypers et al., 2018). Figure 1 shows the biological processes that recycle nitrogen on Earth.

Figure 1.

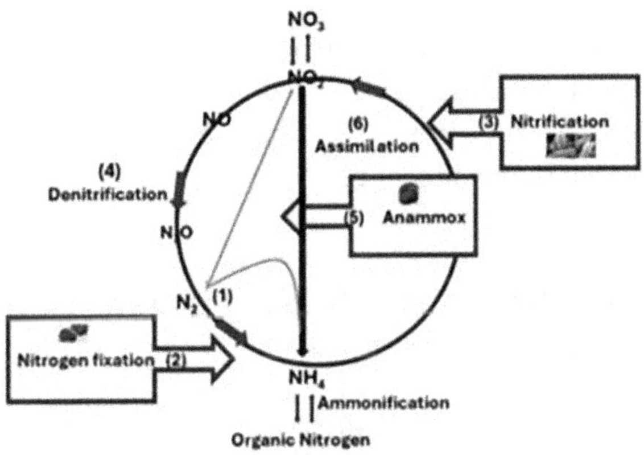

Diverse groups of microorganisms mediate the complex processes of nitrogen-transforming reactions. Amongst which are nitrification and denitrification processes conducted by aerobic ammonia-oxidizing bacteria (AOB), nitrite-oxidizing bacteria (NOB), and denitrifying bacteria (Kuypers et al., 2018; Koch et al., 2019; Daims et al., 2015). During partial nitrification (Equations 1 and 2), NH_3 is converted to nitrite (NO_2) by aerobic ammonia-oxidizing bacteria (AOB), which include *Nitrosomonas* spp., and ammonia-oxidizing archaea (AOA).

$$NH_3 + O_2 + 2H + 2e^- \rightarrow NH_2OH + H_2O \qquad (1)$$

$$NH_2OH + H_2O \rightarrow NO_2^- + 4e^- + 5H^+ \qquad (2)$$

The sequential nitrification reaction is completed by the actions of nitrite-oxidizing bacteria (NOB), e.g., *Nitrobacter sp.*, *Nitrospira* sp., which convert nitrite-nitrogen (NO_2-$N_)$ to nitrate-nitrogen (NO_3-N) (Equations 3 and 4).

$$\frac{1}{2} O_2 + 2\,H^+ + 2\,e^- \rightarrow H_2O \qquad (3)$$

$$NO_2- + H_2O \rightarrow NO_3^{\ -} + 2H^+ + 2e^- \qquad (4)$$

Facultative anaerobic bacteria (FAB) such as *Pseudomonas* and *Paracoccus* Spp conduct the denitrification process, which completes the stepwise reduction of NO_3 to dinitrogen (N_2) gas under anoxic conditions (Equation 5).

$$2NO_3^- + 10\,e^- + 12\,H^+ \rightarrow N_2 + 6\,H_2O \qquad (5)$$

The aerobic oxidizers involved in the nitrification pathway require molecular oxygen, and the products (NO_2-N and NO_3-N) are eutrophic pollutants that are toxic to aquatic life when the partially treated wastewater is discharged. Another drawback is that the denitrification process requires organic carbon input. Complete nitrification and denitrification processes consume large amounts of energy, which could increase the overall operational costs (Wang et al., 2017; Liwarska-Bizukojc et al., 2018).

1.2 Position of the Anammox Process within Biological Nitrogen Transformation

The limitations and demands of nitrification-denitrification processes led researchers to intensify efforts to develop more sustainable alternative WWT treatment methods that could effectively remove nitrogen from wastewater. The new pathway includes the environment-friendly anaerobic ammonia oxidation (Anammox) process (Equation 6).

$$NH_3 + NO_2^- + H^+ \rightarrow N_2 + 2H_2O \qquad (6)$$

Figure 2 shows the Anammox wastewater treatment process.

Figure 2.

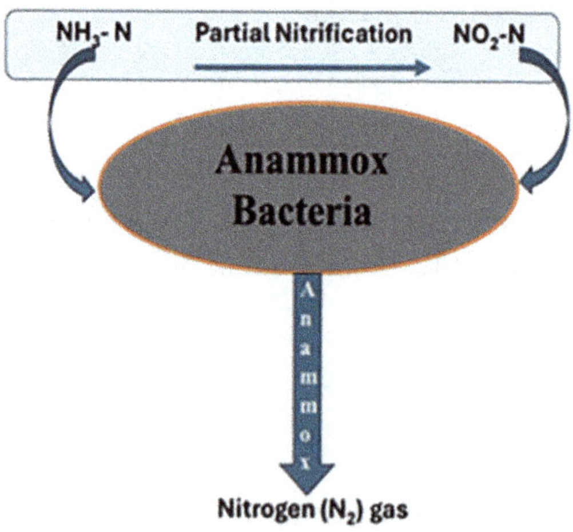

The Anammox process WWT pathway is mediated by the anaerobic ammonia-oxidizing (AnAOB) bacteria, which are obligate anaerobes that produce nitrogen from ammonia and nitrite under anoxic conditions, with hydroxylamine and hydrazine as intermediates (Wang et al., 2022). The Anammox process (Fig. 2), accidentally discovered in a yeast factory in the Netherlands in the 1990s, anaerobically transforms ammonia and nitrite to atmospheric dinitrogen (N_2) gas (Kuenen, 2020). The autotrophic Anammox bacteria, belonging to the phylum Planctomycetes, derive energy from inorganic carbon sources, including carbon dioxide and carbonates, for their metabolic activity. Species of the five genera (*Anammoxoglobus, Brocadia, Kuenenia, Jettenia, and Scalindua) of anammox bacteria have been* identified for Anammox WWT using only molecular techniques to date.

Unlike the conventional methods of nitrogen removal from wastewater (nitrification and denitrification), Anammox offers significant operational and environmental advantages as a WWT method; it bypasses the need for oxygen and external organic carbon. This unique reaction reduces aeration costs by 60%, energy consumption by 63%, and produces 70 – 80% less sludge (González-Martínez et al., 2018; Magrí et al., 2023).

Additionally, the Anammox process can remove approximately 90% of the nitrogen in wastewater when NH_3-N and NO_2-N are simultaneously metabolized to nitrogen gas. The release of nitrogen gas, trapped in NH3-N and NO2-N, into the atmosphere reduces the negative consequences (e.g., eutrophication) of overloading natural water bodies with nitrogen-rich raw or poorly treated wastewater. The Anammox process is a beneficial, eco-friendly, cost-effective, energy-efficient, and sustainable wastewater treatment method; it lowers greenhouse gas emissions - carbon dioxide is not released, and the supply of an external organic carbon source is not required (Wongkiew et al., 2017). The Anammox process also generates wastewater useful for agricultural practices, such as irrigation (Adams et al., 2020).

The application of the Anammox process in engineered systems faced multiple technical challenges, including the very slow growth rate, low biomass yields, and sensitivity of the Anammox bacteria to fluctuations in environmental conditions and a lack of Anammox sludge (Wang et al., 2022; Hendrickx et al., 2014; Ni and Zhang et al., 2013). To overcome the initial problems confronting the Anammox process, researchers enriched Anammox bacterial seed granules to shorten the startup time, designed bioreactors with various configurations, equipped with different support materials to develop stable microbial biomass for the efficient removal of nitrogen.

A variety of biomass growth supports have been developed and used for the Anammox WWT process; these materials include charcoal, porous ceramics, and volcanic rock (Adams et al., 2020; Dai et al., 2019; Lu et al., 2018), amongst others. Critical to the Anammox wastewater treatment process is the availability of suitable biomass growth support materials to provide both a physical surface for attachment and protection for the Anammox bacteria. Such media could shield the bacteria from predation by grazing protozoa (Suarez et al., 2015) and reduce biomass washout during mixing, thereby promoting biofilm formation (Zhang et al., 2013). Additional benefits are improved process stability and protection against the toxic effects of molecular oxygen on Anammox bacteria (Niederdorfer et al., 2021; Mozumder et al., 2014). Furthermore, biomass carriers provide a high surface area for microbial attachment, facilitating the development of dense, stable biofilms (Yang et al., 2012), which are required for efficiently removing eutrophic nitrogen pollutants from wastewater.

1.3 Anammox Bioreactors and Conditions of Operation

Many bioreactors have successfully implemented the Anammox wastewater treatment process under different operating conditions. The operational conditions of some major Anammox bioreactors are summarized in Table 1.

Table 1. Operation conditions of some major Anammox bioreactors

Type of reactor	Working Volume (L)	Duration (d)	Influent NO_2-N/ NH_3-N (mg L^{-1})	N-Removal Rate (kg N L^{-1} d^{-1})	NO_2-N: NH_3-N removal efficiency (%)	References
FBR	23	150	840: 1100	1.5	81 - 99	Adams et al., 2020
GLR	7	200	70:70	2.0	99:88	Kosgey et al., 2021
MBBR	256000	111	580	1.8	> 80	Dimitrova et al., 2020
MBR	8	>250	552:552	-	-	Lackner et al., 2014
RBC	1.7	100	350:350	0.091	100:99	Adams et al., 2020
SBR	12	-	20	0.00004	-	Laureni et al., 2015
UASB	5	280	420:350	-	88	Li and Sung et al., 2015

Key: **FBR** = fluidized bed reactor, **GLR** = gas lift reactor, **MBBR** = moving bed biofilm reactor, **MSBR** = membrane sequencing batch reactor, **MBR** = membrane biofilm reactor, **SBR** = sequencing batch reactor, **RBC** = rotating biological contactor. Adapted from Anyanwu (2023).

The engineered Anammox bioreactors not only vary in configurations, sizes, volumes, and operating conditions (Table 11.1), but also in the type and strength of wastewater treated at laboratory and industrial scales. Wastewater treated by the Anammox process includes that arising from agricultural and landfill runoffs, municipal and industrial sources, to enhance nitrogen removal before discharge. For example, the fluidized bed reactor (FBR) has been applied to treat municipal sewage with low concentrations of ammonia-nitrogen (NH_3-N), and high-strength ammonia-rich wastewater, including industrial wastewater, digester effluent, and landfill leachate (Adams et al., 2020). Deploying the moving bed biofilm reactor (MBBR), Dimitrova et al. (2020) used the partial nitrification-Anammox process to treat a highly rich nitrogen wastewater concentrate that originated from the dewatering stage of an anaerobically digested sludge. Using a high-strength NH_3-N synthetic wastewater, Kosgey et al. (2021) compared the performance of a gas-lift reactor to remove nitrogen via the Anammox process with that of MBBR in a laboratory study.

1.4 Estimated Global Expansion of Anammox Wastewater Treatment Installations

Improvements and technological advancements in the Anammox process have made it a priority wastewater treatment technology for removing nitrogen. After the first full-scale installation of an Anammox plant in Rotterdam, Netherlands (Wang et al., 2022), the last 10 years have witnessed a rapid global expansion in the number of Anammox wastewater treatment (WWT) plants from approximately 40 in 2014 (Lackner et al., 2014) to over 110 (Magrí et al., 2023; Yin et al., 2025).

Figure 3 shows the estimated number of global full-scale Anammox installations.

Figure 3.

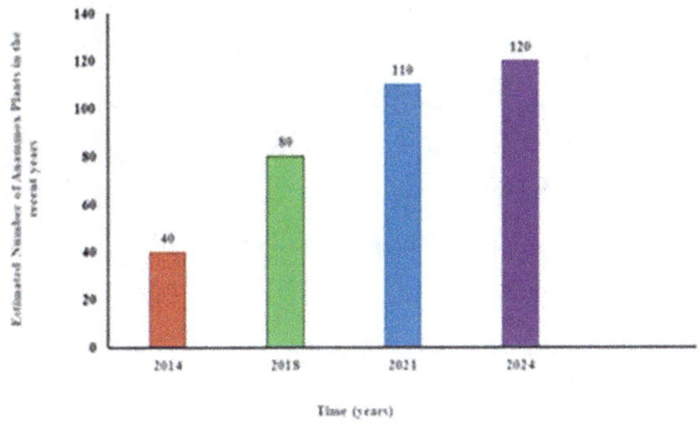

Improved developments, better management of the Anammox process, and the urgent global need to tackle water pollution and increase water security have made Anammox a popular wastewater treatment technology for removing eutrophic nitrogen.

Around the world, there are marked variations in the level of deployment of the Anammox process for wastewater treatment. China has the most (> 50%) operational full-scale Anammox wastewater treatment plants. Europe, particularly Germany, and Asia (Japan) have a significant number of Anammox installations (Zhang et al., 2021). Furthermore, countries such as the USA and the UK have started to deploy the Anammox wastewater treatment technology (Trigo et al., 2006). On the contrary, Africa, which stands to benefit more from the potential of Anammox technology in combating water scarcity and poor sanitary conditions, has no full-scale Anammox installations operating on the continent. However, concerted research efforts to establish the Anammox process are increasing, particularly in South Africa. Recently, a successful lab-scale enrichment of *Candidatus Brocadia species* from sludges, sourced from active municipal wastewater treatment facilities, and pilot reactor studies have been reported (Tikilili et al., 2016; Kosgey et al., 2022).

The Anammox wastewater treatment process has reduced the negative effects of releasing wastewater rich in eutrophic nitrogen into natural aquatic environments. Anammox has successfully treated a wide range of high-strength ammonia wastewaters emanating from reject water, industrial effluents, agricultural, and domestic sources (Adams et al., 2020). Anammox WWT operations have been conducted in various bioreactors fitted with different materials as support biomass development, thereby confirming the important contributions of the Anammox process in wastewater treatment worldwide (Dai et al., 2019; Lu et al., 2018).

A recent addition to the list of Anammox bioreactors is the adapted Laboratory expanded bed biofilm reactor (EBBR). The EBBR, initially engineered for complete oxidation of ammonia to nitrate, achieved notable success due to its unique ability to utilize the 0.7- 1.0 mm ABDite biomass support particles. The ABDite particles provide a large surface area for the development of stable biofilm. Comparatively, the expanded bed biofilm reactor also meets most criteria, as a sustainable WWT technology by promoting effective mass transfer, being simple to set up, low-cost, and energy efficient. The EBBR was deployed for the first laboratory-scale Anammox wastewater treatment process in a research study (Anyanwu et al., 2023).

2 EXPERIMENTAL DESIGN AND SET-UP OF THE EBBR FOR ANAMMOX INVESTIGATION

2.1 Expanded Bed Biofilm Reactor (EBBR) Schematic and Design

A transparent lab-scale (0.7 L) EBBR used in this study - the Anammox wastewater treatment investigation (Fig. 11.4) was made of borosilicate glass (QVF Corning Process Systems, UK) with two vertical columns (C1 and C2). The C1 was used for bed expansion, and C2, for wastewater recirculation, was connected using silicon tubing (Silex, UK) to ensure adequate mixing of the high-strength nitrogen influent from the feed bottle and partially treated (effluent) synthetic wastewater (SWW) leaving the expansion column (Anyanwu et al., 2024).

Figure 4 shows the schematic diagram of the expanded bed biofilm reactor (EBBR).

Figure 4.

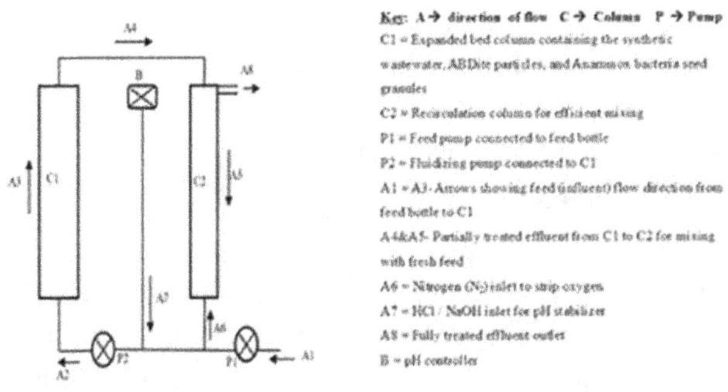

2.2 The EBBR ABDite Biomass Growth Support Particles and Anammox Seed Granules

To start the Anammox process investigation, the EBBR requires a supply of the ABDite biomass support and Anammox seed granules.

Figure 5 shows the ABDite particle and the Anammox seed granule.

Figure 5.

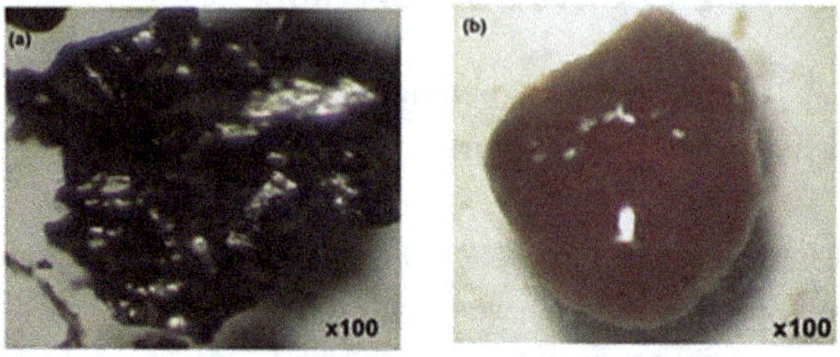

The 0.7–1.0 mm ABDite® biomass particles sourced from ABD Ltd. (Manchester, UK) were used as growth media to support biomass attachment in the expansion column (C1) of the EBBR-Anammox system to fill 75% (12.0 cm) of bed height capacity (Fig. 5a). Bricked red Anammox bacteria (AnAOB) seed granules (9.0 g) supplied by Severn Trent Waters (UK) were inoculated into the EBBR expansion column, continuously fed with high-strength nitrogen SWW containing a high strength (500 mg L^{-1}) NH_3-N and 660 mg L^{-1} (NO_2-N), and trace element solution following the method described in Anyanwu et al. (Anyanwu, 2023).

2.3 EBBR-Anammox Process Setup and Maintenance of Operation

Experimental conditions within the continuous culture EBBR were controlled as follows: a 50% bed expansion from the baseline height was maintained by the upward-flowing SWW, which provided a velocity of 1 cm s^{-1} using peristaltic pump (Watson Marlow 120S, UK); a temperature of 30° was maintained with a thermo-circulator (Thermofisher, UK) connected to a water jacket, and a pH controller (Electrolab 260, UK) added 0.1M HCl or 0.25 M NaOH to achieve pH 8.3. A continuous injection of nitrogen gas flowing at the rate of 100 m^3 min $^{-1}$ at the base of the recirculation column (Fig. 11.4, point A6) was used to reduce dissolved oxygen (DO).

2.4 Physical Observations and Chemical Measurements for Anammox Activity Detection in EBBR

The EBBR (C1) was monitored every 2–3 days to determine the development of brick-red granules or brown bioparticles indicating biofilm formation on the ABDite® biomass support, and the generation of gas bubbles; the physical evidence associated with the Anammox process. Increases in bed height, pH, and temperature measurements were recorded manually. Chemical analysis to determine NH_3-N,

NO_2-N, and NO_3-N concentrations in the influent and effluent was performed using standard spectrophotometric (standard) methods on 5 mL of liquid samples collected at points A1 (influent) and A8 (effluent) (Fig. 4).

The collected data were used to calculate nitrogen removal rate (NRR) and nitrogen removal efficiency (NRE) (Anyanwu, 2023). The NRR was determined as the difference between the influent NH_3-N and NO_2-N loaded into the EBBR and effluent NH_3-N & NO_2-N removed from the Anammox bioreactor based on the stoichiometric reaction and expressed in Kg N m^{-3} d^{-1} (Anyanwu et al., 2024; Wang et al., 2017). ABDite bioparticles rich in Anammox bacteria biofilms were collected by suction from the top of the EBBR and used for scanning electron microscopy (SEM), genomic DNA extraction, and DNA amplification by PCR, as well as the detection of 16S rRNA and NGS sequencing for the identification of the Anammox bacteria community in the EBBR.

3 RESULTS AND DISCUSSION

3.1 Physical Evidence Supporting the Establishment of the Anammox Process in the EBBR

The occurrence of the Anammox process in a bioreactor can be assessed with specific parameters

Figure 6 shows evidence for the establishment of the Anammox process in the EBBR

Figure 6.

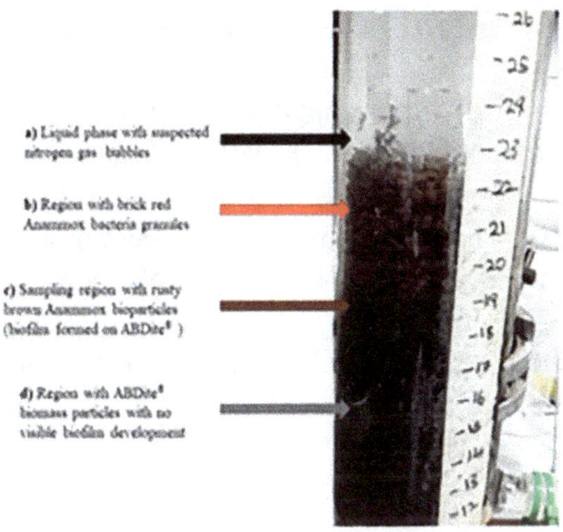

The evidence for Anammox bacterial activity was: (a) Gas bubble production suspected to be atmospheric nitrogen gas, (b) brick-red coloured Anammox bacterial granules, (c) brown-coloured and bioparticles developed on the ABDite®. There are also black-coloured ABDite® particles without observable biofilm attached (d). These physical observations provided compelling evidence for the establishment of the Anammox process within the EBBR (C1, Fig. 4) in expanded mode (Fig.6).

The evolution of unconfirmed bubbles of nitrogen gas within the liquid phase (Fig.6 (a)), the appearance of brick-red Anammox granules (Fig. 6 (b)), and the development of rusty-brown biofilms on the ABDite® support particles in the sampling region (Fig.6 (c)) collectively indicated active Anammox bacterial metabolism. These characteristic features, which are widely recognized as hallmarks of a mature Anammox community, suggest that the operating conditions within the EBBR facilitated the successful enrichment and activity of Anammox bacterial biomass. Comparable observations - including nitrogen gas production, red–orange granule formation, and red–brown biofilm development—have been reported as diagnostic indicators of established Anammox activity in a range of wastewater treatment bioreactors, including SBR, UASB, and MBBR reactors (Kuypers et al., 2018; Li and Sung et al., 2015; Dimitrova et al., 2020).

3.2 Chemical Performance Parameters Supporting the Establishment of the Anammox Process in the EBBR

The performance of the EBBR–Anammox system following the start-up of operations was evaluated using these chemical parameters: concentrations of ammonia-nitrogen (NH_3-N), nitrite nitrogen (NO_2-N), and nitrate nitrogen (NO_3-N), nitrogen loading (NLR) and removal (NRR) rates, and nitrogen removal efficiency (NRE). The start-up period (data not shown) was marked by fluctuations in the concentrations of NH_3-N, NO_2-N, and NO_3-N species recovered from effluent.

3.2.1 Concentrations of Ammonia-Nitrogen, Nitrite Nitrogen, and Nitrate Nitrogen in EBBR Effluent

The establishment of the Anammox WWT process in a system is determined by the simultaneous removal of NH_3-N and NO_2-N, and the production of about 10% NO_3-N.

Figure 7 shows the concentration of NH_3-N and NO_2-N removed from the EBBR.

Figure 7.

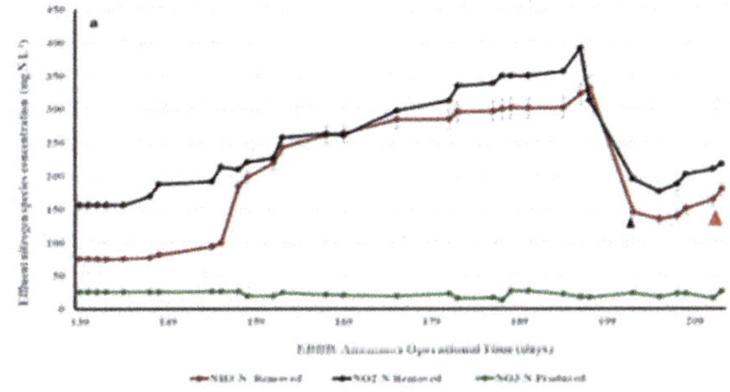

There was a marked change in the concentrations of NH_3-N, NO_2-N, and NO_3-N species after the initial turbulence from day 139 to day 212. During the 73 days of stable operation (fewer fluctuations) after the Anammox bacteria had adjusted to the operating (30 °C, pH 8.3) conditions within the reactor, the EBBR–Anammox system achieved a steady increase in NH_3–N (75 – 313 mg N L^{-1}) and NO_2–N (156 – 392 mg N L^{-1}) concentration removals, respectively. An average molar consumption ratio of 1:1.25 (NH_3-N: NO_2-N) was accompanied by the production of 12 – 27 mg N L^{-1} (NO_3-N), which accounted for ~10% of the total influent nitrogen supply in the synthetic wastewater (Fig.7).

3.2.2 Nitrogen Loading Rate and Nitrogen Removal Rate

Figure 8 shows the Nitrogen loading and removal rates of the EBBR-Anammox Process

Figure 8.

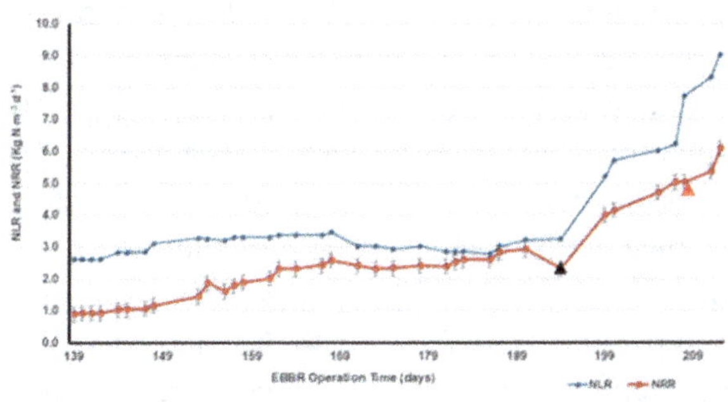

During the period from day 139 to 212 at 30 °C, pH 8. 3 operating conditions, the EBBR- Anammox system recorded a maximum nitrogen removal rate (NRR) of 6.1 kg N m^{-3} d^{-1} (day 212), and a minimum of NRR of 0.9 kg N m^{-3} d^{-1} (Fig.8).

3.2.3 EBBR-Anammox % Nitrogen Removal Efficiency

Figure 9 shows the % nitrogen removal efficiency of the EBBR-Anammox process

Figure 9.

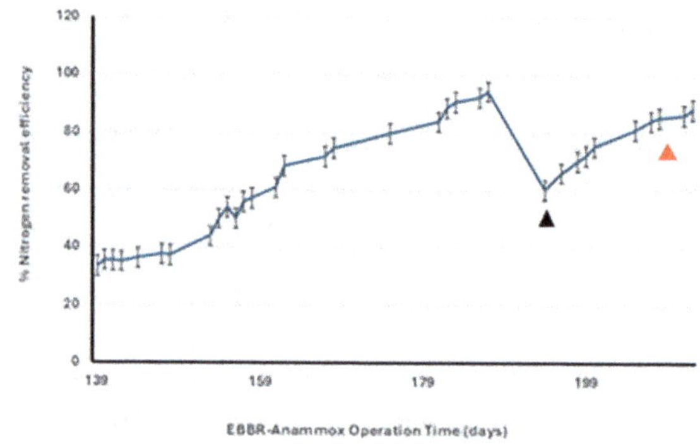

The nitrogen removal efficiency (NRE) of the Anammox wastewater treatment in the EBBR indicated a steady increase between days 139 and 212. The NRE exceeded 90% between 183 and 187 (Fig.9).

A stable operational phase in the EBBR-Anammox system was indicated by a sustained increase in nitrogen removal rate (NRR > 67%) and the development of active Anammox biomass, observed as red granules transitioning to rusty-brown coloured bioparticles with suspected nitrogen gas bubble formation (Fig. 6). This confirmed effective Anammox activity and high NRR performance. A brief disturbance occurred between days 187–194 due to a temperature drop (30 °C → 23 °C) caused by a fault in the temperature regulator, causing >20% and >30% declines in NRR and nitrogen removal efficiency (NRE), respectively (black triangle; Fig.8 & 9). Recovery to normal conditions (red triangle; day 199) demonstrated system and microbial resilience, with NLR and NRR reaching 9.0 kg N m^{-3} d^{-1} (>180%) and 6.1 kg N m^{-3} d^{-1} (>160%) within 20 days (Fig. 8 & 9).

Restoration of temperature from 23 °C to 30 °C reactivated Anammox activity in the EBBR, confirming reports that the optimal temperature for Anammox performance in engineered systems is around 30 °C, although it may vary with reactor configuration (González-Martínez et al., 2018; Langone et al., 2014). Similar temperature sensitivity has been reported in other systems, including reduced efficiency in a lab-scale SBR following temperature controller failure (Fernandez et al., 2012), a 16% loss in activity at 15 °C (Morales et al., 2015), and decreased nitrogen removal below 20 °C in a hybrid SBR–MBBR reactor Laureni et al., 2015). Consistent with these findings, the observed decline in NRE within the EBBR-Anammox system was attributed to the temperature drop to 23 °C.

3.3 SEM Evidence Supporting Anammox Process Establishment in the EBBR

The SEM micrographs reveal distinct stages of biofilm formation, and Anammox bacteria biomass colonization of the ABDite® support particles collected from the EBBR expanded column – C1 (Fig.4)

3.3.1 *SEM Micrographs of Uncolonized and Partially Colonized ABDite Particles With a Thin Biofilm*

The early growth of Anammox biomass in a system is detected by the formation of a thin biofilm, with some growth surfaces uncolonized.

Figure 10 shows early stages of biofilm development on EBBR ABDite particles.

Figure 10.

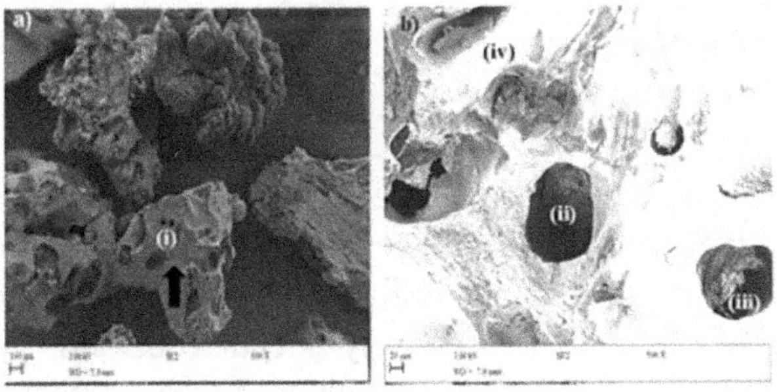

The pattern of biofilm development on the ABDite® confirms that both the internal and external structural features of the support medium provided an extensive attachment surface conducive for the growth of Anammox bacteria in the EBBR (Fig. 10).

At the early stage of biofilm development on the EBBR ABDite, a thin-layered surface biofilm (20 μm, at x500) and a very little extrapolymeric substance were visible in the partially colonized pores and surface; some parts of the ABDite were yet to be fully colonized (Fig. 10b). In contrast, the uncolonized ABDite® particles

(100μm, at x100) exhibited a rough, angular surface morphology devoid of any visible microbial material (Fig.10a, b (iii)), thereby emphasizing the transformation that occurred following biofilm establishment in the EBBR system (Anyanwu, 2023).

3.3.2 Advanced Biofilm Formation and Bioparticle With Extracellular Polymeric Substances on the EBBR ABDite Pores and the Surface With

Advancement in biofilm development on a growth particle is characterized by an increase in biofilm thickness, extrapolymeric substance size, and bioparticle formation (Anyanwu et al., 2024).

Figure 11 shows advanced biofilm development on EBBR ABDite particles.

Figure 11.

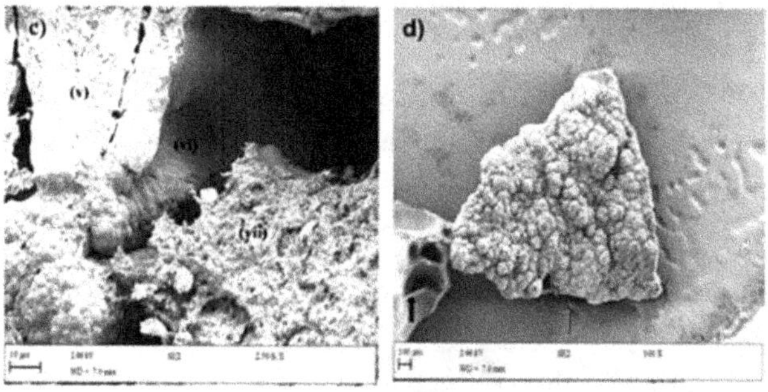

The SEM micrographs revealed biofilm development both within the internal cavities (pores) and across the outer layer (surface) of the EBBR-Anammox ABDite® biomass support (Fig. 11). The colonization pattern indicated that biofilm formation likely initiated in the pores and progressively extended outward, covering the outer layer, with the formation of thicker extracellular polymeric substances and fully formed individual bioparticles (Fig. 11c-vii). This stage of biofilm development indicated a fully colonized pore (internal structure) with a 10 μm biofilm and extensive extrapolymeric material (Fig. 11c). The observed increase in the size of the biofilm formed at this phase is consistent and indicated further development from the thin biofilm layers recorded in the initial stage, where portions of the outer structure and internal channels remained only partially colonized (Fig. 9, 10b iii & iv).

3.3.3 Final Stage of Biofilm With Cauliflower Morphology on the EBBR ABDite Particles

The presence of mature Anammox bacterial biofilm with the cauliflower morphology is evidence that the biofilm is fully formed.

Figure 12 shows a fully biofilm on the EBBR ABDite particles

Figure 12.

The SEM micrographs indicated that the mature biofilms were characterized by dense, irregular "cauliflower-like" structure of biomass layers that completely enveloped the internal cavities and outer surfaces of the ABDite® particles (Fig. 12 The size, shape, and surface structure of the fully formed biofilm contrasted with the thin, patchy biofilm coverage recorded during the early and advanced colonization of the ABDite particles (Fig. 10, Fig. 11). In addition to the fully developed cauliflower -structured Anammox bacteria biofilm, there were identifiable cocci-like bacteria (2µm, x5.00 K) cells seen on the ABDite surface SEM micrograph (Fig. 12e).

3.3.4 Mixed Culture of Distinct Bacterial-Like Cells Developed on the EBBR-ABDite Particles

In engineered systems, Anammox bacteria prefer to grow as a mixed culture alongside other organisms.

Figure 13 shows mixed bacterial-like cells formed on EBBR ABDite.

Figure 13.

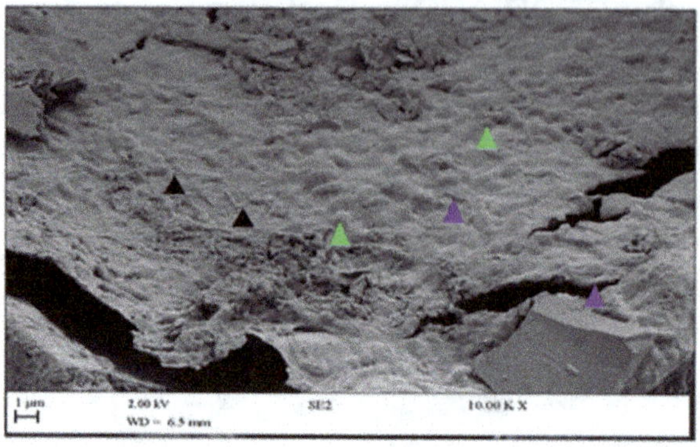

The various bacteria-like morphotypes, which constitute the mixed culture, are differentiated by colour annotations. They include short rods **(green triangles),** long rods with flagella **(black triangles),** and filamentous (**purple triangles**) cells (Fig. 13). The rod-like bacteria, ranging from <1 µm to ~1 µm in length at x10.00K, densely covered the thin layer of a partially formed biofilm surface (Fig. 13), suggesting these cells were the early colonizers. The longer rod cells, which had visible flagella, were more abundant than the short rods and filamentous cells (Fig. 13). The growth of these bacterial cells indicates that the EEBR-Anammox system has the potential to provide a conducive operating environment for the growth of different bacterial biomass suitable for the Anammox WWT process.

3.3.5 Predominantly Cocci-Like Cells Colonization of the EBBR ABDite

In addition to the rods and filamentous bacterial groups previously described, cocci-like bacteria were among the diverse bacterial cells found in Anammox WWT systems.

Figure 14 shows cocci-shaped bacterial cells on the EBBR ABDite particle.

Figure 14.

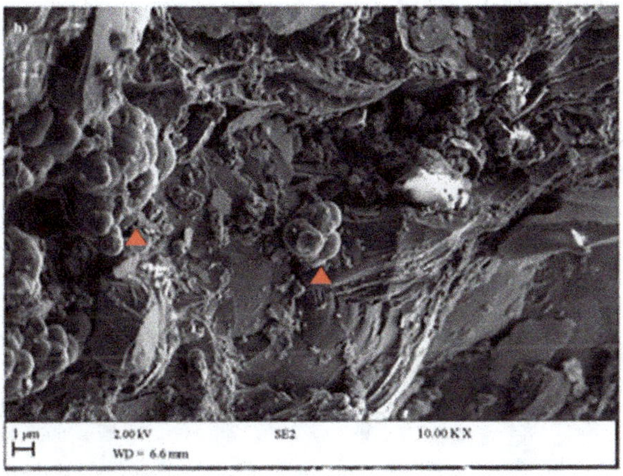

The suspected coccus-shaped bacteria (1 –2µm) were arranged in clusters of four or more cells, dominating the biofilm surface (Fig. 12 & 14) at this stage of the EBBR–Anammox process. The dominance of the coccoid cells in the EBBR over others and the presence of fewer coccoid cells in the (Fig. 12) suggest that variation in operating conditions could favour the growth and abundance of a particular type of bacterial cell.

Comparison of the EBBR biofilm architecture and bacterial morphologies with other established Anammox systems — such as SBR (Ni et al., 2018), ABBR (Wang et al., 2018), and MBBR (Wang et al., 2019) — revealed strong similarities, including the formation of cauliflower-like biofilms and diverse bacterial cell types. These findings confirm that the EBBR ABDite supports Anammox biofilm development, which could be useful for removing nitrogen pollutants from wastewater alongside other functional bioreactor configurations.

3.4 Molecular Biology Evidence Supporting Anammox Process Activity in the EBBR

The establishment of the Anammox wastewater treatment process in the EBBR was confirmed through DNA extraction, PCR amplification, Sanger sequencing, and BLAST search confirmation of bacterial DNA identity using specific primers, and microbial identification via NGS microbiome analysis.

3.4.1 Gel Electrophoresis Images of PCR- Amplified DNAs Extracted From EBBR-Anammox Samples

A successful gel electrophoresis run produces specific bands according to the size of the amplicons. Figure 15 shows the gel electrophoresis result of the EBBR-Anammox bacteria.

Figure 15.

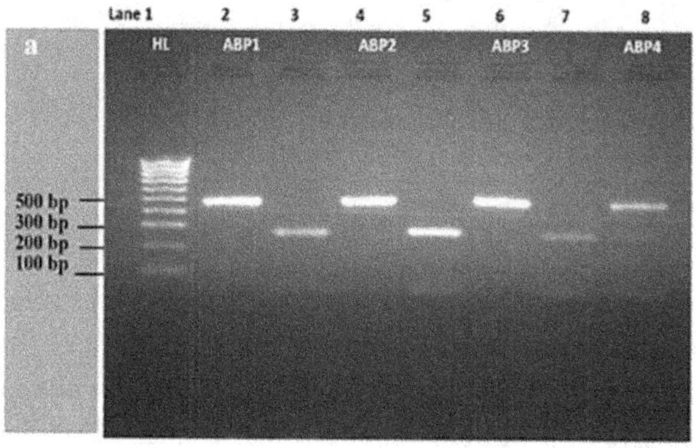

The PCR amplification used Anammox bacterial DNA from ABDite® samples (ABDP1–ABDP4) collected over 13 months (Fig.15). Amplicons generated with the primer pair Amx368F/Amx820R corresponded to species of any of the five known Anammox bacteria genera — *Candidatus Scalindua, Kuenenia, Jettenia, Brocadia, and Anammoxoglobus; and* the primer set - Brod541F/Amx820R specifically targeted *Candidatus Scalindua and Brocadia species*. The amplicon sizes (475–480 bp and ~280 bp) were within the expected range for Anammox bacteria, and the values based on the standard curve and gel image profiles of templates detected with Amx368F/Amx820R in Lanes 2, 4, 6, 8: (Amx368F/Amx820R amplicons; Lanes 3, 5, 7, and 100 bp HyperLadder – Lane 1, respectively. The sampling time (months) vs the electrophoresis gel run lanes were - Lanes 2–3 (1st month), 4–5 (5th month), 6 –7 (7th month), and Lane 8 (13th month).

3.4.2 *Sanger Sequencing – BLAST Identification of Anammox Bacteria Community From EBBR Samples*

The names of the Anammox bacterial community identified by BLAST search using the DNA nucleotide, initially sequenced by the Sanger method, are displayed in Table 2.

Table 2. Names of the Anammox bacterial community identified by Sanger and BLAST analysis

Accession number	Expected Bacterial group	Primer used	Identified Bactria	(%) identity	E-value
CP049055 LT934425 CT573071	AnAOB	Amx820R / Amx368F	*Kuenenia stuttgartiensis sp.* *Candidatus Jettenia sp.*	98.39	0.0
KX027318 KX027317			*Candida Brocadia sp.*	98 - 99	0.0
KX020503 KY399802 KP663625	AnAOB	Brod541F/ Amx820R	*Candidatus Brocadia spp. - sapporoensis, fulgida & caroliniensis*	99 - 100	0.0
MG846296 MG831208 HQ443293 KJ859558	AOB	amoA-1F / amoA-2IR20	*Nitrosomonas sp. clones*	99 - 100	0.0
GU562977	NOB	FGPS872/ FGPS1269	*Nitrobacter sp. FGPS_EFE_3b*	98 - 100	0.0
MH084827.1 MK088094 KU961663.1	Yeast	ITS1 / ITS2	*Filobasidium uniguttulatum sp. strains* *Cryptococcus sp.*	97 - 99 99	2e-79 -

Key: AOB = aerobic ammonia-oxidizing bacteria, **AnAOB** = anaerobic ammonia-oxidizing bacteria, **NOB** = Nitrite-oxidizing bacteria. (Adapted from Anyanwu, 2023).

BLAST analysis was based on > 98% query nucleotide sequence identity using Sanger sequenced DNA extracted from the EBBR-Anammox system **ABDite** bioparticle.

The DNA fragments analyzed via BLAST confirmed that the amplified DNA originated from the targeted bacterial groups, enabling comparison of Anammox populations colonizing the EBBR ABDite® particles at different stages. Bacterial identification was by selection of the top 100 BLAST hits (Anyanwu, 2023).

The BLAST analysis results revealed the presence of *Candidatus Kuenenia stuttgartiensis* and *Candidatus Brocadia* species. (AnAOB), *Nitrosomonas* sp. (AOB, detected with primer amoA-1F/amoA-2R), and *Nitrobacter* sp. (NOB, detected with primer FGPS872/FGPS1269) (Table 2). Two human pathogenic yeast species, *Filobasidium uniguttulatum* and *Cryptococcus* sp., were also identified with 92% query coverage and ≥ 97% identity (Table 3). *F. uniguttulatum* appeared more abundant, and its detection highlights the potential biohazard concerns, emphasizing the need for tertiary treatment to remove nitrogenous contaminants and possibly kill pathogens (e.g., by chlorination) before discharge to reduce negative environmental and health impacts of wastewater (Glibert, 2017; Ngatia et al., 2019; Blaettle, 2018).

The coexistence of Anammox bacteria and yeasts, previously reported in both engineered and natural systems (Kuenen et al., 2020; Tomaszewski et al., 2017), was reaffirmed in this study. Although the precise role of yeasts in an Anammox WWT system remains unclear, yeasts may promote localized micro-anoxic niches that are favourable to Anammox activity (Crouzet et al., 2014; Van Niftrik et al., 2012).

3.4.3 Evolutionary Relationship of EBBR-Anammox Bacterial Community/ Phylogenetic Tree

An evolutionary tree provides a visual representation of the diverse groups of organisms that conduct the Anammox process.

Figure 16 shows the major bacterial groups of the EBBR-Anammox process

Figure 16.

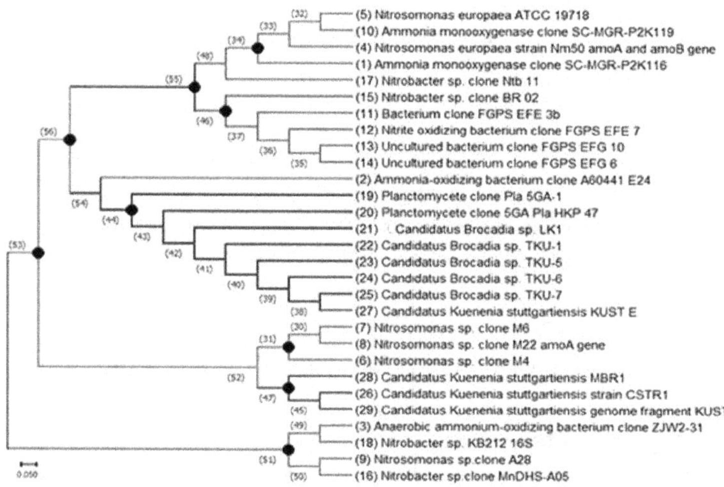

The unrooted phylogenetic tree of the main bacterial types that thrived within the EBBR-Anammox system was constructed using MEGA X (scale bar = 5% substitutions per 1,000 bootstrap replicates) (Kumar et al., 2018), and the bacterial groups identified from the BLAST search. The evolutionary tree indicated 29 bacterial clones grouped into six clades (Fig. 16). The unrooted phylogenetic tree revealed the % contributions (based on the targeted nucleotide sequences) of the various bacterial groups to the EBBR-Anammox community. Anaerobic ammonia-oxidizing bacteria (AnAOB) were the most dominant group (41%), followed by ammonia-oxidizing bacteria (AOB, 31%) and nitrite-oxidizing bacteria (NOB, 28%), with *Nitrosomonas europaea* (7%) distinctly represented. The relatively low abundance of NOB suggests that nitrite oxidizers did not outcompete Anammox bacteria for nitrite. This situation could have created ideal conditions within the reactor, allowing efficient nitrogen removal from the EBBR-Anammox system.

3.4.4 Next-Generation Sequencing (NGS) and Identification of EBBR-Anammox Bacterial Community

The Next-generation sequencing (NGS) microbiome analyses were performed by Eurofins, a company based in Germany, under the identification details (NG-25934; Report No. 2022113-1). The NGS analysis used the DNA samples extracted from ABDite® bioparticles from the EBBR-Anammox system and the V3V4a primer set to enumerate the different groups of bacteria and the % operational taxonomic unit (OUT) contributions made by the groups

3.4.4.1 NGS-Identified Mixed Microbial Community of the EBBR-Anammox system

Next-generation sequencing (NGS) technology has the potential to enumerate all the microbial groups growing in a bioreactor, whether directly or not associated with the Anammox process.

Figure 17 shows the mixed microbial community of the EBBR-Anammox process

Figure 17.

The main Anammox-relevant bacterial groups identified among the mixed bacterial consortium in the EBBR-Anammox system were aerobic ammonia-oxidizing bacteria (AOB), anaerobic ammonia-oxidizing bacteria (AnAOB), and nitrite-oxidizing bacteria (NOB). Equally present were the complete ammonia-oxidizing bacteria (Commamox), ammonia-oxidizing archaea (AOA), and nitrifying heterotrophic bacteria (NPB), amongst others (Figure 17). The species of AnAOB present were *Candidatus Kuenenia* and *Candidatus Brocadia, Nitrosomonas and Nitrosospira spp.* represented the AOB, with *Nitrobacter sp., being the main NOB (*Figure 17).

3.4.4.2 The Operational Taxonomic Unit (OUT) Contributions to the EBBR-Anammox Process

Different groups of bacteria inhabited the EBBR-Anammox system and contributed differently to the total microbial community, quantified as operational taxonomic units (% OTU). The % OTU contributions of major bacterial groups to the EBBR-Anammox WWT process are summarized in Table 3.

Table 3. Contributions (% OTU) of major bacterial groups to the EBBR-Anammox WWT process

Microbial Group	*Species*	*Number of OTU*	*% Identity*	*Sequence length (bp)*
AOB	*Nitrosomonas sp.*	18	95-100	424-425
	Nitrosomonas europaea	7	98	424
	Nitrosospira sp.	1	78	424
AnAOB	*Candidatus Kuenenia sp.*	5	95-97	424
NOB	*Nitrobacter sp. LAMK1242*	2	99	399
	Nitrobacter sp. NS5-5	1	100	404
Commamox	*Nitrospira sp.*	2	98	417
AOA	*Candidatus Nitrosarchaeum limnium*	1	94	400
AHB	*Caldilinea aerophile*	2	88	405
	Anaerolinea thermolimosa	7	83 - 88	402-407

Key: **OTU** = operational taxonomic unit, **AOB** = aerobic ammonia oxidizing bacteria, **AnAOB** = anaerobic ammonia oxidizing bacteria, **NOB** = Nitrite oxidizing bacteria, **Commamox** = complete ammonia oxidizing bacteria, **AOA** = ammonia oxidizing Archaea, **AHB** = anaerobic heterotrophic bacteria

Each cohort of bacteria contributed varying, distinct proportions to the overall microbial community structure of the EBBR-Anammox system. Anammox bacterial types that contributed were AnAOB, represented by *Candidatus Kuenenia* sp. (5% OTU) and *Candidatus Brocadia* sp. Present in abundance were the AOB *Nitrosomonas spp.,* with 18% OTU, and *Nitrobacter sp.* (NOB) had only 2% (Fig. 17 & Table 3). The identification and % OTU contributions of these Anammox bacterial groups corroborated the Sanger sequencing and BLAST search data, confirming the prevalence of the AnAOB and AOB, which play a major role in nitrogen removal, thereby reinforcing the reliability of the EBBR-Anammox findings (Figs. 16 & Table 2).

The lab-scale EBBR-Anammox system supported a diverse microbial consortium comprising AnAOB, Comammox, AOB, AOA, NOB, NPB, AHB, and yeasts. The presence of these organisms in the EBBR was confirmed by NGS microbiome analysis (Fig. 17, Table 3) and supported by Sanger sequencing using species-specific primers and BLAST search (Fig. 16; Table 2). The agreement between molecular results and SEM observations of Anammox biofilm development on ABDite® bioparticles (Figs. 10 – 14) demonstrates that the preceding molecular procedures—DNA extraction, PCR, and sequencing—were effective in identifying key bacterial taxa within the EBBR.

Consistent with reports from previous Anammox wastewater treatment processes conducted using different bioreactors, the EBBR-Anammox data indicated that AnAOB thrives within multispecies biofilms, where heterotrophic bacteria initiate biofilm formation, and coexistence provides protection from protozoan grazing (Mozumder et al., 2014). Similarly, Botchkova et al. (2015) showed that cohabitation with AOB and yeasts in membrane bioreactors enhanced AnAOB resistance to oxygen and nitrite stress (up to 400 mg L^{-1}). The detection of comparable communities in the EBBR-Anammox system suggests analogous ecological interactions that supported stable and sustained Anammox performance.

4 CONCLUSION

This study provides strong evidence for the first successful establishment of the Anammox process in a lab-scale Expanded Bed Biofilm Reactor (EBBR), achieving a maximum nitrogen removal rate (NRR) of 6.1 kg N m^{-3} d^{-1} and a nitrogen removal efficiency (NRE) of 80–100%. The formation of brick-red coloured granules and rusty-brown coloured biofilms—typical of Anammox enrichment—on ABDite® support media, coupled with simultaneous NH_3–N (313 mg L N L^{-1}) and NO_3–N (392 mg N L^{-1}) removal, minimal NO_2–N (27 mg N L^{-1}) accumulation, and sustained production of gas bubbles confirmed active Anammox performance. Molecular analyses (PCR, Sanger sequencing, BLAST, and NGS) identified three Anammox genera—*Candidatus Brocadia*, *Kuenenia*, and *Jettenia*—alongside functionally associated microbial groups. These findings demonstrate that EBBR technology, originally designed for complete nitrification, can be effectively repurposed for the Anammox process to treat eutrophic, high-strength nitrogen wastewater.

Given the increasing nitrogen load from agricultural runoff, fertilizer overuse, aquaculture, and livestock production, Anammox-based systems offer a sustainable and efficient way to mitigate the environmental problems caused by eutrophication from nitrogen-rich wastewater. Sequentially, the EBBR-Anammox WWT system provides a solution by producing wastewater suitable for reuse in irrigation, with a positive impact on running costs.

REFERENCES

Adams, M., Xie, J., Xie, J., Chang, Y., Guo, M., Chen, C., & Zhang, T. C. (2020). The effect of carrier addition on Anammox start-up and microbial community: A review. *Reviews in Environmental Science and Biotechnology, 19*(2), 355–368. DOI: 10.1007/s11157-020-09530-4

Anyanwu, C., Akhidime, D., Dempsey, M., & Melling, L. (2024). Establishment of the Anammox wastewater treatment process in a lab-scale expanded bed biofilm reactor with ABDite®. In: Ekpo, S.C. (ed.) *The Second International Adaptive and Sustainable Science, Engineering and Technology Conference (ASSET 2023).* Signals and Communication Technology. Springer, Cham. https://doi.org/DOI: 10.1007/978-3-031-53935-0_4

Anyanwu, C. C. (2023). Establishment of Anammox wastewater treatment process in the Expanded Bed Biofilm Reactor (EBBR) with ABDite. Available at: https://core.ac.uk/download/568419276.pdf

Blaettle, K. G. (2018). The Effects of Sewage on Aquatic Ecosystems. Sciencing. Available at: https://sciencing.com/effects-sewage-aquatic-ecosystems-21773.html (Accessed 09 A. Nov. 2025).

Botchkova, E. A., Plakunov, V. K., & Nozhevnikova, A. N. (2015). Dynamics of biofilm formation on microscopic slides submerged in an Anammox bioreactor. *Microbiology, 84*(3), 456–460. DOI: 10.1134/S0026261715030029

Boyd, C. E. (2019). Nitrogen. In *Water Quality: An Introduction* (pp. 269–290). Springer International Publishing.

Crouzet, M., Le Senechal, C., Brozel, V. S., Costaglioli, P., Barthe, C., Bonneu, M., Garbay, B., & Vilain, S. (2014). Exploring early steps in biofilm formation: Set-up of an experimental system for molecular studies. *BMC Microbiology, 14*(1), 253. DOI: 10.1186/s12866-014-0253-z PMID: 25266973

Dai, F., De Prá, M. C., Vanotti, M. B., Gilmore, K. R., & Cumbie, W. E. (2019). Microbial characteristics of nitrifiers, denitrifiers, and anammox bacteria on different support media to treat space mission wastewater. *Journal of Environmental Management, 232*, 943–951. DOI: 10.1016/j.jenvman.2018.12.008 PMID: 33395762

Daims, H., Lebedeva, E. V., Pjevac, P., Han, P., Herbold, C., Albertsen, M., Jehmlich, N., Palatinszky, M., Vierheilig, J., Bulaev, A., Kirkegaard, R. H., von Bergen, M., Rattei, T., Bendinger, B., Nielsen, P. H., & Wagner, M. (2015). Complete nitrification by *Nitrospira* bacteria. *Nature, 528*(7583), 504–509. DOI: 10.1038/nature16461 PMID: 26610024

Delkash, M., Al-Faraj, F. A., & Scholz, M. (2018). Impacts of anthropogenic land use changes on nutrient concentrations in surface waterbodies: A review. *Clean (Weinheim)*, *46*(5), 1800051. DOI: 10.1002/clen.201800051

Dimitrova, I., Dabrowska, A., & Ekström, S. (2020). Start-up of a full-scale partial nitritation-anammox MBBR without inoculum at Klagshamn WWTP. *Water Science and Technology*, *81*(9), 2033–2042. DOI: 10.2166/wst.2020.271 PMID: 32666956

Fernández, J. D., Fajardo, C., Campos, J. L., Mosquera-Corral, A., & Méndez, R. (2012). Short- and long-term effects of ammonium and nitrite on the Anammox process. *Journal of Environmental Management*, *95*, S170–S174. DOI: 10.1016/j.jenvman.2010.10.044 PMID: 21074312

Glibert, P. M. (2017). Eutrophication, harmful algae, and biodiversity—Challenging paradigms in a world of complex nutrient changes. *Marine Pollution Bulletin*, *124*(2), 591–606. DOI: 10.1016/j.marpolbul.2017.04.027 PMID: 28434665

González-Martínez, A., Muñoz-Palazon, B., Rodríguez-Sánchez, A., & González-López, J. (2018). New concepts in anammox processes for wastewater nitrogen removal: Recent advances and future prospects. *FEMS Microbiology Letters*, *365*(6), fny031. DOI: 10.1093/femsle/fny031 PMID: 29438563

Hendrickx, T. L. G., Kampman, C., Zeeman, G., Temmink, B. G., Hu, Z., Kartal, B., & Buisman, C. J. N. (2014). High specific activity for Anammox bacteria enriched from activated sludge at 10 °C. *Bioresource Technology*, *163*, 214–222. DOI: 10.1016/j.biortech.2014.04.025 PMID: 24814247

Koch, H., Van Kessel, M. A. H. J., & Lucker, S. (2019). Complete nitrification: Insights into the ecophysiology of comammox *Nitrospira*. *Applied Microbiology and Biotechnology*, *103*(1), 177–189. DOI: 10.1007/s00253-018-9486-3 PMID: 30415428

Kosgey, K., Awolusi, O. O., Kiambi, S. L., Allam, M., Ismail, A., Bux, F., Chandran, K., & Kumari, S. (2021). Process performance and microbial community structures in three ANAMMOX-mediated systems with different mixing conditions. *Journal of Environmental Chemical Engineering*, *9*(6), 106466. DOI: 10.1016/j.jece.2021.106466

Kosgey, K., Zungu, P. V., Bux, F., & Kumari, S. (2022). Biological nitrogen removal from low carbon wastewater. *Frontiers in Microbiology*, *13*, 968812. DOI: 10.3389/fmicb.2022.968812 PMID: 36466689

Kuenen, J. G. (2020). Anammox and beyond. *Environmental Microbiology*, *22*(2), 525–536. DOI: 10.1111/1462-2920.14904 PMID: 31867834

Kumar, S., Stecher, G., Li, M., Knyaz, C., & Tamura, K. (2018). MEGA X: Molecular evolutionary genetics analysis across computing platforms. *Molecular Biology and Evolution, 35*(6), 1547–1549. DOI: 10.1093/molbev/msy096 PMID: 29722887

Kuypers, M. M. M., Marchant, H. K., & Kartal, B. (2018). The microbial nitrogen-cycling network. *Nature Reviews. Microbiology, 16*(5), 263–276. DOI: 10.1038/nrmicro.2018.9 PMID: 29398704

Lackner, S., Glibert, E. M., Vlaeminck, S. E., Joss, A., Horn, H., & Van Loosdrecht, M. C. M. (2014). Full-scale partial nitritation/anammox experiences – An application survey. *Water Research, 55*, 292–303. DOI: 10.1016/j.watres.2014.02.032 PMID: 24631878

Langone, M., Yan, J., Haaijer, S. C. M., Op den Camp, H. J. M., Jetten, M. S. M., & Andreottola, G. (2014). Coexistence of nitrifying, anammox and denitrifying bacteria in a sequencing batch reactor. *Frontiers in Microbiology, 5*, 28. DOI: 10.3389/fmicb.2014.00028 PMID: 24550899

Laureni, M., Weissbrodt, D. G., Szivak, I., Robin, O., Nielsen, J. L., Morgenroth, E., & Joss, A. (2015). Activity and growth of anammox biomass on aerobically pre-treated municipal wastewater. *Water Research, 80*, 325–336. DOI: 10.1016/j.watres.2015.04.026 PMID: 26024830

Li, X., & Sung, S. (2015). Development of the combined nitritation-anammox process in an upflow anaerobic sludge blanket (UASB) reactor with anammox granules. *Chemical Engineering Journal, 281*, 837–843. DOI: 10.1016/j.cej.2015.07.016

Liwarska-Bizukojc, E., Chojnacki, J., Klink, M., & Olejnik, D. (2018). Effect of the type of external carbon source on denitrification kinetics of wastewater. *Desalination and Water Treatment, 101*, 143–150. DOI: 10.5004/dwt.2018.21758

Lu, Y. F., Ma, L. J., Ma, L., Shan, B., & Chang, J. J. (2018). Improvement of start-up and nitrogen removal of the anammox process in reactors inoculated with conventional activated sludge using biofilm carrier materials. *Environmental Technology, 39*(1), 59–67. DOI: 10.1080/09593330.2017.1294624 PMID: 28278777

Magrí, A., & Lotti, T. (2023). Special issue on "Anammox-based processes for wastewater treatment". *Processes (Basel, Switzerland), 11*(5), 1422. DOI: 10.3390/pr11051422

Morales, N., Val del Río, A., Vázquez-Padín, J. R., Gutiérrez, R., Fernández-González, R., Icaran, P., Rogalla, F., Campos, J. L., Méndez, R., & Mosquera-Corral, A. (2015). Influence of dissolved oxygen concentration on the start-up of the anammox-based process: ELAN®. *Water Science and Technology, 72*(4), 520–527. DOI: 10.2166/wst.2015.233 PMID: 26247749

Mozumder, M. S. I., Picioreanu, C., Van Loosdrecht, M. C. M., & Volcke, E. I. P. (2014). Effect of heterotrophic growth on autotrophic nitrogen removal in a granular sludge reactor. *Environmental Technology, 35*(8), 1027–1037. DOI: 10.1080/09593330.2013.859711 PMID: 24645487

Ngatia, L., Grace, J. M., III, & Moriasi, D. (2019). Nitrogen and phosphorus eutrophication in marine environments. *Monitoring of marine pollution*, pp.77-93.

Ni, L., Wang, Y., Lin, X., Yan, Y., Zhang, Y., & Wang, W. (2018). Enhancement of the adaptability of anammox granules to zinc shock by appropriate organic carbon treatment. *Bioresource Technology, 268*, 496–503. DOI: 10.1016/j.biortech.2018.08.035 PMID: 30114669

Ni, S. Q., & Zhang, J. (2013). Anaerobic ammonium oxidation: From laboratory to full-scale application. *BioMed Research International, 2013*, 1–10. Advance online publication. DOI: 10.1155/2013/469360 PMID: 23956985

Niederdorfer, R., Fragner, L., Yuan, L., Hausherr, D., Wei, J., Magyar, P., Joss, A., Lehmann, M. F., Ju, F., & Bürgmann, H. (2021). Distinct growth stages controlled by the interplay of deterministic and stochastic processes in functional anammox biofilms. *Water Research, 200*, 117225. DOI: 10.1016/j.watres.2021.117225 PMID: 34052477

Suarez, C., Persson, F., & Hermansson, M. (2015). Predation of nitritation–anammox biofilms used for nitrogen removal from wastewater. *FEMS Microbiology Ecology, 91*(11), 124. DOI: 10.1093/femsec/fiv124 PMID: 26472578

Tikilili, P. V., & Chirwa, E. (2016). Searching for indigenous anaerobic ammonium oxidizing (anammox) bacteria in South African habitats: Pretoria region. *Biotechnology, Biotechnological Equipment, 30*(6), 1097–1105. DOI: 10.1080/13102818.2016.1225515

Tomaszewski, M., Cema, G., & Ziembińska-Buczyńska, A. (2017). Influence of temperature and pH on the anammox process: A review and meta-analysis. *Chemosphere, 182*, 203–214. DOI: 10.1016/j.chemosphere.2017.05.003 PMID: 28499181

Van Niftrik, L., & Jetten, M. S. M. (2012). Anaerobic ammonium-oxidizing bacteria: Microorganisms with exceptional properties. *Microbiology and Molecular Biology Reviews : MMBR, 76*(3), 585–596. DOI: 10.1128/MMBR.05025-11 PMID: 22933561

Wang, C., Sitong, L., Xu, X., & Zhao, C. (2017). Potential coupling effects of ammonia-oxidizing and anaerobic ammonium-oxidizing bacteria on completely autotrophic nitrogen removal over nitrite biofilm formation induced by the second messenger cyclic diguanylate. *Applied Microbiology and Biotechnology, 101*(9), 101. DOI: 10.1007/s00253-016-7981-y PMID: 28078398

Wang, S., Zhang, D., Wang, T., Wang, X., & Yuan, L. (2018). Preliminary performance and biofilm characteristics of Anammox process in an improved FBR with mixture of honeycomb-like and non-woven carriers. *Desalination and Water Treatment, 116*, 67–74. DOI: 10.5004/dwt.2018.22273

Wang, T., Wang, X., Yuan, L., Luo, Z., & Indira, H. K. (2019). Start-up and operational performance of Anammox process in an anaerobic baffled biofilm reactor (ABBR) at a moderate temperature. *Bioresource Technology, 279*, 1–9. DOI: 10.1016/j.biortech.2019.01.114 PMID: 30710814

Wang, Z., Zheng, M., Duan, H., Yuan, Z., & Hu, S. (2022). A 20-year journey of partial nitritation and anammox (PN/A): From sidestream toward mainstream. *Environmental Science & Technology, 56*(12), 7522–7531. DOI: 10.1021/acs.est.1c06107 PMID: 35657148

Wongkiew, S., Hu, Z., Chandran, K., Lee, J. W., & Khanal, S. K. (2017). Nitrogen transformations in aquaponic systems: A review. *Aquacultural Engineering, 76*, 9–19. DOI: 10.1016/j.aquaeng.2017.01.004

World Water Assessment Programme (WWAP)/UN-Water (2018). *The United Nations World Water Development Report 2018: Nature-based Solutions for Water*. Paris: UNESCO. Available at: https://www.unwater.org/publications/world-water-development-report-2018 (Accessed 7 Aug. 2019).

Yang, G. F., & Jin, R. C. (2012). The joint inhibitory effects of phenol, copper (II), oxytetracycline (OTC) and sulfide on Anammox activity. *Bioresource Technology, 126*, 187–192. DOI: 10.1016/j.biortech.2012.09.023 PMID: 23073107

Yin, X., Li, W., Zhang, M., Du, R., & Chen, C. (2025). Advances in anammox process for municipal wastewater treatment plants: A review focusing on applied study and full-scale implementation cases. *Critical Reviews in Environmental Science and Technology, 55*(13), 1001–1024. DOI: 10.1080/10643389.2025.2483522

Zhang, J., Zhu, T., Meng, T., Zhang, Y., Yang, J., Yang, W., Müller, C., & Cai, Z. (2013). Agricultural land use affects nitrate production and conservation in humid subtropical soils in China. *Soil Biology & Biochemistry*, *62*, 107–114. DOI: 10.1016/j.soilbio.2013.03.006

Zhang, L., Liu, H., & Chen, Q. (2021). Recent development of the ANAMMOX process in China: Challenges and opportunities. *Processes (Basel, Switzerland)*, *9*(7), 1207. DOI: 10.3390/pr9071207

Chapter 6
Biotechnological Approaches for Textile Effluent Treatment:
Microbial Decolourisation and Biodegradation of Synthetic Dyes

Pendo Nandiga Bigambo
http://orcid.org/0000-0002-2365-2015
University of Dar es Salaam, Tanzania

Jerome Michael Bidu
University of Dar es Salaam, Tanzania

ABSTRACT

The textile industry drives economic growth and improves human welfare, yet it remains a significant source of industrial water pollution. Textile effluents contain dyes, salts, and auxiliary chemicals from dyeing, printing, and washing processes, many of which are toxic, carcinogenic, persistent, and resistant to degradation. These pollutants threaten human health and aquatic ecosystems. Conventional physicochemical treatments are often costly, energy-intensive, and generate secondary pollutants, limiting their sustainability. Biotechnological approaches are eco-friendly and cost-effective alternatives, but are limited by the complexity and toxicity of textile wastewater. This chapter reviews current biotechnological strategies for textile effluent management, emphasizing bacterial, fungal, and algal dye-decolourisation, their enzymatic degradation mechanisms, and factors influencing treatment performance. It also discusses biodegradation pathways, key operational challenges, and emerging strategies to improve the efficiency and industrial use of biological treatment technologies.

DOI: 10.4018/979-8-2600-2998-5.ch006

1 INTRODUCTION

The textile industry is one of the largest in the world, playing an important role in the global economy by producing clothing and other everyday materials while also providing employment to a wide range of people. However, the growth of the textile industry has come at a high environmental cost, as it generates large volumes of wastewater from its various processing stages. The textile wet-processing operations, including pre-treatment, dyeing, printing and washing, consume large amounts of water and consequently generate substantial volumes of wastewater. This wastewater contains complex mixtures of dyes, auxiliary chemicals and heavy metals, making it a major source of pollution in receiving water bodies and soil (Kusumlata et al., 2024; Periyasamy, 2025).

Textile effluents are characterised by high colour intensity, elevated chemical and biological oxygen demand (COD and BOD) and the presence of heavy metals and other toxic substances (Periyasamy, 2025). Inefficient dye-fixation processes result in approximately 10–50% of the applied dyes remaining unfixed on fabrics, eventually entering wastewater streams (Bhatia et al., 2017; Holkar et al., 2016; Scholz, 2019). The dyes and chemicals used in the textile wet processes are often toxic, carcinogenic and resistant to natural degradation, leading to persistent contamination of aquatic and terrestrial ecosystems (Wang et al., 2022). Moreover, the accumulation of untreated or inadequately treated textile wastewater discharged into the environment disrupts ecological balance, threatens aquatic biodiversity, reduces sunlight penetration in water bodies, affects photosynthesis in aquatic plants and presents risks to human health through direct exposure or bioaccumulation in food chains (Kocaman, 2025; Madhav et al., 2018; Riza et al., 2021). Conventional physicochemical methods used for the treatment of textile effluent, such as coagulation, flocculation, adsorption and advanced oxidation processes, have shown effectiveness. However, they are often associated with high operating costs, the generation of secondary pollutants and incomplete removal of recalcitrant dyes (Bidu et al., 2021; Donkadokula et al., 2020). These challenges necessitate the exploration of alternative approaches that are sustainable, economical and capable of complete mineralisation of pollutants.

The complex and recalcitrant nature of textile effluents poses significant challenges to conventional treatment technologies, which are often costly and generate secondary pollutants. In response, biotechnology approaches, particularly microbial-based treatments, have emerged as promising, environmentally benign approaches for textile effluent management. Microorganisms, such as bacteria, fungi, and algae, possess diverse metabolic pathways and enzymatic systems capable of degrading and detoxifying a wide range of synthetic dyes and associated contaminants (Ceretta et al., 2021; Periyasamy, 2025). Through microbial degradation, complex dye molecules can be broken down into less toxic products, contributing to the restoration of polluted ecosystems. However, challenges related to dye toxicity, process efficiency, and large-scale applicability remain inadequately addressed. Therefore, this chapter provides a detailed review of the biotechnological approaches for textile effluent management with emphasis on microbial decolouration of dyes in polluted environments. It explores the types of microbes involved, their mechanisms of dye degradation, and the practical applications of these biological systems for treating textile effluent. The chapter also highlights existing challenges and future directions in microbial bioremediation of textile effluents.

2 COMPOSITION AND CHARACTERISTICS OF TEXTILE EFFLUENTS

Textile effluents are complex chemical mixtures generated during various stages of textile manufacturing, including sizing, de-sizing, scouring, bleaching, mercerising, dyeing, printing, and finishing processes (Scholz, 2019). Each of these processes involves a variety of chemical inputs that eventually result in effluents with highly variable yet characteristically high pollutant loads. These effluents typically contain a wide range of chemical substances, such as dyes, salts and surfactants, heavy metals, alkalis, acids, suspended solids and varieties of organic compounds (Wang et al., 2022). Table 1 presents typical textile processes and the wastewater generated.

Table 1. Textile effluent produced from typical textile processes

Process	Wastewater generated
Fibre preparation	Little or no wastewater generated
Yarn spinning	Little or no wastewater generated
Slashing/sizing	BOD, COD; metals, cleaning waste, size
Weaving	Little or no wastewater generated
De-sizing	BOD from water-soluble sizes, synthetic sizes, lubricants, biocides; anti-static compounds
Singeing	Little or no wastewater generated
Scouring	Sodium hydroxide, detergents, fats, oils, pectin, wax, spin finishes
Bleaching	Hydrogen peroxide, sodium silicate, high pH
Mercerizing	High pH and Sodium hydroxide
Heat setting	Little or no wastewater generated

continued on following page

Table 1. Continued

Process	Wastewater generated
Dyeing and Printing	Colour, metals, salt, surfactants, toxins, auxiliary, BOD, alkalinity
Finishing	BOD, COD, suspended solids, toxins, spent Solvents

(Holkar et al., 2016)

The heterogeneous composition of textile wastewater poses significant treatment challenges because many pollutants exhibit high chemical stability, low biodegradability, toxicity to microorganisms, and resistance to conventional treatment processes. In addition, the simultaneous presence of dyes, salts, surfactants, heavy metals, and auxiliary chemicals creates synergistic inhibitory effects, thereby reducing the effectiveness of biological and physicochemical treatments (Al-Tohamy et al., 2022; Bhatia et al., 2017; Donkadokula et al., 2020; Holkar et al., 2016). Synthetic dyes, in particular, are among the most critical pollutant groups in textile wastewater, as approximately 10–50% of the dyes do not bind to fibres during dyeing and printing and are therefore discharged in wastewater (Ali, 2010; Scholz, 2019). They are classified by chemical structure (e.g., azo, anthraquinone, phthalocyanine, indigoid, triphenylmethane, xanthene, nitro, and heterocyclic dyes) and application method (e.g., acid dyes, basic dyes, direct dyes, reactive dyes, disperse dyes, vat dyes, and sulfur dyes). The molecular structures of most synthetic dyes are complex and contain several aromatic rings, such as benzene or naphthalene, arranged in a conjugated structure. Among these, azo dyes are the most prevalent class, accounting for approximately 60–70% of dyes produced and used globally (Saratale et al., 2011). Their chromophoric azo bonds (-N=N-) produce a vibrant colour range, are economically efficient to produce, and can be easily manufactured (Periyasamy, 2025). In addition, most synthetic dyes have high molecular weights, which resist chemical and biological degradation and persist in the environment, accumulating over time. Even low dye concentrations (<1 mg/L) can produce visible colouration in receiving water bodies due to the strong chromophoric structures and high tinctorial strengths of synthetic dyes (Holkar et al., 2016; Periyasamy, 2025).

Apart from dyes, textile effluents also contain elevated salt concentrations, particularly sodium chloride and sodium sulphate, which are commonly used during the dyeing processes. However, high salt concentrations increase the salinity and total dissolved solids (TDS) of textile wastewater. Elevated salinity imposes osmotic stress on microorganisms, inhibits enzymatic activity, reduces microbial diversity, and lowers the efficiency of biological treatment systems. Excessive salts also interfere with membrane-based treatment technologies by increasing fouling and scaling tendencies (Aragaw, 2024; Ceretta et al., 2021; Donkadokula et al., 2020; Holkar et al., 2016).

Levelling agents, dispersing agents and fixing agents are also commonly used as textile dyeing auxiliaries. However, they further complicate the wastewater treatment because many are non-biodegradable or toxic to aquatic organisms. Surfactants, such as nonylphenol ethoxylates, linear alkylbenzene sulfonates, sodium dodecyl sulfate, fatty alcohol ethoxylates, and cationic quaternary ammonium compounds, which are commonly used in wet processing of textiles, can stabilise emulsified pollutants, increase foaming and interfere with oxygen transfer and microbial activity during biological treatment (Arora et al., 2022). Ultimately, this can reduce degradation efficiency and affect reactor performance.

Other hazardous components include heavy metals such as chromium, copper, nickel, cobalt and zinc, which are found in dyes, pigment formulations, mordants and finishing agents (Md. Islam et al., 2025). These compounds are non-biodegradable and may accumulate in sludge, sediments, and biological tissues (Jaishankar et al., 2014). In addition, heavy metals can inhibit microbial growth, deactivate key degradative enzymes and suppress metabolic pathways involved in the biodegradation of dyes, thereby reducing the efficiency of biological treatment systems (Ali et al., 2009; Mudhoo & Kumar, 2013).

High COD and BOD as well as high pH values are other characteristic parameters of textile industry effluents, as clearly shown in Table 2. High COD values indicate substantial organic pollutant loads that require extensive oxidation or biodegradation, while elevated BOD reflects high concentrations of biodegradable organic matter that can rapidly deplete dissolved oxygen in receiving water bodies, thereby threatening aquatic life and increasing stress on effluent treatment systems (Bhatia et al., 2017). Similarly, fluctuating pH conditions and high temperatures may destabilise microbial communities, inhibit enzyme activity and reduce overall treatment efficiency (Jamee & Siddique, 2019; Mishra & Maiti, 2018). Collectively, these complex pollutant compositions in textile effluent affect both the environment and human health, necessitating integrated treatment approaches that simultaneously address colour removal, toxicity reduction, organic load degradation, and salinity management.

Table 2. General Characteristics of textile wastewater

Parameter	Value
pH	6–10
Temperature (°C)	35–45
Biochemical oxygen demand (mg/L)	80–6,000

continued on following page

Table 2. Continued

Parameter	Value
Chemical oxygen demand (mg/L)	150–12,000
Oil and grease (mg/L)	10–30
Total suspended solids (mg/L)	15–8,000
Free ammonia	<10
Total dissolved solids (mg/L)	2,900–3,100
Chloride (mg/L)	1,000–1,600
Sodium (mg/L)	70
Trace elements (mg/L)	<10
Silica (mg/L)	<15
Total Kjeldahl nitrogen (mg/L)	70–80
Color (Pt-Co)	50–2,500

(Wang et al., 2022)

3 ENVIRONMENTAL IMPACTS OF THE TEXTILE EFFLUENT

Textile effluents contain large amounts of pollutants from textile wet-processing, presenting significant environmental and public health concerns. Synthetic dyes have particularly complex chemical structures that are chemically stable and resistant to degradation. Azo dyes, which constitute the majority group of dyes in textile dyeing, however, have adverse environmental and public health effects due to their persistence, toxic nature and ability to release aromatic amines upon breakdown. When disposed of without proper treatment, they persist in soil and aquatic ecosystems, significantly disrupting the ecological processes and accumulating through food chains. In addition, the colour present in textile effluent is also detrimental to the environment. The intense colouration of textile effluents significantly reduces sunlight penetration into water bodies, thereby inhibiting photosynthesis in phytoplankton and submerged aquatic vegetation (Holkar et al., 2016; Madhav et al., 2018). Furthermore, their transformation products, such as aromatic amines from azo dyes, are often more toxic than the parent dye molecules, amplifying the environmental impacts. Numerous studies have reported a correlation between particular textile dyes and the occurrence of various diseases in both humans and animals, including mutagenic and carcinogenic effects (Kocaman, 2025; Mani et al., 2023; Ramamurthy et al., 2024).

3.1 Impact on Land and Soil

When untreated or partially treated textile effluents are discharged onto land or used for irrigation, they adversely impact the soil quality and biodiversity (Kocaman, 2025). Dye pollutants present in wastewater have been reported to interfere with seed germination, reduce seedling height, and affect the survival of soil microorganisms (Kumar et al., 2025). Studies report that irrigating with untreated textile wastewater can reduce seed germination rates by 20–60%, decrease chlorophyll levels by up to 45%, and suppress root and shoot development, depending on dye concentration and exposure duration (Kumar et al., 2025; Panhwar et al., 2024). High concentrations of salts and other dyeing auxiliaries introduced during the dyeing process, particularly total dissolved solids (TDS), reduce chlorophyll content and inhibit plant growth. These impacts occur because textile chemical pollutants alter soil pH, decrease essential nutrients such as nitrogen, phosphorus, potassium, and calcium, increase sodium and chloride concentrations, and disrupt the physicochemical properties necessary for plant growth (Panhwar et al., 2024).

Dye and chemical contaminants present in textile effluents can also affect the soil microbial communities. Chemical pollutants reduce the populations of beneficial microorganisms and impair the activities of important bacterial and fungal colonies, such as urease, arginine ammonification, and nitrification (Riza et al., 2021). These organic contaminants directly affect soil fertility and subsequently hinder plant blooming and germination processes (Islam & Mostafa, 2018; Nirmala et al., 2021).

Gradually, the direct discharge of untreated or poorly treated textile wastewater into the land results in soil contamination, which subsequently reduces agricultural yields (Madhav et al., 2018). Chemicals in textile wastewater present in soil can also accumulate in crops and then be transferred to humans and livestock through the food chain (Periyasamy, 2025).

3.2 Impact on Aquatic Systems

In water bodies, textile effluent constitutes a major source of water pollution that affects aquatic ecosystems, fish populations, and human health (Periyasamy, 2025). Dye-containing effluents enter aquatic systems through multiple channels, including industrial effluents, domestic wastewater, and runoff from dyeing facilities worldwide. The problem is particularly severe in rapidly industrialising textile-producing regions, including parts of Asia, Africa and Latin America, where wastewater treatment infrastructure is often inadequate (Kavitha et al., 2025).

The presence of synthetic dyes in textile effluents has raised particular concerns in aquatic ecosystems because they resist natural degradation processes and persist in water bodies due to their complex chemical compositions (Madhav et al., 2018).

Once released into aquatic environments, textile dyes can undergo transformation through photodegradation, hydrolysis, and microbial degradation (Alzain et al., 2023). However, these reactions may generate intermediate compounds, including aromatic amines from azo dyes, phenols, and other low-molecular-weight aromatic derivatives, some of which are harmful and more toxic than the original dye molecules.

Dye-containing textile effluent presents ecotoxicological effects to aquatic fauna, especially fish, due to the presence of hazardous substances, including synthetic dyes, heavy metals, surfactants, and textile industrial by-products (Periyasamy, 2025). Fish exposed to such contaminated water experience respiratory distress from reduced dissolved oxygen levels, disrupted endocrine function, impaired swimming capability, reduced feeding efficiency, altered behaviour, and organ damage (Carney et al., 2021). Contaminants such as heavy metals and aromatic chemicals may accumulate in fish tissues over time, resulting in bioaccumulation and subsequent chronic toxicity and disrupted reproduction (Periyasamy, 2025; Qiu et al., 2019; Unal et al., 2024). It may also impair the availability and visibility of prey, resulting in diminished feeding efficiency and growth rates (Hossain et al., 2018; Khattab et al., 2020). Reactive dyes have been reported to induce DNA damage, oxidative stress, and reduced survival rates in zebrafish and tilapia (Qiu et al., 2019; Unal et al., 2024).

The presence of dyes in aquatic ecosystems may disrupt the entire food chain, affecting phytoplankton, zooplankton, and fish that depend on these creatures for sustenance (Tigini et al., 2011). Dye pollution may accumulate in sediments, resulting in habitat destruction, especially for benthic fish species (Periyasamy, 2025). Modified habitats are impacted by the dark colour of textile wastewater, which obstructs sunlight penetration into the aquatic environment, disrupts aquatic vegetation, and diminishes the availability of shelter and breeding grounds for fish (Carney et al., 2021). In the long run, these cumulative impacts lower biodiversity, reduce fish populations and increase the dominance of pollution-tolerant species, thereby diminishing overall ecosystem resilience.

Despite the degradation that can be observed in water bodies, the majority of these dyes persist due to their complex chemical structures and resistance to decomposition (Kallawar & Bhanvase, 2024). Their resistance continues to make textile dye pollution a major challenge for the health of aquatic ecosystems (Tyagi et al., 2025).

3.3 Impact on Human Health

Human exposure to pollutants in textile effluent primarily occurs through the food chain, as pollutants accumulate in food crops, fish, and other plants and animals that are consumed by humans. Additionally, direct contact with contaminated water or chemicals can lead to skin irritation, allergic reactions, dermatitis, respiratory tract challenges and mutagenicity (Kavitha et al., 2025). Inhalation of textile dyes may also cause breathing difficulties, nausea, vomiting, diarrhoea, gastritis, and cognitive impairments (Das et al., 2023). Communities near industrial areas or among workers in textile dyeing and finishing processes are especially at risk of exposure from textile effluents.

Many synthetic dyes, especially those containing the azo group, pose significant health hazards to humans because their degradation produces carcinogenic serotonin metabolites that are reported to pose health risks, including skin irritation and allergic reactions (Periyasamy, 2025). Other classes of dyes used in textiles, including heterocyclic dyes, anthraquinone dyes, triarylmethane dyes, naphthalene-based dyes, and indigoid-based dyes, have also been associated with skin and eye irritation, respiratory complications, and probable carcinogenic effects (Mani et al., 2023; Ramamurthy et al., 2024). Textile dyes are also linked with possible liver and kidney damage in humans (Periyasamy, 2025). Various studies have particularly reported connections between prolonged exposure to such dye compounds and increased incidences of bladder and liver cancer (Kavitha et al., 2025; Ramamurthy et al., 2024).

Heavy metals, particularly copper, cobalt, nickel, and chromium, which are commonly present in textile wastewater, present a substantial genotoxic risk and, over time, can accumulate in human tissues (Jorge et al., 2023). The release of textile dyes and heavy metals into aquatic environments results in their absorption by fish and other organisms, which are subsequently consumed by humans, further increasing health risks (Periyasamy, 2025). Figure 1 summarises the root of the flow of textile wastewater pollutants to organisms such as plants and animals.

Figure 1. Schematic representation of textile dye effluent contamination, environmental dispersion and associated effects on human health

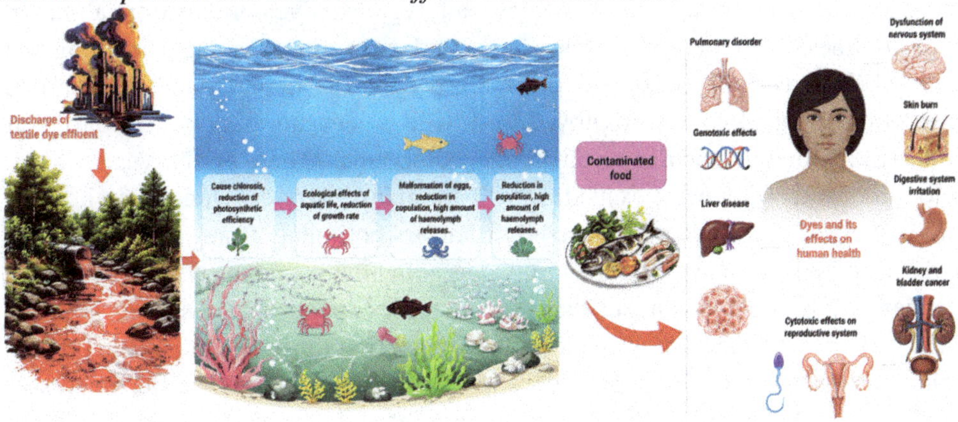

(Periyasamy, 2025)

4. APPROACHES TO TEXTILE EFFLUENT MANAGEMENT

Textile wastewater treatment techniques differ significantly in terms of pollutant removal efficiency, operational cost, sludge generation, energy requirements and environmental sustainability (Donkadokula et al., 2020; Holkar et al., 2016). Physical and chemical methods generally achieve rapid colour removal and high treatment efficiency; however, they are often associated with high chemical consumption, secondary sludge generation, and elevated operational costs (Holkar et al., 2016). Conversely, biological treatments are more environmentally sustainable and cost-effective but may require longer retention times and are sensitive to wastewater composition, salinity, pH and operational conditions (Aragaw, 2024; Bhatia et al., 2017; Ceretta et al., 2021).

Several technologies and treatment strategies are employed to treat textile-dye-containing effluents, particularly those characterised by high colour intensity, elevated COD and BOD, toxic compounds, and salinity. Textile wastewater treatment is typically implemented in three main stages: primary, secondary, and tertiary (Jahan et al., 2022). Each stage is designed to progressively reduce pollutant concentrations and improve water quality.

Primary treatments involve processes, such as sedimentation, filtration (including membrane filtration), adsorption, precipitation, and oxidation. Primary methods serve as the initial textile effluent treatment step, where coarse solids, suspended particles/fibres and some dissolved pollutants are removed to make effluent amenable to secondary treatment (Bhatia et al., 2017; Holkar et al., 2016). Coagulation and flocculation processes are also commonly employed in this stage to stabilise colloidal particles in textile effluent before sedimentation (Carrascal-Hernández et al., 2025). These processes are reported to remove approximately 60–90% of colour and suspended solids depending on coagulant type and dye composition, although large volumes of chemical sludge are generated, requiring secondary handling and disposal (Holkar et al., 2016). Adsorption processes, particularly activated carbon adsorption, are also widely employed for textile wastewater treatment due to their high affinity for dyes and dissolved organic compounds. These systems can achieve colour removal efficiencies exceeding 90% for many azo dyes (Donkadokula et al., 2020). Nevertheless, adsorption mainly transfers pollutants from the liquid phase to the adsorbent surface rather than fully degrading them, and the ned for adsorbent regeneration or replacement substantially increases operational costs.

Secondary treatments include biological transformation or degradation of organic pollutants that are resistant to physical and chemical treatments. The step addresses BOD, COD, phenol, oil, and colour reduction through biological treatments, such as aerobic, anaerobic, and fungal treatments, which are central to this stage (Das et al., 2023; Rathour et al., 2024). However, other advanced technologies, such as advanced oxidation processes, membrane bioreactors, enzymatic treatments, activated carbon adsorption, and oxidative biotransformation, also fall under this step. Biological systems typically achieve 50–90% COD removal depending on the wastewater composition, microbial consortium, and operating conditions (Bhatia et al., 2017; Ceretta et al., 2021).

Tertiary processes represent the final stage of textile effluent treatments. They are designed to remove residual pollutants and colours that remain after secondary treatment (Samuchiwal et al., 2021). They include membrane filtration, reverse osmosis, ion exchange, activated carbon adsorption, electrochemical treatment, and constructed wetland treatment systems. Membrane technologies such as ultrafiltration, nanofiltration, and reverse osmosis are particularly effective at producing high-quality effluent suitable for reuse in textile processing. In many cases, membrane systems achieve dye and salt removal efficiencies of over 95% (Jallouli et al., 2023; Uddin et al., 2024). However, membrane fouling, concentrate disposal, high-pressure requirements, and elevated operational costs remain major limitations affecting their long-term sustainability and industrial-scale application.

Textile effluent treatment techniques can also be grouped into physical, chemical, physicochemical and biological processes as presented in Figure 2. These techniques differ in terms of pollutant removal efficiency, operational cost, sludge generation, energy requirements and environmental sustainability (Donkadokula et al., 2020; Holkar et al., 2016). Physical and chemical methods generally achieve rapid colour removal and high treatment efficiency; however, they are often associated with high chemical consumption, secondary sludge generation, and elevated operational costs (Holkar et al., 2016). Conversely, biological treatments are more environmentally sustainable and cost-effective but may require longer retention times and are sensitive to wastewater composition and operational conditions (Aragaw, 2024; Bhatia et al., 2017; Ceretta et al., 2021).

Examples of physical techniques include membrane separation, evaporation, screening, flotation, and sedimentation-based processes, while chemical and physicochemical techniques include adsorption, coagulation-flocculation, oxidation, ion exchange, and electrochemical processes such as electrodialysis (Bhatia et al., 2017; Carrascal-Hernández et al., 2025; Holkar et al., 2016). Under optimised conditions, the physical and chemical methods of textile effluent treatment can achieve colour removal exceeding 80–95%. For instance, coagulation-flocculation processes are reported to remove approximately 60–90% of colour and suspended solids depending on coagulant type and dye composition, although large volumes of chemical sludge are generated (Holkar et al., 2016). Similarly, activated carbon adsorption has demonstrated colour removal efficiencies exceeding 90% for many reactive and azo dyes (Donkadokula et al., 2020), while advanced oxidation processes, including ozonation and Fenton oxidation, can achieve over 80–95% colour removal and substantial COD reduction (Holkar et al., 2016; Samuchiwal et al., 2021). However, both physical and chemical treatment processes are expensive, consume more energy and are not environmentally friendly. (Das et al., 2023; Donkadokula et al., 2020; Holkar et al., 2016). Coagulation and flocculation generate substantial volumes of chemical sludge requiring secondary disposal and increase overall treatment cost (Holkar et al., 2016), while advanced oxidation processes require substantial energy input and chemical reagents, limiting their economic feasibility for large-scale applications (Donkadokula et al., 2020; Samuchiwal et al., 2021).

Figure 2. Major technologies employed in textile wastewater treatment

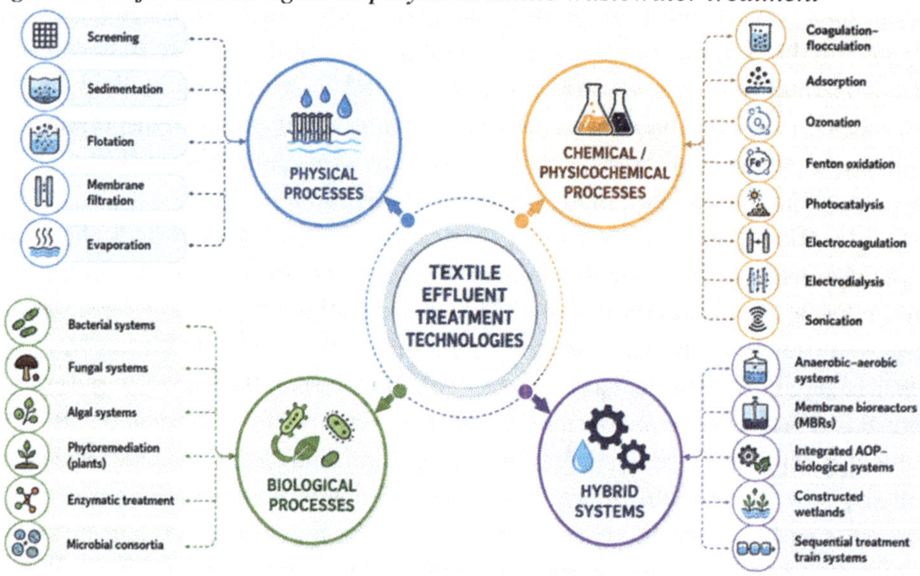

Although physicochemical methods have often achieved rapid pollution removal, their application is limited by high energy demand, chemical consumption, sludge generation and secondary pollution concerns (Holkar et al., 2016). In many cases, these methods transfer contaminants from one phase to another rather than achieving complete mineralisation (Ceretta et al., 2021; Donkadokula et al., 2020). Hybrid treatment systems that integrate physical, chemical and biological processes have also emerged to overcome the challenge that no single treatment technology can effectively address the complexity of textile wastewater (Al-Tohamy et al., 2022; Ceretta et al., 2021). However, such systems that combine the rapid colour removal capacity of physicochemical methods with the sustainability and mineralisation potential of biological treatments have increased operational costs.

Bioremediation technologies that use microbial agents such as bacteria, fungi, algae, yeast, and a consortium of these microorganisms to degrade dyes and other hazardous compounds in textile effluent are promising potential for sustainable degradation of dyes in textile effluents (Al-Tohamy et al., 2022; Aragaw, 2024; Jamee & Siddique, 2019). These biological methods have been identified as the most non-hazardous, cost-effective and environmentally-friendly among the available treatment options due to their lower energy needs, reduced chemical use and ability to mineralise varieties of pollutants rather than merely transfer them from one phase to another (Bener et al., 2020; Donkadokula et al., 2020; Jamee & Siddique, 2019).

4.1 Biotechnological Approaches to Textile Effluent Management

Biological approaches for the treatment of textile wastewater are increasingly recognised as more promising alternatives compared to conventional physical and chemical techniques (Bhatia et al., 2017; Ceretta et al., 2021; Hossain et al., 2022; Shoukat et al., 2019). The approach involves using a wide range of microorganisms, including bacteria, algae, yeast, and fungi, as well as enzyme-based systems, which are viable biological candidates for converting dye molecules present in textile effluent into non-toxic or fully mineralised products (Ali, 2010; Al-Tohamy et al., 2022). The treatment may occur even without supplementing with other easily biodegradable carbon sources, since most microbes are capable of directly utilising dyes or dye intermediates as carbon sources. Biological treatment systems are used individually or in combination and may operate in anaerobic, aerobic, anoxic or sequential conditions. Examples of biological systems available are aerobic activated-sludge processes, rotating biofilm reactors, packed-bed and fluidised-bed reactors, sequential batch and continuous-flow reactors (Das et al., 2023; Rathour et al., 2024). These systems apply several microbial cultures, including bacteria, fungi and algae for textile dye decolourisation (Gonçalves et al., 2000).

Biological treatments are easier to apply, produce less sludge, require fewer chemical reagents, are more cost-effective, have energy-saving features, are safer for the environment, and often generate non-toxic byproducts during microbial metabolic activities (Bhatia et al., 2017; Ceretta et al., 2021). Another advantage of the biological treatment is the ability of microorganisms, either singly or in consortia, to biodegrade dye-containing wastewater with high metabolic efficiency (Shoukat et al., 2019). Depending on the microbial species, dye class and operational conditions, biological systems can achieve colour removal efficiencies ranging from 60% to above 95%, along with substantial reductions in COD and toxicity levels (Bhatia et al., 2017; Ceretta et al., 2021). Compared to conventional approaches, biotechnological methods provide innovative and sustainable solutions for managing complex textile effluents (Gonçalves et al., 2000). Their ability to degrade and detoxify hazardous compounds more efficiently while minimising secondary pollution has gained them attention as eco-friendly alternatives to traditional physicochemical treatment methods (Jamee & Siddique, 2019).

4.2 Processes Used for Biological Decolorization of Textile Effluents

There are two main processes involved in pollutant removal in biological systems for textile wastewater treatment: adsorption or biosorption using microbial biomass and biodegradation by microbes (Al-Tohamy et al., 2022; Rane & Joshi, 2021). Through biosorption, dyes are passively adsorbed onto microbial biomass with potential for subsequent degradation, while in biodegradation, active enzymatic breakdown occurs, leading to mineralisation or conversion into non-toxic metabolites.

Biosorption and biodegradation processes occur under aerobic, anaerobic or anoxic conditions; thus, biological treatments are also classified as aerobic, anaerobic or anoxic. The treatment of textile effluents mainly depends on anaerobic methods, with consortia of microorganisms (anaerobic sludge or biofilm) serving as active agents. The process can take higher organic loads, does not require aeration and produces less sludge (Azanaw et al., 2022). The byproducts of aerobic treatment are biomass, carbon dioxide, and water, while those from anaerobic treatment include methane gas and other chemical compounds (Gonçalves et al., 2000; Panhwar et al., 2024; Punzi, 2015). Biomass generated under aerobic conditions can also function as an adsorbent material for dyes and associated chemical pollutants.

4.3 Adsorption/Desorption in Decolorizing Textile Dyes

Adsorption is a non-metabolic surface-based process in which dye molecules are physically or chemically bound to the cell walls of the microorganisms. When dye molecules and other pollutants are absorbed by biological materials, such as the biomass of microorganisms, including bacteria, yeasts, fungi, and algae, this process is referred to as biosorption (Danouche et al., 2021). Compared to synthetic adsorbents that are used in dye adsorption, biosorption is a relatively new, low-cost and effective strategy.

The microbial cell walls are rich in functional groups such as hydroxyl, carboxyl, and amino groups, which allow interactions with dye molecules through various mechanisms, including physical and chemical adsorption, electrostatic interactions, ion exchange, and complexation (Danouche et al., 2021). For instance, anionic dyes can readily bind to positively charged amino groups present on microbial surfaces such as those of *Bacillus*, *Aspergillus* or *Saccharomyces*. However, the adsorption process does not degrade dye structure; instead, it uses various microbial biomass to absorb or accumulate the dyes, thereby removing them from water (Das et al., 2023; Rathour et al., 2024). Nonetheless, the absorbed dye remains chemically intact on the biomass surface and can potentially be recovered using appropriate solvents. Consequently, biosorption is commonly considered a preliminary or complementary step to enhance dye removal before the actual biodegradation process. But since no metabolic activity is required, adsorption is a rapid process that can occur with both living and dead biomass (Rane & Joshi, 2021).

4.4 Biodegradation of Textile Dyes

Biodegradation is an active, enzyme-driven process in which microorganisms chemically transform or completely mineralise dye molecules into simpler and less toxic compounds. Unlike adsorption, biodegradation requires metabolic activity and is facilitated by a diverse array of oxidoreductive enzymes, such as azoreductase, laccases, catalases, oxidases, and peroxidases (Rane & Joshi, 2021). The biodegradation pathways of dyes depend on the molecular structures of dyes being degraded, the microorganisms, and the enzymes involved. Studies have reported increased biodegradation efficiency of dyes in the presence of a consortium of microorganisms compared to individual pure strains, with efficiencies of approximately 50–70% achieved in single strains and efficiencies above 90% upon applying mixed microbial cultures (Ali et al., 2009; Rane & Joshi, 2021). Although the mechanism involved in the biodegradation of dyes by microbial consortia is difficult to establish, co-metabolic activities within the microbial community are thought to enhance the likelihood of biodegradation and mineralisation (Rane & Joshi, 2021). Through a consortium, multiple groups of microorganisms with different metabolic pathways and complementary enzyme systems can work together more efficiently to fully degrade complex dye molecules and dye-transformed chemicals by reducing the buildup of toxic intermediates (Punzi, 2015; Samuchiwal et al., 2021). In addition, within the consortium, some microorganisms could detoxify and neutralise toxic chemicals produced during the degradation process, while others could perform further degradation. For example, under aerobic conditions, bacteria such as *Pseudomonas* and *Bacillus* initiate azo bond cleavage through azoreductase activity, producing aromatic amines as intermediates. The resulting intermediates can subsequently be mineralised under aerobic conditions by organisms such as *Sphingomonas* and *Phanerochaete chrysosporium*, thereby reducing toxicity and achieving more complete degradation (Saratale et al., 2011). Fungal-bacterial consortia comprising *Trametes versicolor* and *Pseudomonas putida* have demonstrated synergistic degradation, with fungal enzymes partially oxidising complex dye molecules and bacteria further metabolising the resulting intermediates (Ali, 2010). Similarly, in algal-bacterial systems, algae such as *Chlorella vulgaris* provide oxygen through photosynthesis, supporting aerobic bacterial degradation, while bacteria release carbon dioxide and nutrients that promote algal growth (Rane & Joshi, 2021). These synergistic interactions enhance both resilience and efficiency in degrading dye compounds (Al-Tohamy et al., 2022; Samuchiwal et al., 2021).

4.5 Existing Biological Systems for Decolourisation of Textile Effluents

Various biological systems have been developed for textile wastewater treatment using bacteria, fungi, algae, enzymes, and microbial consortia. The ability of these systems to treat textile wastewater depends heavily on the ability of the organisms to resist the dye toxicity and dye intermediates, such as aromatic amines and phenolic compounds (Aragaw, 2024). Consequently, selection of biological treatment systems is usually done depending on factors such as dye composition, wastewater characteristics, reactor conditions, and the desired level of pollutant mineralisation. Because of the structural diversity of textile dyes, effluent treatment can also be achieved using a mixture of microbial communities, often described as a consortium.

4.6 Enzymatic Treatment of Textile Wastewater

Enzymatic treatment of textile effluent has recently emerged as a green technology, offering promising alternatives to conventional methods for dye removal. They can efficiently cleave chromophoric groups present in dyes, break down aromatic rings and transform complex dye structures under mild operational conditions (Khan, 2025). Several enzymes of microbial and plant origin can be used to treat textile effluents, either alone or in combination with microorganisms. Peroxidase, azoreductase, and laccase are examples of microbial enzymes commonly used in the biodegradation of synthetic dyes present in textile effluents (Pandey et al., 2017). These enzymes are produced by bacteria, fungi, and algae, including *Bacillus*, *Pseudomonas*, *Streptomyces*, *Trametes*, *Pleurotus*, and *Aspergillus* species and have demonstrated effectiveness against a broad range of azo, anthraquinone, and reactive dyes. Nevertheless, their actions in dye degradation differ; for example, peroxidase exhibits broad substrate specificity and mild reaction conditions. Lignin peroxidase and manganese peroxidase produced by the fungus *P. chrysosporium* are especially valuable in degrading resistant dyes (Floros et al., 2012). Conversely, azoreductase specialises in the degradation of azo dyes by targeting azo bonds (-N=N-), initiating reductive cleavage (Pandey et al., 2017). Laccases oxidise phenolic and non-phenolic compounds and are known to be highly effective against reactive dyes and dye intermediates. Plant-derived enzymes, particularly peroxidases and lignin peroxidase, laccases, and polyphenol oxidases from species such as *Saccharum spontaneum*, *Brassica rapa*, *Rhus vernicifera*, *Blumea malcolmii*, and *Salvinia molesta*, have also shown significant potential in dye decolourisation (Rane & Joshi, 2021). Nevertheless, the combination of various enzymes used to treat textile wastewater can significantly enhance the biodegradation of textile dyes (Das et al., 2023). Mixed enzyme systems comprising laccase, lignin peroxidase, manganese peroxidase, tyrosinase, azoreductase, and reductases have been reported from diverse microbial sources, enhancing the degradation of structurally complex dyes (Rane & Joshi, 2021). It is also important to note that the choice of enzyme combination and the specific enzymes used depend on the dye type, its chemical structure, and the microbial species involved (Singh et al., 2016).

4.7 Treatment of Textile Wastewater With Bacterial Systems

Bacteria have been shown to degrade synthetic dyes, a process aided by their oxidases (Jamee & Siddique, 2019). To date, several bacteria, including *Bacillus subtilis*, *Aeromonas hydrophila*, *Bacillus cereus*, *Pseudomonas*, *Rhodococcus*, and *Achromobacter* species, have been reported to possess high dye decolourisation capacity from textile effluents (Das et al., 2023; Rathour et al., 2024). Strains such as *Micrococcus glutamicus* and *Lysinibacillus* sp. have been reported to achieve nearly complete removal of reactive dyes under optimised laboratory conditions, whereas most *Pseudomonas* and *Bacillus* strains achieve 80%–95% efficiency (Al-Tohamy et al., 2022). Several other actinomycetes stains have been reported to decolourise reactive dyes, including anthraquinone, phthalocyanine and azo dyes, through adsorption onto the cellular biomass without any degradation (Moyo et al., 2022).

Azoreductase enzymes have been reported to initiate the degradation of azo dyes, particularly Orange II, where anaerobic cleavage of the azo bond is followed by aerobic mineralization (Khan et al., 2025; Moyo et al., 2022). Complete degradation of azo dyes requires bacterial strains that work in alternating anaerobic and aerobic conditions (Bidu, 2024; Das et al., 2023; Rathour et al., 2024). This is because some bacteria can degrade dyes under aerobic conditions, while others do so under anaerobic conditions (Jamee & Siddique, 2019). Thus, alternating anaerobic and aerobic conditions are important for enabling different members of the consortium, which require varying conditions to perform their optimal reactions sequentially (Das et al., 2023; Rathour et al., 2024). Most bacteria that degrade dyes under aerobic conditions, i.e. in the presence of oxygen, require an additional carbon source since they are unable to utilize the dye as a carbon source. Very few bacteria, such as *Pigmentiphaga kullae* K24 and *Xenophilus azovorans* KF 46 can cleave the -N=N- bonds present in azo dye groups, and utilize the resultant amines for their growth (Jamee & Siddique, 2019).

Bacterial species have been particularly reported to be more efficient than other microbial strains in degrading dyes in textile wastewater (Zhuang et al., 2020). Their efficacy in dye degradation is determined by the bacteria's adaptability and their ability to degrade dyes under the prevailing environmental conditions (Al-Tohamy et al., 2022). The main advantages of using bacteria for dye degradation are their ease of cultivation and their high specific growth rates when compared to other microorganisms (Pearce et al., 2003; Y. Wang et al., 2020). A list of selected bacterial strains and their respective dye removal performance is reported in Table 3.

Table 3. Bacterial Degradation of Textile Dyes

Bacteria	Dye	Dye Removal (%)	References
Micrococcus glutamicus NCIM-2168	C.I. Reactive Green 19A	100	(Saratale et al., 2010)
Pseudomonas sp. SUK1	Reactive Red 2	>80	(Hossen et al., 2020)
Bacillus sp. VUS	Navy Blue 2GL	~90	(Dawkar et al., 2009)
Kocuria rosea MTCC 1532	Methyl Orange	100	(Parshetti et al., 2010)
Proteus mirabilis LAG	Reactive Blue 13	87.91	(Olukanni et al., 2010)
Lysinibacillus sp. RGS	C.I. Remazol Red	100	(Srinivasan & Bankole, 2022)
Lysinibacillus sp. KMK-A	Reactive Orange M2R	98	(Chaudhari et al., 2013)
Bacillus cereus	Orange II / Acid Orange 7	52.5	(Tripathi et al., 2018)
Pseudomonas extremorientalis BU118	Congo Red	36–94	(Neifar et al., 2020)

4.8 Textile Wastewater Treatment With Fungal Systems

Fungi play an important role in the degradation and mineralisation of a wide range of textile dyes (Anastasi et al., 2011; Blánquez et al., 2008; Pundir et al., 2024; Yang et al., 2009). The first strains of fungi discovered for dye decolouration were *P. chrysosporium*, which was found to degrade polymeric lignin molecules. The discovery led to intensive research into the degradation of a wide range of recalcitrant xenobiotic compounds such as textile dyes. Nowadays, fungi are employed in the treatment of textile wastewater using various fungal strains or consortia (Anastasi et al., 2011; Blánquez et al., 2008). Table 4 presents the selected list of fungal strains and the dyes they degrade.

Table 4. Fungal strains reported in degrading textile dyes

Fungal Strain	Dyes Degraded	References
Fusarium solani	Crystal Violet, Malachite Green	(H. Ali, 2010)
Myrothecium sp. IMER1	Remazol Brilliant Blue R (RBBR)	(Huang et al., 2014)
Coriolus versicolor	Vat dyes (e.g., Cibanon Blue)	
Aspergillus niger	Reactive Black 5; various azo dyes	(Saratale et al., 2011)
Armillaria sp. F022	Azo, triphenylmethane, anthraquinone dyes	(Rodriguez-couto, 2013)
Schizophyllum commune IBL-06	Solar Brilliant Red 80	(Asgher et al., 2013)
Piptoporus betulinus IEBL-3	Industrial textile effluents	(Pundir et al., 2024)
Aspergillus flavus & *Fusarium oxysporium* (consortium)	Real textile effluent	(Nadu, 2023)
Penicillium chrysogenum	Real textile dyeing wastewater	(Lanfranconi, 2022)

The primary advantage of using fungi for dye-containing wastewater treatment is their ability to accelerate their metabolism to achieve optimal environmental conditions (Pundir et al., 2024; Yang et al., 2009). White-rot fungi are among the strains identified as capable of degrading recalcitrant organic pollutants, such as textile dyes, including azo dyes. The ability originates from their non-specific lignin-modifying enzymes (Al-Tohamy et al., 2022). Fungi such as *Phanerochaete chrysosporium* and *Trametes* species have demonstrated dye degradation efficiencies of 85–90% for azo, anthraquinone, and triphenylmethane dyes owing to their powerful extracellular oxidative enzymes (Pundir et al., 2024).

Similar to bacteria, fungi use enzymes to degrade complex chemical structures, such as dye molecules (Chakraborty et al., 2025; Yang et al., 2009). Fungal enzymes, such as laccases, peroxidases, xylanases, cellulases, and proteases (Al-Tohamy et al., 2022), have already been effectively used for the degradation and decolourisation of azo dyes (Anastasi et al., 2011). A combination of several fungal enzymes, such as manganese peroxidase, lignin peroxidase, and laccase, has been reported to have high degradation efficiency (Pundir et al., 2024). Depending on the processing medium, fungi can produce a wide range of oxidative enzymes (Al-Tohamy et al., 2022). Azo dye decolourisation by fungal enzymes, such as azoreductases and laccases, is commonly carried out using the pathway shown in Figure 3.

Figure 3. Mechanism of fungal azoreductase enzyme-mediated reductive cleavage of azo dyes

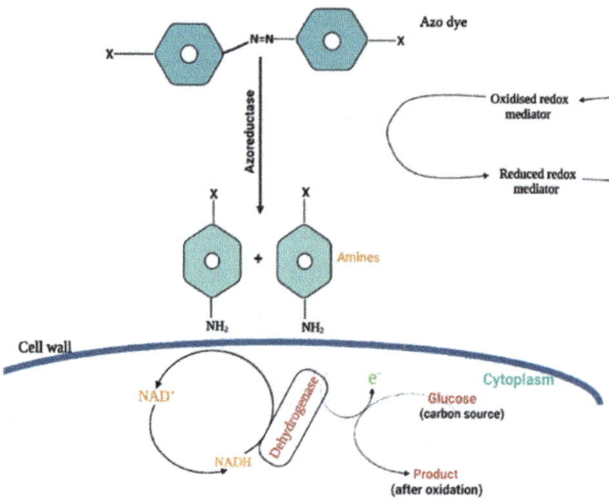

(Pundir et al., 2024)

12.4.9 Textile Wastewater Treatment With Algal Cultures

Algae are another group of organisms that are capable of degrading azo dyes, particularly through azoreductase enzymes (Das et al., 2023; Rathour et al., 2024). The growing interest in the use of algae as bio-coagulants for textile dye biodegradation arises from their ideal cell wall properties, large surface, high capacity, and affinity for binding (Khan et al., 2022; Vijayaraghavan & Shanthakumar, 2020). Examples of algae species that have been reported to degrade azo dyes include *Chlorella* and *Oscillatoria*. The reduction of azo dyes produces aromatic amines, which are then reduced to simpler compounds (Periyasamy, 2025). Some algal cultures have been using azo dyes as their sole sources of carbon and nitrogen. Using such algae was proposed by the authors for use in stabilisation ponds, as they can play a role in aromatic amine removal (Periyasamy, 2025). Table 5 shows the performance of the selected algae species for specific dye removal.

Table 5. Selected algae species for dye degradation (Al-Tohamy et al., 2022)

Algae Species	Dye	% Removal	References
Chlorella vulgaris	Remazol	53.2	(Sani et al., 2021)
Chlorella vulgaris	Brilliant Blue R	—	(Nazlı & Karatay, 2024)
Coelastrella sp.	Rhodamine B	80	(Lúcia et al., 2018)
Dermatocarpon vellereceum	Navy Blue HE22	95	(Kulkarni et al., 2018)
Chara vulgaris L.	Methyl Red	70–100	(Mahajan & Kaushal, 2020)

4.10 Textile Wastewater Treatment With Yeast Systems

Yeasts present another potential system employed for textile wastewater remediation. They remove dyes from textile effluents using biosorption and reductive cleavage, and perform well in hybrid reactors (Danouche et al., 2021; Mendes, Moreira et al., 2023). The biosorption of dyes using yeast is facilitated by the numerous functional groups, such as proteins and nitrogen-containing groups, present on the cell wall of the yeast strains. Other active groups, such as acidic polysaccharides, lipids and amino acids, may also be involved (Danouche et al., 2021).

The common yeast used for textile wastewater bioremediation is *Saccharomyces cerevisiae*, also known as baker's yeast (Al-Najar et al., 2021). This yeast strain has been extensively studied as a biosorbent (Danouche et al., 2021). Yeast uses azoreductases to facilitate the transformation of azo dyes containing -N=N- groups, leading to the generation of amines and decolourisation (Ertu rul et al., 2009). Many studies report that yeast can efficiently digest dye effluents (Al-Najar et al., 2021; Ertu rul et al., 2009; Mendes, Cassoni, et al., 2023). For example, yeast *Rhodotorula mucilaginosa* has been used for the degradation of Remazol Blue (Ertu rul et al., 2009). Table 6 shows various yeasts involved in dye decolourisation.

Table 6. Selected yeast strains for textile dye degradation

Yeast / Fungus Species	Dye	% Removal	References
Galactomyces geotrichum MTCC1360	Methyl Red	100	(Waghmode et al., 2011)
Trichosporon beigelii NCIM-3326	Navy Blue HER	100	(Qu et al., 2010)
Galactomyces geotrichum MTCC1360	Mixture of dyes	88	(Waghmode et al., 2011)
Trametes trogii BAFC463	Fast Blue RR	> 85	(Levin et al., 2012)
Magnusiomyces ingens LH-F1	Acid Red B	97.37	(Tan et al., 2014)
Scheffersomyces spartinae TLHS-SF1	Acid Scarlet 3R	98.14	(Tan et al., 2016)

continued on following page

Table 6. Continued

Yeast / Fungus Species	Dye	% Removal	References
Trichosporon akiyoshidanum HP2023	Reactive Black 5	89	(Al-tohamy et al., 2020)
Sterigmatomyces halophilus SSA1575	Reactive Black 5	100	

Overall, using yeast for the remediation of textile wastewater and for mitigating the environmental impact of dye production shows potential (Kurade et al., 2017; Mendes, Moreira, et al., 2023; Nicula et al., 2023). Compared to bacteria and fungi, yeasts are more advantageous when it comes to degrading textile dyes due to their rapid growth rates and ability to tolerate a wide range of environmental conditions (Al-Tohamy et al., 2022). Furthermore, yeasts have a special flocculating characteristic that enables them to aggregate into multicellular flakes (masses), facilitating their recovery after treatment of coloured effluent (Danouche et al., 2021).

4.11 Phytoremediation

Plants are naturally absorbers of pollutants from wastewater, especially in swampy areas. The process of cleaning wastewater by plants and their associated microbial communities is called phytoremediation. It is an environmentally friendly and cost-effective process. Phytoremediation is an example of a synergistic relationship where both species act together for a better result (Adelodun et al., 2021). Its mechanism for removing the toxicity of pollutants involves the combined action of some plants and their root-associated microbes (Adelodun et al., 2021; Chandanshive et al., 2018; Rizwana et al., 2014). In particular, it involves the combined actions of the plant's phytodegradation and phytoaccumulation, and the enzymatic action of bacteria that drive the whole remediation process (Chandanshive et al., 2018; Rizwana et al., 2014).

Textile dye phytoremediation has been researched and reported as a promising approach, particularly the use of constructed wetland systems (Bidu, 2025). These systems are said to be economically viable and ecologically sustainable, making them attractive for low-resource settings (Bharathiraja et al., 2018). Microbial-assisted phytoremediation offers several advantages, such as low cost, sustainability, and the potential to treat large areas of contaminated soil or water. Previous studies have reported that bacterial species of *Ochrobactrum* sp., *Pseudomonas aeruginosa*, and *Providencia vermicola* can enhance textile dye decolouration and reduce the toxicity of textile-emitted effluent dyes when used with suitable plant hosts (Adelodun et al., 2021; Bharathiraja et al., 2018; Chandanshive et al., 2018).

Commonly used phytoremediation practices include constructed wetland systems. They are effective in removing organic matter, nutrients, heavy metals, organic and inorganic chemicals, and pathogens from textile wastewater. The removal of pollutants in wetlands involves interconnected mechanisms, including physical, chemical, and biological processes that remove various contaminants (Sharma & Malaviya, 2022). The rhizosphere zone plays an important role in these reactions, serving as a hotspot for physico-chemical and biological reactions driven by interactions of macrophytes, microorganisms, substrate materials, and contaminants (Sharma & Malaviya, 2022). The zones in a constructed wetland for wastewater treatment sectioning of the substrate, different mechanisms involved in pollutant removal, and the location where the removal phenomenon is shown in Figure 4.

Figure 4. Graphical representation of phytoremediation mechanisms in constructed wetlands for textile wastewater treatment

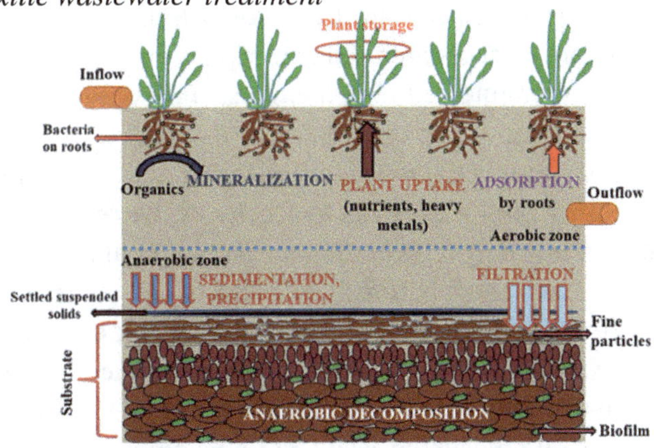

(Sharma & Malaviya, 2022)

Despite the advantages of phytoremediation, the performance of dye degradation during treatment depends on the success of microbial phytoremediation for textile dye removal. This depends on various factors, including the choice of suitable plant species, the presence of specific dye-degrading microorganisms, and optimisation of environmental conditions (Das et al., 2023). Suboptimal environmental conditions or high pollutant levels can prevent plant growth and microbial activities, thereby affecting the overall phytoremediation performance.

4.12 Advanced Bioremediation Methods

Apart from the traditional bioremediation systems, encompassing bacteria, yeast, fungi, algae and enzymes, innovative and advanced bioremediation methods such as microbial fuel cells (MFCs), genetically engineered microorganisms, and nanoparticle-based bioremediation are gaining prominence (Kumar et al., 2025).

These advanced technologies address challenges related to conventional biological dye removal, including slow degradation rates, incomplete dye removal or breakdown, formation of toxic aromatic amines and sensitivity to biodegradation conditions such as pH, temperature and dye concentration.

Microbial fuel cells are devices that use microorganisms as biocatalysts to remove dyes from wastewater while simultaneously generating electricity (Rane & Joshi, 2021), a process that is not possible in conventional anaerobic or aerobic reactors. Electroactive microbes such as *Bacillus*, *Geobacter*, *Enterobacter* and *Shewanella bacteria* are commonly used in modern MFCs (Karthika et al., 2024; Khan et al., 2015; Solanki et al., 2013).

Genetically engineered microorganisms are bioengineered strains that have been modified to improve their dye-degrading or transforming efficiency. Studies have reported significant improvements in the decolourisation efficiency of the acid violet dye when mutant laccases are used compared to wild-type enzymes (Kumar et al., 2025). Similarly, engineered *E. coli* expressing azoreductase genes found in *Pseudomonas luteola* removed up to 95% of azo dyes (Dixit & Garg, 2018).

Nanoparticle-based bioremediation is an emerging field that combines nanotechnology with microorganisms or enzymes to accelerate dye removal through adsorption, redox reactions or catalytic degradation (Kumar et al., 2025). Nanoparticles (NPs) such as Fe_2O_3, CuO, and functionalized metal oxides are characterised by high surface area, reactivity, and adjustable surface properties, enabling their effective interactions with dye molecules. Studies have demonstrated high decolourisation efficiencies when using NPs-based remediation, with Fe_2O_3-NPs rapidly removing up to 95% of dyes, while biosynthesised CuO-NPs embedded in beads removing up to 90% of dyes (Kumar et al., 2025).

Membrane technologies, including both chemical membrane reactors and membrane bioreactors, represent another important advancement in textile wastewater treatment techniques and are effective in producing high-quality effluent suitable for reuse (Jallouli et al., 2023; Uddin et al., 2024). However, membrane bioreactors are limited by longer treatment times, sensitivity to toxic compounds, and the need to maintain microbial communities (Aragaw, 2024).

Hybrid treatment systems integrating biological, physical and chemical processes have increasingly been adopted to overcome the limitations of individual treatment technologies. For example, sequential anaerobic–aerobic bioreactors combine anaerobic azo bond reduction with aerobic mineralisation of aromatic amines, thereby improving overall dye degradation efficiency (Bidu, 2024). Similarly, membrane bioreactors integrated with activated sludge systems have demonstrated high colour, COD, and suspended solids removal efficiencies while producing reusable effluent (Jallouli et al., 2023). Constructed wetlands coupled with microbial degradation processes have also shown promise for low-cost treatment of textile wastewater in resource-limited settings (Bidu, 2025). These integrated systems enhance treatment stability, reduce toxicity, and improve pollutant mineralisation compared with single-stage biological treatments.

5 FACTORS AFFECTING BIODEGRADATION EFFICIENCY

Biodegradation of synthetic dyes and other chemicals in textile effluent depends on the physical, chemical, and biological factors (Ali, 2010; Singh, 2015). These factors influence microbial growth, enzymatic activity and the metabolic pathways involved in dye transformation. Process parameters, such as the ratio of organic to microbial charge, temperature, and the oxygen present in the system, all influence the efficiency of biological effluent treatment methods (Jamee & Siddique, 2019; Panhwar et al., 2024). Dye structure is particularly important, with simpler structures and low molecular weights exhibiting higher rates of biodegradation. The nature of substituents on the aromatic ring is also important in determining oxidation behaviour (Jamee & Siddique, 2019). In azo dyes specifically, studies have demonstrated that electron-donating groups such as methyl and methoxy substituents enhance oxidative enzymatic degradation of azo phenols, whereas electron-withdrawing groups, including chloro, fluoro, and nitro substituents, inhibit oxidative enzymatic breakdown (Jamee & Siddique, 2019). Anthraquinone and phthalocyanine dyes are generally more resistant than azo dyes because of their highly condensed aromatic structures and greater chemical stability.

Apart from factors associated with the dye structure, biodegradation is also influenced by biotic and abiotic parameters (Mishra & Maiti, 2018). Their effects come from how these factors affect the biochemical reactions and functional aspects of the microbial communities responsible for dye degradation (Krishnan et al., 2017; Mishra & Maiti, 2018). Additionally, pretreatment techniques such as chemical degradation can significantly affect microbial decolourisation efficiency by providing simple nitrogen and carbon sources (Saratale et al., 2011).

5.1 Abiotic Factors

Abiotic factors are environmental and operational conditions that affect the efficiency of dye biodegradation (Aragaw, 2024). The key parameters include reaction temperature, pH, oxygen level, mixing intensity, retention time, dye concentration, dye class or structure, and the quantity and quality of nitrogen and carbon sources (Krishnan et al., 2017; Mishra & Maiti, 2018). Microbial growth and enzymatic activity are strongly affected by reaction temperature. Low temperatures reduce metabolic activity and slow down biodegradation rates, whereas excessively high temperatures may denature degradative enzymes and inhibit microbial survival. Similarly, pH influences membrane transport, enzyme structure, nutrient solubility and redox reactions during dye degradation. Most bacterial dye-degrading systems perform optimally under neutral to slightly alkaline conditions, while fungal systems often tolerate a wider pH range (Al-Tohamy et al., 2022).

Oxygen availability also determines the biodegradation pathway. Anaerobic conditions favour reductive cleavage of azo bonds through azoreductase activity, whereas aerobic conditions are essential for the subsequent oxidation and mineralisation of aromatic amines produced during azo dye reduction (Bidu, 2024; Das et al., 2023). Consequently, sequential anaerobic–aerobic systems often achieve more complete dye degradation than single-stage systems. The availability of nutrients such as carbon and nitrogen sources is equally important since many microorganisms require co-metabolism for effective dye degradation. Inadequate nutrient supply limits microbial growth and enzyme production (Saratale et al., 2011).

Mixing intensity and retention time further influence biodegradation efficiency by affecting contact between microorganisms and dye molecules. Proper mixing enhances oxygen transfer, nutrient distribution and microbial-pollutant interactions within the reactor system while adequate retention time allows for sufficient microbial growth and enzymatic degradation (Krishnan et al., 2017; Mishra & Maiti, 2018). However, excessive mixing may disrupt microbial flocs and biofilms, reduce biomass retention, and negatively affect treatment efficiency, particularly in biofilm-based and activated sludge systems (Punzi, 2015; Rathour et al., 2024).

Apart from reaction conditions, bioreactor type, and characteristics of the wastewater to be treated are also important when it comes to biodegradation systems (Mishra & Maiti, 2018). Many studies have reported that such higher dye concentrations in textile effluents tend to decrease the rate of microbial decolourisation (Aragaw, 2024; El Awady et al., 2024). For instance, Guadie and colleagues (Guadie et al., 2018) evaluated the decolourisation efficiency of *Halomonas* sp. across dye concentrations ranging from 50 to 250 ppm for different dye types (reactive yellow, reactive red, vat blue, and reactive blue) and real textile effluent. The applied microorganism (*Halomonas* sp.) achieved a decolourisation efficiency greater than 85% after 24 hours of incubation at 50 to 150 ppm, but the efficiency decreased to between 25% and 55% at higher dye concentrations (200-250 ppm). Although the study showed decreased decolourisation efficiency at higher dye concentrations, it also reported an increased efficiency of over 98% across all concentrations with extended incubation time, suggesting the importance of reaction time in the biodegradation of dyes present in textile effluent. These findings further demonstrate that a combination of dye concentration and incubation time can influence the overall efficiency of microbial degradation of dyes present in textile effluents (Aragaw, 2024).

5.2 Biotic Factors

Biotic factors involve common characteristics and interactions of microbial communities involved in biological wastewater treatment. These include enzyme activity, food competition, community growth, sufficient biomass, composition, and genetic factors such as mutation and horizontal gene transfer (Aragaw, 2024).

Competition for limited carbon and nitrogen sources among bacterial species and antagonistic interactions between bacterial species and microbial predators can highly reduce the biodegradation efficiency of the biological treatment of textile effluent (Ajaz et al., 2020; Singh, 2015). Furthermore, the rate of dye biodegradation in wastewater depends on the availability of organic load within the process, and on the number and type of microbial communities capable of metabolising specific dye compounds (Ajaz et al., 2020). Additionally, the capacity of microbes to produce specific enzymes during biodegradation and the food-to-microbe ratio also affect the rate of biodegradation (Singh, 2015). Inhibition of enzyme activity or microbial proliferation resulting from toxic intermediates, limited nutrients, or predator effect may further hinder effective dye biodegradation (Mohanty & Kumar, 2025).

Successful biodegradation of dyes present in textile effluent requires both adequate biomass and sufficient enzyme production with a corresponding number of target microbial cells. Since dyes contained in wastewater are highly dependent on the specificity of enzymes excreted and the availability of food sources, the extent of contaminant metabolism by specific enzymes must have contributed to their specific contaminant affinity and this availability is highly needed (Aragaw, 2024). Generic adaptability further affects biodegradation performance. Mutation and horizontal gene transfer may enhance the acquisition of degradative pathways and improve microbial tolerance to toxic textile pollutants. Such adaptive responses are highly important in long-term treatment systems exposed to fluctuating wastewater composition and dye concentrations (Ajaz et al., 2020; Singh, 2015).

6 CHALLENGES AND FUTURE PROSPECTS

The application of biotechnology for managing textile effluent is promising due to its environmental friendliness and cost-effectiveness compared to conventional physicochemical treatments. However, it faces several technical and operational challenges that must be addressed in order to realise its full potential in textile wastewater treatment. These challenges are particularly severe due to the complex and variable nature of textile wastewater, characterised by mixtures of dye classes, heavy metals, salts and auxiliary chemicals. Microbial sensitivity and practical operational concerns further complicate the microbial degradation processes. One of the significant challenges in microbial decolouration of textile effluents is the incomplete degradation of dyes and associated pollutants. This is because most synthetic dyes require specific biodegradation conditions for complete degradation or mineralisation (Ali, 2010). However, achieving these conditions consistently may be difficult in typical bioreactor systems, such as activated sludge reactors, membrane bioreactors, packed-bed bioreactors, fluidised-bed reactors and sequential anaerobic-aerobic systems. Fluctuations in hydraulic loading, oxygen transfer, salinity, pH, and dye concentration within these systems can significantly affect microbial activity and degradation efficiency (Rathour et al., 2024; Yang et al., 2020). In addition, during microbial degradation, particularly under anaerobic conditions, some dye compounds, metabolites or intermediates are generated. These intermediates, such as azo-based aromatic amines including aniline, benzidine, naphthylamines, and sulfanilic acid derivatives, are more toxic and mutagenic, and can be more recalcitrant than the parent dye molecules, posing secondary environmental risks if discharged without being fully mineralised. Some aromatic amines, such as aniline, benzidine, and 4-aminobenzenesulfonic acid, can inhibit microbial enzymes, suppress biomass growth, and pose carcinogenic risks to aquatic organisms and humans (Saratale et al., 2011).

Microbes are also highly sensitive to most compounds commonly present in textile effluents. High concentrations of toxic substances, such as high salinity, can induce osmotic stress and reduce microbial respiration, while heavy metals such as chromium and copper may inhibit oxidoreductive enzymes involved in dye degradation (Aragaw, 2024). Similarly, extreme pH and toxic dye aromatic intermediates may inhibit or suppress microbial growth and disrupt enzymatic activity, thereby reducing the overall treatment efficiency. Maintaining optimal conditions such as pH, temperature, oxygen levels, and nutrient balance is therefore important, although challenging, especially in fluctuating industrial wastewater streams. Long-term stability and activity of microbial populations, particularly during continuous or large-scale operations, further require robust process control and monitoring.

Upscaling biological treatment processes from laboratory or pilot scale to full industrial-scale systems remains challenging. Although many laboratory-scale studies report dye removal efficiencies above 90%, translating these results to industrial-scale applications is still difficult. This is due to the complexity of bioreactor design, biomass retention, and process stability that are difficult to maintain under fluctuating industrial wastewater composition compared to laboratory experiments, which are typically conducted under controlled conditions. Economically, biological wastewater treatment systems may require extended retention times, careful nutrient balancing, and additional polishing stages to ensure consistent effluent quality. These requirements can increase operational complexity and costs as demonstrated by a comparative study of various biological treatments, including conventional activated sludge, membrane bioreactor and moving bed biofilm reactor (Yang et al., 2020). Consequently, the long-term sustainability of biological treatment of textile wastewater depends on improving reactor stability, enhancing microbial adaptability, minimising the accumulation of toxic intermediates, and integrating complementary treatment technologies capable of achieving complete mineralisation.

To address these limitations, various emerging technological approaches, including advanced enzymatic techniques, microbial consortia engineering and nanotechnology-based biocatalyst immobilisation, are being explored. The integration of microbial treatment with advanced chemical and physical methods, such as membrane filtration, adsorption and advanced oxidation processes, can create hybrid systems that improve the overall efficiency and treatment outcomes. Such integrated approaches combine the rapid action of chemical or physical processes with the environmental and cost benefits of biological degradation. Furthermore, emerging biotechnological tools for real-time monitoring, as well as process control and optimisation are also important to enable precise control of the treatment parameters and microbial activity, thereby enhancing process reliability.

Advanced technologies, such as genetic engineering and synthetic biology, present another promising avenue, since these technologies can lead to the development of genetically engineered bacteria with greater tolerance and a wider range of pollutant degradation (Kumar et al., 2025). In addition, enzyme engineering techniques can improve the activity, stability and substrate range of key-degrading enzymes such as laccases, peroxidases and azoreductases. Extensive research is also ongoing to establish methods for utilising waste biomass, such as agro-industrial residuals, as a low-cost substrate to support microbial growth and reduce treatment costs. Such strategies support the overall sustainability of wastewater treatment systems.

Moving forward, the future of textile effluent waste management lies in developing scalable, cost-effective and environmentally sustainable biotechnological solutions. This will require interdisciplinary collaborations among microbiologists, environmental engineers, chemists, textile dyers and colourists, and overall industrial stakeholders. Alternative dyeing techniques, such as natural dyes that can be easily degraded, waterless dyeing, dyeing without dyes, and intrinsic dyeing, should be studied, and their potential for sustainable textile dyeing should be emphasised.

7 CONCLUSION

Textile effluents present significant environmental and health risks due to their complex mixture of dyes, dye auxiliaries, salts and other toxic chemical compounds. These effluents remain among the most complex and environmentally hazardous pollutants of water bodies and land. The pollution caused by textile wastewater is related to its high colour intensity, elevated chemical oxygen demand, salinity, and the presence of synthetic dyes that are resistant to degradation. Conventional physicochemical treatment technologies are energy-intensive, generate secondary pollutants and have proven inefficient at removing contaminants from textile effluent.

Biotechnological systems, such as microbial and enzyme-mediated dye-degrading systems, offer viable and environmentally sustainable solutions for effectively managing textile effluents and dye decolouring. While challenges remain in terms of effluent variability, incomplete mineralisation of dyes, sensitivity of microorganisms to toxic compounds and scalability, advancements in microbial biotechnology and continuous exploitation of effective dye-degrading enzymes have enhanced the efficiency of dye decolouration and detoxification. Moreover, hybrid treatment systems such as sequential anaerobic–aerobic bioreactors, membrane bioreactors, and microbial assisted constructed wetlands have demonstrated superior dye decolourisation, pollutant removal efficiencies and effluent detoxification. These hybrid treatment systems combine the operational advantages of physical and chemical processes, making them particularly attractive for industrial-scale applications. These efforts suggest potential for environmentally friendly, efficient industrial-scale implementation.

Future research should focus on the development of strong microbial strains through genetic and metabolic engineering, optimisation of reactor designs, and the application of optimisation tools to better understand microbial dye degradation pathways. Additionally, comprehensive techno-economic assessments and life-cycle analyses are essential to support the large-scale implementation of biotechnological solutions in the textile industry. Coordinated efforts in biotechnology and proper environmental management are also needed for effective detoxification and the restoration of dye-polluted ecosystems.

DECLARATION OF CONFLICT OF INTEREST

The authors declare to have no conflict of interest.

REFERENCES

Adelodun, A. A., Olajire, T., Afolabi, N. O., Akinwumiju, A. S., Akinbobola, E., & Hassan, U. O. (2021). Phytoremediation potentials of *Eichhornia crassipes* for nutrients and organic pollutants from textile wastewater. *International Journal of Phytoremediation, 23*(13), 1333–1341. DOI: 10.1080/15226514.2021.1895719 PMID: 33788648

Ajaz, M., Shakeel, S., & Rehman, A. (2020). Microbial use for azo dye degradation—A strategy for dye bioremediation. *International Microbiology, 23*(2), 149–159. DOI: 10.1007/s10123-019-00103-2 PMID: 31741129

Al-Najar, J. A., Lutfee, T., & Alwan, N. F. (2021). The action of yeast as an adsorbent in wastewater treatment: A Brief Review. *IOP Conference Series. Earth and Environmental Science, 779*(1), 012054. Advance online publication. DOI: 10.1088/1755-1315/779/1/012054

Al-Tohamy, R., Ali, S. S., Li, F., Okasha, K. M., Mahmoud, Y. A. G., Elsamahy, T., Jiao, H., Fu, Y., & Sun, J. (2022). A critical review on the treatment of dye-containing wastewater: Ecotoxicological and health concerns of textile dyes and possible remediation approaches for environmental safety. *Ecotoxicology and Environmental Safety, 231*, 113160. DOI: 10.1016/j.ecoenv.2021.113160 PMID: 35026583

Al-tohamy, R., Kenawy, E., Sun, J., Ali, S. S., & Di Gregorio, S. (2020). Performance of a Newly Isolated Salt-Tolerant Yeast Strain *Sterigmatomyces halophilus* SSA-1575 for Azo Dye Decolorization and Detoxification. *Frontiers in Microbiology, 11*(June), 1–19. DOI: 10.3389/fmicb.2020.01163 PMID: 32595618

Ali, H. (2010). Biodegradation of Synthetic Dyes—A Review. *Water, Air, and Soil Pollution, 213*(1–4), 251–273. DOI: 10.1007/s11270-010-0382-4

Ali, N., Hameed, A., & Ahmed, S. (2009). Physicochemical characterization and Bioremediation perspective of textile effluent, dyes and metals by indigenous Bacteria. *Journal of Hazardous Materials, 164*(1), 322–328. DOI: 10.1016/j.jhazmat.2008.08.006 PMID: 18818017

Alzain, H., Kalimugogo, V., & Hussein, K. (2023). A Review of Environmental Impact of Azo Dyes. *International Journal of Research and Review, 10*(6), 64–689. DOI: 10.52403/ijrr.20230682

Anastasi, A., Parato, B., Spina, F., Tigini, V., Prigione, V., & Varese, G. C. (2011). Decolourisation and detoxification in the fungal treatment of textile wastewaters from dyeing processes. *New Biotechnology, 29*(1), 38–45. DOI: 10.1016/j.nbt.2011.08.006 PMID: 21911089

Aragaw, T. A. (2024). A review of dye biodegradation in textile wastewater, challenges due to wastewater characteristics, and the potential of alkaliphiles. *Journal of Hazardous Materials Advances, 16*(August), 100493. DOI: 10.1016/j.hazadv.2024.100493

Arora, J., Ranjan, A., Chauhan, A., Biswas, R., Rajput, V. D., Sushkova, S., Mandzhieva, S., Minkina, T., & Jindal, T. (2022). Surfactant pollution, an emerging threat to ecosystem: Approaches for effective bacterial degradation. *Journal of Applied Microbiology, 133*(3), 1229–1244. DOI: 10.1111/jam.15631 PMID: 35598183

Asgher, M., Yasmeen, Q., Muhammad, H., & Iqbal, N. (2013). Enhanced decolorization of Solar brilliant red 80 textile dye by an indigenous white rot fungus *Schizophyllum commune* IBL-06. *Saudi Journal of Biological Sciences, 20*(4), 347–352. DOI: 10.1016/j.sjbs.2013.03.004 PMID: 24235871

Azanaw, A., Birlie, B., Teshome, B., & Jemberie, M. (2022). Textile effluent treatment methods and eco-friendly resolution of textile wastewater. *Case Studies in Chemical and Environmental Engineering, 6*, 100230. DOI: 10.1016/j.cscee.2022.100230

Bener, S., Atalay, S., & Ersöz, G. (2020). The hybrid process with eco-friendly materials for the treatment of the real textile industry wastewater. *Ecological Engineering, 148*(December 2019). DOI: 10.1016/j.ecoleng.2020.105789

Bharathiraja, B., Jayamuthunagai, J., Praveenkumar, R., & Iyyappan, J. (2018). Phytoremediation Techniques for the Removal of Dye in Wastewater. *Energy. Environmental Sustainability*, •••, 243–252. DOI: 10.1007/978-981-10-7485-1_12

Bhatia, D., Sharma, N. R., Singh, J., & Kanwar, R. S. (2017). Biological methods for textile dye removal from wastewater: A review. *Critical Reviews in Environmental Science and Technology, 47*(19), 1836–1876. DOI: 10.1080/10643389.2017.1393263

Bidu, J. M. (2024). Sequential anaerobic–intermittently aerated treatment of textile and domestic wastewater. *Water Practice & Technology, 19*(12), 4875–4888. DOI: 10.2166/wpt.2024.283

Bidu, J. M. (2025). Integrated anaerobic reactor and intermittently aerated constructed wetland for enhanced textile wastewater treatment. *Cleaner Water, 4*(November 2024). DOI: 10.1016/j.clwat.2025.100093

Bidu, J. M., Bruggen, B. Van der, Rwiza, M. J., & Njau, N. K. (2021). *Current status of textile wastewater management practices and ef fl uent characteristics in Tanzania J. M. Bidu, B. Van der Bruggen, M. J. Rwiza and K. N. Njau*. 2363–2376. DOI: 10.2166/wst.2021.133

Blánquez, P., Sarrà, M., & Vicent, T. (2008). Development of a continuous process to adapt the textile wastewater treatment by fungi to industrial conditions. *Process Biochemistry (Barking, London, England), 43*(1), 1–7. DOI: 10.1016/j.procbio.2007.10.002

Carney, A. B., Cartine, J., Jönander, C., Karlsson, M., Langlois, J., Lindström, M., Lundin, J., Melander, N., Pesqueda, A., Rahmqvist, I., Renaux, J., Roos, J., Spilsbury, F., Svalin, J., Vestlund, H., Zhao, L., Asker, N., Ašmonaitė, G., Birgersson, L., … Sturve, J. (2021). Assessing the effects of textile leachates in fish using multiple testing methods: From gene expression to behavior. *Ecotoxicology and Environmental Safety, 207*(July 2020). DOI: 10.1016/j.ecoenv.2020.111523

Carrascal-Hernández, D. C., Orozco-Beltrán, E. J., Insuasty, D., Márquez, E., & Grande-Tovar, C. D. (2025). Systematic Evaluation of Biodegradation of Azo Dyes by Microorganisms: Efficient Species, Physicochemical Factors, and Enzymatic Systems. *International Journal of Molecular Sciences, 26*(16), 1–37. DOI: 10.3390/ijms26167973 PMID: 40869295

Ceretta, M. B., Nercessian, D., & Wolski, E. A. (2021). Current Trends on Role of Biological Treatment in Integrated Treatment Technologies of Textile Wastewater. *Frontiers in Microbiology, 12*(March), 1–7. DOI: 10.3389/fmicb.2021.651025 PMID: 33841377

Chakraborty, S., Talukdar, A., Dey, S., & Bhattacharya, S. (2025). Role of fungi, bacteria and microalgae in bioremediation of emerging pollutants with special reference to pesticides, heavy metals and pharmaceuticals. In *Discover Environment* (Vol. 3, Number 1). Springer International Publishing. DOI: 10.1007/s44274-025-00217-7

Chandanshive, V. V., Kadam, S. K., Khandare, R. V., Kurade, M. B., Jeon, B. H., Jadhav, J. P., & Govindwar, S. P. (2018). In situ phytoremediation of dyes from textile wastewater using garden ornamental plants, effect on soil quality and plant growth. *Chemosphere, 210*, 968–976. DOI: 10.1016/j.chemosphere.2018.07.064 PMID: 30208557

Chaudhari, A. U., Tapase, S. R., Markad, V. L., & Kodam, K. M. (2013). Simultaneous decolorization of reactive Orange M2R dye and reduction of chromate by Lysinibacillus sp. KMK-A. *Journal of Hazardous Materials, 262*, 580–588. https://doi.org/dx.doi.org/10.1016/j.jhazmat.2013.09.006. DOI: 10.1016/j.jhazmat.2013.09.006 PMID: 24095998

Danouche, M., El Arroussi, H., Bahafid, W., & El Ghachtouli, N. (2021). An overview of the biosorption mechanism for the bioremediation of synthetic dyes using yeast cells. *Environmental Technology Reviews*, *10*(1), 58–76. DOI: 10.1080/21622515.2020.1869839

Das, S., Cherwoo, L., & Singh, R. (2023). Decoding dye degradation: Microbial remediation of textile industry effluents. *Biotechnology Notes*, *4*(August), 64–76. DOI: 10.1016/j.biotno.2023.10.001 PMID: 39416919

Dawkar, V. V., Jadhav, Æ. U. U., Ghodake, G. S., & Govindwar, Æ. S. P. (2009). Effect of inducers on the decolorization and biodegradation of textile azo dye Navy blue 2GL by *Bacillus sp.* VUS. *Biodegradation*, *20*(6), 777–787. DOI: 10.1007/s10532-009-9266-y PMID: 19468842

Dixit, S., & Garg, S. (2018). Biodegradation of Environmentally Hazardous Azo Dyes and Aromatic Amines Using *Klebsiella pneumoniae. Journal of Environmental Engineering*, *144*(6), 04018035. Advance online publication. DOI: 10.1061/(ASCE)EE.1943-7870.0001353

Donkadokula, N. Y., Kola, A. K., Naz, I., & Saroj, D. (2020). A review on advanced physico-chemical and biological textile dye wastewater treatment techniques. *Reviews in Environmental Science and Biotechnology*, *19*(3), 543–560. DOI: 10.1007/s11157-020-09543-z

El Awady, M. E., El-Shall, F. N., Mohamed, G. E., Abd-Elaziz, A. M., Abdel-Monem, M. O., & Hassan, M. G. (2024). Exploring the decolorization efficiency and biodegradation mechanisms of different functional textile azo dyes by *Streptomyces albidoflavus* 3MGH. *BMC Microbiology*, *24*(1), 1–22. DOI: 10.1186/s12866-024-03347-9 PMID: 38877404

Ertu rul, S., San, N. O., & Dönmez, G. (2009). Treatment of dye (Remazol Blue) and heavy metals using yeast cells with the purpose of managing polluted textile wastewaters. *Ecological Engineering*, *35*(1), 128–134. DOI: 10.1016/j.ecoleng.2008.09.015

Floros, M., Hojabri, L., Abraham, E., Jose, J., Thomas, S., Pothan, L., Leao, A. L., & Narine, S. (2012). Enhancement of thermal stability, strength and extensibility of lipid-based polyurethanes with cellulose-based nanofibers. *Polymer Degradation & Stability*, *97*(10), 1970–1978. DOI: 10.1016/j.polymdegradstab.2012.02.016

Gonçalves, I. M. C., Gomes, A., Brás, R., Ferra, M. I. A., Amorim, M. T. P., & Porter, R. S. (2000). Biological treatment of effluent containing textile dyes. *Coloration Technology*, *116*(12), 393–397. DOI: 10.1111/j.1478-4408.2000.tb00016.x

Guadie, A., Gessesse, A., & Xia, S. (2018). Halomonas sp. strain A55, a novel dye decolorizing bacterium from dye-uncontaminated Rift Valley Soda lake. *Chemosphere, 206*, 59–69. DOI: 10.1016/j.chemosphere.2018.04.134 PMID: 29730566

Holkar, C. R., Jadhav, A. J., Pinjari, D. V., Mahamuni, N. M., & Pandit, A. B. (2016). A critical review on textile wastewater treatments: Possible approaches. *Journal of Environmental Management, 182*, 351–366. DOI: 10.1016/j.jenvman.2016.07.090 PMID: 27497312

Hossain, L., Sarker, S. K., & Khan, M. S. (2018). Evaluation of present and future wastewater impacts of textile dyeing industries in Bangladesh. *Environmental Development, 26*, 23–33. DOI: 10.1016/j.envdev.2018.03.005

Hossain, M. S., Sarker, P., Rahaman, M. S., Ahmed, F. F., Shaibur, M. R., & Khabir Uddin, M. (2022). Biological Treatment of Textile Wastewater by Total Aerobic Mixed Bacteria and Comparison with Chemical Fenton Process. *Pollution, 8*(4), 1418–1433. DOI: 10.22059/POLL.2022.340753.1408

Hossen, Z., Akhter, S., Rahman, T., & Khatun, M. (2020). Bacterial Degradation of Synthetic Dye by *Pseudomonas sp.* Obtained from Dyeing Mill Effluent. *American Journal of Pure and Applied Biosciences, 2(6), 192-199, 2020 Publisher, 2*(6), 192–199. https://doi.org/doi.org/10.34104/ajpab.020.01920199

Huang, J., Fu, Y., & Liu, Y. (2014). Comparison of Alkali-Tolerant Fungus *Myrothecium Sp.* IMER1 and White-Rot Fungi for Decolorization of Textile Dyes and Dye Effluents Bioremediation & Biodegradation. *Journal of Bioremediation & Biodegradation, 5*(3). Advance online publication. DOI: 10.4172/2155-6199.1000221

Islam, M., Aidid, A. R., Mohshin, J. N., Mondal, H., Ganguli, S., & Chakraborty, A. K. (2025). A critical review on textile dye-containing wastewater: Ecotoxicity, health risks, and remediation strategies for environmental safety. *Cleaner Chemical Engineering, 11*(January), 100165. DOI: 10.1016/j.clce.2025.100165

Islam, M. R., & Mostafa, M. G. (2018). Textile Dyeing Effluents and Environment Concerns -. *RE:view, 11*, 131–144.

Jahan, N., Tahmid, M., Shoronika, A. Z., Fariha, A., Roy, H., Pervez, M., Cai, Y., Naddeo, V., & Islam, M. S. (2022). A Comprehensive Review on the Sustainable Treatment of Textile Wastewater: Zero Liquid Discharge and Resource Recovery Perspectives. *Sustainability (Basel), 14*(22), 15398. DOI: 10.3390/su142215398

Jaishankar, M., Tseten, T., Anbalagan, N., Mathew, B. B., & Beeregowda, K. N. (2014). Toxicity, mechanism and health effects of some heavy metals. *Interdisciplinary Toxicology, 7*(2), 60–72. DOI: 10.2478/intox-2014-0009 PMID: 26109881

Jallouli, S., Buonerba, A., Borea, L., Hasan, S. W., Belgiorno, V., Ksibi, M., & Naddeo, V. (2023). Living membrane bioreactor for highly effective and eco-friendly treatment of textile wastewater. *The Science of the Total Environment*, *871*(January), 161963. DOI: 10.1016/j.scitotenv.2023.161963 PMID: 36737022

Jamee, R., & Siddique, R. (2019). Biodegradation of synthetic dyes of textile effluent by microorganisms: An environmentally and economically sustainable approach. *European Journal of Microbiology & Immunology*, *9*(4), 114–118. DOI: 10.1556/1886.2019.00018 PMID: 31934362

Jorge, A. M. S., Athira, K. K., Alves, M. B., Gardas, R. L., & Pereira, J. F. B. (2023). Textile dyes effluents: A current scenario and the use of aqueous biphasic systems for the recovery of dyes. *Journal of Water Process Engineering*, *55*, 104125. DOI: 10.1016/j.jwpe.2023.104125

Kallawar, G. A., & Bhanvase, B. A. (2024). A review on existing and emerging approaches for textile wastewater treatments: Challenges and future perspectives. *Environmental Science and Pollution Research International*, *31*(2), 1748–1789. DOI: 10.1007/s11356-023-31175-3 PMID: 38055170

Karthika, A., Seenivasagan, R., Kasimani, R., Sudhakara Rao, J., & Poonkuzhali, K. (2024). Microbial technologies for sustainable textile effluent treatment: A review. *Journal of Environmental Chemical Engineering*, *12*(5), 113275. DOI: 10.1016/j.jece.2024.113275

Kavitha, G., Govindhan, M., & Premkumar, S. (2025). Dye pollution and its implications for human health, aquatic ecosystems, and sustainable wastewater treatment: A comprehensive review. *Journal of Water Process Engineering*, *80*, 109071. DOI: 10.1016/j.jwpe.2025.109071

Khan, A., Nayarisseri, A., & Singh, S. K. (2025). Characterization and optimization of azo dyes degrading microbes isolated from textile effluent. *Scientific Reports*, *15*(1), 1–23. DOI: 10.1038/s41598-025-95359-2 PMID: 40175441

Khan, A. A., Gul, J., Naqvi, S. R., Ali, I., Farooq, W., Liaqat, R., AlMohamadi, H., Štěpanec, L., & Juchelková, D. (2022). Recent progress in microalgae-derived biochar for the treatment of textile industry wastewater. *Chemosphere*, *306*(July), 135565. Advance online publication. DOI: 10.1016/j.chemosphere.2022.135565 PMID: 35793745

Khan, M. D., Abdulateif, H., Ismail, I. M., Sabir, S., & Khan, M. Z. (2015). Bioelectricity Generation and Bioremediation of an Azo-Dye in a Microbial Fuel Cell Coupled Activated Sludge Process. *PLoS One*, *10*(10), e0138448. DOI: 10.1371/journal.pone.0138448 PMID: 26496083

Khan, M. F. (2025). Recent Advances in Microbial Enzyme Applications for Sustainable Textile Processing and Waste Management. *Sci*, *7*(2), 46. Advance online publication. DOI: 10.3390/sci7020046

Khattab, T. A., Abdelrahman, M. S., & Rehan, M. (2020). Textile dyeing industry: Environmental impacts and remediation. *Environmental Science and Pollution Research International*, *27*(4), 3803–3818. DOI: 10.1007/s11356-019-07137-z PMID: 31838699

Kocaman, S. (2025). Impacts of dye-contaminated effluents on the environment. In *Engineered Biocomposites for Dye Adsorption* (Number 2022, pp. 11–22). INC. DOI: 10.1016/B978-0-443-29877-6.00002-0

Krishnan, J., Arvind Kishore, A., Suresh, A., Madhumeetha, B., & Gnana Prakash, D. (2017). Effect of pH, inoculum dose and initial dye concentration on the removal of azo dye mixture under aerobic conditions. *International Biodeterioration & Biodegradation*, *119*, 16–27. DOI: 10.1016/j.ibiod.2016.11.024

Kulkarni, A. N., Watharkar, A. D., Rane, N. R., & Jeon, B. (2018). Ecotoxicology and Environmental Safety Decolorization and detoxification of dye mixture and textile effluent by lichen *Dermatocarpon vellereceum* in fixed-bed upflow bioreactor with subsequent oxidative stress study. *Ecotoxicology and Environmental Safety*, *148*(September 2017), 17–25. DOI: 10.1016/j.ecoenv.2017.10.001

Kumar, M., Mishra, A., Patel, S. K., Kushwaha, J., Singh, S., Mishra, V., Singh, D., Singh, V., Giri, B. S., Singhania, R. R., & Singh, D. (2025). Environmental Impacts and Strategies for Bioremediation of Dye-Containing Wastewater. *Bioengineering (Basel, Switzerland)*, *12*(10), 1043. DOI: 10.3390/bioengineering12101043 PMID: 41155042

Kurade, M. B., Waghmode, T. R., Patil, S. M., Jeon, B. H., & Govindwar, S. P. (2017). Monitoring the gradual biodegradation of dyes in a simulated textile effluent and development of a novel triple layered fixed bed reactor using a bacterium-yeast consortium. *Chemical Engineering Journal*, *307*, 1026–1036. DOI: 10.1016/j.cej.2016.09.028

Kusumlata, A., Ambade, B., Kumar, A., & Gautam, S. (2024). Sustainable Solutions: Reviewing the Future of Textile Dye Contaminant Removal with Emerging Biological Treatments. *Limnological Review*, *24*(2), 126–149. DOI: 10.3390/limnolrev24020007

Lanfranconi, I., Ceretta, M. B., Bertola, N., Wolski, E. A., & Durruty, I. (2022). Textile dyeing wastewater treatment by *Penicillium chrysogenum*: Design of a sustainable process. *Water Science and Technology*, *86*(2), 292–301. DOI: 10.2166/wst.2022.204 PMID: 35906908

Levin, L., Grassi, E., & Carballo, R. (2012). International Biodeterioration & Biodegradation Efficient azoic dye degradation by *Trametes trogii* and a novel strategy to evaluate products released. *International Biodeterioration & Biodegradation*, *75*, 214–222. DOI: 10.1016/j.ibiod.2012.10.005

Lúcia, A., Carissimi, E., Dotto, G. L., & Feris, L. A. (2018). *Biosorption of rhodamine B dye from dyeing stones effluents using the green microalgae Chlorella pyrenoidosa*. Cleaner Production., DOI: 10.1016/j.jclepro.2018.07.128

Madhav, S., Ahamad, A., Singh, P., & Mishra, P. K. (2018). A review of textile industry: Wet processing, environmental impacts, and effluent treatment methods. *Environmental Quality Management*, *27*(3), 31–41. DOI: 10.1002/tqem.21538

Mahajan, P., & Kaushal, J. (2020). Phytoremediation of azo dye methyl red by macroalgae *Chara vulgaris L:* kinetic and equilibrium studies. *Environmental Science and Pollution Research.*

Mani, M., Pachaiappan, R., Mahendra Gowda, R. V., & Aroulmoji, V. (2023). Environmental and Health Effects of Emerging Contaminants – A Critical Review. *International Journal of Advanced Science and Engineering*, *10*(2), 3449–3470. DOI: 10.29294/IJASE.10.2.2023.3449-3470

Mendes, M., Cassoni, A. C., Alves, S., Moreira, P., Castro, P. M., & Pintado, M. E. (2023). Assessment of drying conditions for storage of a yeast-based decolorization solution for application in textile industrial wastewater treatment plants. *Fungal Biology*, *127*(7–8), 1111–1117. DOI: 10.1016/j.funbio.2023.06.002 PMID: 37495301

Mendes, M., Moreira, I. S., Moreira, P., Pintado, M., & Castro, P. M. L. (2023). Bioaugmentation of Aerobic Granular Sludge with Dye-Decolorizing Yeast for Textile Industrial Wastewater. *Processes (Basel, Switzerland)*, *11*(6), 1–16. DOI: 10.3390/pr11061654

Mishra, S., & Maiti, A. (2018). The efficacy of bacterial species to decolourise reactive azo, anthroquinone and triphenylmethane dyes from wastewater: A review. *Environmental Science and Pollution Research International*, *25*(9), 8286–8314. DOI: 10.1007/s11356-018-1273-2 PMID: 29383646

Mohanty, S. S., & Kumar, A. (2025). Microbial decolorization of anthraquinone dyes: Batch and continuous treatment systems - a mini-review. *Frontiers in Environmental Engineering*, *4*(June), 1–21. DOI: 10.3389/fenve.2025.1553712

Moyo, S., Makhanya, B. P., & Zwane, P. E. (2022). Use of bacterial isolates in the treatment of textile dye wastewater: A review. *Heliyon*, *8*(6), e09632. DOI: 10.1016/j.heliyon.2022.e09632 PMID: 35677403

Mudhoo, A., & Kumar, S. (2013). Effects of heavy metals as stress factors on anaerobic digestion processes and biogas production from biomass. *International Journal of Environmental Science and Technology*, *10*(6), 1383–1398. DOI: 10.1007/s13762-012-0167-y

Nadu, T.Sanmathi K R. (2023). Biodegradation on the Removal of Dyes from Textile Effluent. *International Journal of Advanced Research in Science. Tongxin Jishu*, *4–8*, 4–8. Advance online publication. DOI: 10.48175/IJARSCT-13102

Nazlı, S. B., & Karatay, S. E. (2024). Evaluation of Food Wastes in Chlorella vulgaris Cultivation for Remazol Brilliant Blue R Biosorption. *Aquatic Sciences and Engineering*, *39*(1), 1–7. DOI: 10.26650/ASE20241305101

Neifar, M., Chouchane, H., Mahjoubi, M., Jaouani, A., & Cherif, A. (2020). *Pseudomonas extremorientalis* BU118 : A new salt-tolerant laccase-secreting bacterium with biotechnological potential in textile azo dye decolourization. *Biotech*, *6*(1), 1–9. DOI: 10.1007/s13205-016-0425-7 PMID: 35822774

Nicula, N. O., Lungulescu, E. M., Rîmbu, G. A., Marinescu, V., Corbu, V. M., & Csutak, O. (2023). Bioremediation of Wastewater Using Yeast Strains: An Assessment of Contaminant Removal Efficiency. *International Journal of Environmental Research and Public Health*, *20*(6), 4795. Advance online publication. DOI: 10.3390/ijerph20064795 PMID: 36981703

Nirmala, A., Soundariya, S., & Praveena, A. R. (2021). *On the Phytotoxicity of Waste-Water from Textile Industry on Selected Crops Seed Germination and its Treatment Using Bacteria and Zinc Oxide Nanoparticles. 14*(December), 1564–1570.

Olukanni, O. D., Osuntoki, A. A., Kalyani, D. C., Gbenle, G. O., & Govindwar, S. P. (2010). Decolorization and biodegradation of Reactive Blue 13 by *Proteus mirabilis* LAG. *Journal of Hazardous Materials*, *184*(1–3), 290–298. DOI: 10.1016/j.jhazmat.2010.08.035 PMID: 20832936

Pandey, K., Singh, B., Pandey, A. K., Badruddin, I. J., Pandey, S., Mishra, V. K., & Jain, P. A. (2017). Application of Microbial Enzymes in Industrial Waste Water Treatment. *International Journal of Current Microbiology and Applied Sciences*, *6*(8), 1243–1254. DOI: 10.20546/ijcmas.2017.608.151

Panhwar, A., Sattar Jatoi, A., Ali Mazari, S., Kandhro, A., Rashid, U., & Qaisar, S. (2024). Water resources contamination and health hazards by textile industry effluent and glance at treatment techniques: A review. *Waste Management Bulletin*, *1*(4), 158–163. DOI: 10.1016/j.wmb.2023.09.002

Parshetti, G. K., Telke, A. A., Kalyani, D. C., & Govindwar, S. P. (2010). Decolorization and detoxification of sulfonated azo dye methyl orange by *Kocuria rosea* MTCC 1532. *Journal of Hazardous Materials*, *176*(1-3), 503–509. DOI: 10.1016/j.jhazmat.2009.11.058 PMID: 19969416

Pearce, C. I., Lloyd, J. R., & Guthrie, J. T. (2003). The removal of colour from textile wastewater using whole bacterial cells: A review. *Dyes and Pigments*, *58*(3), 179–196. DOI: 10.1016/S0143-7208(03)00064-0

Periyasamy, A. P. (2025). A review of bioremediation of textile dye containing wastewater. *Cleaner Water*, *4*(June), 100092. DOI: 10.1016/j.clwat.2025.100092

Pundir, A., Thakur, M. S., Prakash, S., Kumari, N., Sharma, N., Parameswari, E., He, Z., Nam, S., Thakur, M., Puri, S., Puranik, S., Kumar, S., Madhu, , & Kumar, M. (2024). Fungi as versatile biocatalytic tool for treatment of textile wastewater effluents. *Environmental Sciences Europe*, *36*(1), 185. Advance online publication. DOI: 10.1186/s12302-024-01007-3

Punzi, M. (2015). *Treatment of textile wastewater by combining biological processes and advanced oxidation* [Lund University]. https://portal.research.lu.se/portal/files/6076976/5367871.pdf

Qiu, W., Zheng, M., Sun, J., Tian, Y., Fang, M., Zheng, Y., Zhang, T., & Zheng, C. (2019). Photolysis of enrofloxacin, pefloxacin and sulfaquinoxaline in aqueous solution by UV/H2O2, UV/Fe(II), and UV/H2O2/Fe(II) and the toxicity of the final reaction solutions on zebrafish embryos. *The Science of the Total Environment*, *651*, 1457–1468. DOI: 10.1016/j.scitotenv.2018.09.315 PMID: 30360275

Qu, Y., Shi, S., Ma, F., & Yan, B. (2010). Bioresource Technology Decolorization of Reactive Dark Blue K-R by the synergism of fungus and bacterium using response surface methodology. *Bioresource Technology*, *101*(21), 8016–8023. DOI: 10.1016/j.biortech.2010.05.025 PMID: 20566285

Ramamurthy, K., Priya, P. S., Murugan, R., & Arockiaraj, J. (2024). Hues of risk: Investigating genotoxicity and environmental impacts of azo textile dyes. *Environmental Science and Pollution Research International, 31*(23), 33190–33211. DOI: 10.1007/s11356-024-33444-1 PMID: 38676865

Rane, A., & Joshi, S. J. (2021). Biodecolorization and Biodegradation of Dyes: A Review. *The Open Biotechnology Journal, 15*(1), 97–108. DOI: 10.2174/1874070702115010097

Rathour, R. K., Sharma, D., Ullah, S., Mahmoud, E.-H. M., Sharma, N., Kumar, P., Bhatt, A. K., Ahmad, I., & Bhatia, R. K. (2024). Bacterial–microalgal consortia for bioremediation of textile industry wastewater and resource recovery for circular economy. *Biotechnology for the Environment, 1*(1), 1–21. DOI: 10.1186/s44314-024-00005-2

Riza, M., Ehsan, M. N., & Hoque, S. (2021). *Portrayal of Textile Based Pollutants and its Impact on Soil, Plants and Fisheries. 20.*

Rizwana, M., Darshan, M., & Nilesh, D. (2014). Phytoremediation of Textile Waste Water Using Potential Wetland Plant: Eco Sustainable Approach. *International Journal of Interdisciplinary and Multidisciplinary Studies, 1*(4), 130–138. http://www.ijims.com

Rodriguez-couto, S. (2013). Treatment of Textile Wastewater by White-rot Fungi: Still a Far Away Reality? *Textiles and Light Industrial Science and Technology, 2*(3), 113–119.

Samuchiwal, S., Gola, D., & Malik, A. (2021). Decolourization of textile effluent using native microbial consortium enriched from textile industry effluent. *Journal of Hazardous Materials, 402*(September 2020), 123835. DOI: 10.1016/j.jhazmat.2020.123835

Sani, Z. M., Dalhatu, A. S., Adam, B. S., Mohammed, K., Muhammad, Y. Y., & Ibrahim, S. (2021). Bioremediation of Some Reactive Dyes Commonly used in Fabric Re-dyeing by *Chlorella vulgaris. Asian Journal of Environment & Ecology, 15*(December 2019), 10–19. DOI: 10.9734/ajee/2021/v15i430233

Saratale, R. G., Saratale, G. D., Chang, J. S., & Govindwar, S. P. (2010). Decolorization and biodegradation of reactive dyes and dye wastewater by a developed bacterial consortium. *Biodegradation, 21*(6), 999–1015. DOI: 10.1007/s10532-010-9360-1 PMID: 20407917

Saratale, R. G., Saratale, G. D., Chang, J. S., & Govindwar, S. P. (2011). Bacterial decolorization and degradation of azo dyes: A review. *Journal of the Taiwan Institute of Chemical Engineers, 42*(1), 138–157. DOI: 10.1016/j.jtice.2010.06.006

Scholz, D. A. Y. M. (2019). Textile dye wastewater characteristics and constituents of synthetic effluents: a critical review. In *International Journal of Environmental Science and Technology* (Vol. 16, Number 2). Springer Berlin Heidelberg. DOI: 10.1007/s13762-018-2130-z

Sharma, R., & Malaviya, P. (2022). Constructed wetlands for textile wastewater remediation: A review on concept, pollutant removal mechanisms, and integrated technologies for efficiency enhancement. *Chemosphere, 290*(December 2021), 133358. DOI: 10.1016/j.chemosphere.2021.133358

Shoukat, R., Khan, S. J., & Jamal, Y. (2019). Hybrid anaerobic-aerobic biological treatment for real textile wastewater. *Journal of Water Process Engineering, 29*(July 2018), 100804. DOI: 10.1016/j.jwpe.2019.100804

Singh, J. S. (2015). Microbes: The chief ecological engineers in reinstating equilibrium in degraded ecosystems. *Agriculture, Ecosystems & Environment, 203*, 80–82. DOI: 10.1016/j.agee.2015.01.026

Singh, R., Kumar, M., Mittal, A., & Mehta, P. K. (2016). Microbial enzymes: industrial progress in 21st century. *3 Biotech, 6*(2), 1–15. DOI: 10.1007/s13205-016-0485-8

Singh, S. N. (2015). Microbial degradation of synthetic dyes in wastewaters. *Environmental Science and Engineering (Subseries: Environmental Science)*, (9783319109411), iii–iv. DOI: 10.1007/978-3-319-10942-8

Solanki, K., Subramanian, S., & Basu, S. (2013). Microbial fuel cells for azo dye treatment with electricity generation: A review. *Bioresource Technology, 131*, 564–571. DOI: 10.1016/j.biortech.2012.12.063 PMID: 23403060

Srinivasan, S., Bankole, P. O., & Sadasivam, S. K. (2022). Biodecolorization and degradation of textile azo dyes using *Lysinibacillus sphaericus* MTCC 9523. *Frontiers in Environmental Science, 10*(October), 1–13. DOI: 10.3389/fenvs.2022.990855

Tan, L., He, M., Song, L., Fu, X., & Shi, S. (2016). Bioresource Technology Aerobic decolorization, degradation and detoxification of azo dyes by a newly isolated salt-tolerant yeast *Scheffersomyces spartinae* TLHS-SF1. *Bioresource Technology, 203*, 287–294. DOI: 10.1016/j.biortech.2015.12.058 PMID: 26744802

Tan, L., Li, H., Ning, S., & Xu, B. (2014). Bioresource Technology Aerobic decolorization and degradation of azo dyes by suspended growing cells and immobilized cells of a newly isolated yeast *Magnusiomyces ingens* LH-F1. *Bioresource Technology, 158*, 321–328. DOI: 10.1016/j.biortech.2014.02.063 PMID: 24632410

Tigini, V., Giansanti, P., Mangiavillano, A., Pannocchia, A., & Varese, G. C. (2011). Evaluation of toxicity, genotoxicity and environmental risk of simulated textile and tannery wastewaters with a battery of biotests. *Ecotoxicology and Environmental Safety, 74*(4), 866–873. DOI: 10.1016/j.ecoenv.2010.12.001 PMID: 21176963

Tripathi, M., Pandey, R., & Kumar, S. (2018). Biodecolorization of Orange II dye by native *Bacillus sp* and *Staphylococcus sp* in simulated medium. *Journal of Pharmacognosy and Phytochemistry*, •••, 1366–1368.

Tyagi, S., Kapoor, R. T., Singh, R., & Shah, M. P. (2025). Insights on microbial enzymes mediated biodegradation of azo dyes: A sustainable strategy for environment clean up. *Bioremediation Journal, 0*(0), 1–43. DOI: 10.1080/10889868.2025.2498695

Uddin, M., Islam, M. K., & Dev, S. (2024). Investigation of the performance of the combined moving bed bioreactor-membrane bioreactor (MBBR-MBR) for textile wastewater treatment. *Heliyon, 10*(10), e31358. DOI: 10.1016/j.heliyon.2024.e31358 PMID: 38813235

Unal, Y. E., Gundogdu, H., Bilgin, O. N., Yokus, S., Turk, O. K., Yazici Guvenc, S., Can-Güven, E., & Varank, G. (2024). Effective treatment of textile industry wastewater by combined ultraviolet assisted and Fe2+ catalyzed percarbonate oxidation. *Journal of Dispersion Science and Technology, 45*(8), 1585–1598. DOI: 10.1080/01932691.2023.2222823

Vijayaraghavan, G., & Shanthakumar, S. (2020). Removal of crystal violet dye in textile effluent by coagulation using algal alginate from brown algae *Sargassum sp. Desalination and Water Treatment, 196*(September 2019), 402–408. DOI: 10.5004/dwt.2020.25569

Waghmode, T. R., Kurade, M. B., & Govindwar, S. P. (2011). International Biodeterioration & Biodegradation Time dependent degradation of mixture of structurally different azo and non azo dyes by using *Galactomyces geotrichum* MTCC 1360. *International Biodeterioration & Biodegradation, 65*(3), 479–486. DOI: 10.1016/j.ibiod.2011.01.010

Wang, X., Jiang, J., & Gao, W. (2022). *Reviewing textile wastewater produced by industries: characteristics, environmental impacts, and treatment strategies. 85*(7), 2076–2096. DOI: 10.2166/wst.2022.088

Wang, Y., Wang, H., Wang, X., Xiao, Y., Zhou, Y., Su, X., Cai, J., & Sun, F. (2020). Resuscitation, isolation and immobilization of bacterial species for efficient textile wastewater treatment: A critical review and update. *The Science of the Total Environment, 730*, 139034. DOI: 10.1016/j.scitotenv.2020.139034 PMID: 32416505

Yang, Q., Li, C., Li, H., Li, Y., & Yu, N. (2009). Degradation of synthetic reactive azo dyes and treatment of textile wastewater by a fungi consortium reactor. *Biochemical Engineering Journal, 43*(3), 225–230. DOI: 10.1016/j.bej.2008.10.002

Yang, X., López-Grimau, V., Vilaseca, M., & Crespi, M. (2020). Treatment of Textile Wastewater by CAS, MBR, and MBBR: A Comparative Study from Technical, Economic, and Environmental Perspectives. *Water (Basel), 12*(5), 1306. DOI: 10.3390/w12051306

Zhuang, M., Sanganyado, E., Zhang, X., Xu, L., Zhu, J., Liu, W., & Song, H. (2020). Azo dye degrading bacteria tolerant to extreme conditions inhabit nearshore ecosystems: Optimization and degradation pathways. *Journal of Environmental Management, 261*(October 2019), 110222. DOI: 10.1016/j.jenvman.2020.110222

Chapter 7

Microbial Strategies for Industrial Waste Bioremediation: Insights from Mining, Paint, and Detergent Industries

Muibat Omotola Fashola
http://orcid.org/0000-0001-5233-0042
Lagos State University, Nigeria

Olajide Solomon Anagun
http://orcid.org/0000-0002-5492-2516
Lagos State University, Nigeria

Timileyin Emmanuel Olufisoye
http://orcid.org/0009-0003-6101-3198
Lagos State University, Nigeria

Olorunjuwon Omolaja Bello
University of Medical Sciences, Ondo, Nigeria

Kehinde Oreoluwa Adebiyi
http://orcid.org/0000-0002-6172-4075
Indiana University, Bloomington, USA

ABSTRACT

Rapid industrialization has accelerated economic growth but also increased hazardous industrial wastes that threaten environmental sustainability and public health. Mining, paint, and detergent industries release pollutants such as heavy

DOI: 10.4018/979-8-2600-2998-5.ch007

metals, dyes, surfactants, volatile organic compounds, and acid-forming substances. Conventional physicochemical treatment methods are often costly, energy-intensive, and may generate secondary pollution, creating the need for eco-friendly alternatives. This chapter examines the characteristics, environmental fate, and health effects of wastes from these industries while highlighting microbial remediation strategies. Techniques including biosorption, bioaccumulation, bioprecipitation, bioleaching, biomineralization, bioaugmentation, and biostimulation are discussed for heavy metal removal and degradation of organic pollutants. Emerging technologies such as microbial fuel cells, nano-bioremediation, and genetic engineering are also explored for improving treatment efficiency and supporting resource recovery.

7.1 INTRODUCTION

Socioeconomic development and the pursuit of better livelihoods are essential to a country's progress. The globalisation of trade, commerce, and tourism is directly linked with the socio-economic status and lifestyle of the people. Industrialisation and urbanisation are the common features of financial development and economic growth. It leads to a dramatic increase in the generation of hazardous industrial wastes (Prempeh, 2024). Modern society is fundamentally reliant on products derived from the mining, paint, and petrochemical sectors, from structural metals and pharmaceuticals to modern plastics and pigments. This industrial activity, however, generates an environmental burden on a scale that is difficult to comprehend (Ameen, 2023). The mining sector is the dominant contributor by sheer volume and has undergone rapid growth in the past two decades. Specifically, global production of mineral fuels, metal ores, and industrial minerals increased by 52% from 11.3 billion tonnes in 2000 to 17.3 billion tonnes in 2020 (Reichl and Schatz, 2022). Some commodities, such as iron ore (151%) and aluminium (166%), substantially outpaced this growth rate during the same period (Reichl and Schatz, 2022). Industrial waste from mining, often laden with heavy metals and acid-generating sulfides (Ojonimi et al., 2021; Nayak, 2023; Moyo, 2024), dwarfs the other industrial sectors in this review. The paint and petrochemical industries are principal sources of a lower volume, but highly toxic, waste stream (Nair et al., 2021; Adebiyi, 2022). These sectors are major contributors to the estimated 400 million tonnes of hazardous waste generated each year, a staggering rate that averages 12.7 tonnes per second and corresponds to approximately 48.8 kg per capita annually (Rosli et al., 2023). Industrial waste from these sectors is not only defined by its mass and volume, but by its complex chemical composition, including volatile organic compounds (VOCs), persistent organic pollutants (POPs), hydrocarbons, dyes, and toxic heavy metal sludge, which pose immediate and direct threats to public health and environmental stability (Aznar-Sánchez et al., 2018; Nair et al., 2021; Keawboonchu et al., 2023; Kumar et al., 2023).

While this technological progress drives economic prosperity and improved global living standards, it concurrently results in contamination that threatens the very ecosystems and populations it claims to serve. Therefore, this review will critically evaluate the literature on the public health impacts, focusing on the environmental and health implications of the major pollutants from the waste generated by these industries.

7.2 MINING INDUSTRY: WASTE GENERATION AND ENVIRONMENTAL IMPACT

7.2.1 Overview of Waste Generation

Global economic growth and a notable increase in population have resulted in a rapid rise in demand for mineral commodities. This surge in mining activity has brought environmental concerns to the forefront, with a growing social demand for the sustainable development of all mining-related activities. A primary challenge is the adequate management of the large quantity of residues, or waste products, generated during each phase of the mining process. This requires combining economic efficiency with demands for environmental sustainability, considering energy requirements, water resources, and human health risks (Aznar-Sánchez et al., 2018).

The selection of mining techniques depends on the morphology of the natural deposit and the type of mineral substance being extracted. Regardless of the method, each step can generate large quantities of unmarketable or uneconomic materials (Wieszczycka, 2018). These wastes may be in the form of solids, tailings, or slurry, with the most common being tailings, waste rock, slag, and tail ends. These wastes are typically maintained in tailing ponds, dams, or tips, in accordance with local legislation (Cebada, 2016). These structures may be classified as inert if they pose no danger, or as "dangerous" when they cause negative effects to the soil, ground and surface water, vegetation, and even the local fauna and population (Aznar-Sánchez et al., 2018).

7.2.2 The Problem of Acid Mine Drainage (AMD)

A primary environmental threat from mining effluent is AMD. The effluent discharge from mining causes serious environmental pollution, such as suspended solids, heavy metals, and salts, which in turn spoils surface and groundwater quality. The generation of AMD occurs at both operating and abandoned polymetallic mine sites that contain heavy metals and sulphate. The process is often biogeochemical: iron/sulphur oxidising bacteria oxidise iron and sulfur compounds present in these pyritic/sulphidic minerals, liberating protons (H^+), sulfur, and other metals, which lowers the pH to below 3 (Fashola et al., 2015; Singh et al., 2020).

This acidic water then begins to dissolve other metal sulfidic minerals, contributing additional metal ions to the solution. Over time, as water evaporates from the surface of these acidic pools, it results in the formation of a yellow-red solid precipitate called iron hydroxide (Wang, 2022). This process is visualised in Figure 1. The resulting leachate is highly acidic, laden with high concentrations of metal, and its storage and disposal become indispensable to curtail long-term consequences (Fashola et al., 2015). Despite containment efforts in tailing ponds, spillage and leakage are common, allowing this leachate to enter nearby aqueous ecosystems and contaminate surface water bodies as shown in Figure 2.

Figure 1. Visual evidence of acid mine drainage and iron precipitation at the Keonjhar iron ore mine tailing facilities, Odisha

(Singh et al., 2020). **(A)** *A panoramic view of a tailing dam at the Keonjhar iron ore mines.* **(B)** *Yellow-red solid precipitates of iron hydroxide formed due to water evaporation from acidic pools.* **(C)** *The flow of highly acidic leach liquor from the tailing impoundment areas into nearby surface water ecosystems.*

Figure 2. Visual documentation of leach liquor migrating from an iron ore tailing dam into surrounding ecosystems

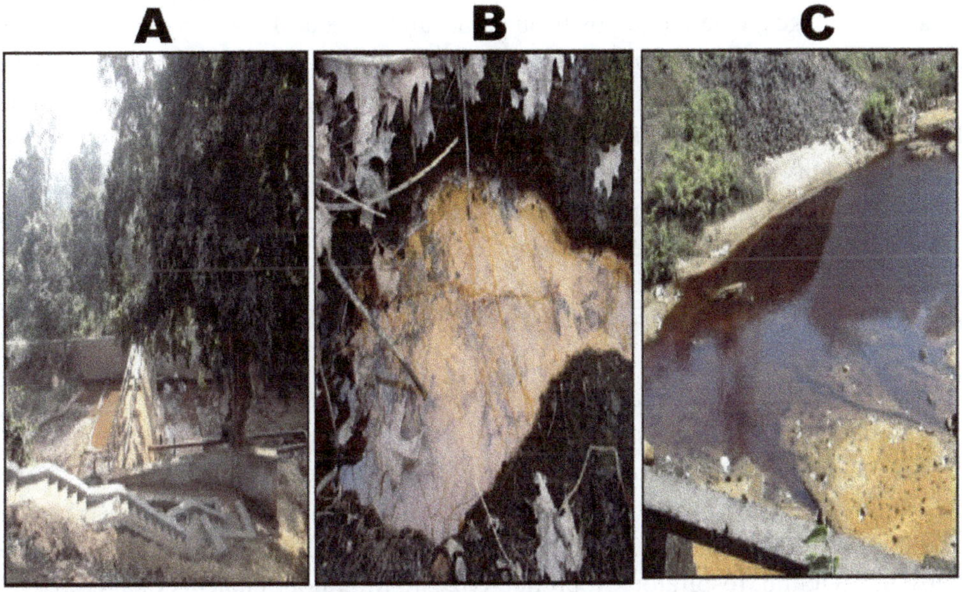

(Singh et al., 2020). **(A)** *Structural pathways and spillage facilitating the flow of leachate away from the primary tailing impoundment area.* **(B)** *Localised precipitation of yellow-red iron hydroxides resulting from the highly acidic, metal-laden runoff interacting with the surrounding soil and organic matter.* **(C)** *The ultimate accumulation of the contaminated, acidic leach liquor in a nearby surface water body, demonstrating severe environmental contamination.*

7.2.3 Impacts on Environmental Quality and Public Health

The impact of mining and subsequent dumping of fine tailings results in the leaching of heavy metals and acidic solution, with severe impacts on the surrounding water ecosystem and local inhabitants (Fashola et al., 2020a).

7.2.4. Ecosystem and Environmental Impacts

The most noticeable impact of mine waste is the sparse or complete absence of flora and fauna in the surrounding areas and water bodies (Moyo, 2024). This is a direct result of the high acidity, elevated salinity, and heavy metal concentrations, coupled with a deficiency in soil nutrients (N and P) and organic matter (Fashola et al., 2020a). The lack of vegetation cover exacerbates soil erosion, which allows

AMD and contaminated sediments to be transported even further downstream (Moyo, 2024).

Fine dust from the waste heaps is raised during summer and transported by wind to nearby villages, where it deposits on plants and objects. This dust can cause skin itching and laboured breathing (Prasad, 2023), while the acidic water contaminates farmlands and water bodies, leading to other skin diseases (Ojonimi et al., 2021). The acidic runoff also makes the water "harder," which leads to slow growth, poor aesthetic quality, and even the gradual death of crops (Shanmukha et al., 2024). High iron deposition on the soil forces plants to uptake excessive amounts, leading to leaf deformation, nutrient deficiencies, and reduced crop productivity (Musah, 2024).

7.2.5. Public Health Impacts and Bioaccumulation

The heavy metals in mine containing wastes are not biodegradable and may be accumulated in living tissues (Moyo, 2024). This process of bioaccumulation occurs when plants use substrates from contaminated soil and water (Nnaji et al., 2023). The heavy metals are then passed up the food chain as animals consume the contaminated plants or water directly (Singh et al., 2022). Bioaccumulation can cause behavioural, respiratory, reproductive, and migratory changes in animals, or even the annihilation of sensitive species (Rambabu et al., 2020).

Heavy metals enter the human food chain, having serious effects on humans and animals (Rambabu et al., 2020). Health impacts of heavy metals are very significant. For instance, while the EPA sets the standard for permissible limits of iron in drinking water at 0.3 mg/L, high levels from mine waste can cause hemochromatosis in humans (Singh et al., 2020). In the long run, this can lead to liver, heart, and pancreatic damage, as well as increase the chances of diabetes (Singh et al., 2020). Other heavy metals have severe, well-documented repercussions on human health and plant physiology (Rambabu et al., 2020), as summarised in Table 1.

Table 1. Effects of heavy metals on human health and plant physiology.

Heavy metals	Potential repercussions on human health	Potential plant physiological response
Arsenic	Bronchitis, skin and bladder cancer, kidney failure, haemolysis, and bone marrow depression	Growth inhibition, loss of yield and fruit production, food chain poisoning
Cadmium	Renal dysfunction, lung disease, lung cancer	Decrease in seed germination and lipid content
Chromium	Liver/kidney necrosis, skin ulcers	Decreased plant growth through membrane damage, chlorosis and root damage

continued on following page

Table 1. Continued

Heavy metals	Potential repercussions on human health	Potential plant physiological response
Copper	Anaemia, liver and kidney damage	Inhibits photosynthesis and the reproductive process
Lead	Mental retardation in children, developmental delay	Decreased chlorophyll production and plant growth
Manganese	Damage to the central nervous system	-
Mercury	Impaired neurodevelopment, decreases in memory	Decreased photosynthetic activity, water uptake and antioxidant
Nickel	Allergic contact dermatitis, chronic bronchitis, lung and nasal cancer	Decreased seed germination, protein and enzyme production
Zinc	Damage to the nervous membrane	Lowers Ni toxicity and promotes plant growth

(Adapted From: Rambabu et al., 2020).

7.2.6 Mining Waste Management and Resource Recovery

The ecological and health crises caused by heavy metals, particularly toxic ones like, copper, nickel, mercury, cadmium, and lead, justify the urgent need for their recovery and treatment from industrial wastewaters (Fu and Wang, 2011).

Considering the high market price of metals and the environmental impact of mismanagement, resource recovery is an increasingly viable option (Huntington et al., 2022). Since mine tailings have already been part-processed, the recovery of remaining metals can be economical (Wang and Ren, 2014). Several mineral processing and separation technologies are available to recycle these untapped metals (Shirani Faradonbeh et al., 2025). For example, work has been initiated to recover high-grade Fe_2O_3 (pure red iron oxide) from iron ore tailing ponds, which can then be sold as a raw material, allowing the mining industry to offset its processing costs. As shown in Table 2, mining wastewater can contain a variety of materials with significant estimated value (Wang and Ren, 2014).

Table 2. Metals concentration in mining wastewater.

Metal	Mining Wastewater (mg/L)
Aluminum	0.161
Cadmium	0.004
Calcium	548
Cobalt	0.0126

continued on following page

Table 2. Continued

Metal	Mining Wastewater (mg/L)
Copper	0.244
Iron	0.033
Magnesium	29.52
Nickel	0.142
Potassium	34.9
Sodium	100.05
Zinc	0.023
Estimated value (USD)*	61.56

(Adapted From: Wang and Ren, 2014).

This approach aligns with the "3Rs" (Reuse, Reduce and Recycle), which requires that all stakeholders, government, mining industries, and private sectors collaborate for the exploration, extraction, and recovery of this untapped wealth (Mahartin, 2023).

7.3 PAINT INDUSTRY WASTES

7.3.1 Generation and Environmental Impact

The paint industry, a prominent sub-sector of the chemical industry, has witnessed a surge through large-scale development. Despite facing economic downturns, the global coating market reached $106.7 billion in 2011, a figure attributed to increased demand from the construction, industrial, and transportation sectors. This market is particularly robust within the Asia-Pacific region, which accounts for 37% of the total economic growth in the paint industry (Noruzman and Apandi, 2020). Paints and surface coating industries have become major economic drivers today (Khan, 2023). Consequently, a considerable amount of paint is utilised universally for protection against mechanical, chemical, and atmospheric weathering, as well as for decorative purposes and to meet aesthetic requirements driven by urbanisation. Environmental pollution continues to be one of the critical issues of the modern world. The high dependence on industrial products leads to the release of waste, which fuels environmental pollution; this remains the centre of the discussion (Ahmed, 2023). Specifically, the increased demand for paint and coating materials has led to high amounts of industrial wastewater from this sector (Viktoryová et al., 2022).

7.3.2 Paint Composition and Manufacturing Processes

Most paints consist of similar core components: a mixture of pigments (suspended solids) in a vehicle (liquid medium), a volatile solvent, a binder (polymeric material), extenders, and suitable additives (Kakoi et al., 2017; Aniyikaiye et al., 2019). Modern paint formulations are broadly categorised into two matrices: the water-based (architectural/decorative) and the solvent-based (industrial/protective) (Fan et al., 2024). While their dispersion media differ, both share four core components: pigments for colour and opacity, binders (resins) for film formation, solvents as vehicles for viscosity control, and additives such as biocides, defoamers, and plasticisers (Christie and Abel, 2021; Müller and Schackmann, 2023). The production of paint is a complex process involving the dispersion of pigments and additives into a solution of resin and solvent, followed by relatively simple mixing operations (Doroszkowski, 2020). An important step in this process is the initial pigment dispersion, known as grinding. The raw materials used are mainly resins, solvents, pigments, extenders, and fillers (Maile, 2021). Water-based paints are typically composed of water, pigments, resins, extenders, ammonia, dispersants, antifoam additives, polyvinyl alcohol emulsion, and a preservative (Rehan and Usman, 2023). Their manufacturing begins by mixing water, ammonia, and dispersant in a tank, after which ground pigment and extenders are added (Bhavsar and Sardesai, 2022). Conversely, solvent-based paints are dominated by solvents, pigments, extenders, resins, plasticisers, and drying oils (Baij et al., 2020). Their production involves mixing ground pigments, resins, and extenders in a high-speed mixer before adding solvents and plasticisers (Pfaff, 2022). Pigments are central to this process; these natural or synthetic compounds are either organic or inorganic powdered colour substances dissolved in water, alcohol, or oils to colour the substrate. Synthetic organic pigments, such as dioxazine, beta naphthol, azo, and phthalocyanine, are hydrocarbon compounds made from petroleum and chemicals, primarily by exposure to high temperatures or pressure (Kumar et al., 2021).

7.3.3 Waste Generation and Characterisation

The manufacturing process of paint uses a large volume of water and chemical reagents, resulting in the release of large volumes of heavily polluted wastewater (Viktoryová and Hrouzková, 2022). Paint-producing industries use approximately 284–321 million litres of water per day, of which only about 4% is recycled. Shockingly, approximately 70% of this wastewater is discharged untreated into natural river basins. Waste generation occurs primarily during the cleaning of dispersion mills, mixing tanks, and packaging lines. Empirical studies indicate that approximately 80% of the total waste load from this sector originates from equipment washing

(Noruzman and Apandi, 2020; Szarka et al., 2022). Since paint manufacturing is essentially a blending operation, the compounds present in the wastes are the same compounds present in the input material (Ruffino et al., 2023).

Liquid waste is a combination of residue from production and the mixture of materials, containing washing water, oils, chemical dyes, ammonia, and calcium carbonate. Paint industries produce tons of solid waste, such as paper, cardboard, plastic waste, and traces of damaged raw materials. These industries have been showed to release volatile organic compounds and dust into the air (Ahmed, 2023). Water-based paints are characterised by high volumes of water containing suspended organic pigments, polymer binder, latex resins, dyes, extenders, surfactants and cellulosic thickeners (Obidi et al., 2022a). Therefore, effluents generated from their production are prone to microbial contamination, necessitating the use of biocides, which further complicate toxicity (Pandey and Kiran, 2020). In contrast, solvent-based wastes are dominated by Volatile Organic Compounds (VOCs) and semi-volatile compounds, often containing hazardous solvents such as toluene, xylene, and various ketones (Aydemir and Özsoy, 2020).

7.3.4 Wastewater Quality Analysis From Some Paint Industries in Sudan

The following data (Table 3) illustrates the severe non-compliance of paint factory effluents compared to standard instructions.

Table 3. Quality of wastewater generated from some paint factories in Sudan.

Isotropic	pH	Total dissolved solids (mg/L)	Total suspended solids (mg/L)	Biochemical oxygen demand (mg/L)	Chemical oxygen demand (mg/L)	Oil and grease (mg/L)	Lead (mg/L)	Manganese (mg/L)	Iron (mg/L)	Copper (mg/L)	Sulphides (mg/L)
Standard	6.9	2500	400	350	250	20	0.1	0.1	2	-	2.5
Factory 1	7.9	720	-	80	6500	10000	9232	0.1353	0.349	9.86	0.022
Factory 2	7.3	840	240	1250	2200	140	0.1522	0.128	0.465	0.027	472
Factory 3	7.6	140	340	600	1600	220	0.0993	0.144	1.96	0.03	11.6

(Data Adapted from: Ahmed et al., 2023)

Factory 1 represents a hazard, with lead concentrations reaching 9232 mg/L, orders of magnitude above the 0.1 mg/L standard, and oil/grease levels at 10,000 mg/L compared to the limit of 20 mg/L. Meanwhile, Factory 2 exhibits massive sulphide contamination (472 mg/L against a 2.5 mg/L standard). All three factories significantly exceed chemical oxygen demand (COD) and biological oxygen demand (BOD) limits, indicating a high load of organic pollutants that deplete oxygen in receiving waters.

7.3.5 Environmental Toxicity and Ecological Impact

The discharge of untreated paint industry effluents precipitates a cascade of deleterious effects that pose high environmental and health challenges (Yadav and Dutta, 2024). While legacy regulations focus on heavy metals, modern formulations have shifted the pollutant burden toward high concentrations of metallic oxides and minerals, specifically Titanium Dioxide (TiO_2), Iron Oxides (Fe_2O_3), Aluminium (Al), and Mica flakes (Arslan Younas et al., 2024). The most immediate impact of effluents containing TiO_2 and Iron Oxides is a drastic alteration in water column transparency (Kolya and Kang, 2024). TiO_2, used universally as a white pigment, results in effluents with extreme turbidity, which scatter and absorb sunlight, inhibiting photosynthesis in phytoplankton and submerged macrophytes (Haynes, 2020). As primary productivity collapses, dissolved oxygen levels drop, destabilising the food web (Mishra, 2025). Unlike dissolved organic pollutants, mineral components like Mica flakes and coarse pigment particles are chemically inert but physically destructive (Pfaff, 2022). These suspended solids settle out of the water column, forming a sludge blanket over the riverbed in a process known as benthic smothering (Weis, 2024). This destroys the spawning grounds of fish and habitats of benthic macroinvertebrates, while suspended Mica flakes act as abrasive agents, causing physical micro-trauma to gill tissues of aquatic organisms (Weis, 2024). Chemical ecotoxicity further exacerbates these physical impacts. Azo dyes, which constitute the largest class of organic colourants, are characterised by azo bonds (-N=N-) (Alzain et al., 2023). These structures are recalcitrant (resistant to light and water), making them desirable for durability but disastrous for environmental degradation (Hoxha et al., 2019). Untreated effluents with dyes are potentially hazardous to freshwater ecosystems as they block sunlight and disrupt the food web (Kumari, 2024). Solvents such as benzene, toluene, ethylbenzene, and xylene (BTEX) are ubiquitous in solvent-based waste; these are potent neurotoxins and precursors to ground-level ozone (Salihoglu and Salihoglu, 2016). Aluminium presents a complex hazard as it is amphoteric. In mixing zones where alkaline effluent meets neutral river water, aluminium can precipitate onto fish gills, causing ion regulatory failure and respiratory distress (Fensham et al., 2022). This dissolved aluminium is also phytotoxic, inhibiting root growth in aquatic and riparian vegetation (Ahmed et al., 2023).

7.3.6 Public Health Implications

These effects not only negatively impact the environment but also pose a significant threat to human health. The effluents contain diluted paints, azo dyes, heavy metals, and oils, which can irritate the skin and eyes or cause headaches (Khan and Malik, 2013). In addition, the occurrence of components such as phthalocyanines in paint effluents are very toxic. These are pigments that do not degrade easily and decompose only at elevated temperatures. They are very stable up to 410°C and only begin decomposing at about 425°C which is not obtainable in paints (Obidi et al., 2022b)

Toxins present in this wastewater can cause respiratory problems, muscle weakness, and organ damage, such as to the liver or kidneys (Alsukaibi, 2022). Specifically, the primary toxicological risk of azo dyes arises when they enter anaerobic environments, such as the human intestinal tract (Chavan and Kaur, 2024). Reductive cleavage of the azo bond releases aromatic amines, which are established mutagens and carcinogens (Ahmed et al., 2021). Beyond dyes, plasticisers like phthalates are known endocrine disruptors, capable of interfering with hormonal systems in humans (Chavan and Kaur, 2024). Toxic heavy metals discharged into the environment can cause detrimental health problems via the food chain, leading to oxidative stress, nervous system impairments, and reduced growth (Ahmed et al., 2021). Concerns have been raised regarding nano-scale Titanium Dioxide ($nTiO_2$), which can generate reactive oxygen species (ROS) via photocatalysis under UV light, potentially causing oxidative stress in human tissues (Carmen and Daniela, 2012). Consequently, sufficient treatment of paint industry effluents is critical to remove colour and microorganisms and reduce the levels of COD and BOD to mitigate these severe risks (Hrouzková and Szarka, 2021).

7.4 DETERGENT INDUSTRY WASTES

7.4.1 Chemical Characterisation and Effluent Composition

The waste stream produced by the detergent industry forms a complex, high-strength matrix composed of active cleaning agents, builders, additives, and synthesis by-products (Ahmed et al., 2018; Dolganova et al., 2018). Understanding the physicochemical profile of this effluent is essential for assessing its environmental behaviour (Ohanwemadu and Nwakanma, 2025).

7.4.2 Surfactant Composition and Environmental Fate: LAS, NPEs, and Biosurfactants

The organic burden surfactants constitute the primary organic load in detergent effluents. Linear Alkylbenzene Sulfonates (LAS) remain the dominant anionic surfactant globally, accounting for approximately 40% of the market share (Khan, 2022). Chemically, commercial LAS comprises a mixture of homologues with linear alkyl chains ranging from 10 to 14 carbon atoms attached to a sulfonated benzene ring (Larson and Woltering, 2020). While introduced to replace recalcitrant branched alkylbenzene sulfonates (BAS), LAS remains pseudo-persistent in high-load environments.

In contrast, Nonylphenol Ethoxylates (NPEs) represent a significant class of nonionic surfactants. Unlike LAS, NPEs degrade into persistent, lipophilic metabolites such as Nonylphenol (NP), creating a toxic reservoir in aquatic sediments (Bernal-Jácome et al., 2024). The industry has increasingly adopted biosurfactants (e.g., rhamnolipids), which offer superior biodegradability and lower toxicity profiles compared to synthetic counterparts (Farias et al., 2021).

7.4.3 Evolution of Detergent Builders

Phosphate-Zeolite Shift, Sodium Triphosphate (STP) served as the standard builder for sequestering calcium and magnesium ions. However, the direct correlation between phosphate release and the eutrophication of stagnant surface waters necessitated a fundamental industry shift (Schwuger and Smulders, 2020). Modern eco-friendly formulations have largely replaced STP with Zeolite A (Sodium Aluminosilicate), a water-insoluble ion exchanger. A representative "green" detergent formulation consists of 15% Zeolite A, 55% Sodium Sulfate, and 15% LAS (Ahmed et al., 2018). Zeolite A is characterised by a cubic morphology with a mean particle size of 2.47 µm, designed to optimise calcium binding without depositing residues on fabrics. Recently, Hierarchical Sodium P Zeolite (Na-P) has emerged as an alternative, offering a higher calcium binding capacity (328 mg $CaCO_3$/g) due to its unique morphology and high specific surface area (Ahmed et al., 2018; Koohsaryan et al., 2021).

7.4.4 Industrial vs. Laundry Effluent Profiles

Physicochemical characteristics of detergent waste vary significantly by source (Table 4). Manufacturing effluent from production plants acts as an extreme pollutant, with COD levels reaching up to 23,800 mg/L and surfactant concentrations exceeding 3.4 g/L (Mousavi and Khodadoost, 2019). Industrial textile laundry effluent, on the other hand, is characterised by high alkalinity (pH 9.5–10.5), elevated temperature, and a specific "oil and grease" load of 45–120 mg/L derived from soiled textiles (de Santana da Silva et al., 2018).

Table 4. Physicochemical Profile of High-Strength Detergent

Parameter	Detergent Manufacturing Effluent	Industrial Textile Laundry Effluent
Primary Source	Production Line Wash-down	Soiled Textile Cleaning
pH	Highly Variable	Alkaline (9.5 – 10.5)
Chemical Oxygen Demand (COD)	Up to 23,800 mg/L	645 – 1,350 mg/L
Biochemical Oxygen Demand (BOD)	Up to 3,670 mg/L	280 – 450 mg/L
Surfactant Concentration	> 3,400 mg/L	~45 – 120 mg/L (as Oil/Grease)
Total Suspended Solids (TSS)	Up to 2,230 mg/L	350 – 650 mg/L
Key Pollutants	Sulfones, Tetralins, Phosphorus	Soil, Lint, Surfactants, Bleach

(Adapted from: Centurion et al., 2018; de Santana da Silva et al., 2018; Sharghi et al., 2022).

7.4.5 Characterisation of Non-Surfactant Pollutants and Residues

Detergent effluent is frequently co-contaminated with pollutants generated during synthesis or sequestered during cleaning. The sulfonation of alkylbenzenes generates specific high-viscosity by-products, namely sulfones and tetralins, which constitute up to 2 wt % of the final product as "unsulfonated matter" (Dolganova et al., 2018). Untreated wastewater often carries a hazardous burden of heavy metals. Analysis of laundry sediments has quantified significant concentrations of iron (13.22 mg/L), lead (2.14 mg/L), and cadmium (0.05 mg/L), far exceeding international permissible limits (Ohanwemadu and Nwakanma, 2025). These metals accumulate in bottom sediments, creating a long-term toxicological risk. The effluent also typically contains a high sulfate load (60-90 mg/L), which biotransforms into toxic hydrogen sulfide (H_2S) during anaerobic treatment (Centurion et al., 2018; Dolganova et al., 2018; Ohanwemadu and Nwakanma, 2025). Further co-contamination examples are shown in Table 5.

Table 5. Heavy Metal Burden in Detergent Industry Wastewater

Heavy Metal	Concentration (Wet Season) (mg/L)	Concentration (Dry Season) (mg/L)	Permissible Limit (WHO/ NESREA) (mg/L)	Health Implication
Iron (Fe)	13.22	15.28	0.3	Tissue damage, neurodegeneration
Lead (Pb)	1.16	2.14	0.01	Neurotoxicity, growth retardation
Zinc (Zn)	5.39	6.30	3.0	Immune system disruption

Heavy Metal	Concentration (Wet Season) (mg/L)	Concentration (Dry Season) (mg/L)	Permissible Limit (WHO/ NESREA) (mg/L)	Health Implication
Copper (Cu)	3.40	4.49	1.0	Gastrointestinal distress, liver damage
Cadmium (Cd)	0.02	0.05	0.003	Kidney damage, carcinogenicity
Chromium (Cr)	0.07	0.06	0.05	Carcinogenic, liver/ kidney damage

(Adapted from: Ohanwemadu and Nwakanma, 2025)

7.4.6 Interfacial Phenomena and Contaminant Fate

Surfactants accumulate at the air-water interface, lowering surface tension and causing foam formation. This foaming physically blocks oxygen transfer (reaeration), leading to hypoxic conditions in receiving waters (Arora et al., 2023). The fate of builders is dictated by solubility, while phosphates dissolve, insoluble Zeolites (containing ~5% insoluble matter) primarily partition into sewage sludge via sedimentation, effectively removing them from the aqueous phase (Ahmed et al., 2018).

7.4.7 Biodegradation Kinetics and Ecotoxicological Risk of Industrial Detergent Waste

Biodegradability varies by surfactant class. Alcohol Ethoxylates (AE) and Alcohol Ethoxysulfates (AES) demonstrated rapid mineralisation in river die-away tests (Talmage, 2020). Linear Alkylbenzene Sulfonates (LAS) is biodegradable under aerobic conditions but degrades slowly in anaerobic environments, leading to accumulation in anoxic sediments. A national-scale risk assessment in Spain revealed that LAS concentrations in surface waters frequently exceed safety thresholds, with 17.5% of sampling points showing a high-risk quotient (RQ > 1) for aquatic life (Luo et al., 2023). Biosurfactants, however, have demonstrated 100% removal rates in remediation applications, highlighting superior environmental compatibility (Farias et al., 2021).

7.4.8 Advanced Treatment of Detergent Manufacturing Effluents and Laundry Wastewater

Conventional biological treatment is often insufficient for high-strength detergent waste due to surfactant toxicity, necessitating the development of advanced treatment technologies (Table 6). However, these engineering solutions must be tailored to the specific waste profile, with examples including;

i. **Physicochemical Coagulation:** Ferric Chloride ($FeCl^{2+}$) or Lime removes up to 88% of suspended solids and 76% of COD from manufacturing effluent, effectively breaking stable emulsions (Sharghi et al., 2022).
ii. **Membrane Bioreactors (MBR):** For water reuse, MBRs using ultrafiltration eliminate 90% of COD and 84% of anionic surfactants, although the removal of non-ionic surfactants is less efficient (approx. 66%) due to adsorption properties (Bering et al., 2018).
iii. **Advanced Oxidation Processes (AOPs):** For recalcitrant loads, the Fenton process (H_2O_2/Fe^{2+}) chemically degrades LAS with 86.4% efficiency within 60 minutes, significantly improving effluent biodegradability (Mousavi and Khodadoost, 2019).

Table 6. Efficiency of Advanced Treatment Technologies

Treatment Technology	Target Pollutant	Removal Efficiency	Mechanism
Chemical Coagulation (Lime/Alum)	Manufacturing Effluent (COD, P)	67% COD, 92% Phosphorus	Destabilisation of emulsions
Membrane Bioreactor (MBR)	Laundry Wastewater (Surfactants)	90% COD, 84% Anionic Surfactants	Physical filtration + Biological degradation
Fenton Process (H_2O_2/Fe^{2+})	Recalcitrant LAS	86.4% LAS removal (60 min)	Advanced Oxidation (Hydroxyl radicals)
Dissolved Air Flotation (DAF)	Suspended Solids	< 10% COD (Ineffective alone)	Physical separation

(Adapted from: Bering et al., 2018; Mousavi and Khodadoost, 2019)

7.4.9 Microbiological Impact of Waste from Detergent Industries and Degradation Pathways

The influx of detergents into ecosystems acts as a potent selective pressure, fundamentally altering the structure and function of microbial communities. The biodegradation of LAS is a deterministic process requiring a specialised microbial consortium. The metabolic pathway proceeds via ω-oxidation (oxidation of the terminal methyl group), followed by β-oxidation (stepwise shortening of the alkyl chain), and finally desulfonation and aromatic ring cleavage (Mousavi and Khodadoost, 2019). Specific bacterial strains, such as *Ochrobactrum anthropi* and *Pseudomonas aeruginosa*, possess the genetic machinery (e.g., the *sdsA* gene) to utilise surfactants as a sole carbon source. The regulation of this gene involves a positive regulator (*SdsB*) and a negative regulator (*SdsR*) that ensure enzymes are only produced when surfactant is the primary fuel source (Mousavi and Khodadoost, 2019).

7.4.10 Resilience of Microbial Consortia to Industrial Detergent Waste and Sulfate Loads

In anaerobic systems, high surfactant loads reduce microbial diversity, selecting for a resistant "functional core." Acidogenic bacteria like *Clostridium* and *Geobacter* drive the initial degradation steps, while methanogenic archaea (*Methanosaeta*) are highly sensitive to surfactant toxicity and survive only within protective granular biomass (Centurion et al., 2018). The high sulfate load in detergent waste promotes the growth of Sulfate-Reducing Bacteria (SRB), generating toxic hydrogen sulfide (H_2S). This creates a competitive environment between sulfidogens and methanogens. However, acclimated microbial consortia demonstrate resilience, maintaining high COD removal efficiency despite sulfide accumulation (Centurion et al., 2018). Certain metal-tolerant bacteria, such as *Paenibacillus amylolyticus* and *Bacillus lentus*, have demonstrated "dual tolerance," degrading surfactants while simultaneously removing heavy metals from the environment (Adekanmbi et al., 2018).

7.4.11 Industrial Detergent Waste and the Co-selection of Antibiotic Resistance

A critical emerging concern is the link between detergent pollution and Antimicrobial Resistance (AMR). Detergent biocides, such as Quaternary Ammonium Compounds (QACs), co-select for antibiotic resistance genes through mechanisms of cross-resistance (e.g., multi-drug efflux pumps) and co-resistance (genetic linkage on plasmids) (Mousavi and Khodadoost, 2019; Andrews et al., 2024). Research indicates that exposure to common surfactant residues facilitates the horizontal transfer of resistance genes, effectively turning detergent-polluted environments into breeding grounds for multi-drug-resistant bacteria (Mousavi and Khodadoost, 2019).

7.4.12 Ecological and Human Health Risks Associated with Industrial Detergent Waste

The impact of detergent waste extends beyond immediate environmental degradation to chronic ecological and human health risks.

a. Aquatic Toxicity

The primary mechanism of aquatic toxicity is the disruption of biological membranes, leading to gill damage, swelling, and osmoregulatory failure in fish (Wood and Eom, 2021). Acute toxicity assays reveal that cationic surfactants are the most lethal to aquatic life (LC_{50} as low as 0.35 mg/L for *Daphnia magna*), followed by anionics, while non-ionics are generally the least toxic ($LC_{50} > 100$ mg/L) (Mousavi and Khodadoost, 2019). This is further delineated in Table 7.

Table 7. Comparative Aquatic Toxicity of Surfactant Classes

Surfactant Class	Representative Compound	Test Organism	Acute Toxicity (LC_{50})	Toxicity Rating
Cationic	Dodecyl Trimethyl Ammonium Chloride (DTAC)	*Daphnia magna*	0.35 mg/L	Highly Toxic
Anionic	Sodium Dodecyl Sulfate (SDS)	*Danio rerio*	3.6 – 12.5 mg/L	Moderately Toxic
Non-ionic	Alkyl Polyglucosides (APG)	*Daphnia magna*	> 100 mg/L	Low Toxicity
Biosurfactant	Rhamnolipids	*Artemia salina*	> 500 mg/L	Non-Toxic

(Data adapted from: Farias et al., 2021; Takada et al., 2025)

b. Genotoxicity and Endocrine Disruption

Beyond acute lethality, "standard" commercial detergents induce genotoxicity, causing chromosomal aberrations and micronucleus formation in plant assays (*Vicia faba*), whereas eco-friendly formulations show negligible genetic effects (Horozal and Aksoy, 2020). The metabolite Nonylphenol (NP) acts as a potent Endocrine Disrupting Chemical (EDC). By mimicking the hormone 17β-estradiol, NP induces feminisation in male fish and poses a potential risk to human reproductive health due to its persistence and bioaccumulation in the food web (Mousavi and Khodadoost, 2019).

c. Human Health Risks

The accumulation of heavy metals (lead, cadmium, arsenic) derived from detergent waste in water sources presents a direct threat to human health. Chronic exposure to these bio-accumulative toxins is linked to neurodevelopmental disorders in children, renal failure, and carcinogenesis, underscoring the urgent need for rigorous management of detergent industry effluents (Okechukwu and Onwukeme, 2025).

7.4.13 Microbial Remediation Strategies for Industrial Wastes

Bioremediation is defined as the biotechnological process utilising biological agents, principally bacteria, fungi, and algae, to eliminate, mineralise, degrade, or neutralise pollutants of organic and inorganic origin (Saini et al., 2023). Unlike conventional physicochemical treatments, which are often energy-intensive and generate secondary toxic sludge, microbial remediation offers a cost-effective and eco-friendly alternative for purifying industrial effluents (Radojević and Jakovljević, 2025).

The remediation of industrial effluents is governed by two distinct mechanisms, dictated by the pollutant's toxicity and bioavailability (Panasia et al., 2019). Passive physicochemical interactions (e.g., biosorption) are thermodynamically spontaneous and independent of cellular respiration. This independence renders them the optimal strategy for handling high-toxicity effluents that would otherwise be lethal to living biomass.

In contrast, metabolically dependent processes (e.g., bioaccumulation, biomineralisation) require cellular energy (ATP) but offer a critical advantage: the ability to enzymatically transform recalcitrant organic xenobiotics into harmless by-products (mineralisation) rather than merely sequestering them (Radojević and Jakovljević, 2025).

Bridging the gap between cellular mechanisms and field application requires ecological engineering, specifically through bioaugmentation (introducing absent metabolic pathways) or biostimulation (overcoming nutrient limitations). The urgency of these strategies is driven by the sheer volume of surfactant production; anionic surfactants (such as SDS and LAS) now account for 60% of the global market, making them a primary target for remediation (Figure 3).

Figure 3. Global production share of surfactant classes. Anionic surfactants, such as LAS and SDS, constitute the majority (60%) of global production, underscoring the environmental urgency of their remediation

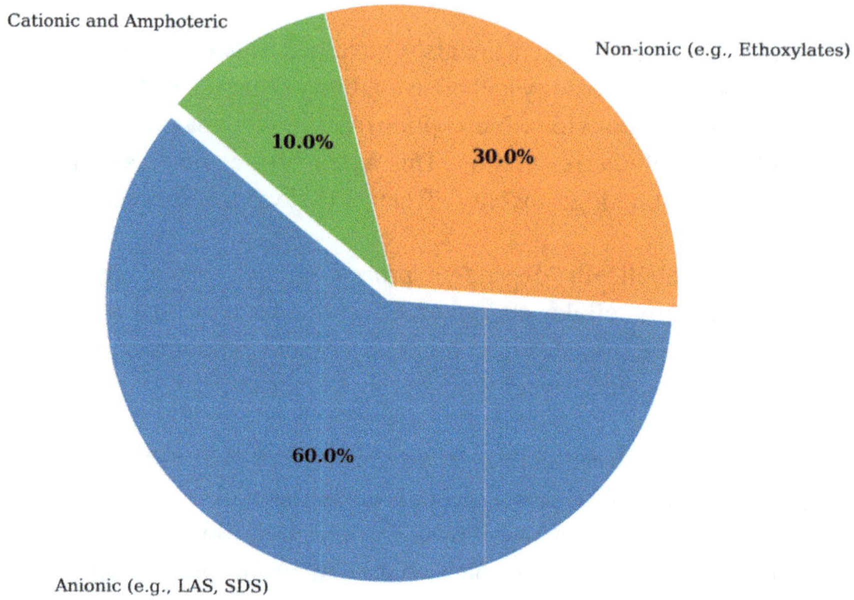

(Data adapted from Arora et al., 2022).

7.5. CONVENTIONAL MICROBIAL REMEDIATION STRATEGIES

To optimise remediation, it is necessary to distinguish between cellular mechanisms (how the cell works) and field application strategies (how the site is managed). Cellular mechanisms refer to the biochemical or physicochemical processes carried out by microorganisms, such as adsorption, transformation, precipitation, or intracellular accumulation of metals. In contrast, field application strategies involve the practical implementation of these processes through approaches such as bioaugmentation, biostimulation etc.

7.5.1. Biosorption

Biosorption is a passive mechanism involving the physicochemical binding of pollutants to functional groups (e.g., carboxyl, amine, and phosphate) on the biomass cell wall (Ayele et al., 2021). Its primary advantage is the independence from metabolism, allowing the use of dead biomass in toxic environments where active growth is inhibited (Monteiro et al., 2012).

Operational parameters heavily influence efficiency. Recent studies (Figure 4) utilising *Vitreoscilla sp.* ENSG301 demonstrated removal efficiencies of 97.1% for Cu, significantly outperforming other strains like *Bacillus sp.* (75.7%) (Radojević and Jakovljević, 2025). In the context of Nigerian laundry wastewater, *Bacillus lentus* BAL2 demonstrated a simultaneous capacity to remove 3.1% of Cu and 39% of Cd while actively degrading surfactants (Adekanmbi et al., 2018).

Figure 4. Heavy metal removal efficiency by distinct microbial strains. Vitreoscilla sp. ENSG301 exhibits superior removal of Cu (97.1%), whereas P. amylolyticus BAL1 provides moderate removal of Cd (39.0%) in complex laundry effluents

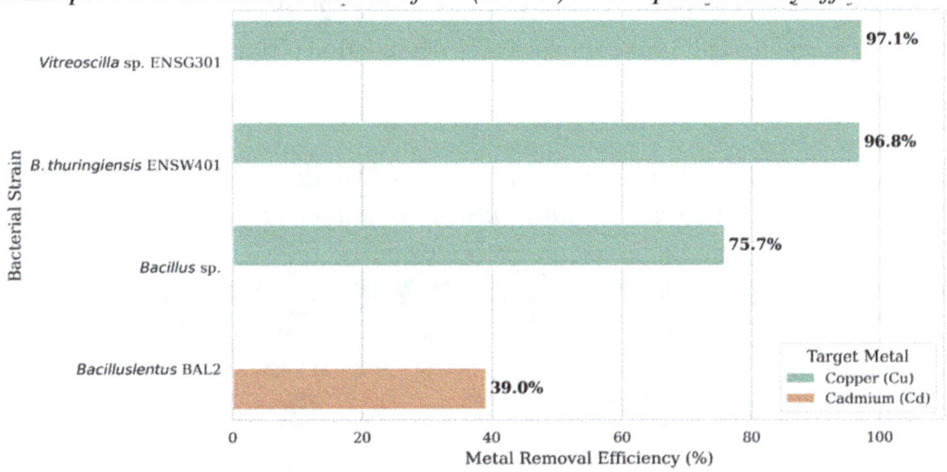

(Data adapted from Radojević and Jakovljević, 2025; Adekanmbi et al., 2018).

7.5.1.2 Bioaccumulation

Unlike passive biosorption, bioaccumulation is an active, metabolically driven process where living microorganisms transport contaminants across the cell membrane for intracellular sequestration (Fashola et al., 2016; Parmar and Patel, 2025). This is often mediated by ATP-dependent transport systems that import heavy metals via pathways intended for essential nutrients like magnesium or calcium.

While bioaccumulation enables the remediation of low-concentration pollutants, it requires the organism to possess high tolerance to toxicity (Panasia et al., 2019). Resilient strains such as *Pseudomonas putida* can accumulate Cd at high concentrations, with some strains tolerating up to 50 mg/L (Adekanmbi et al., 2018). Sequestration is often facilitated by metal-binding proteins, such as metallothioneins, primarily during the culture's exponential growth phase (Fashola et al., 2016).

7.5.1.3 Bioaugmentation

Bioaugmentation involves inoculating contaminated sites with specific, allochthonous microbial strains possessing superior degradation capabilities (Muter, 2023). This targets recalcitrant xenobiotics for which indigenous populations lack the necessary metabolic pathways.

Applications of bioaugmentation have targeted complex pollutants such as heavy oils and motor oils. A notable success was achieved using *Pseudomonas aeruginosa* ATCC 10145, which produced a biosurfactant capable of removing 100% of heavy oil (OCB1) from contaminated surfaces (Farias et al., 2021). Consortia containing *Pseudomonas putida* SP3 have shown 100% degradation efficiency for SDS within 24 hours, as shown in Figure 5

Figure 5. Comparative biodegradation efficacy of Sodium Dodecyl Sulfate (SDS) by selected bacterial strains. Strains such as P. putida SP3 and S. marcescens demonstrate complete (100%) degradation capability

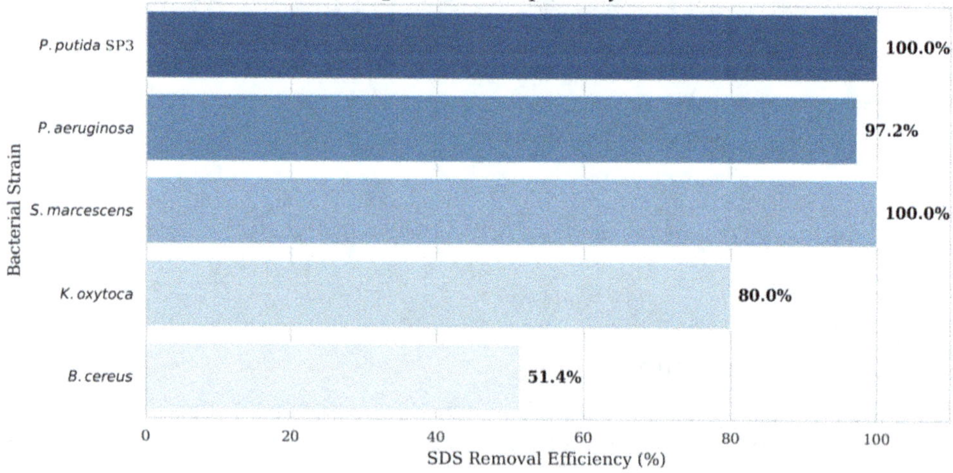

(Data adapted from Radojević and Jakovljević, 2025).

Bioaugmentation approach had also been successfully used in immobilization of heavy metals in mine tailings. Fashola et al. (2020b), reported the ability of two indigenous heavy metal tolerant *Bacillus cereus* OMF 003 and *Enterobacter asburiae* OMF 532 previously isolated from gold mine tailings in reducing the mobility of

Ni, Pb and Zn in gold mine tailings collected from three abandoned sites in Krugersdorp South Africa. Mobility factor of 35%, 90% and 60% was recorded for Ni, Pb and Zn present in the mine tailings.

However, reviews of field applications highlight that success is not guaranteed; the efficacy of introduced strains is often compromised by ecological competition with native flora or predation by protozoa, leading to a rapid decline in the population of the inoculant (Radojević and Jakovljević, 2025).

7.5.1.4 Biostimulation

Biostimulation modifies environmental conditions to enhance the metabolic activity of *indigenous* microbial populations rather than introducing foreign strains (Adams et al., 2015). Strategies include adding rate-limiting nutrients or terminal electron acceptors. Standard protocols involve adding carbon sources (e.g., sugarcane extract or L-cysteine) to support activity in nutrient-poor effluents. In electrochemical systems, adding sugarcane extract to the anode resulted in 82% removal efficiency of Pb(II) (Aleid et al., 2023).

7.5.1.5 Bioprecipitation

Bioprecipitation converts soluble, toxic contaminants into insoluble, stable mineral phases (e.g., sulfides or carbonates), reducing their mobility (Hoffmann et al., 2021). Validated applications in microbial fuel cells (MFCs) using substrates like sugarcane extract achieved 82% removal for Pb. Non-electrochemical studies with *Vitreoscilla sp.* reported 97.1% removal of Cu, confirming efficacy in high-concentration effluents (Haque et al., 2021).

7.5.1.6 Biomineralization

Biomineralization involves the conversion of dissolved metal ions into stable mineral phases through microbial metabolic or extracellular activities. This mechanism plays a crucial role in the remediation of mining-impacted environments, particularly in the stabilization of heavy metals and mitigation of AMD. One of the most studied biomineralization pathways in mining environments is sulfate reduction by SRB, which leads to the precipitation of metals as metal sulfides (e.g., PbS, ZnS, FeS). Chen et al. (2016), demonstrated that *Bacillus cereus* isolated from lead-zinc mine tailings exhibited a high Pb (II) uptake capacity (up to 340 mg/g) and facilitated its transformation into stable phosphate minerals. Further, a study on acid mine drainage (AMD) reported the complete removal of zinc and arsenic using a consortium of indigenous SRB with Zn precipitating as ZnS nanoparticles and arsenic forming sulfide minerals such as amorphous orpiment (As_2S_3) and realgar (La Pape et al., 2017).

7.5.1.7 Bioleaching

Bioleaching is a microbial-mediated process that involves the extraction of metals from ores and waste materials through biochemical oxidation and solubilization. This technique is widely applied in the mining industry for the recovery of valuable metals such as copper, gold, nickel, and cobalt, particularly from low-grade ores and mine tailings. (Jerez, 2017). This bioprocess has gained traction in the mining industry as it has a cost-effective metal recovery rate. (Rendón-Castrillón et al., 2023; Jia et al., 2024). A mixed microbial consortium comprising *Acidithiobacillus ferrooxidans* and *Sulfobacillus thermosulfidooxidans* has been used in zinc bioleaching from lead-zinc mine tailings, achieving a 94% recovery while reducing the leaching cycle by 30% (Liao et al., 2021). In fact, bioleaching has been implicated as a sustainable microbial remediation in mine waste and space mining (Abraham and Schopf, 2026).

7.5.1.8 Enzymatic Degradation

Enzymatic degradation is a highly specific, metabolically driven conventional strategy crucial for treating complex effluents from both the paint and detergent industries. For paint effluents, microbial consortia deploy extracellular enzymes such as laccases, peroxidases, and azo-reductases to cleave the chromophoric bonds of synthetic dyes (Rajendran et al, 2025).

Simultaneously, these industrial effluents carry a heavy burden of anionic surfactants, predominantly LAS, which are notoriously recalcitrant and form stable foams in receiving waters. The microbial breakdown of LAS relies heavily on a sequential, multi-enzyme pathway that systematically degrades the surfactant structure. This pathway (Figure 6) occurs in four primary stages (Balson and Felix, 1995):

1. **ω-Oxidation:** The degradation initiates at the terminal methyl group of the straight alkyl chain. Catalyzed by an alkane monooxygenase and two dehydrogenases, the terminal carbon is oxidized sequentially into an alcohol, an aldehyde, and finally a carboxylic acid.
2. **β-Oxidation:** The resulting carboxylic acid undergoes β-oxidation, entering the tricarboxylic acid cycle. This process systematically shortens the alkyl chain by cleaving two-carbon fragments to produce acetyl-CoA.
3. **Desulfonation:** To break down the remaining aromatic ring structure, the sulfonated component must be removed. Microorganisms achieve this through three potential enzymatic mechanisms:

a) Hydroxylative desulfonation

$RSO_3H + H_2O ROH + 2H+ + SO_3^{2-}$ (i)

b) **Monooxygenase catalysis under acidic conditions.**

$RSO_3H + O2 + 2NADH ROH + H_2O + SO_3^{2-} + 2NAD^+$ (ii)

c)Reductive desulfonation

$RSO_3H + NADH + H+ RH + NAD+ + H_2SO_4$ (iii)

4. **Aromatic Ring Cleavage:** The removal of the alkyl and sulfonate groups leaves behind benzoic and phenylacetic acids. Microbial oxidation further cleaves these rings, converting them into innocuous intermediates such as fumaric acid, acetoacetic acid, and catechol benzene.

Figure 6. Enzymatic pathway for the biodegradation of Linear Alkylbenzene Sulfonates (LAS)

Linear Alkylbenzene Sulfonate (LAS)

↓ (Alkane monooxygenase / ω -Oxidation)

Sulfophenyl Carboxylic Acids (SPCs)

↓ (β -Oxidation / successive removal of 2 carbon units)

Short-chain SPCs

↓ (Desulfonation)

Aromatic Ring Intermediates (e.g., Benzoic and Phenylacetic acids)

↓ (Microbial Ring Cleavage)

Harmless Byproducts (Fumaric acid, Acetoacetic acid, CO_2, H_2O)

(Adapted from Kaida et al., 2021).

7.5.2 Advanced Microbial Remediation Strategies

As industrial waste becomes increasingly complex, containing nanomaterials, pharmaceutical residues, and synthetic polymers, traditional strategies often fall short. Recent advancements integrate microbiology with nanotechnology, genetic engineering, and electrochemistry.

7.5.2.1 Nano-Bioremediation and Hybrid Bio-Nano Systems

Integrating nanotechnology with microbial processes shifts the paradigm from simple degradation to catalytic synergy (Haider et al., 2024). Hybrid systems use inorganic catalysts to overcome the kinetic bottlenecks of traditional bioremediation.

A prime example is the Photo-Microbial Fuel Cell (PMFC). Standard MFCs are often limited by the cathode reaction. Recent advancements incorporate Titanium Dioxide (TiO_2) nanoparticles as a photocathodic catalyst (Nasr et al., 2018). In systems treating Sodium Dodecyl Sulfate (SDS), the TiO_2-coated cathode generates reactive hydroxyl radicals under UV irradiation (Vera et al., 2023).

This creates a dual-oxidation mechanism: microbes oxidise organic matter at the anode, while nanoparticles photochemically degrade resistant surfactants at the cathode. This "pincer attack" achieved an SDS removal efficiency of >96% and a COD removal of >71%, significantly outperforming microbial treatment alone. Crucially, the system generated a maximum power density of 0.73 W/m^3, validating its potential as a self-sustaining technology (Sathe et al., 2020).

7.5.2.2 Genetically Engineered Microorganisms (GEMs)

Metabolic engineering facilitates the creation of microorganisms with optimised degradation pathways (Park et al., 2018). By employing genome-editing tools, researchers can enhance the expression of catabolic enzymes. Specific breakthroughs include the regulation of the *sdsA1* gene (encoding alkylsulfatase) in *Pseudomonas aeruginosa* PAO1. Research has shown that overexpression or specific induction of this gene by SDS allows the bacterium to utilise toxic detergents as a sole carbon source (Radojević and Jakovljević, 2025). However, the deployment of GEMs remains constrained by regulatory concerns regarding biosafety (Hoffmann et al., 2021). To mitigate risks such as horizontal gene transfer, recent developments include the incorporation of biocontainment strategies (e.g., suicide genes or auxotrophy) to prevent the escape of engineered traits into the environment (Sebesta et al., 2022; Radojević and Jakovljević, 2025).

7.5.2.3 Microbial Fuel Cells (MFCs)

Microbial Fuel Cells (MFCs) represent a paradigm shift from energy-consuming wastewater treatment to energy-generating systems (Rani et al., 2022). In these bioelectrochemical devices, electroactive bacteria (exoelectrogens) oxidise organic matter at the anode, transferring electrons to an external circuit to generate electricity while simultaneously treating the effluent (Garbini et al., 2023; Mahto and Das, 2025).

Key performance metrics for MFCs include the power density and coulombic efficiency (Zafar et al., 2022). Recent configurations utilising a biochar-based anode (derived from coconut shells) inoculated with *Geobacter sulfurreducens* and *Shewanella oneidensis* have achieved power densities of 1.72 W/m^2 while removing 92.3% of COD from industrial wastewater. This performance significantly exceeds traditional systems, verifying the efficacy of carbon-based electrodes in enhancing electron transfer (Mkilima and Baimukasheva, 2025).

7.5.2.4 Microbial Decolorization

Microbial decolorization has emerged as an effective and sustainable strategy for treating dye-containing effluents from industries such as paint manufacturing, where synthetic dyes are widely used for pigmentation. Microbial decolorization often involve the use of various oxidative and reductive enzymes such as laccase and azo reductase (Mohanty and Kumar, 2025). Different studies have isolated microbes that have shown promising decolorization efficacy. These include a study by Lade et al. (2012), wherein the isolate *Rhodocyclus gelatinosus* XL-1 from contaminated wastewater and attained ≥93% decolorization efficacy of the reactive brilliant blue (KN-R). Other microbes such as *Shewanella decolorationis* S12 and *Galactomuces geotrichum* have also been implicated in decolourization of dyes yielding 99% and 55.5% efficacy respectively (Xu et al., 2006).

7.6. CONCLUSIONS

Wastes generated from mining, paint and detergents industries represent a major and rising environmental problem because of the various composite chemical composition, toxicity and persistence nature of these wastes. The numerous components of the wastes from these industries have been linked to various health hazards in humans, animals as well as degradation of soil and water quality when these wastes are released into the environment.

The introduction of antimicrobial resistance genes from the wastes into the environment is another disturbing problem that needs urgent attention. Heavy metals and detergent biocides such as QACs are known to co-select for antibiotics resistance genes in the polluted environment. Effective and efficient bioremediation strategies that offer a viable and sustainable alternative to conventional treatment technologies such as biosorption, bioaccumulation, bioaugmentation and bioprecipitation were discussed. The use of recent advances in bioremediation such as microbial fuel cells, nano-bioremediation, and genetically engineered microorganisms further enhance remediation efficiency while enabling resource recovery and energy generation.

Despite these promising developments, the successful field-scale application of microbial technologies is limited by challenges including non-uniformity of the waste composition, extreme environmental conditions that can hinder microbial survival, potential ecological risks, and regulatory limitations. Microbial remediation holds significant potential, prevent industrial pollution and support environmental protection, public health, and long-term industrial sustainability. The integration of microbial remediation with conventional physicochemical methods and development of more robust microbial consortia are needed in future to enhance the effectiveness of the advanced methods.

REFERENCES

Abraham, A. P., & Schopf, S. (2026). Bioleaching as a biotechnological tool for metal recovery: From sewage to space mining. *Frontiers in Bioengineering and Biotechnology*, *13*, 1712157. DOI: 10.3389/fbioe.2025.1712157 PMID: 41625076

Adams, G. O., Fufeyin, P. T., Okoro, S. E., & Ehinomen, I. (2015). Bioremediation, biostimulation and bioaugmention: A review. *International Journal of Environmental Bioremediation & Biodegradation*, *3*(1), 28–39. DOI: 10.12691/ijebb-3-1-5

Adebiyi, F. M. (2022). Air quality and management in petroleum refining industry: A review. *Environmental Toxicology and Chemistry*, *4*, 89–96.

Adekanmbi, A. O., Oyeladun, W. O., & Olaposi, A. V. (2018). Degradation of surfactant and metal-removal by bacteria from a Nigerian laundry environment. *European Journal of Biological Research*, *8*(4), 243.

Ahmed, J., Thakur, A., and Goyal, A. (2021). Industrial wastewater and its toxic effects. https://doi.org/DOI: 10.1039/9781839165399-00001

Ahmed, M. A. M. (2023). Evaluation of Industrial Waste in the Paints Industry Sector: Khartoum North Industrial Area. *Journal of Nature. Life and Applied Sciences*, *7*(4), 56–64. DOI: 10.26389/AJSRP.R080622

Ahmed, S., Sanni, K. A., Azeez, O. S., Saka, A. A., & Oluwafemi, S. (2018). Production of 100 kg/day of zeolite a as a builder for powdered detergent from Nigerian Ahoko Kaolin using locally fabricated mini zeolite plant. [IOP Publishing.]. *IOP Conference Series. Materials Science and Engineering*, *413*, 012064. DOI: 10.1088/1757-899X/413/1/012064

Aleid, G. M., Alshammari, A. S., Ahmad, A. R. D., Hussain, F., Oh, S. E., Ahmad, A., & Umar, K. (2023). Advancement in microbial fuel cells technology by using waste extract as an organic substrate to produce energy with metal removal. *Processes (Basel, Switzerland)*, *11*(8), 2434. DOI: 10.3390/pr11082434

Alsukaibi, A. K. (2022). Various approaches for the detoxification of toxic dyes in wastewater. *Processes (Basel, Switzerland)*, *10*(10), 1968. DOI: 10.3390/pr10101968

Alzain, H., Kalimugogo, V., Hussein, K., & Karkadan, M. (2023). A review of environmental impact of azo dyes. *International Journal of Research and Review*, *10*(6), 673–689. DOI: 10.52403/ijrr.20230682

Ameen, G. (2023). Industrial Chemistry: Chemicals And Technology In Everyday Life. *Shengming De Huaxue*, •••, 213.

Andrews, K., Landeryou, T., & Nale, J. Y. N. (2024). Fleas of fleas: The potential role of bacteriophages in *Salmonella* diversity and pathogenicity. In *Microbiology Society Annual Conference 2024*.

Aniyikaiye, T. E., Oluseyi, T., Odiyo, J. O., & Edokpayi, J. N. (2019). Physico-chemical analysis of wastewater discharge from selected paint industries in Lagos, Nigeria. *International Journal of Environmental Research and Public Health, 16*(7), 1235. DOI: 10.3390/ijerph16071235 PMID: 30959965

Arora, U., Khuntia, H. K., Chanakya, H. N., & Kapley, A. (2023). Surfactants: Combating the fate, impact, and aftermath of their release in the environment. *International Journal of Environmental Science and Technology, 20*(10), 11551–11574. DOI: 10.1007/s13762-022-04549-2

Arslan Younas, A. S., Maqbool, H., Qamar, M. S., Parveen, A., Afzal, R., & Saleem, A., and Ullah10, Q. (2024). Nanoparticle Innovations in Wastewater Treatment: A Comparative study of titanium dioxide, gold, and nano zero-valent iron nanoparticles for lead, chromium, and mercury removal. *Global Scientific and Academic Research Journal of Multidisciplinary Studies*, *2*(8), 15–31.

Aydemir, C., & Özsoy, S. A. (2020). Environmental impact of printing inks and printing process. *Journal of Graphic Engineering and Design, 11*(2), 11–18. DOI: 10.24867/JGED-2020-2-011

Ayele, A., Suresh, A., & Benor, S. (2021). Phycoremediation of Heavy Metals, Factors Involved and Mechanisms Related to Functional Groups in the Algae Cell Surface – A Review. In Aravind, J., Kamaraj, M., Prashanthi Devi, M., & Rajakumar, S. (Eds.), *Strategies and Tools for Pollutant Mitigation* (pp. 269–289). Springer., DOI: 10.1007/978-3-030-63575-6_13

Aznar-Sánchez, J. A., García-Gómez, J. J., Velasco-Muñoz, J. F., & Carretero-Gómez, A. (2018). Mining Waste and Its Sustainable Management: Advances in Worldwide Research. *Minerals (Basel)*, *8*(7), 284. DOI: 10.3390/min8070284

Baij, L., Hermans, J., Ormsby, B., Noble, P., Iedema, P., & Keune, K. (2020). A review of solvent action on oil paint. *Heritage Science*, *8*(1), 1–23. DOI: 10.1186/s40494-020-00388-x

Balson, T., & Felix, M. S. B. (1995). Biodegradability of non-ionic surfactants. In Karsa, D. R., & Porter, M. R. (Eds.), *Biodegradability of Surfactants* (pp. 204–230). Springer Netherlands. DOI: 10.1007/978-94-011-1348-9_7

Bering, S., Mazur, J., Tarnowski, K., Dąbkowska, N., Janus, M., Mozia, S., & Morawski, A. W. (2018). Removal of organic pollutants and surfactants from laundry wastewater in membrane bioreactor (MBR). *Desalination and Water Treatment, 134*, 281–288. DOI: 10.5004/dwt.2018.23207

Bernal-Jácome, L. A., Izar-Landeta, J. M., Flores-Ramírez, R., González-Salazar, L. F., & Vargas-Berrones, K. X. (2024). Nonylphenol ethoxylate degradation in detergents during shelf time, a new exposure pathway, and a perspective on their substitution. *Environmental Science and Pollution Research International, 31*(21), 30497–30508. DOI: 10.1007/s11356-024-33260-7 PMID: 38607492

Bhavsar, R. A., & Sardesai, A. (2022). Investigation of effect of type of pigment/extender on the stability of high pigment volume concentration water-based architectural paint. *Journal of Coatings Technology and Research, 19*(3), 919–930. DOI: 10.1007/s11998-021-00568-9

Carmen, Z., & Daniela, S. (2012). *Textile organic dyes-characteristics, polluting effects and separation/elimination procedures from industrial effluents-a critical overview* (Vol. 3). IntechOpen. DOI: 10.5772/32373

Cebada, J. D. P. (2016). Mining corporations and air pollution science before the Age of Ecology. *Ecological Economics, 123*, 77–83. DOI: 10.1016/j.ecolecon.2015.12.001

Centurion, V. B., Moura, A. G. L., Delforno, T. P., Okada, D. Y., Dos Santos, V. P., Varesche, M. B. A., & Oliveira, V. M. (2018). Anaerobic co-digestion of commercial laundry wastewater and domestic sewage in a pilot-scale EGSB reactor: The influence of surfactant concentration on microbial diversity. *International Biodeterioration & Biodegradation, 127*, 77–86. DOI: 10.1016/j.ibiod.2017.11.017

Chavan, A. R., and Kaur, G. (2024). Keeping Food Safe: Tackling The Health Risks Of Phthalates And Azo Dyes.

Chen, Z., Pan, X., Chen, H., Guan, X., & Lin, Z. (2016). Biomineralization of Pb (II) into Pb-hydroxyapatite induced by *Bacillus cereus* 12-2 isolated from Lead–Zinc mine tailings. *Journal of Hazardous Materials, 301*, 531–537. DOI: 10.1016/j.jhazmat.2015.09.023 PMID: 26468754

Christie, R., & Abel, A. (2021). Organic pigments: General principles. *Physical Sciences Reviews, 6*(12), 807–834. DOI: 10.1515/psr-2020-0187

de Santana da Silva, J., Heck, M. C., Buzo, M. G., Almeida, I. V., & Vicentini, V. E. P. (2018). Evaluation of textile laundry effluents and their cytotoxic effects on *Allium cepa*. *Environmental Science and Pollution Research International, 25*(28), 27890–27898. DOI: 10.1007/s11356-018-2813-5 PMID: 30056543

Dolganova, I., Dolganov, I., Ivanchina, E., & Ivashkina, E. (2018). Alkylaromatics in detergents manufacture: Modeling and optimizing linear alkylbenzene sulfonation. *Journal of Surfactants and Detergents*, *21*(1), 175–184. DOI: 10.1002/jsde.12009

Doroszkowski, A. (2020). Paints. In *Technological Applications of Dispersions* (pp. 1–68). CRC Press. DOI: 10.1201/9781003067252-1

Fan, Y., Song, Z., Wu, Y., Ren, X., Bi, C., Ye, W., & Xu, Y. (2024). Chemicals of emerging concern in water-based paint products. *Environmental Science & Technology Letters*, *11*(5), 445–452. DOI: 10.1021/acs.estlett.4c00052

Farias, C. B. B., Rita de Cássia, F., Almeida, F. C. G., Santos, V. A., & Sarubbo, L. A. (2021). Removal of heavy oil from contaminated surfaces with a detergent formulation containing biosurfactants produced by *Pseudomonas* spp. *PeerJ*, *9*, e12518. DOI: 10.7717/peerj.12518 PMID: 34900433

Fashola, M. O., Ngole-Jeme, V. M., & Babalola, O. O. (2015). Diversity of acidophilic bacteria and archaea and their roles in bioremediation of acid mine drainage. *British Microbiology Research Journal*, *8*(3), 443–456. DOI: 10.9734/BMRJ/2015/14365

Fashola, M. O., Ngole-Jeme, V. M., & Babalola, O. O. (2016). Heavy metal pollution from gold mines: Environmental effects and bacterial strategies for resistance. *International Journal of Environmental Research and Public Health*, *13*(11), 1047. DOI: 10.3390/ijerph13111047 PMID: 27792205

Fashola, M. O., Ngole-Jeme, V. M., & Babalola, O. O. (2020a). Physicochemical properties, heavy metals, and metal-tolerant bacteria profiles of abandoned gold mine tailings in Krugersdorp, South Africa. *Canadian Journal of Soil Science*, *100*(3), 217–233. DOI: 10.1139/cjss-2018-0161

Fashola, M. O., Ngole-Jeme, V. M., & Babalola, O. O. (2020b). Heavy metal immobilization potential of indigenous bacteria isolated from gold mine tailings. *International Journal of Environmental Research*, *14*(1), 71–86. DOI: 10.1007/s41742-019-00240-6

Fensham, E., Ling, N., and Tempero, G. W. (2022). Toxicological effects of aluminium in relation to diel pH changes on fish and kōura.

Fu, F., & Wang, Q. (2011). Removal of heavy metal ions from wastewaters: A review. *Journal of Environmental Management*, *92*(3), 407–418. DOI: 10.1016/j.jenvman.2010.11.011 PMID: 21138785

Garbini, G. L., Barra Caracciolo, A., & Grenni, P. (2023). Electroactive bacteria in natural ecosystems and their applications in microbial fuel cells for bioremediation: A review. *Microorganisms*, *11*(5), 1255. DOI: 10.3390/microorganisms11051255 PMID: 37317229

Haider, H. U., Arshad, E., Amir, M. A., Zafar, N., Akhtar, I., Ahmad, M. T., and Ali, M. (2024). Integrating Nanotechnology and Bacteria-Based Remediation: A Synergistic Approach to Mitigating Climate Change and Soil Degradation in Agriculture. *Global Scientific and Academic Research Journal of Multidisciplinary Studies. 1-13*

Haque, M. M., Mosharaf, M. K., Haque, M. A., Tanvir, M. Z. H., & Alam, M. K. (2021). Biofilm formation, production of matrix compounds and biosorption of copper, nickel and lead by different bacterial strains. *Frontiers in Microbiology*, *12*, 615113. DOI: 10.3389/fmicb.2021.615113 PMID: 34177820

Haynes, V. (2020). The Interactive Effects of Titanium Dioxide Nanoparticles and Light on Coastal Planktonic Organisms. Doctoral dissertation

Hoffmann, T. D., Reeksting, B. J., & Gebhard, S. (2021). Bacteria-induced mineral precipitation: A mechanistic review. *Microbiology (Reading, England)*, *167*(4), 001049. DOI: 10.1099/mic.0.001049 PMID: 33881981

Horozal, A., & Aksoy, Ö. (2020). Evaluation of the genotoxicity of some standard and eco-friendly detergents with Vicia faba. *Caryologia*, *73*(4), 129–139. DOI: 10.13128/caryologia-856

Hoxha, D., Belayachi, N., Brunetaud, X., & Rémond, S. (2019). A Critical Review on Durability of Sustainable Materials and Structures. In *International Conference on Critical Thinking in Sustainable Rehabilitation and Risk Management of the Built Environment* (181-199). Cham: Springer International Publishing.

Hrouzková, S., & Szarka, A. (2021). Development of a modified QuEChERS procedure for the isolation of pesticide residues from textile samples, followed by GC–MS determination. *Separations*, *8*(8), 106. DOI: 10.3390/separations8080106

Huntington, V. E., Coulon, F., & Wagland, S. T. (2022). Innovative resource recovery from industrial sites: A critical review. *Sustainability (Basel)*, *15*(1), 489. DOI: 10.3390/su15010489

Jerez, C. A. (2017). Biomining of metals: How to access and exploit natural resource sustainably. *Microbial Biotechnology*, *10*(5), 1191–1193. DOI: 10.1111/1751-7915.12792 PMID: 28771998

Jia, Y., Ruan, R., Qu, J., Tan, Q., Sun, H., & Niu, X. (2024). Multi-scale and trans-disciplinary research and technology developments of heap bioleaching. *Minerals (Basel), 14*(8), 808. DOI: 10.3390/min14080808

Kaida, H., Syed, M. A., Shukor, M. Y., & Othman, A. R.da. (2021). Biodegradation of linear alkylbenzene sulfonates (LAS): A mini review. *Bioremediation Science and Technology Research, 9*(1), 1–6. DOI: 10.54987/bstr.v9i1.590

Kakoi, B., Kaluli, J. W., Ndiba, P., & Thiong'o, G. (2017). Optimization of Maerua Decumbent bio-coagulant in paint industry wastewater treatment with response surface methodology. *Journal of Cleaner Production, 164*, 1124–1134. DOI: 10.1016/j.jclepro.2017.06.240

Keawboonchu, J., Thepanondh, S., Kultan, V., Pinthong, N., Malakan, W., & Robson, M. G. (2023). Integrated Sustainable Management of Petrochemical Industrial Air Pollution. *International Journal of Environmental Research and Public Health, 20*(3), 2280. DOI: 10.3390/ijerph20032280 PMID: 36767648

Khan, N. (2023). *Analysis of Market Preference in Paint Industry* (Doctoral dissertation, BIITM).

Khan, R. A. (2022). Detergents. In *Environmental micropollutants* (pp. 117–130). Elsevier., DOI: 10.1016/B978-0-323-90555-8.00014-3

Khan, S., & Malik, A. (2013). Environmental and health effects of textile industry wastewater. In *Environmental deterioration and human health: Natural and anthropogenic determinants* (pp. 55–71). Springer Netherlands.

Kolya, H., & Kang, C. W. (2024). Toxicity of metal oxides, dyes, and dissolved organic matter in water: Implications for the environment and human health. *Toxics, 12*(2), 111. DOI: 10.3390/toxics12020111 PMID: 38393206

Koohsaryan, E., Anbia, M., Sepehrian, M., & Maghsoodlu, M. (2021). Facile hydrothermal synthesis of hierarchical sodium P zeolite as a non-phosphate detergent builder. *Journal of Surfactants and Detergents, 24*(1), 85–97. DOI: 10.1002/jsde.12452

Kumar, A. Utkarsh, D., Kaman, S., Satya Prakash, G., and Mirza S. Jamal Beg. (2021). "Structure and properties of dyes and pigments." In *Dyes and pigments-novel applications and waste treatment.* IntechOpen, Available at DOI: 10.5772/intechopen.97104

Kumar, A., Thakur, A. K., Gaurav, G. K., Klemeš, J. J., Sandhwar, V. K., Pant, K. K., & Kumar, R. (2023). A critical review on sustainable hazardous waste management strategies: A step towards a circular economy. *Environmental Science and Pollution Research International, 30*(48), 105030–105055. DOI: 10.1007/s11356-023-29511-8 PMID: 37725301

Kumari, U. (2024). Textile dyes and their impact on the natural environment. In *Dye Pollution from Textile Industry: Challenges and Opportunities for Sustainable Development* (pp. 17–30). Springer Nature Singapore. DOI: 10.1007/978-981-97-5341-3_2

Lade, H. S., Waghmode, T. R., Kadam, A. A., & Govindwar, S. P. (2012). Enhanced biodegradation and detoxification of disperse azo dye Rubine GFL and textile industry effluent by defined fungal-bacterial consortium. *International Biodeterioration & Biodegradation, 72*, 94–107. DOI: 10.1016/j.ibiod.2012.06.001

Larson, R. J., & Woltering, D. M. (2020). Linear alkylbenzene sulfonate (LAS). In *Fundamentals of Aquatic Toxicology* (pp. 859–882). CRC Press. DOI: 10.1201/9781003075363-35

Le Pape, P., Battaglia-Brunet, F., Parmentier, M., Joulian, C., Gassaud, C., Fernandez-Rojo, L., Guigner, J. M., Ikogou, M., Stetten, L., Olivi, L., Casiot, C., & Morin, G. (2017). Complete removal of arsenic and zinc from a heavily contaminated acid mine drainage via an indigenous SRB consortium. *Journal of Hazardous Materials, 321*, 764–772. DOI: 10.1016/j.jhazmat.2016.09.060 PMID: 27720469

Liao, X., Ye, M., Li, S., Liang, J., Zhou, S., Fang, X., Gan, Q., & Sun, S. (2021). Simultaneous recovery of valuable metal ions and tailings toxicity reduction using a mixed culture bioleaching process. *Journal of Cleaner Production, 316*, 128319. DOI: 10.1016/j.jclepro.2021.128319

Luo, Y., Jin, X., Xie, H., Ji, X., Liu, Y., Guo, C., & Xu, J. (2023). Linear alkylbenzene sulfonate threats to surface waters at the national scale: A neglected traditional pollutant. *Journal of Environmental Management, 342*, 118344. DOI: 10.1016/j.jenvman.2023.118344 PMID: 37320921

Mahartin, T. L. (2023). Waste management plan with reduce, reuse, recycle (3r) method. *Journal of Sustainability, Society, and eco-welfare*, 1 (1).

Mahto, K. U., & Das, S. (2025). Electroactive biofilm communities in microbial fuel cells for the synergistic treatment of wastewater and bioelectricity generation. *Critical Reviews in Biotechnology, 45*(2), 434–453. DOI: 10.1080/07388551.2024.2372070 PMID: 39009474

Maile, F. J. (2021). Colorants in coatings. *Physical Sciences Reviews, 6*(11), 707–789. DOI: 10.1515/psr-2020-0160

Mishra, S. (2025). Warming Seas: The Impact of Climate Change on Ocean Life and Ecosystems. In *Navigating Climate Change: Impacts on Biodiversity and Ecosystem Resilience* (pp. 467–501). Springer Nature Singapore. DOI: 10.1007/978-981-95-0409-1_24

Mkilima, T., & Baimukasheva, S. (2025). Performance of a multi-chamber microbial fuel cell with biochar anode for industrial wastewater treatment and energy recovery. *Journal of Ecological Engineering, 26*(7), 367–380. DOI: 10.12911/22998993/203696

Mohanty, S. S., & Kumar, A. (2025). Microbial decolorization of anthraquinone dyes: Batch and continuous treatment systems-a mini-review. *Frontiers in Environmental Engineering, 4*, 1553712. DOI: 10.3389/fenve.2025.1553712

Monteiro, C. M., Castro, P. M., & Malcata, F. X. (2012). Metal uptake by microalgae: Underlying mechanisms and practical applications. *Biotechnology Progress, 28*(2), 299–311. DOI: 10.1002/btpr.1504 PMID: 22228490

Mousavi, S. A., & Khodadoost, F. (2019). Effects of detergents on natural ecosystems and wastewater treatment processes: A review. *Environmental Science and Pollution Research International, 26*(26), 26439–26448. DOI: 10.1007/s11356-019-05802-x PMID: 31352596

Moyo, A. (2024). Controlling acid and metalliferous drainage generated by legacy mine wastes in Tasmania using industrial wastes (Doctoral dissertation, University of Tasmania).

Müller, B., & Schackmann, M. (2023). *Coatings formulation.* Lack in Vincentz GmbH and Company KG.

Musah, B. I. (2024). Effects of heavy metals and metalloids on plant-animal interaction and biodiversity of terrestrial ecosystems - An overview. *Environmental Monitoring and Assessment, 197*(1), 12. DOI: 10.1007/s10661-024-13490-5 PMID: 39623084

Muter, O. (2023). Current trends in bioaugmentation tools for bioremediation: A critical review of advances and knowledge gaps. *Microorganisms, 11*(3), 710. DOI: 10.3390/microorganisms11030710 PMID: 36985282

Nair, S., Manu, B., & Azhoni, A. (2021). Sustainable treatment of paint industry wastewater: Current techniques and challenges. *Journal of Environmental Management, 296*, 113105. DOI: 10.1016/j.jenvman.2021.113105 PMID: 34216906

Nasr, M., Eid, C., Habchi, R., Miele, P., & Bechelany, M. (2018). Recent progress on titanium dioxide nanomaterials for photocatalytic applications. *ChemSusChem*, *11*(18), 3023–3047. DOI: 10.1002/cssc.201800874 PMID: 29984904

Nayak, N. P. (2023). Microorganisms and their application in mining and allied industries. *Materials Today: Proceedings*, *72*, 2886–2891. DOI: 10.1016/j.matpr.2022.07.392

Nnaji, N. D., Onyeaka, H., Miri, T., & Ugwa, C. (2023). Bioaccumulation for heavy metal removal: A review. *SN Applied Sciences*, *5*(5), 125. DOI: 10.1007/s42452-023-05351-6

Noruzman, A. H., & Apandi, N. M. (2020). Paint waste management in industry. *Journal of Advanced Research in Business and Management Studies*, *20*(1), 28–33. DOI: 10.37934/arbms.20.1.2833

Obidi, O. F., Kade, E. A., & Fadipe, O. T. (2022a). A first report on the identification of a novel archaea, *Methanospirillum lacunae* from spoilt paints in Lagos, Nigeria, using a metagenomics approach. *Scientific African*, *15*, 2468–2276.

Obidi, O. F., Olushina, O. A., Peter, V., Weinan, C., Matthew, C., Bright, O., Adetunji, A., Damilare, O., Kehinde, A., Malak, A., Idowu, S., Favour, O., & Obaloluwa, S. (2022b). Vibrational spectroscopy and non-parametric analysis for biodegradation evaluation: The acrylic-emulsion paint scenario. *Vibrational Spectroscopy*, *123*, 103427. DOI: 10.1016/j.vibspec.2022.103427

Ohanwemadu, O. S., & Nwakanma, C. (2025). Ecological Risk Evaluation of Heavy Metal Accumulation in Aba River Bottom Sediments: Implications for Soap and Detergent Industry Wastewater Management. *Journal of Agricultural Science and Technology A*, *15*(1). Advance online publication. DOI: 10.17265/2161-6256/2025.01.002

Ojonimi, T. I., Chanda, T. P., & Ameh, E. G. (2021). Acid mine drainage (AMD) contamination in coal mines and the need for extensive prediction and remediation: A review. *Journal of Degraded and Mining Lands Management*, *9*(1), 3129–3136. DOI: 10.15243/jdmlm.2021.091.3129

Okechukwu, V. U., & Onwukeme, V. (2025). Health Risk Assessment of Selected Heavy Metals in Surface Water and Sediments from Aba River in Abia state, Nigeria. *Chemical Research and Technology*, *2*(3), 115–127.

Panasia, G., Oetermann, S., Steinbüchel, A., & Philipp, B. (2019). Sulfate ester detergent degradation in *Pseudomonas aeruginosa* is subject to both positive and negative regulation. *Applied and Environmental Microbiology*, *85*(23), e01352–e19. DOI: 10.1128/AEM.01352-19 PMID: 31540990

Pandey, P., & Kiran, U. V. (2020). Degradation of paints and its microbial effect on health and environment. *Journal of Critical Reviews*, *7*(19), 4879–4884.

Park, S. Y., Yang, D., Ha, S. H., & Lee, S. Y. (2018). Metabolic engineering of microorganisms for the production of natural compounds. *Advanced Biosystems*, *2*(1), 1700190. DOI: 10.1002/adbi.201700190

Parmar, K. S., & Patel, K. M. (2025). Biosorption and bioremediation of heavy metal ions from wastewater using algae: A comprehensive review. *World Journal of Microbiology & Biotechnology*, *41*(7), 262. DOI: 10.1007/s11274-025-04424-5 PMID: 40640643

Pfaff, G. (2022). The world of inorganic pigments. *Chemtexts*, *8*(3), 15. DOI: 10.1007/s40828-022-00166-1

Prasad, M. N. V. (2023). Minerogenic dust and human health. *Medical Geology: En route to One Health*, pp.65-82.

Prempeh, K. B. (2024). The role of economic growth, financial development, globalization, renewable energy and industrialization in reducing environmental degradation in the economic community of West African States. *Cogent Economics & Finance*, *12*(1), 2308675. DOI: 10.1080/23322039.2024.2308675

Radojević, I. D., & Jakovljević, V. D. (2025). Bacterial Potential for Bioremediation of Surfactants and Heavy Metals: Current Knowledge and Trends in Wastewater Treatment Processes. *Separations*, *12*(11), 308. DOI: 10.3390/separations12110308

Rajendran, S., Kalairaj, A., & Senthilvelan, T. J. B. C. (2025). A comprehensive review on enzymatic decolorization of various azo dyes using laccase for the abatement of industrial pollution. *Biomass Conversion and Biorefinery*, *15*(9), 13079–13101. DOI: 10.1007/s13399-024-06104-0

Rambabu, K., Banat, F., Pham, Q. M., Ho, S. H., Ren, N. Q., & Show, P. L. (2020). Biological remediation of acid mine drainage: Review of past trends and current outlook. *Environmental Science and Ecotechnology*, *2*, 100024. DOI: 10.1016/j.ese.2020.100024 PMID: 36160925

Rani, A., Snyder, S. W., Kim, H., Lei, Z., & Pan, S. Y. (2022). Pathways to a net-zero- carbon water sector through energy-extracting wastewater technologies. *NPJ Clean Water*, *5*(1), 49. DOI: 10.1038/s41545-022-00197-8

Rehan, Z. A., & Usman, A. (2023). Polymeric paints and coatings. In *Advanced Functional Polymers: Synthesis to Applications* (pp. 49–76). Springer Nature Singapore. DOI: 10.1007/978-981-99-0787-8_4

Reichl, C., & Schatz, M. (2022). *World Mining Data 2022* (Vol. 37). Austrian Federal Ministry of Agriculture, Regions and Tourism.

Rendón-Castrillón, L., Ramírez-Carmona, M., Ocampo-López, C., & Gómez-Arroyave, L. (2023). Bioleaching techniques for sustainable recovery of metals from solid matrices. *Sustainability (Basel)*, *15*(13), 10222. DOI: 10.3390/su151310222

Rosli, N. A., Aziz, H. A., Pueh, L. L. L., Othman, I. B., Zawawi, M. H., & Hung, Y. T. (2023). Management of various sources of hazardous waste. In *Waste Treatment in the Biotechnology, Agricultural and Food Industries* (Vol. 2, pp. 19–64). Springer International Publishing.

Ruffino, B., Campo, G., Idris, S. S., Salihoğlu, G., & Zanetti, M. (2023). Automotive paint sludge: A review of pretreatments and recovery options. *Resources*, *12*(4), 45. DOI: 10.3390/resources12040045

Saini, S., Tewari, S., Dwivedi, J., & Sharma, V. (2023). Biofilm-mediated wastewater treatment: A comprehensive review. *Materials Advances*, *4*(6), 1415–1443. DOI: 10.1039/D2MA00945E

Salihoglu, G., & Salihoglu, N. K. (2016). A review on paint sludge from automotive industries: Generation, characteristics and management. *Journal of Environmental Management*, *169*, 223–235. DOI: 10.1016/j.jenvman.2015.12.039 PMID: 26773426

Sathe, S. M., Bhowmick, G. D., Dubey, B. K., & Ghangrekar, M. M. (2020). Surfactant removal from wastewater using photo-cathode microbial fuel cell and laterite-based hybrid treatment system. *Bioprocess and Biosystems Engineering*, *43*(11), 2075–2084. DOI: 10.1007/s00449-020-02396-4 PMID: 32596770

Schwuger, M. J., & Smulders, E. J. (2020). Inorganic builders. In *Detergency* (pp. 371–439). CRC Press. DOI: 10.1201/9781003065494-6

Sebesta, J., Xiong, W., Guarnieri, M. T., & Yu, J. (2022). Biocontainment of genetically engineered algae. *Frontiers in Plant Science*, *13*, 839446. DOI: 10.3389/fpls.2022.839446 PMID: 35310623

Shanmukha, N. T., Vinayaka, M., & Lokeshappa, B. (2024). Biodiversity loss due to mining activities. In *Impact of societal development and infrastructure on biodiversity decline* (pp. 166–191). IGI Global.

Sharghi, E. A., & Davarpanah, L. (2022). Optimization of chemical coagulation–flocculation process of detergent manufacturing plant wastewater treatment for full scale applications: A case study. *Desalination and Water Treatment*, *262*, 38–53. DOI: 10.5004/dwt.2022.28494

Shirani Faradonbeh, R., Shah, M. I., Bahadori, M., & Jang, H. (2025). Toward Sustainable Mining: Exploring Alternative Mineral Resources and Innovative Extraction Techniques. *Mining (Basel)*, *5*(4), 66. DOI: 10.3390/mining5040066

Singh, A., Sharma, A., Verma, R. K., Chopade, R. L., Pandit, P. P., Nagar, V., & Sankhla, M. S. (2022). Heavy metal contamination of water and their toxic effect on living organisms. In *The toxicity of environmental pollutants*. IntechOpen. DOI: 10.5772/intechopen.105075

Singh, S., Sukla, L. B., & Goyal, S. K. (2020). Mine waste and circular economy. *Materials Today: Proceedings*, *30*, 332–339. DOI: 10.1016/j.matpr.2020.01.616

Szarka, A., Viktoryová, N., Horváth, G., Szalay, Z., Šimo, F., & Hrouzková, S. (2022). GC–MS methods for the evaluation of the performance of electrochemical water treatment for the degradation of pollutants from paint industry effluents. *Monatshefte für Chemie*, *153*(2), 161–169. DOI: 10.1007/s00706-022-02890-3

Takada, N., Kobayashi, T., & Oya, M. (2025). Interfacial Activity and Acute Toxicity of Surfactants to Aquatic Animals. *Journal of Water and Environment Technology*, *23*(3), 133–144. DOI: 10.2965/jwet.24-058

Talmage, S. S. (2020). *Environmental and human safety of major surfactants: alcohol ethoxylates and alkylphenol ethoxylates*. CRC Press. DOI: 10.1201/9780138757199

Vera, M. L., Traid, H. D., Dwojak, A. N., Rosenberger, M. R., & Schvezov, C. E. (2023). Advances in nanostructured TiO2 coatings for industrial applications. In *Industrial Applications of Nanoparticles* (pp. 228–255). CRC Press. DOI: 10.1201/9781003183525-14

Viktoryová, N., Szarka, A., & Hrouzková, S. (2022). Recent developments and emerging trends in paint industry wastewater treatment methods. *Applied Sciences (Basel, Switzerland)*, *12*(20), 10678. DOI: 10.3390/app122010678

Wang, H., & Ren, Z. J. (2014). Bio electrochemical metal recovery from wastewater: A review. *Water Research*, *66*, 219–232. DOI: 10.1016/j.watres.2014.08.013 PMID: 25216302

Wang, Z. (2022). *Geological and hydrgeological characteristics of acid mine drainage from an abandoned coal mine: A case study of Shandi coal mine in Niangziguan spring catchment*.

Weis, J. S. (2024). *Marine pollution*. Oxford University Press. DOI: 10.1093/wentk/9780197753804.001.0001

Wieszczycka, K. (2018). Wastes generated by mineral extraction industries. *Physical Sciences Reviews*, *3*(6). Advance online publication. DOI: 10.1515/psr-2018-0026

Wood, C. M., & Eom, J. (2021). The Osmo respiratory compromise in the fish gill. *Comparative Biochemistry and Physiology. Part A, Molecular & Integrative Physiology*, *254*, 110895. DOI: 10.1016/j.cbpa.2021.110895 PMID: 33429056

Xu, M., Guo, J., Zeng, G., Zhong, X., & Sun, G. (2006). Decolorization of anthraquinone dye by *Shewanella decolorationis* S12. *Applied Microbiology and Biotechnology*, *71*(2), 246–251. DOI: 10.1007/s00253-005-0144-1 PMID: 16160829

Yadav, D., & Dutta, J. (2024). Advancing environmental sustainability: Recent trends and developments in treatment methods for paint industry wastewater. *Journal of Water Process Engineering*, *61*, 105290. DOI: 10.1016/j.jwpe.2024.105290

Zafar, H., Ishaq, S., Peleato, N., & Roberts, D. (2022). Meta-analysis of operational performance and response metrics of microbial fuel cells (MFCs) fed with complex food waste. *Journal of Environmental Management*, *315*, 115152. DOI: 10.1016/j.jenvman.2022.115152 PMID: 35525044

Chapter 8
Microbial Bioremediation in Circular Economy: Integrating Waste-to-Resource Strategies

Elijah I. Ohimain
http://orcid.org/0000-0002-5491-6271
Niger Delta University, Nigeria

Oladele A. Ogunseitan
http://orcid.org/0000-0003-1317-6219
University of California, Irvine, USA

ABSTRACT

Beginning in the 18th century, the industrial revolution transformed societies with the mass production of manufactured products that also generated a legacy of environmental pollution worldwide. The revolution's environmental footprint has expanded to include the digital electronic revolution and a rapidly globalized supply chain characterized by linear "take-make-dispose" economic models. The adverse impacts of wastes including novel entities on planetary health are expected to transcend planetary boundaries if effective solutions are not implemented. The aspiration of circular economy frameworks presents unprecedented opportunities to transform waste management including harnessing of microbial bioremediation technologies through genetic engineering and improved knowledge of microbial diversity and ecosystem science. This chapter focuses on the integration of microbial bioremediation processes within circular economy framework, emphasizing waste-to-resource transformation strategies that simultaneously address environmental contamination and resource recovery.

DOI: 10.4018/979-8-2600-2998-5.ch008

1 INTRODUCTION

This chapter explores how microbial diversity can be harnessed within circular economy frameworks to simultaneously address environmental pollution and create valuable resources. We examine how the vast metabolic capabilities of Earth's microorganisms can transform waste streams into useful products, creating closed-loop systems that benefit both environmental quality and economic sustainability. We first address some of the pressing opportunities for harnessing microbial diversity in industrial applications, followed by an assessment of innovative approaches for artificially enhancing microbial properties, and risk-benefit considerations.

1.1 Bridging Microbial Solutions and Circular Systems

Microbial diversity defines the variety of microorganisms including bacteria, archaea, fungi plus yeasts, protists and viruses in an ecosystem. It encompasses their species, genetic makeup, morphological differences, landscape and micro-niches, and metabolic and functional profiles. Knowledge of microbial diversity is crucial for elucidating ecosystem functions and services, including climate regulation and nutrient cycling, and their vital contributions to the morbidity and mortality of macroorganisms across a broad phylogenetic spectrum (Ogunseitan, 2005, 2026).

Microbial diversity is the foundation for bioremediation in a circular economy where the byproducts of manufacturing and industrial processes are not designated as waste, but instead are recycled or repurposed as raw materials, and inadvertently polluted environments are cleaned. The broad range of metabolic capabilities, rapid rates of reproduction, brief generation time periods, and efficient genetic innovation and adaptive strategies of microorganisms form the foundation of circular economy. Circular economy has arisen as a vital strategy for sustainable development, especially because it results in waste reduction, reuse and upscaling, which can promote resource efficiency, conserve natural resources and mitigate environmental impact, potentially resulting in economic growth (Ohimain, 2025). Microorganisms, particularly bacteria, archaea, and fungi, are amazingly diverse and have been crucial in numerous biotechnological applications including the manufacture of biofuels from lignocellulosic wastes such as rice straws and bagasse in biorefineries (Yadav and Yadav, 2025).

Microorganisms which perform important functions in the biogeochemical cycling of elements, are vital for maintaining ecosystem services, and can be harnessed to develop innovative solutions for waste management, resource recovery, and sustainable production. By leveraging microbial diversity, we can create closed-loop systems that diminish waste generation and encourage more efficient use of resources.

There is a long history of solutions to environmental problems for which microorganisms are particularly suited, although the scale of implementation and sustainability of such solutions vary widely, especially in the following contexts:

- Sustainable Waste Management: Microorganisms can be used to break down organic substances especially food and other household wastes, which diminishes environmental impact and the volume of wastes destined for landfills. Microbial treatment of wastewater for instance, can result in the recovery of useful resources such as water, nutrients, and energy, which promotes sustainable waste management and waste-to wealth (Parde et al. 2024; Vidanage, 2024).
- Biodegradation and Bioremediation: Microorganisms can biodegrade and clean organic pollutants such as crude oil spills from contaminated environments. This practice reduces waste, promotes environmental sustainability and restores ecosystem health and vitality (Asim et al., 2021; Corrochano et al., 2025)
- Microbial-based ecosystem maintenance and restoration: Microbial diversity associated with sensitive features like mangrove ecosystems have been shown to contribute to their health, climate resilience and vitality, which have found applications in microbial-based maintenance of intact forest and restoration of impacted sites (Allard et al., 2020).
- Materials Resource Recovery: Microbes can recover useful substances such as nutrients, metals, energy, and other valuable materials from diverse waste streams, minimizing the need for primary production from virgin ingredients and minimizing waste disposal into landfills (Ohimain, 2025; Yadav and Yadav 2025).
- Sustainable/Bio-based Production: Microbial conversion technology such as hydrolysis, fermentation and anaerobic digestion can be used to produce biofuels, bioplastics, biofertilizer, biopesticides and other biochemicals, which can displace or decrease dependence on fossil fuels and chemicals. The spectrum of biochemicals and other bio-products can further increase, when biochemical methods are combined with thermochemical methods. Thus, promoting clean and sustainable production processes (Ohimain, 2025).

The development of microorganism-based solutions can create a multiproduct economy, with opportunities for employment in the fields of biotechnology, process engineering, renewable energy, environmental science and waste management with multiple industrial applications (Ohimain, 2025). While microbial diversity holds enormous potential for driving circular economy, there are potential challenges that need to be addressed such as the problem of scalability, insufficient regulatory framework and underdeveloped public awareness of benefits and risks (Arya and Kumar 2020; Naik and Eswari 2022). For instance, scaling up microbial-based processes to industrial levels while maintaining efficiency and cost-effectiveness is still a challenge (Kapoor et al., 2021; Ayub et al., 2025; Ferreira-Filipe et al. 2025). In the meantime, small -scale processes can be commercialized especially in high populated low-income countries including India and Nigeria (Yadav and Yadav 2025). Acceptability and public awareness of recycled or upcycled products through microbial technology is still low. Hence, increasing public consciousness pertaining to the safety and benefits of emerging products is urgently needed. The formulation of regulatory frameworks that support the advancement, commercialization and deployment of microbial-based products from circular economy is also essential (Naik and Eswari 2022; Sniatala et al., 2024).

1.2 The Circular Economy Paradigm in Environmental Microbiology

Traditionally, waste management used to be linear; make-use-dispose, but has now evolved to waste recovery in a circular fashion. The linear model is quite problematic because, wastes get to landfills prematurely before reaching its end-of-life. This practice is unsustainable because it does not conserve resources nor optimize inputs in the production of the product such as labour, energy and basic materials; thus, fueling the exploration of virgin materials. This practice has environmental impacts because it generates large volumes of wastes, mount pressure on landfills, cause emissions including greenhouse gases (GHGs) and impacts the environment from mineral extraction.

The circular economy paradigm is increasingly being applied to environmental science and waste management, offering a promising approach to sustainable development. The circular economy paradigm signifies a fundamental shift from conventional waste disposal methods to resource recovery, emphasizing the recycling and reuse of resources. This style purposes to lessen waste, conserve input, and encourage sustainable development, which aligns with global development agenda. Circular economy is a key mechanism to achieving net-zero emissions and United Nations Sustainable Development Goals (SDGs) especially SDG 13 (climate action) and other goals including SDG1 (no poverty), SDG2 (zero hunger), SDG3 (good health and well-being), SDG 6 (clean water and sanitation), SDG7 (affordable and clean energy), SDG8 (decent work and economic growth), SGD9 (industry, innovation and infrastructure), and SDG 12 (responsible consumption and production).

Kinejara (2023) discussed the history of waste management from 3Rs model (reduce, reuse, and recycle) to 4Rs model (Reduce, Reuse, Recycle, Repurpose), while Fisher (2024) discussed the evolution of the three Rs model (reuse, recycling, and recovery) of waste management to 5Rs (refuse, reduce, reuse, repurpose, and recycle), which is quite similar to refuse, reduce, reuse, recycle, rot model. Ohimain (2025) recently applied a 6-model approach (avoidance, minimize, reuse, recycle, recover, and disposal) for the handling and valorization of food wastes. Here, we present a 7Rs model integrating all the aforementioned models (Table 1). The first step in this hierarchy is to AVOID waste generation, which can be achieved by making informed choices such as refusing substances that has large environmental footprints, containing non-recyclables or banned substances such as single-use plastics. The second R is REDUCE, which involve minimizing wastes at source by reducing the volume and diversity of wastes, reduce emissions especially GHGs, reduce energy and material input. Also, you can choose to process agricultural produces in the farm to minimize waste generation at home. Waste reduction requires making sustainable choices and lifestyle. For instance, by using renewable energy such as solar instead of diesel generators can minimize emissions. The third and fourth Rs are REUSE and REPURPOSE respectively. Reuse involves using the item multiple times instead of single use, while repurpose involved the use of the item for alternative purposes. For instance, pet bottles can be reused for storing drinking water in the fridge or repurposed for storing peanuts. The fifth R is RECYCLE, where the wastes are processed using diverse techniques such as physical, biochemical or thermochemical to convert the waste to alternative products. For instance, some farmers partially process their products in the farm, can use the wastes for compost or manure production. The sixth R is RECOVERY, where energy and other materials are recovered from the wastes. The 6Rs model results in the maximizing resources salvage from wastes, before disposal (ROT), which is the 7th R.

Table 1. The 7Rs model of decreasing waste management hierarchy

R concept	Synonym	Definition and implications	Examples
Refuse	Avoid, prevent	Avoid the use of certain products that generate wastes. This could include substances that have large carbon footprint, non-recyclables, banned substances	Avoid the use of single use plastics
Reduce	Minimize	Minimize waste resulting in reduced consumption of materials and resources. It involves reducing the volume and diversity of wastes, air emissions particularly GHGs and energy and material input. Minimize waste generation at source.	Use of solar power instead of diesel generators can minimize emissions
Reuse	Repeated use	Repeated use of the material before disposal.	Reuse of pet bottles for water
Repurpose	Alternative use	Alternative use of the material	Repurposing of pet bottles to package peanut

continued on following page

Table 1. Continued

R concept	Synonym	Definition and implications	Examples
Recycle	Upcycling	Conversion to a new product using physical, biochemical or thermochemical processes	Conversion of plastics to bioplastics
Recover	Energy recovery	Conversion to energy using biochemical and thermochemical processes	Conversion of food wastes to biofuels
Rot	Dumping, dispose	Remnants after 6Rs are disposal in landfill	Reduced wastes

Microbial metabolism is a fundamental component of nature's circular system, serving critical functions in maintaining ecosystem balance and enhancing sustainability. One of the tools available to microbial diversity is their ability to exhibit diverse metabolic profile that play key roles in biodegradation and bioremediation, decomposition and nutrient cycling, and biogeochemical cycles of elements in nature (Atlas and Bartha, 1998; Madigan et al., 2012). Microbial decomposition and degradation of organic matter and other organic materials including wastes facilitates nutrient release and their reuse by other organisms resulting in nutrient cycling in nature. This process of waste reuse minimizes waste accumulation, can clean pollutants from contaminated environments, restoring ecosystem health and promoting circular economy. Microbes perform important functions in biogeochemical cycles, such as the carbon, nitrogen, iron, and sulphur cycles. Microbial metabolism provides crucial ecosystem services, such as decomposition and nutrient cycling that support ecosystem health and biodiversity (Ogunseitan 2026).

Microbes exhibit diverse metabolism that enables them to inhabit different habitats and metabolise a wide range of substrates including carbohydrates especially sugars, proteins including amino acids, lipids including fatty acids, hydrocarbons and other macromolecules, which increases their resilience and ability to live in extreme environments. The metabolic diversity of microbes is not paralleled by any other domain or kingdom of living things. Microbes can obtain nutrients and energy from diverse sources including organic and inorganic sources. For instance, some microbes are living chemotrophically from chemical substances, while others phototrophically from light. Some microbes obtain reducing equivalents used either for energy conservation or biosynthetic reactions lithotrophically from inorganic substances or organotrophically from organic substances. Some microbes can obtain carbon for synthesizing cell mass autotrophically by fixing atmospheric carbon dioxide or heterotrophically from organic substances, yet others called mixotrophs combine both approaches. Note that autotrophic fixation of carbon dioxide by microbes especially algae and cyanobacteria have a net positive effect on climate change. Some microbes called photolithoautotrophs such as Cyanobacteria, Chromaticaceae and Chlorobiaceae obtain carbon through CO_2 fixation, energy from light and using reducing equivalents from inorganic substances. Chemolithoautotrophs such as sulphur-oxidising bacteria, nitrifying bacteria, Knallgas-bacteria and iron-oxidising bacteria, obtain carbon from the fixation of CO_2 and energy from the oxidation of inorganic substances. Another group of microbes called chemolithoheterotrophs such as *Nitrobacter* spp. and *Wolinella*, cannot fix CO_2, but obtain energy from the oxidation of inorganic substances (Madigan et al., 2012).

Microbes exhibit diverse oxygen requirements ranging from anaerobic, microaerophilic and aerobic. Anaerobes also have subgroups such as strict anaerobes and facultative anaerobes. Microbes like other living things use oxygen as terminal electron acceptor but in addition, use many other substances such as nitrogen compounds (nitrogen gas, ammonia, nitrate, nitrite, dinitrogen oxide), CO_2, sulphur and sulphates, heavy metals (Fe^{3+}, Mn^{4+}, arsenate) during anaerobic respiration and many organic substances during fermentation. Fermentation provides a flexible strategy for the microbial use of carbon for the synthesis of diverse important organic biochemicals such as methanol, ethanol, propanol, isopropanol, butanol, butanediol, succinic acid, butyric acid, propionic acid, lactic acid, hydrogen, acetone, acetic acid, CO_2, formic acid. Microbes also synthesize diverse enzymes that make them to be versatile in their nutrition, respiration and energy usage, all of which play major role in nature's circular economy and provide opportunities for industrial biomimetic approaches (Atlas and Bartha, 1998; Madigan et al., 2012).

Industrial biomimetic approaches by imitating natural processes, designs and efficiencies can play important roles in the development of circular economy. It can assist to minimize waste, preserve resources, and enhance sustainable production, which can reduce environmental impact and improve product efficiencies, all of which are important steps toward net-zero. Key biomimetic concept and approaches can involve applying biologically inspired designs using natural processes for the development of biologically inspired materials. Bio-inspired design can involve the application of design principles from nature to develop products and systems. Industrial processes can be designed that mimic natural processes, such as photosynthesis, respiration, fermentation, bioprecipitation or other processes (Mitikie and Elsaigh, 2025; Sandbhor et al., 2025). Finally, biologically inspired materials that mimic nature's structures and properties, such as self-healing materials and nanostructured materials can be manufactured sustainably (Amran et al., 2022; Rollakanti and Srinivasu 2022; Uddin et al., 2023).

2 CUTTING-EDGE TECHNOLOGIES AND INNOVATIONS

To keep novel entities within safe operating zones of planetary boundaries requires reducing production, use, and disposal of hazardous chemicals and materials while developing innovative technologies for cleaning polluted environments (Ogunseitan, 2023). The convergence of microbial bioremediation with circular economy principles represents one of the most promising frontiers in environmental biotechnology, offering transformative solutions to the interconnected challenges of waste management, resource depletion, and environmental pollution (Sharma et al., 2021; Koul & Taak, 2018). The transition from linear "take-make-dispose" models to regenerative systems, positions microorganisms as ultimate circular economy engineers—organisms that have perfected waste-to-resource conversion over billions of years of evolution (MacArthur, 2013; Singh & Singh, 2019). These microscopic entities possess metabolic capabilities that far exceed current chemical engineering approaches, operating at ambient temperatures and pressures while achieving remarkable specificity and efficiency in transforming complex organic and inorganic materials (Lovley, 2003).

Integration of microbial processes into circular economy frameworks simultaneously addresses multiple sustainability challenges, including climate change mitigation, biodiversity conservation, pollution reduction, and sustainable resource management (Ghisellini et al., 2016; Kirchherr et al., 2017). Unlike conventional waste treatment approaches that focus solely on pollution mitigation through destruction or containment, microbial bioremediation in circular economy contexts transforms waste from environmental liability into economic asset, creating value streams that incentivize proper waste management while reducing dependence on virgin resource extraction (Kirchherr et al., 2017; Stahel, 2016). This paradigm shift fundamentally alters how industries, municipalities, and societies conceptualize waste management and resource utilization, moving from end-of-pipe solutions to integrated systems that prevent waste generation while maximizing resource productivity (Geissdoerfer et al., 2017; Lacy & Rutqvist, 2015).

The economic implications of this transition are profound. The Ellen MacArthur Foundation estimates that circular economy transitions could generate $4.5 trillion in economic benefits by 2030 through reduced material costs, innovation opportunities, and new business models (Ellen MacArthur Foundation, 2015). Microbial technologies are positioned to capture significant portions of this value, particularly in sectors such as waste management, chemical production, materials recovery, and energy generation (Philp & Winickoff, 2018). Moreover, circular economy approaches create employment opportunities across skill levels, from community-based waste collection and sorting to advanced biotechnology research and development, potentially generating millions of jobs globally while addressing environmental challenges (Stahel, 2016; Lacy & Rutqvist, 2015).

The environmental benefits extend beyond waste reduction. Microbial circular economy systems can significantly reduce greenhouse gas emissions by displacing fossil fuel-based energy and materials, preventing methane emissions from landfills, and sequestering carbon in stable organic matter (Paritosh et al., 2017; Smith et al., 2014). Life cycle assessments consistently demonstrate that biological waste treatment approaches have lower environmental footprints than thermal or chemical alternatives, particularly when resource recovery is incorporated (Laurent et al., 2014). Additionally, these systems can reduce pressure on natural ecosystems by decreasing demand for virgin resource extraction, thereby contributing to biodiversity conservation and ecosystem health (Ghisellini et al., 2016).

2.1 Synthetic Biology Applications

Synthetic biology has revolutionized our ability to design and engineer microorganisms with precisely tailored capabilities for waste transformation, moving beyond traditional genetic modification approaches to enable systematic, predictable engineering of biological systems (Benner, 2005; Cameron et al., 2014; Church et al., 2014). This engineering approach draws upon principles from electrical engineering, computer science, and systems biology to create biological systems with predictable, controllable behaviors analogous to electronic circuits (Khalil & Collins, 2010). The field has matured from proof-of-concept demonstrations to practical applications addressing real-world environmental challenges, with engineered organisms now operating in industrial facilities worldwide (Nielsen & Keasling, 2016; Chae et al., 2017).

The conceptual foundation of synthetic biology rests on several key principles: standardization, decoupling, and abstraction (Endy, 2005). Standardization involves creating interchangeable biological parts with well-characterized functions, analogous to standardized electronic components. Decoupling separates the design process from fabrication, allowing researchers to focus on system-level functions rather than molecular details. Abstraction organizes biological complexity into hierarchical levels—from DNA sequences to genetic circuits to cellular behaviors—enabling engineers to work at appropriate levels without being overwhelmed by underlying complexity (Endy, 2005; Canton et al., 2008).

Standardized biological molecular components—promoters that control gene expression, ribosome binding sites that regulate translation, coding sequences that specify protein functions, and terminators that define transcript boundaries—have dramatically accelerated synthetic biology applications by enabling modular assembly of genetic circuits (Canton et al., 2008; Salis et al., 2009). These standardized components, catalogued in repositories like the Registry of Standard Biological Parts maintained by the iGEM Foundation, enable researchers worldwide to build upon previous work rather than starting from scratch, accelerating innovation and reducing duplication of effort (Smolke, 2009; Casini et al., 2015). The registry now contains over 20,000 characterized biological parts, with detailed documentation of their functions, performance characteristics, and optimal applications (iGEM Foundation, 2021).

Computational tools for genetic circuit design have automated much of the process of converting desired functions into DNA sequences, dramatically reducing the time and expertise required for strain engineering (Nielsen et al., 2016; Appleton et al., 2017). Software platforms like Cello, developed at MIT, can translate high-level functional specifications into complete genetic circuit designs, selecting appropriate parts from libraries and optimizing circuit architecture for desired performance characteristics (Nielsen et al., 2016). These tools incorporate machine learning algorithms that improve predictions based on experimental data, creating positive feedback loops that continuously enhance design accuracy (Hossain et al., 2020). Additionally, DNA synthesis costs have plummeted from over $10 per base pair in the 1980s to less than $0.10 per base pair today, making it economically feasible to construct large, complex genetic circuits (Carlson, 2016).

2.2 Engineered Microorganisms for Specific Waste Streams

Designer microbes for targeted pollutant degradation represent perhaps the most direct application of synthetic biology to circular economy objectives, offering precision tools for addressing specific waste streams that resist conventional treatment (Strong et al., 2015; Dvořák et al., 2017). Researchers have developed engineered strains capable of degrading polyethylene terephthalate (PET) plastics, breaking down per- and polyfluoroalkyl substances (PFAS), metabolizing complex pharmaceutical residues that resist conventional treatment, and transforming industrial chemicals into valuable products (Wierckx et al., 2015; Gong et al., 2022; Elmore et al., 2023).

The discovery of *Ideonella sakaiensis* in 2016 marked a watershed moment in biological plastic degradation research, demonstrating that microorganisms could evolve novel capabilities to metabolize synthetic polymers that have existed for less than a century (Yoshida et al., 2016). This bacterium, isolated from sediments outside a plastic bottle recycling facility in Japan, produces two enzymes—PETase and MHETase—that work synergistically to break down PET into its constituent monomers (Tanasupawat et al., 2016; Yoshida et al., 2016). PETase hydrolyzes PET into mono(2-hydroxyethyl) terephthalic acid (MHET) and other intermediates, while MHETase further degrades MHET into terephthalic acid and ethylene glycol, which the bacterium can then metabolize as carbon and energy sources (Austin et al., 2018).

Through directed evolution—systematically introducing mutations and selecting variants with improved performance—and rational protein design based on structural analysis, researchers have created enzyme variants operating efficiently at higher temperatures and broader pH ranges, making them suitable for industrial-scale plastic recycling (Tournier et al., 2020; Lu et al., 2022; Bell et al., 2022). French company

Carbios has developed an engineered enzyme variant that degrades PET at 72°C with reaction rates 10,000 times faster than the native enzyme, achieving near-complete depolymerization within 10 hours (Tournier et al., 2020). These engineered systems can depolymerize PET into constituent monomers—terephthalic acid and ethylene glycol—which can be purified and repolymerized into virgin-quality plastic, creating truly circular material flows that avoid the quality degradation associated with mechanical recycling (Austin et al., 2018; Zimmermann & Dombrowski, 2020).

The economic and environmental implications are substantial. Global PET production exceeds 30 million metric tons annually, with significant portions ending up in landfills or natural environments (Geyer et al., 2017). Enzymatic recycling could recover these materials while using 30-50% less energy than producing virgin PET from petroleum feedstocks (Zimmermann & Dombrowski, 2020). Carbios has announced plans for a demonstration plant capable of processing 40,000 metric tons of PET waste annually, with commercial operations beginning in 2025 (Carbios, 2022). Major brands including PepsiCo, L'Oréal, and Nestlé have partnered with Carbios to incorporate enzymatically recycled PET into their packaging, signaling growing commercial acceptance (Carbios, 2022).

Beyond PET, synthetic biology approaches are addressing other problematic plastics. Researchers have engineered strains of *Pseudomonas putida* capable of metabolizing polyurethane, a versatile polymer used in foams, coatings, and adhesives that is particularly difficult to recycle through conventional means (Magnin et al., 2019). Other teams have developed microbial systems for degrading polystyrene, polyethylene, and mixed plastic waste streams (Wierckx et al., 2015; Amobonye et al., 2021). While these technologies remain at earlier development stages than PET degradation, they demonstrate the breadth of synthetic biology's potential for addressing plastic waste challenges.

Synthetic biology has also enabled creation of microbial consortia specifically designed for electronic waste processing, one of the fastest-growing and most problematic waste streams globally (Ilyas et al., 2007; Işıldar et al., 2016; Kaya, 2016). Electronic devices contain complex mixtures of valuable metals (gold, silver, copper, palladium), rare earth elements (neodymium, dysprosium, europium), and hazardous substances (lead, mercury, brominated flame retardants). Conventional e-waste recycling relies primarily on mechanical shredding followed by pyrometallurgical or hydrometallurgical processing, approaches that are energy-intensive, generate toxic emissions, and often fail to recover many valuable materials (Forti et al., 2020).

Engineered microbial systems offer more selective, environmentally benign alternatives. Bioleaching strains can solubilize precious metals from circuit boards through production of organic acids and oxidizing agents, achieving recovery rates exceeding 95% for gold and copper under optimized conditions (Ilyas & Lee, 2014a, 2014b; Pollmann et al., 2018). Researchers have engineered strains with enhanced metal tolerance and acid production capabilities, improving performance in the harsh chemical environments of e-waste leachates (Kaya, 2016; Işıldar et al., 2016). Additionally, synthetic biology approaches have created strains capable of biosorbing and bioaccumulating specific metals from dilute solutions, enabling recovery of rare earth elements that are critical for renewable energy technologies but difficult to extract through conventional means (Hennebel et al., 2015; Pollmann et al., 2018).

Engineered organisms are also addressing pharmaceutical waste, an emerging environmental concern with significant ecological and human health implications (Larsson et al., 2007; Larsson et al., 2018; Gong et al., 2022). Pharmaceutical residues enter aquatic environments through wastewater treatment plant effluents, agricultural runoff, and improper disposal, where they can disrupt endocrine systems, promote antimicrobial resistance, and harm aquatic organisms at concentrations as low as nanograms per liter (Daughton & Ternes, 1999; Kümmerer, 2009). Conventional wastewater treatment removes only 30-90% of pharmaceutical compounds, with highly variable efficiency depending on compound properties and treatment conditions (Verlicchi et al., 2012).

Synthetic biology approaches have created microbial strains with enhanced capabilities for degrading specific pharmaceutical compounds. Researchers have engineered *Sphingobium* species with expanded substrate ranges for degrading nonsteroidal anti-inflammatory drugs, engineered *Rhodococcus* strains capable of metabolizing steroid hormones, and developed consortia that cooperatively degrade complex antibiotic molecules (Gong et al., 2022; Larsson et al., 2018; Nguyen et al., 2020). These systems can be integrated into wastewater treatment processes as tertiary treatment steps, significantly reducing pharmaceutical loads in effluents (Gong et al., 2022).

The development of biosensors represents another important synthetic biology application for circular economy systems (Gui et al., 2017; Wan et al., 2019). Engineered microorganisms can be designed to produce detectable signals—fluorescence, color changes, electrical currents—in response to specific compounds, enabling real-time monitoring of waste streams and treatment processes (Gui et al., 2017). These living sensors offer advantages over conventional analytical methods in terms of cost, speed, and ability to assess bioavailability rather than just total concentrations (Wan et al., 2019). Whole-cell biosensors have been developed for detecting heavy metals, organic pollutants, nutrients, and process inhibitors, providing valuable tools for optimizing bioremediation systems (Gui et al., 2017; Bereza-Malcolm et al., 2015).

2.3 Biosafety Considerations and Containment Strategies

Biosafety considerations and containment strategies remain paramount as synthetic biology creates engineered organisms for transitioning from laboratory to field applications (Wright et al., 2013; Trump et al., 2020). The Cartagena Protocol on Biosafety provides an international framework for safe transfer, handling, and use of genetically modified organisms (Secretariat of the Convention on Biological Diversity, 2000).

Multiple containment strategies have been developed to address biosafety concerns (Moe-Behrens et al., 2013; Wright et al., 2013). Auxotrophic strains engineered to require synthetic amino acids unavailable in natural environments provide metabolic containment, ensuring organisms cannot survive outside controlled systems (Mandell et al., 2015). Genetic kill switches trigger cell death in response to specific environmental signals or absence of particular compounds (Chan et al., 2016). Genomic recoding creates organisms genetically isolated from natural populations (Fredens et al., 2019).

Regulatory pathways for engineered organism deployment vary significantly across jurisdictions (Marchant & Stevens, 2015; Trump et al., 2020). Progressive regulatory frameworks are emerging that recognize distinctions between contained use in industrial bioreactors and environmental release, applying proportionate oversight based on actual risk profiles (Trump et al., 2020).

2.4 Bioelectrochemical Systems

Bio-electrochemical systems (BES) represent a paradigm shift in microbial waste processing, transforming microorganisms from passive degraders into active participants in energy conversion and resource recovery (Logan & Rabaey, 2012; Lovley & Nevin, 2013). These systems exploit electroactive bacteria that exchange electrons directly with electrodes, effectively wiring microbial metabolism to electrical circuits (Lovley, 2012; Kumar et al., 2017).

Microbial fuel cells (MFCs) generate electricity from organic waste through microbial oxidation at anodes (Logan & Rabaey, 2012). While power densities remain modest—typically 1-10 watts per square meter of electrode surface—MFCs offer advantages for remote or resource-limited applications (Slate et al., 2019). Microbial electrolysis cells (MECs) reverse this process, applying small voltage to drive reactions that would otherwise be thermodynamically unfavorable (Logan et al., 2008).

2.5 Biohydrogen Production from Organic Waste

MECs operating on organic substrates require only 0.2-0.8 volts because microorganisms provide electrons from organic oxidation rather than water splitting (Logan et al., 2008). This dramatic reduction in energy input makes hydrogen production from waste economically attractive, particularly when coupled with renewable electricity sources (Kadier et al., 2016).

Microbial electrosynthesis systems use electroactive microorganisms at cathodes to accept electrons and reduce CO_2 into organic compounds (Nevin et al., 2010; Bajracharya et al., 2017). This technology could enable carbon-negative chemical production, capturing CO_2 from industrial emissions and converting it into valuable products using renewable electricity (Prévoteau et al., 2020).

2.6 Advanced Monitoring and Control

Advanced monitoring and control technologies are transforming microbial waste treatment, providing unprecedented insight into microbial processes and enabling precision management (Vanwonterghem et al., 2014; Bouchez et al., 2016). Next-generation sequencing technologies enable continuous, automated monitoring of microbial populations and activities (Knight et al., 2018). Portable sequencing devices like Oxford Nanopore's MinION enable on-site DNA and RNA sequencing with results available within hours (Jain et al., 2016).

2.7 Artificial Intelligence-Driven Process Optimization

AI-driven process optimization leverages massive data streams from advanced monitoring systems to identify patterns, predict outcomes, and optimize operations (Corominas et al., 2018; Newhart et al., 2019). Machine learning algorithms analyze sensor data and equipment performance metrics to forecast component failures, enabling scheduled maintenance before breakdowns occur (Newhart et al., 2019). These AI systems continuously improve as they accumulate data, identifying optimal strategies that maximize resource recovery while minimizing energy inputs.

2.8 Predictive Modeling for System Performance

Predictive modeling integrates mechanistic understanding of microbial processes with data-driven approaches to forecast system behavior (Batstone et al., 2015; Henze et al., 2000). The Anaerobic Digestion Model No. 1 (ADM1) exemplifies mechanistic modeling approaches, describing biochemical and physicochemical processes in anaerobic digesters (Batstone et al., 2002). Digital twin technology creates virtual replicas of physical bioremediation systems that mirror real-world performance in real-time (Negri et al., 2017).

2.9 Systems Integration and Optimization

Integrated biorefineries combining multiple microbial processes maximize resource recovery and economic value (Kamm & Kamm, 2004; Cherubini, 2010). Cascading value recovery—sequentially extracting different value streams from waste materials—maximizes economic returns (Mussatto, 2014). For example, food waste might first undergo extraction of high-value compounds, then protein recovery, followed by anaerobic digestion for biogas production, with final digestate used as fertilizer (Paritosh et al., 2017).

Hybrid systems combining microbial processes with chemical, thermal, or electrochemical treatments often outperform single-technology approaches (Sharma et al., 2021). Process intensification—achieving greater productivity in smaller reactor volumes—reduces capital costs and physical footprints (Stankiewicz & Moulijn, 2000). High-rate anaerobic digesters using granular sludge achieve organic loading rates 5-10 times higher than conventional systems (Paritosh et al., 2017).

3 CONCLUDING THOUGHTS

The chapter argues that combining microbial bioremediation with circular economy principles is more than a waste-management fix. It represents a shift from linear, extractive systems that deplete resources and create pollution to regenerative systems that keep materials in use and restore environmental health. By treating waste as "misplaced resources" and using microbes that have evolved metabolic capabilities over billions of years, we can simultaneously tackle environmental degradation, resource scarcity, and climate change while creating economic value and improving quality of life. Four converging trends make this transition possible now. Advances in biotechnology provide stronger tools for engineering organisms, understanding ecosystems, and optimizing processes. Rising public concern about climate change, pollution, and resource depletion creates social and political pressure for change. Policy support is growing through circular economy strategies, climate commitments, and the Sustainable Development Goals, which set frameworks and incentives. Economic drivers are also aligning, as resource price volatility, high waste-management costs, and new business opportunities make circular models more attractive. Realizing this vision requires coordinated action. Policymakers need enabling frameworks with regulations, incentives, public investment, and institutional reform to ensure equitable outcomes. Businesses must adopt circular business models that create value through resource productivity rather than extraction, which can improve competitiveness while lowering impacts. Researchers should push technological boundaries and bridge the gap between lab demos and real-world, context-appropriate solutions. Communities must help shape transitions to reflect local values and ensure cultural and social benefits.

DECLARATION OF CONFLICT-OF-INTEREST STATEMENT

The authors have no conflict of interest in this work.

ACKNOWLEDGEMENT

OAO is deeply grateful for all my mentors and mentees on the quest to understand microbial diversity and the contributions of the "unseen world" to our visible world; and for the resources available for current academic work through the University of California Presidential Chair funds; and the World Institute for Sustainable Development of Materials (WISDOM).

REFERENCES

Allard, S. M., Costa, M. T., Bulseco, A. N., Helfer, V., Wilkins, L. G., Hassenrück, C., & Bowman, J. (2020). Introducing the mangrove microbiome initiative: Identifying microbial research priorities and approaches to better understand, protect, and rehabilitate mangrove ecosystems. *mSystems*, *5*(5), e00658–e20. DOI: 10.1128/mSystems.00658-20 PMID: 33082281

Amobonye, A., Bhagwat, P., Singh, S., & Pillai, S. (2021). Plastic biodegradation: Frontline microbes and their enzymes. *The Science of the Total Environment*, *759*, 143536. DOI: 10.1016/j.scitotenv.2020.143536 PMID: 33190901

Amran, M., Onaizi, A. M., Fediuk, R., Vatin, N. I., Muhammad Rashid, R. S., Abdelgader, H., & Ozbakkaloglu, T. (2022). Self-healing concrete as a prospective construction material: A Review. *Materials (Basel)*, *15*(9), 3214. DOI: 10.3390/ma15093214 PMID: 35591554

Appleton, E., Madsen, C., Roehner, N., & Densmore, D. (2017). Design automation in synthetic biology. *Cold Spring Harbor Perspectives in Biology*, *9*(4), a023978. DOI: 10.1101/cshperspect.a023978 PMID: 28246188

Arya, S., & Kumar, S. (2020). Bioleaching: Urban mining option to curb the menace of E-waste challenge. *Bioengineered*, *11*(1), 640–660. DOI: 10.1080/21655979.2020.1775988 PMID: 32538256

Asim, N., Badiei, M., Torkashvand, M., Mohammad, M., Alghoul, M. A., Gasaymeh, S. S., & Sopian, K. (2021). Wastes from the petroleum industries as sustainable resource materials in construction sectors: Opportunities, limitations, and directions. *Journal of Cleaner Production*, *284*, 125459. DOI: 10.1016/j.jclepro.2020.125459

Atlas, R. M., & Bartha, R. (1998). *Microbial ecology: Fundamentals and applications* (4th ed.). Benjamin/Cummings.

Austin, H. P., Allen, M. D., Donohoe, B. S., Rorrer, N. A., Kearns, F. L., Silveira, R. L., Pollard, B. C., Dominick, G., Duman, R., El Omari, K., Mykhaylyk, V., Wagner, A., Michener, W. E., Amore, A., Skaf, M. S., Crowley, M. F., Thorne, A. W., Johnson, C. W., Woodcock, H. L., & Beckham, G. T. (2018). Characterization and engineering of a plastic-degrading aromatic polyesterase. *Proceedings of the National Academy of Sciences of the United States of America, 115*(19), E4350–E4357. DOI: 10.1073/pnas.1718804115 PMID: 29666242

Ayub, A., Wani, A. K., Malik, S. M., Ayub, M., Singh, R., Chopra, C., & Malik, T. (2025). Green nanoscience for healthcare: Advancing biomedical innovation through eco-synthesized nanoparticle. *Biotechnology Reports (Amsterdam, Netherlands), 00913*, e00913. Advance online publication. DOI: 10.1016/j.btre.2025.e00913 PMID: 40895589

Bajracharya, S., ter Heijne, A., Dominguez Benetton, X., Vanbroekhoven, K., Buisman, C. J. N., Strik, D. P. B. T. B., & Pant, D. (2017). Carbon dioxide reduction by mixed and pure cultures in microbial electrosynthesis using an assembly of graphite felt and stainless steel as a cathode. *Bioresource Technology, 195*, 14–24. DOI: 10.1016/j.biortech.2015.05.081 PMID: 26066971

Batstone, D. J., Keller, J., Angelidaki, I., Kalyuzhnyi, S. V., Pavlostathis, S. G., Rozzi, A., Sanders, W. T. M., Siegrist, H., & Vavilin, V. A. (2002). The IWA Anaerobic Digestion Model No 1 (ADM1). *Water Science and Technology, 45*(10), 65–73. DOI: 10.2166/wst.2002.0292 PMID: 12188579

Batstone, D. J., Puyol, D., Flores-Alsina, X., & Rodríguez, J. (2015). Mathematical modelling of anaerobic digestion processes: Applications and future needs. *Reviews in Environmental Science and Biotechnology, 14*(4), 595–613. DOI: 10.1007/s11157-015-9376-4

Bell, E. L., Smithson, R., Kilbride, S., Foster, J., Hardy, F. J., Ramachandran, S., Tedstone, A. A., Haigh, S. J., Garforth, A. A., Day, P. J. R., Levy, C., Shaver, M. P., & Green, A. P. (2022). Directed evolution of an efficient and thermostable PET depolymerase. *Nature Catalysis, 5*(8), 673–681. DOI: 10.1038/s41929-022-00821-3

Benner, S. A. (2005). Synthetic biology: Act natural. *Nature, 421*(6919), 118. DOI: 10.1038/421118a PMID: 12520282

Bereza-Malcolm, L. T., Mann, G., & Franks, A. E. (2015). Environmental sensing of heavy metals through whole cell microbial biosensors: A synthetic biology approach. *ACS Synthetic Biology, 4*(5), 535–546. DOI: 10.1021/sb500286r PMID: 25299321

Bouchez, T., Blieux, A. L., Dequiedt, S., Domaizon, I., Dufresne, A., Ferreira, S., Godon, J. J., Hellal, J., Joulian, C., Quaiser, A., Martin-Laurent, F., Mauffret, A., Monier, J. M., Peyret, P., Schmitt-Koplin, P., Sibourg, O., D'oiron, E., Bispo, A., Deportes, I., & Ranjard, L. (2016). Molecular microbiology methods for environmental diagnosis. *Environmental Chemistry Letters*, *14*(4), 423–441. DOI: 10.1007/s10311-016-0581-3

Cameron, D. E., Bashor, C. J., & Collins, J. J. (2014). A brief history of synthetic biology. *Nature Reviews. Microbiology*, *12*(5), 381–390. DOI: 10.1038/nrmicro3239 PMID: 24686414

Canton, B., Labno, A., & Endy, D. (2008). Refinement and standardization of synthetic biological parts and devices. *Nature Biotechnology*, *26*(7), 787–793. DOI: 10.1038/nbt1413 PMID: 18612302

. Carbios. (2022). *Annual Report 2021*. Carbios SA.

Carlson, R. (2016). Estimating the biotech sector's contribution to the US economy. *Nature Biotechnology*, *34*(3), 247–255. DOI: 10.1038/nbt.3491 PMID: 26963545

Casini, A., Storch, M., Baldwin, G. S., & Ellis, T. (2015). Bricks and blueprints: Methods and standards for DNA assembly. *Nature Reviews. Molecular Cell Biology*, *16*(9), 568–576. DOI: 10.1038/nrm4014 PMID: 26081612

Chae, T. U., Choi, S. Y., Kim, J. W., Ko, Y. S., & Lee, S. Y. (2017). Recent advances in systems metabolic engineering tools and strategies. *Current Opinion in Biotechnology*, *47*, 67–82. DOI: 10.1016/j.copbio.2017.06.007 PMID: 28675826

Chan, C. T. Y., Lee, J. W., Cameron, D. E., Bashor, C. J., & Collins, J. J. (2016). 'Deadman' and 'Passcode' microbial kill switches for bacterial containment. *Nature Chemical Biology*, *12*(2), 82–86. DOI: 10.1038/nchembio.1979 PMID: 26641934

Cherubini, F. (2010). The biorefinery concept: Using biomass instead of oil for producing energy and chemicals. *Energy Conversion and Management*, *51*(7), 1412–1421. DOI: 10.1016/j.enconman.2010.01.015

Church, G. M., Elowitz, M. B., Smolke, C. D., Voigt, C. A., & Weiss, R. (2014). Realizing the potential of synthetic biology. *Nature Reviews. Molecular Cell Biology*, *15*(4), 289–294. DOI: 10.1038/nrm3767 PMID: 24622617

Corominas, L., Garrido-Baserba, M., Villez, K., Olsson, G., Cortés, U., & Poch, M. (2018). Transforming data into knowledge for improved wastewater treatment operation: A critical review of techniques. *Environmental Modelling & Software*, *106*, 89–103. DOI: 10.1016/j.envsoft.2017.11.023

Corrochano, N., de Tuesta, J. D., de Mora, A., Pariente, M. I., Segura, Y., Molina, R., & Martínez, F. (2025). Valorization of refinery sludges by hydrothermal carbonization towards circular economy. *Waste Management, 204*, 114949. DOI: 10.1016/j.wasman.2025.114949 PMID: 40527174

Daughton, C. G., & Ternes, T. A. (1999). Pharmaceuticals and personal care products in the environment: Agents of subtle change? *Environmental Health Perspectives, 107*(Suppl 6), 907–938. DOI: 10.1289/ehp.99107s6907 PMID: 10592150

Dvořák, P., Nikel, P. I., Damborský, J., & de Lorenzo, V. (2017). Bioremediation 3.0: Engineering pollutant-removing bacteria in the times of systemic biology. *Biotechnology Advances, 35*(7), 845–866. DOI: 10.1016/j.biotechadv.2017.08.001 PMID: 28789939

Ellen MacArthur Foundation. (2015). *Towards a circular economy: Business rationale for an accelerated transition.* Ellen MacArthur Foundation. https://www.ellenmacarthurfoundation.org/towards-a-circular-economy-business-rationale-for-an-accelerated-transition

Elmore, J. R., Dexter, G. N., Salvachúa, D., Martinez-Baird, J., Hatmaker, E. A., Huenemann, J. D., Klingeman, D. M., Peabody, G. L., Peterson, D. J., Singer, C., Beckham, G. T., & Guss, A. M. (2023). Production of itaconic acid from alkali pretreated lignin by dynamic two stage bioconversion. *Nature Communications, 14*, 2494. DOI: 10.1038/s41467-023-38012-5 PMID: 33859194

Endy, D. (2005). Foundations for engineering biology. *Nature, 438*(7067), 449–453. DOI: 10.1038/nature04342 PMID: 16306983

Ferreira-Filipe, D. A., Duarte, A. C., Hursthouse, A. S., Rocha-Santos, T., & Silva, A. L. P. (2025). Biobased Strategies for E-Waste Metal Recovery: A Critical Overview of Recent Advances. *Environments (Basel, Switzerland), 12*(1), 26. DOI: 10.3390/environments12010026

Fisher, C. (2024). What are the 5 R's of Waste Management — Refuse, Reduce, Reuse, Repurpose, Recycle. https://www.rts.com/blog/the-5-rs-refuse-reduce-reuse-repurpose-recycle/. Published on March 10, 2024. Downloaded 12 October 2025

. Forti, V., Baldé, C. P., Kuehr, R., & Bel, G. (2020). *The global e-waste monitor 2020: Quantities, flows and the circular economy potential.* United Nations University/United Nations Institute for Training and Research, International Telecommunication Union, and International Solid Waste Association.

Fredens, J., Wang, K., de la Torre, D., Funke, L. F. H., Robertson, W. E., Christova, Y., Chia, T., Schmied, W. H., Dunkelmann, D. L., Beránek, V., Uttamapinant, C., Llamazares, A. G., Elliott, T. S., & Chin, J. W. (2019). Total synthesis of *Escherichia coli* with a recoded genome. *Nature, 569*(7757), 514–518. DOI: 10.1038/s41586-019-1192-5 PMID: 31092918

Geissdoerfer, M., Savaget, P., Bocken, N. M. P., & Hultink, E. J. (2017). The circular economy: A new sustainability paradigm? *Journal of Cleaner Production, 143*, 757–768. DOI: 10.1016/j.jclepro.2016.12.048

Geyer, R., Jambeck, J. R., & Law, K. L. (2017). Production, use, and fate of all plastics ever made. *Science Advances, 3*(7), e1700782. DOI: 10.1126/sciadv.1700782 PMID: 28776036

Ghisellini, P., Cialani, C., & Ulgiati, S. (2016). A review on circular economy: The expected transition to a balanced interplay of environmental and economic systems. *Journal of Cleaner Production, 114*, 11–32. DOI: 10.1016/j.jclepro.2015.09.007

Gong, H., Gao, J., & Jin, Z. (2022). Biodegradation of pharmaceutical compounds in wastewater: Mechanisms and applications. In Varjani, S., Pandey, A., Gnansounou, E., Khanal, S. K., & Raveendran, S. (Eds.), *Bioremediation for environmental sustainability* (pp. 393–421). Elsevier., DOI: 10.1016/B978-0-323-89874-4.00010-4

Gui, Q., Lawson, T., Shan, S., Yan, L., & Liu, Y. (2017). The application of whole cell-based biosensors for use in environmental analysis and in medical diagnostics. *Sensors (Basel), 17*(7), 1623. DOI: 10.3390/s17071623 PMID: 28703749

Hennebel, T., Boon, N., Maes, S., & Lenz, M. (2015). Biotechnologies for critical raw material recovery from primary and secondary sources: R&D priorities and future perspectives. *New Biotechnology, 32*(1), 121–127. DOI: 10.1016/j.nbt.2013.08.004 PMID: 23994422

Henze, M., Gujer, W., Mino, T., & van Loosdrecht, M. (2000). *Activated sludge models ASM1, ASM2, ASM2d and ASM3*. IWA Publishing.

Hossain, A., Lopez, E., Halper, S. M., Cetnar, D. P., Reis, A. C., Strickland, D., Klavins, E., & Tabor, J. J. (2020). Automated design of thousands of nonrepetitive parts for engineering stable genetic systems. *Nature Biotechnology, 38*(12), 1466–1475. DOI: 10.1038/s41587-020-0584-2 PMID: 32661437

iGEM Foundation. (2021). *Registry of standard biological parts*. http://parts.igem.org

Ilyas, S., Anwar, M. A., Niazi, S. B., & Ghauri, M. A. (2007). Bioleaching of metals from electronic scrap by moderately thermophilic acidophilic bacteria. *Hydrometallurgy, 88*(1-4), 180–188. DOI: 10.1016/j.hydromet.2007.04.007

Ilyas, S., & Lee, J. C. (2014a). Biometallurgical recovery of metals from waste electrical and electronic equipment: A review. *ChemBioEng Reviews*, *1*(4), 148–169. DOI: 10.1002/cben.201400001

Ilyas, S., & Lee, J. C. (2014b). Bioleaching of metals from electronic scrap in a stirred tank reactor. *Hydrometallurgy*, *149*, 50–62. DOI: 10.1016/j.hydromet.2014.07.004

Işıldar, A., Rene, E. R., van Hullebusch, E. D., & Lens, P. N. L. (2016). Electronic waste as a secondary source of critical metals: Management and recovery technologies. *Resources, Conservation and Recycling*, *135*, 296–312. DOI: 10.1016/j.resconrec.2017.07.031

Jain, M., Olsen, H. E., Paten, B., & Akeson, M. (2016). The Oxford Nanopore MinION: Delivery of nanopore sequencing to the genomics community. *Genome Biology*, *17*(1), 239. DOI: 10.1186/s13059-016-1103-0 PMID: 27887629

Kadier, A., Simayi, Y., Abdeshahian, P., Azman, N. F., Chandrasekhar, K., & Kalil, M. S. (2016). A comprehensive review of microbial electrolysis cells (MEC) reactor designs and configurations for sustainable hydrogen gas production. *Alexandria Engineering Journal*, *55*(1), 427–443. DOI: 10.1016/j.aej.2015.10.008

Kamm, B., & Kamm, M. (2004). Principles of biorefineries. *Applied Microbiology and Biotechnology*, *64*(2), 137–145. DOI: 10.1007/s00253-003-1537-7 PMID: 14749903

Kapoor, R. T., Salvadori, M. R., Rafatullah, M., Siddiqui, M. R., Khan, M. A., & Alshareef, S. A. (2021). Exploration of Microbial Factories for Synthesis of Nanoparticles – A Sustainable Approach for Bioremediation of Environmental Contaminants. *Frontiers in Microbiology*, *12*, 658294. DOI: 10.3389/fmicb.2021.658294 PMID: 34149647

Kaya, M. (2016). Recovery of metals and nonmetals from electronic waste by physical and chemical recycling processes. *Waste Management*, *57*, 64–90. DOI: 10.1016/j.wasman.2016.08.004 PMID: 27543174

Khalil, A. S., & Collins, J. J. (2010). Synthetic biology: Applications come of age. *Nature Reviews. Genetics*, *11*(5), 367–379. DOI: 10.1038/nrg2775 PMID: 20395970

Kinejara, A. (2023). The 4 Rs of Recycling: Reduce, Reuse, Recycle, Repurpose. https://www.reelpaper.com/blogs/reel-talk/4rs-of-recycling. Published January 11, 2023. Downloaded 12 October 2025

Kirchherr, J., Reike, D., & Hekkert, M. (2017). Conceptualizing the circular economy: An analysis of 114 definitions. *Resources, Conservation and Recycling*, *127*, 221–232. DOI: 10.1016/j.resconrec.2017.09.005

Knight, R., Vrbanac, A., Taylor, B. C., Aksenov, A., Callewaert, C., Debelius, J., Gonzalez, A., Kosciolek, T., McCall, L. I., McDonald, D., Melnik, A. V., Morton, J. T., Navas, J., Quinn, R. A., Sanders, J. G., Swafford, A. D., Thompson, L. R., Tripathi, A., Xu, Z. Z., & Dorrestein, P. C. (2018). Best practices for analysing microbiomes. *Nature Reviews. Microbiology, 16*(7), 410–422. DOI: 10.1038/s41579-018-0029-9 PMID: 29795328

Koul, B., & Taak, P. (2018). *Biotechnological strategies for effective remediation of polluted soils*. Springer. DOI: 10.1007/978-981-13-2420-8

Kumar, R., Singh, L., & Zularisam, A. W. (2017). Exoelectrogens: Recent advances in molecular drivers involved in extracellular electron transfer and strategies used to improve it for microbial fuel cell applications. *Renewable & Sustainable Energy Reviews, 56*, 1322–1336. DOI: 10.1016/j.rser.2015.12.029

Kümmerer, K. (2009). The presence of pharmaceuticals in the environment due to human use: Present knowledge and future challenges. *Journal of Environmental Management, 90*(8), 2354–2366. DOI: 10.1016/j.jenvman.2009.01.023 PMID: 19261375

Lacy, P., & Rutqvist, J. (2015). *Waste to wealth: The circular economy advantage*. Palgrave Macmillan.

Larsson, D. G. J., Andremont, A., Bengtsson-Palme, J., Brandt, K. K., de Roda Husman, A. M., Fagerstedt, P., Fick, J., Flach, C. F., Gaze, W. H., Kuroda, M., Kvint, K., Laxminarayan, R., Manaia, C. M., Nielsen, K. M., Plant, L., Ploy, M. C., Segovia, C., Simonet, P., Smalla, K., & Wernersson, A. S. (2018). Environmental factors influencing the development and spread of antibiotic resistance. *FEMS Microbiology Reviews, 42*(1), 68–80. DOI: 10.1093/femsre/fux053 PMID: 29069382

Larsson, D. G. J., de Pedro, C., & Paxeus, N. (2007). Effluent from drug manufactures contains extremely high levels of pharmaceuticals. *Journal of Hazardous Materials, 148*(3), 751–755. DOI: 10.1016/j.jhazmat.2007.07.008 PMID: 17706342

Laurent, A., Bakas, I., Clavreul, J., Bernstad, A., Niero, M., Gentil, E., Hauschild, M. Z., & Christensen, T. H. (2014). Review of LCA studies of solid waste management systems: Part I—Lessons learned and perspectives. *Waste Management, 34*(3), 573–588. DOI: 10.1016/j.wasman.2013.10.045 PMID: 24369845

Logan, B. E., Call, D., Cheng, S., Hamelers, H. V. M., Sleutels, T. H. J. A., Jeremiasse, A. W., & Rozendal, R. A. (2008). Microbial electrolysis cells for high yield hydrogen gas production from organic matter. *Environmental Science & Technology, 42*(23), 8630–8640. DOI: 10.1021/es801553z PMID: 19192774

Logan, B. E., & Rabaey, K. (2012). Conversion of wastes into bioelectricity and chemicals by using microbial electrochemical technologies. *Science, 337*(6095), 686–690. DOI: 10.1126/science.1217412 PMID: 22879507

Lovley, D. R. (2003). Cleaning up with genomics: Applying molecular biology to bioremediation. *Nature Reviews. Microbiology, 1*(1), 35–44. DOI: 10.1038/nrmicro731 PMID: 15040178

Lovley, D. R. (2012). Electromicrobiology. *Annual Review of Microbiology, 66*(1), 391–409. DOI: 10.1146/annurev-micro-092611-150104 PMID: 22746334

Lovley, D. R., & Nevin, K. P. (2013). Electrobiocommodities: Powering microbial production of fuels and commodity chemicals from carbon dioxide with electricity. *Current Opinion in Biotechnology, 24*(3), 385–390. DOI: 10.1016/j.copbio.2013.02.012 PMID: 23465755

Lu, H., Diaz, D. J., Czarnecki, N. J., Zhu, C., Kim, W., Shroff, R., Acosta, D. J., Alexander, B. R., Cole, H. O., Zhang, Y., Lynd, N. A., Ellington, A. D., & Alper, H. S. (2022). Machine learning-aided engineering of hydrolases for PET depolymerization. *Nature, 604*(7907), 662–667. DOI: 10.1038/s41586-022-04599-z PMID: 35478237

MacArthur, E. (2013). Towards the circular economy. *Journal of Industrial Ecology, 2*, 23–44.

Madigan, M. T., Martinko, J. M., Stahl, D. A., & Clark, D. P. (2012). *Brock Biology of Microorganisms* (13th ed.). Pearson Education, Inc.

Magnin, A., Pollet, E., Phalip, V., & Avérous, L. (2019). Evaluation of biological degradation of polyurethanes. *Biotechnology Advances, 39*, 107457. DOI: 10.1016/j.biotechadv.2019.107457 PMID: 31689471

Mandell, D. J., Lajoie, M. J., Mee, M. T., Takeuchi, R., Kuznetsov, G., Norville, J. E., Gregg, C. J., Stoddard, B. L., & Church, G. M. (2015). Biocontainment of genetically modified organisms by synthetic protein design. *Nature, 518*(7537), 55–60. DOI: 10.1038/nature14121 PMID: 25607366

Marchant, G. E., & Stevens, Y. A. (2015). A new window of opportunity to reject process-based biotechnology regulation. *GM Crops & Food, 6*(4), 233–242. DOI: 10.1080/21645698.2015.1134406 PMID: 26930116

Mitikie, B. B., & Elsaigh, W. A. (2025). Bio-based bacterial concrete: A sustainable alternative in construction. *Innovative Infrastructure Solutions, 10*(6), 234. DOI: 10.1007/s41062-025-02058-y

Moe-Behrens, G. H. G., Davis, R., & Haynes, K. A. (2013). Preparing synthetic biology for the world. *Frontiers in Microbiology*, *4*, 5. DOI: 10.3389/fmicb.2013.00005 PMID: 23355834

Mussatto, S. I. (2014). Brewer's spent grain: A valuable feedstock for industrial applications. *Journal of the Science of Food and Agriculture*, *94*(7), 1264–1275. DOI: 10.1002/jsfa.6486 PMID: 24254316

Naik, S., & Eswari, J. S. (2022). Electrical waste management: Recent advances challenges and future outlook. *Total Environment Research Themes*, *1*, 100002. DOI: 10.1016/j.totert.2022.100002

Negri, E., Fumagalli, L., & Macchi, M. (2017). A review of the roles of digital twin in CPS-based production systems. *Procedia Manufacturing*, *11*, 939–948. DOI: 10.1016/j.promfg.2017.07.198

Nevin, K. P., Woodard, T. L., Franks, A. E., Summers, Z. M., & Lovley, D. R. (2010). Microbial electrosynthesis: Feeding microbes electricity to convert carbon dioxide and water to multicarbon extracellular organic compounds. *mBio*, *1*(2), e00103–e00110. DOI: 10.1128/mBio.00103-10 PMID: 20714445

Newhart, K. B., Holloway, R. W., Hering, A. S., & Cath, T. Y. (2019). Data-driven performance analyses of wastewater treatment plants: A review. *Water Research*, *157*, 498–513. DOI: 10.1016/j.watres.2019.03.030 PMID: 30981980

Nguyen, L. N., Vu, M. T., Johir, M. A. H., Pernice, M., Ngo, H. H., Zdarta, J., Jesionowski, T., Nghiem, L. D., & Hai, F. I. (2020). Promotion of direct interspecies electron transfer and potential impact of conductive materials in anaerobic digestion and its downstream processing: A critical review. *Bioresource Technology*, *316*, 123925. DOI: 10.1016/j.biortech.2020.123925 PMID: 34467893

Nielsen, A. A. K., Der, B. S., Shin, J., Vaidyanathan, P., Paralanov, V., Strychalski, E. A., Ross, D., Densmore, D., & Voigt, C. A. (2016). Genetic circuit design automation. *Science*, *352*(6281), aac7341. Advance online publication. DOI: 10.1126/science.aac7341 PMID: 27034378

Nielsen, J., & Keasling, J. D. (2016). Engineering cellular metabolism. *Cell*, *164*(6), 1185–1197. DOI: 10.1016/j.cell.2016.02.004 PMID: 26967285

Ogunseitan, O. (2005). *Microbial Diversity: Form and Function in Prokaryotes*. John Wiley & Sons.

Ogunseitan, O. A. (2023). Planetary boundaries and safe operating space for novel entities. In Hester, R., & Harrison, R. (Eds.), *Environmental impacts of modern agriculture* (pp. 1–24). Royal Society of Chemistry.

Ogunseitan, O. A. (2026). Bacterial Diversity, Introduction to. In Wolf, J. B., & Russo, C. A. M. (Eds.), *Encyclopedia of Evolutionary Biology* (Vol. 5, pp. 42–47). Elsevier, Academic Press., DOI: 10.1016/B978-0-443-15750-9.00073-2

Ohimain, E. I. (2025). Microbial Biotechnology for Food Waste Reduction and Upcycling. Babalola O. O. et al. (eds.), Ecofriendly Frontiers: Harnessing Microbial Applications for Food Security. https://doi.org/. Springer Nature SwitzerlandDOI: 10.1007/978-3-031-98700-7_12

Parde, D., Ghosh, R., Rajpurohit, P., Bhaduri, S., & Behera, M. (2024). Nutrient Retrieval Techniques in Wastewater Treatment. In Ghangrekar, M. M., Yadav, S., & Yadava, R. N. (Eds.), *Biological and Hybrid Wastewater Treatment Technology. Earth and Environmental Sciences Library*. Springer., DOI: 10.1007/978-3-031-63046-0_7

Paritosh, K., Kushwaha, S. K., Yadav, M., Pareek, N., Chawade, A., & Vivekanand, V. (2017). Food waste to energy: An overview of sustainable approaches for food waste management and nutrient recycling. *BioMed Research International, 2370927*, 1–19. Advance online publication. DOI: 10.1155/2017/2370927 PMID: 28293629

Philp, J., & Winickoff, D. E. (2018). *Realising the circular bioeconomy. OECD Science, Technology and Industry Policy Papers No. 60*. OECD Publishing.

Pollmann, K., Kutschke, S., Matys, S., Raff, J., Hlawacek, G., & Lederer, F. L. (2018). Bio-recycling of metals: Recycling of technical products using biological applications. *Biotechnology Advances, 36*(4), 1048–1062. DOI: 10.1016/j.biotechadv.2018.03.006 PMID: 29555455

Prévoteau, A., Carvajal-Arroyo, J. M., Ganigué, R., & Rabaey, K. (2020). Microbial electrosynthesis from CO_2: Forever a promise? *Current Opinion in Biotechnology, 62*, 48–57. DOI: 10.1016/j.copbio.2019.08.014 PMID: 31593911

Rollakanti, C. R., & Srinivasu, K. (2022). Self-healing Concrete Based on Bacteria for Sustainable Infrastructure: A State-of-the-Art Review. In: Kumar, P.G., Subramaniam, K.V.L., Santhakumar, S.M., Satyam D., N. (eds) Recent Advances in Civil Engineering. Lecture Notes in Civil Engineering, vol 233. Springer, Singapore. DOI: 10.1007/978-981-19-0189-8_15

Salis, H. M., Mirsky, E. A., & Voigt, C. A. (2009). Automated design of synthetic ribosome binding sites to control protein expression. *Nature Biotechnology, 27*(10), 946–950. DOI: 10.1038/nbt.1568 PMID: 19801975

Sandbhor, S., Apte, S., Puranik, S., Asati, M., & Rai, S. (2025). Technique for improving concrete characteristics with bacillus: Insights on strength and durability. *MethodsX, 14*, 103151. DOI: 10.1016/j.mex.2024.103151 PMID: 39866201

Secretariat of the Convention on Biological Diversity. (2000). *Cartagena Protocol on Biosafety to the Convention on Biological Diversity*. Secretariat of the Convention on Biological Diversity.

Sharma, P., Gaur, V. K., Sirohi, R., Varjani, S., Hyoun Kim, S., & Wong, J. W. C. (2021). Sustainable processing of food waste for production of bio-based products for circular bioeconomy. *Bioresource Technology, 325*, 124684. DOI: 10.1016/j.biortech.2021.124684 PMID: 33493748

Singh, J., & Singh, S. (2019). Biochar applications in bioremediation. In Prasad, M., Favas, M., & Maiti, S. (Eds.), *Bio-geotechnologies for mine site rehabilitation* (pp. 417–433). Elsevier., DOI: 10.1016/B978-0-12-812986-9.00023-7

Slate, A. J., Whitehead, K. A., Brownson, D. A. C., & Banks, C. E. (2019). Microbial fuel cells: An overview of current technology. *Renewable & Sustainable Energy Reviews, 101*, 60–81. DOI: 10.1016/j.rser.2018.09.044

Smith, K. R., Bruce, N., Balakrishnan, K., Adair-Rohani, H., Balmes, J., Chafe, Z., Dherani, M., Hosgood, H. D., Mehta, S., Pope, D., & Rehfuess, E. (2014). Millions dead: How do we know and what does it mean? Methods used in the comparative risk assessment of household air pollution. *Annual Review of Public Health, 35*(1), 185–206. DOI: 10.1146/annurev-publhealth-032013-182356 PMID: 24641558

Smolke, C. D. (2009). Building outside of the box: iGEM and the BioBricks Foundation. *Nature Biotechnology, 27*(12), 1099–1102. DOI: 10.1038/nbt1209-1099 PMID: 20010584

Stahel, W. R. (2016). The circular economy. *Nature, 531*(7595), 435–438. DOI: 10.1038/531435a PMID: 27008952

Stankiewicz, A. I., & Moulijn, J. A. (2000). Process intensification: Transforming chemical engineering. *Chemical Engineering Progress, 96*(1), 22–34.

Strong, P. J., Xie, S., & Clarke, W. P. (2015). Methane as a resource: Can the methanotrophs add value? *Environmental Science & Technology, 49*(7), 4001–4018. DOI: 10.1021/es504242n PMID: 25723373

Tanasupawat, S., Takehana, T., Yoshida, S., Hiraga, K., & Oda, K. (2016). *Ideonella sakaiensis* sp. nov., isolated from a microbial consortium that degrades poly(ethylene terephthalate). *International Journal of Systematic and Evolutionary Microbiology, 66*(8), 2813–2818. DOI: 10.1099/ijsem.0.001058 PMID: 27045688

Tournier, V., Topham, C. M., Gilles, A., David, B., Folgoas, C., Moya-Leclair, E., Kamionka, E., Desrousseaux, M. L., Texier, H., Gavalda, S., Cot, M., Guémard, E., Dalibey, M., Nomme, J., Cioci, G., Barbe, S., Chateau, M., André, I., Duquesne, S., & Marty, A. (2020). An engineered PET depolymerase to break down and recycle plastic bottles. *Nature*, *580*(7802), 216–219. DOI: 10.1038/s41586-020-2149-4 PMID: 32269349

Trump, B. D., Cummings, C. L., Klasa, K., Galaitsi, S. E., & Linkov, I. (2020). A decision analytic model to guide early-stage government regulatory action: Applications for synthetic biology oversight. *Regulation & Governance*, *14*(2), 251–272. DOI: 10.1111/rego.12245

Uddin, M. N., Tafsirojjaman, T., Shanmugasundaram, N., Praveenkumar, S., & Li, L. (2023). Smart self-healing bacterial concrete for sustainable goal. *Innovative Infrastructure Solutions*, *8*(1), 46. DOI: 10.1007/s41062-022-01020-6

Vanwonterghem, I., Jensen, P. D., Dennis, P. G., Hugenholtz, P., Rabaey, K., & Tyson, G. W. (2014). Deterministic processes guide long-term synchronised population dynamics in replicate anaerobic digesters. *The ISME Journal*, *8*(10), 2015–2028. DOI: 10.1038/ismej.2014.50 PMID: 24739627

Verlicchi, P., Al Aukidy, M., & Zambello, E. (2012). Occurrence of pharmaceutical compounds in urban wastewater: Removal, mass load and environmental risk after a secondary treatment—A review. *The Science of the Total Environment*, *429*, 123–155. DOI: 10.1016/j.scitotenv.2012.04.028 PMID: 22583809

Vidanage, P. W. (2024). State of Art Technologies for Nutrient Removal from Wastewater. In Yadav, S., Ghangrekar, M. M., & Yadava, R. N. (Eds.), *Emerging Technologies in Biological and Hybrid Wastewater Treatment: Lessons from Developed to Enhancing Practices in Developing Countries*. Springer., DOI: 10.1007/978-3-031-74515-7_7

Wan, X., Volpetti, F., Petrova, E., French, C., Maerkl, S. J., & Wang, B. (2019). Cascaded amplifying circuits enable ultrasensitive cellular sensors for toxic metals. *Nature Chemical Biology*, *15*(5), 540–548. DOI: 10.1038/s41589-019-0244-3 PMID: 30911179

Wierckx, N., Prieto, M. A., Pomposiello, P., de Lorenzo, V., O'Connor, K., & Blank, L. M. (2015). Plastic waste as a novel substrate for industrial biotechnology. *Microbial Biotechnology*, *8*(6), 900–903. DOI: 10.1111/1751-7915.12312 PMID: 26482561

Wright, O., Stan, G. B., & Ellis, T. (2013). Building-in biosafety for synthetic biology. *Microbiology (Reading, England)*, *159*(Pt 7), 1221–1235. DOI: 10.1099/mic.0.066308-0 PMID: 23519158

Yadav, A., & Yadav, K. (2025). Circular biorefineries for rural India: Turning rice straw and bagasse into biofuels. *Academia Green Energy*, *2*(4). Advance online publication. DOI: 10.20935/AcadEnergy7949

Yoshida, S., Hiraga, K., Takehana, T., Taniguchi, I., Yamaji, H., Maeda, Y., Toyohara, K., Miyamoto, K., Kimura, Y., & Oda, K. (2016). A bacterium that degrades and assimilates poly(ethylene terephthalate). *Science*, *351*(6278), 1196–1199. DOI: 10.1126/science.aad6359 PMID: 26965627

Zimmermann, W., & Dombrowski, A. (2020). Enzymatic degradation and recycling of poly(ethylene terephthalate): A critical review. In Kabasci, S. (Ed.), *Bio-based Plastics: Materials and Applications* (pp. 227–242). John Wiley & Sons., DOI: 10.1002/9781119592525.ch10

Chapter 9
The Future of Microbial Environmental Biotechnology:
Integrating Artificial Intelligence (AI), CRISPR, and Systems Biology Approaches

Obinna Chukwuemeka Nwinyi
http://orcid.org/0000-0001-9314-6460
Covenant University, Nigeria

Olugbenga Oluwagbemi
http://orcid.org/0000-0001-9085-3370
Rochester Institute of Technology, USA

ABSTRACT

This review examines the integration of Artificial Intelligence (AI), CRISPR-based genome editing, and Systems Biology in transforming microbial environmental biotechnology. It highlights how AI and machine learning improve the discovery of microbial functions and predictive modeling of microbial communities beyond traditional omics analyses. The review also discusses the role of CRISPR-Cas technologies in engineering microbial strains and synthetic microbial consortia (SynComs) with enhanced abilities for pollutant degradation, biosensing, and carbon sequestration. Furthermore, Systems Biology is presented as a key approach for integrating multi-omics data to design robust microbial ecosystems for environmental applications. While these technologies offer promising solutions for precision bioremediation

DOI: 10.4018/979-8-2600-2998-5.ch009

and sustainable environmental restoration, the review also addresses challenges such as off-target effects, delivery mechanisms, biosafety concerns, ethical issues, regulatory barriers, and public acceptance.

1 INTRODUCTION TO MICROBIAL ENVIRONMENTAL BIOTECHNOLOGY

Humankind has witnessed tremendous progress in the last century due to technological advancement in different industrial sectors. These include but not limited to industries such as, mining, health and medicine (especially therapeutics-antibiotics), pesticides, petrochemicals and refineries (Sheng et al., 2019). However, the rapid progress has also led to elevated and uncontrolled discharge of highly toxic chemicals and intermediates, which contribute to soil, water and air pollution (Wang et al., 2024a). As the burden of toxic pollutants continues to intensify, attention has shifted toward biological systems capable of counteracting their impact (Ayilara et al., 2023).

Microbial biotechnology offers environmentally friendly approaches that can be implemented effectively for identifying the indicator microbial strains – causative agent for the problem and the ones useful for environmental bioremediation (Ahmed, 2022). Microbial-driven bioremediation process can benefit the community by exploiting the metabolomics capabilities of microorganisms that enable the utilization of toxic compounds and biotransformation to utilizable intermediates and products (Ayilara et al., 2023) (See Figure 1).

Figure 1. Bioremediation of pollutants by different applications

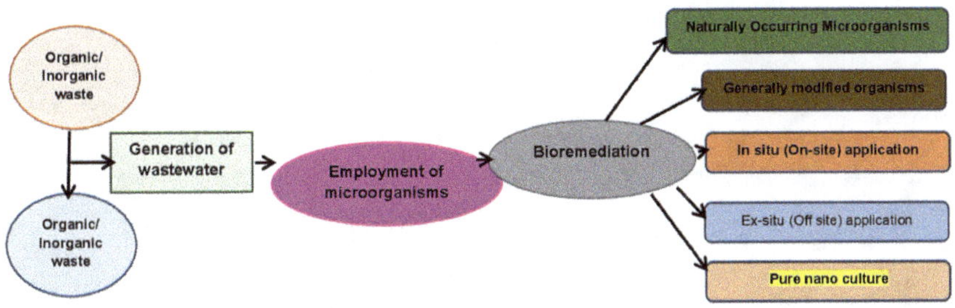

The molecular biology revolution of the 1970s reformed Environmental Biotechnology. The recombinant DNA technology enabled scientists to create new abilities in microbes. Since the 1980s, genetics, genomics and metabolic engineering have allowed microbes to do exponentially more. More possibilities have been presented within the recent decades when systems biology and synthetic biology became present. These methods enable scientists to build new genetic networks and pathways within microbes or study a whole microbial community through the knowledge of multi-omics and computational modelling and simulations (Oluwagbemi et al., 2022; Al-Salihi et al., 2025). Figure 2 depicts of how microbial environmental biotechnology has advanced from the traditional fermentation which was used to produce food and chemicals; to large-scale biochemical production began (citric acid, penicillin, amongst others). Presently, with the help of genetic, computational modeling and computational engineering of microbes, targeted solutions to environmental problems are created.

Figure 2.

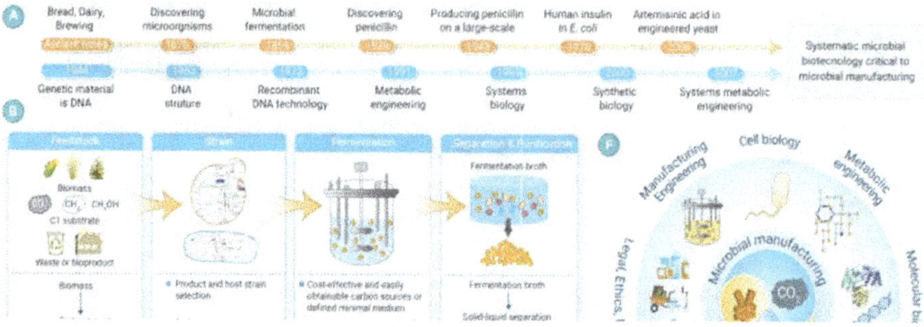

Microbial communities in natural environments rarely act alone but function as complex consortia where species interact synergistically to accomplish tasks unachievable by single strains (Haruta & Yamamoto, 2018). Recent innovations have focused on engineering synthetic microbial consortia designed to optimize environmental functions. Synthetic consortia are carefully selected or genetically modified microbes with complementary metabolic pathways (Wang et al., 2025).

To further optimize their performance and stability, Synthetic biology has emerged as a transformative approach enabling the design and construction of tailored microbial strains with enhanced environmental functionalities. It is the application of engineering, science, and technology to edit the genetic material of living organisms in order to enable them to perform new functions (Cameron *et al.,* 2014). Such engineered organisms can provide robust solutions to our current environmental challenges. Using synthetic biology tools, microorganisms have been engineered to sense, quantify, and report the presence of specific environmental pollutants (Brutesco *et al.,* 2017). By editing microbial genomes, researchers can introduce or optimize metabolic pathways for targeted pollutant degradation, biosynthesis of bioplastics, or biofuel production. For example, engineered *Pseudomonas* and *Escherichia coli* strains have shown improved degradation of aromatic hydrocarbons. Genetically modified microbes can function as biosensors that detect and report the presence of specific pollutants, facilitating real-time environmental monitoring and management (Rafeeq et al., 2022).

As these microbial tools gain traction, the market for environmental biotechnology products is growing, reflecting this promise. For example, bioremediation services are widely offered and are projected to form a multibillion-dollar market. Commercial products include enzyme formulations, biosurfactants (for oil dispersion), bioaugmentation consortia, and bioplastics (e.g. PHA biopolymers). While many of these still use naturally occurring microbes, the research and startup ecosystem are pushing toward more sophisticated, engineered solutions (De Almeida et al., 2016).

In sum, the field today combines established microbial processes with cutting-edge research in synthetic biology, systems biology, and environmental engineering. This synergy is laying the groundwork for the next generation of technologies.

2 ARTIFICIAL INTELLIGENCE IN ENVIRONMENTAL BIOTECHNOLOGY

Artificial Intelligence (AI) has revolutionized the field of genomic research to the point of automation of enormous quantities of sequencing data (Wang *et al.,* 2024b). Artificial Intelligence mimicks human intelligence by reasoning, learning from data, performing problem-solving tasks, demonstrating perception, and engaging decision-making (Shajil et al., 2025). More recent sequencing technologies NGS (next-generation sequencing) yield access to high-dimensional data on the genomes of microbes, at billions of base pairs per sample, far exceeding the capability of manual inspection (Satam et al., 2023). The use of AI in NGS pipelines has transformed the entire field of genomics providing us with the capabilities regarding data analysis, precision, and scalability. Practically, processes in genomic workflow, such as experimental design, image processing, variant calling, and sequence assembly, can be optimized using AI-driven tools (such as machine learning and deep learning models). More importantly, Machine learning based methods are highly effective at discovering previously unknown patterns in either an experimental dataset or one where an expert rule cannot be applied, or the data are too large to be assessed manually (Dada et al., 2022). This capability renders AI highly suitable to decode complex genomic data like microbial sequence sequencing as illustrated in Figure 3 (Wang *et al.,* 2024b). Figure 3 also depicts AI-enhanced microbial genomics workflow. The workflow consists of different layers ranging from data preprocessing, variant calling, annotation to functional prediction, which all lead to enhanced genomic insights.

Figure 3.

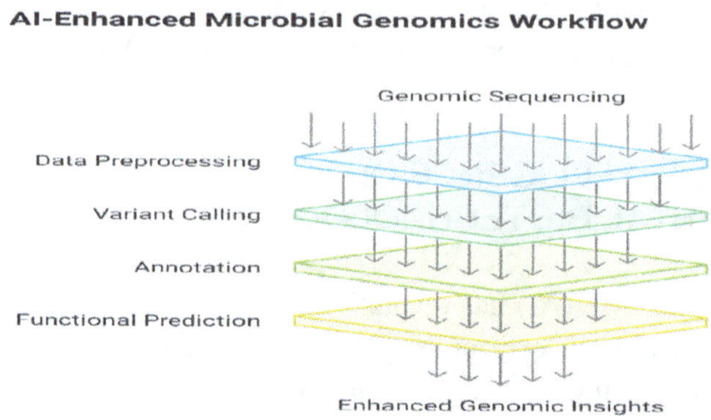

3 GENE DISCOVERY AND FUNCTIONAL PREDICTION

Genome annotation determines genes (and other functional components), regulatory elements (promoters, operators, motifs), and other functional components within the DNA, a central task in microbial genomics. AI tools significantly fast-track this activity as known genes or regulatory motifs are used to train ML models (Support vector machines (SVM), decision trees, random forests, and neural networks) to scan new sequences and look for similar patterns. The decision-tree and the SVM classifiers can predict bacterial phenotypes and gene functions based on genome sequences (classic studies) (Pilalis et al., 2025). Essentially, ML approaches memorize the statistical patterns of genes (open reading frames, coding potential, splice signals) and hence report candidate genes or single-nucleotide polymorphisms (SNPs) throughout a genome in less time than the rule-based approaches. Through this, AI can facilitate the determination of all the functional components present in the genome, including fresh genes and variant loci, with minimum human guidance (Lee et al., 2025). Deep learning techniques have been applied to the classification of datasets of health medications and to predict intramolecular coupling constants of molecules of potential health medications (Oyewola et al., 2022a and Oyewola et al., 2022b). Deep learning algorithms (Google Deep Variant) are now used to detect SNPs or carry out other SNP tasks and have surpassed traditional heuristics, making them more accurate to use, as shown in Figure 4. Unsupervised algorithms (clustering, dimensionality reduction) are helpful in clustering together unlabeled sequences based on a notion of similarity in cases where no labeling is known, to classify microbial strains or operon structures (Mlost et al., 2025). Comparative studies, in which 14 bacterial species were used, showed that Clair3 has a median F1 score of ~99.99% of SNPs and ~99.53% of indels, superior to Illumina-loaded pipelines by Hall *et al.* (2024). Other comparisons reveal that DeepVariant using context-aware models provides the best F1 scores (96-99%) in Illumina and PacBio HiFi data, but can be poorly resourced compared to DNAscope, which provides similar results within 30 minutes on 32-thread machines. Machine learning offers, in general, a data-guided method for gene discovery, trained on labeled instances to generalize to unseen genes, ORFs, or regulatory modules, even in ill characterized species (Markus & Steven, 2021).

Figure 4.

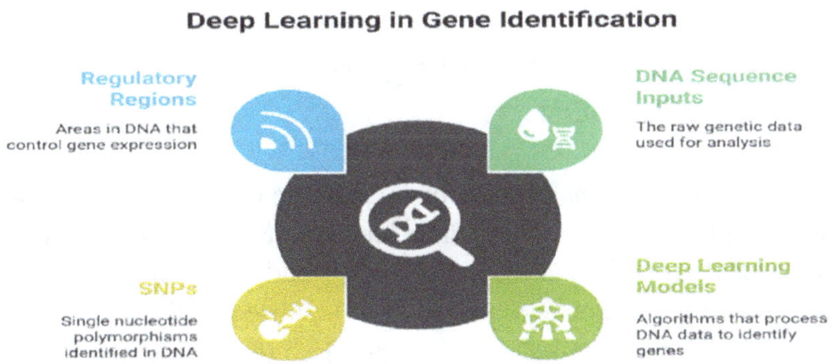

4 ARTIFICIAL INTELLIGENCE (AI) IN MICROBIAL COMMUNITIES AND COMPARATIVE GENOMICS

AI is also critical in the intraspecific and interspecific comparisons of genomes, mainly in microbes. Comparative genomics is an essential aspect of bioinformatics where genomic sequences of different species or organisms are analyzed and compared to discover similarities and differences in DNA; this has found application in providing insights for the control of a virus of global concern such as Ebola (Oluwagbemi and Awe, 2018). Unsupervised machine learning methods (k-means clustering, hierarchical clustering, and latent Dirichl*et al*location) identify new community types or subspecies among the microbes or strains based on genomic signatures in microbiome or population studies. The clustering techniques are also used on presence/absence matrices of gene expression, k-mer frequencies, or functional annotations to stratify microbial communities by habitat, host association, or biochemical completeness (Shi et al., 2022). The AI is especially helpful in sequence alignment, phylogenetic inference, and genome clustering in comparative genomics. It enables large-scale comparative analysis of metagenome-assembled genomes (MAGs) of bacteria in thousands of publicly available gut and soil microbiome data sets possible using neural-network-assisted alignment algorithms and graph-based genome representations (Zhou *et al.*, 2024). Also, the supervised classification and dimensionality reduction provide a framework that can be used to assign newly sequenced genomes to taxonomic and functional hierarchies based on AI-assisted tools, including GTDB-Tk and Anvi'o. These pipelines allow the determination of conserved areas in the genome, horizontal gene transfers, and gene gain/loss patterns that could reflect an area of functional specialization (Chaumeil et al., 2022).

5 MACHINE LEARNING FOR MICROBIAL FUNCTION PREDICTION

Prediction of the functions of microbial genes and organisms using sequencing data is increasingly dependent on machine learning (ML). Conventional sequence alignment-based tools or domain-based databases tend to overlook novel or remote-homologous functions (Katoh & Standley, 2016). Subsequently, ML has a different principle applicable in ML models, which learn rich representations on sequences and can generalize new sequences by learning with protein language models (PLMs). New ML frameworks that are customized precisely to the microbe have recently shown high fidelity in predicting functional genes of our traits, and in multi-omics data integration (Kulikova et al., 2023).

6 DEEP LEARNING TOOLS FOR ENVIRONMENTAL BIOTECHNOLOGY

6.1 DeepGOMeta

The deep learning tool, DeepGOMeta by Tawfiq *et al.* (2024), is a model that infers the protein-level (Gene Ontology proteomics) of microbial genomes (i.e., assigns Gene Ontology (GO) terms). Noting that most of the function predictors are trained on eukaryotic data, DeepGOMeta is instead trained on a selected prokaryotic Swiss-Prot corpus (~ 10,000 reviewed bacterial and archaeal proteins) whose functions are annotated experimentally with GO terms. It uses the Evolutionary Scale Modeling 2 (ESM2) protein language model embeddings as embeddings. The protein sequences are passed into ESM2 for a fixed-length embedding, and afterwards neural classifiers are trained independently per sub-ontology of GO (Molecular Function, Biological Process, Cellular Component) (Tawfiq *et al.,* 2024). DeepGOMeta has been used on the 16S amplicon, metagenome (paired), analysis (human gut, soil), where it predicts functions of 16S using the nearest-sequence approach, but contrasts with the species-level clustering. Across various case studies (e.g., phenotypes of human diet), DeepGOMeta functional profiles commonly led to the clustering of conditions more effectively than did taxonomy alone (Boadu *et al.,* 2023). DeepGOMeta was more wide-ranging in its GO annotation of organisms and produced a more discriminative clustering in some of the datasets compared to metabolic profiling tools.

6.2 Micropherret

The model is an ensemble supervised classifier called MICROPHERRETxthat predicts microbes' phenotypic properties and metabolic abilities based on their genome annotations (Bizzotto *et al.*, 2024). It uses the FAPROTAX database (a literature-derived mapping of taxa to 92 ecological functions). It trains binary classifiers to predict each of the 92 functions, based on the gene content of a genome. Bizzotto *et al.* (2024) constructed a training set by assimilating FAPROTAX functional classes to genomes of pertinent taxa (=14,364 genomes). Individual tuned models were forecasted against each of the classes of functions. They tested each trait with logistic regression, random forests, SVMs, and a neural network (Keras). In each case, they ran cross-validation to identify the best algorithm (Rengasamy *et al*, 2022).

Pragmatically, MICROPHERRET encompasses 86 performed functions (post filter), which are metabolisms (carbon/nitrogen/sulfur cycling), trophic functions, and habitat-related properties (plastic degradation, methane oxidation). It uses feature vectors as genome annotations (KEGG Ortholog counts by eggNOG-mapper). Remarkably, micropherret withstands incomplete genomes on simulated fragments of genomes (down to 70% completeness). Most of the classifiers performed reliably (Bizzotto *et al.*, 2024). The high-quality reference genomes and the abundant type of anaerobic digestion metagenome-assembled genomes (MAMs) in the Biogas Microbiome database confirmed the tool's predictions. The inferred community function profiles generated during the biogas test concurred with known community-level biology and the fine-scale characteristics of archaeal genomes (including methanogenesis). A case study about acetoclastic methanogenesis revealed that the regime of training new trait models can be modified in a modular fashion as necessary. Examples of AI tools for Microbial functional annotation is shown in Table 1.

6.3 AI-Guided Metabolic Pathway Engineering

AI-driven approaches are revolutionizing metabolic engineering by combining mechanistic modeling and machine learning to optimize pathways, predict enzyme function, and guide strain design Figure 5.

Table 1. AI Tools for Microbial Functional Annotation

Tool	Core Algorithm	Input	Output/Function	Performance Metrics
FunGeneTyper	Protein Language Model + Neural Network	Protein sequences	Functional categorization (ARG, VFG)	Accuracy >99%, F1 >0.97
DeepGOMeta	ESM2 + GO classification	Protein sequences	GO term prediction	Outperformed OTU clustering in 5/9 phenotypes (Tawfiq *et al.,* 2024)
MICROPHERRET	Ensemble (SVM, RF, LR, NN)	Genome annotation (KO counts)	Phenotypic trait classification	High accuracy across 86 traits (Bizzotto *et al.*, 2024)
FGBERT	Transformer PLM	Gene sequences (metagenomics)	Functional predictions in microbiomes	Not specified (preprint model)

Figure 5.

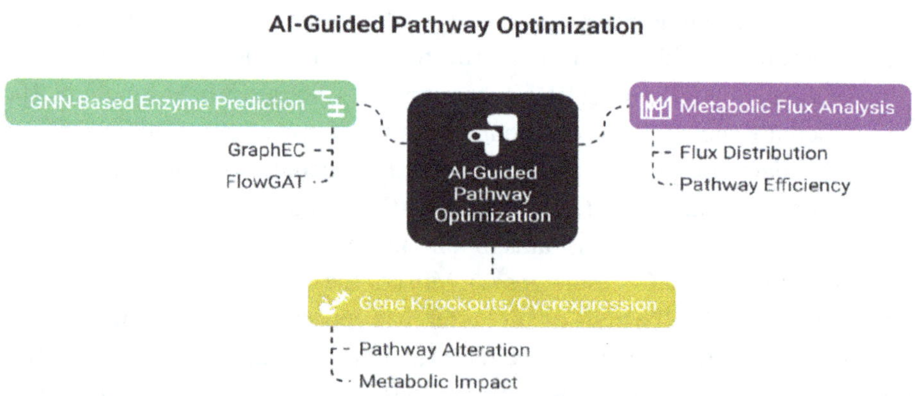

6.4 GraphEC

The GraphEC combines predicted protein structures based on ESMFold with the embeddings obtained with the help of pretrained PLMs (ProtTrans) (Song *et al.,* 2024). It generates geometric graphs of protein structures, finds active-site residues, and proceeds with geometric graph learning to type EC numbers and identify an enzyme's optimal pH. GraphEC consists of combining structural and sequence characteristics into a geometric graph. Attention-based localization of the active site is made before predicting EC. This promotes the diffusion of labels with the use of homology context. Enhances the capability for optimal pH prediction. GraphEC performed well on benchmarks: Area Under the Curve (AUC) was approximately 0.96 on active-site prediction; the EC numbers predicted higher precision than traditional methods, and pH can be predicted accurately. This combines Pre-trained Language Models PLM, structure, and graph learning, bringing substantial breakthroughs in enzyme annotation and silico pathway design (Song *et al.*, 2024).

6.5 FlowGAT

The FlowGAT is a combination of genome-scale Flux Balance analysis (FBA) with a Graph Attention Network (GAT). Nodes are constructed from reactions, and edges carry computed flux relationships calculated by FBA. The GAT acquires patterns on the wild-type flux distributions by using the information on the gene knockout assays (Hasibi, *et al.*, 2024). It avoids the optimality assumption of FBA of knockouts. GAT depicts local dependence on fluxes between reactions and the unification of mechanistic modeling of fluxes and data-driven learning.

6.6 AI in Microbiome Community Modeling and Simulation

Recent advances in artificial intelligence (AI) have profoundly impacted CRISPR guide design, off-target prediction, and synthetic microbiome engineering.

6.7 AI for CRISPR Guide Design

The development of single-guide RNA (sgRNA) that can be used in CRISPR applications has high specificity and efficiency. Deep learning has found a primary place in optimizing such processes as DeepCRISPR 2.0 (Liu *et al.*,2023), an improved convolutional neural network (CNN) with neuron dimensions and a topology capable of integrating the sequence context and chromatin accessibility data to predict sgRNA on efficiency and off-target cleavages. It had a Pearson correlation of 0.85 on on-target efficiency, and an AUC of 0.96 in off-target prediction across human cell lines. Al prediction Scorer 2.0 (Kim *et al.,* 2018) is also a gradient-boosted tree where features are calculated based on the sequence and epigenetically relevant tags. It performed better than the earlier versions using the independent test sets and attained more than 0.93 in efficiency and off-target detection AUROC. Such tools as DeepCRISPR 2.0 implement epigenetic information, increasing prediction accuracy compared to sequence-only models. Pretrained genomic word embeddings, DNABERT, or genomic transformers make it simple to retrain them with new cell types and species quickly. Both tools systematically tested cross-cell-line generalization and performed state-of-the-art on held-out datasets.

6.8 Microbiome–Host Interaction Modeling

The biological complexities in the form of microbial communities and heterogeneity along the host axis are problematic to analyze. A key new role in examining these complex relationships is played by artificial intelligence, especially machine learning and deep learning models, being used to model correlations of microbial composition, gene expression, and metabolic functions against host phenotypes and environmental factors (Przymus et al., 2025). Among the prominent tools of this domain facilitated by AI, one can distinguish gRodon. The model uses the supervised machine learning technique to forecast the growth rate in microbial communities using metagenomic data. Training on more than 1,200 metagenomes concerning known in situ replication rates in various ecosystems yields a coefficient of determination of about 0.76. Such predictions help to understand the mechanics of microbial community functions in various host and environmental conditions, where measurements in the wet labs are labor-intensive (Weissman et al., 2021).

A further tool is revolutionary; MiMeNet (Microbiome-Metabolome Neural Network) reported by Reiman *et al.* (2021). The MiMeNet model is based on a deep neural network with an attention approach to predicting metabolite abundance using microbiome gene relative abundance profiles. The model can learn complex, non-linear relationships between microbial community functions and host-related metabolites with Spearman correlation coefficients between 0.70 and 0.85 in multiple disease cohorts. This method can be used to identify the most important microbial taxa or genes that impact metabolic outputs, and these outputs.

7 SYNTHETIC MICROBIAL COMMUNITY DESIGN

The synthesis of synthetic microbial consortia with a specific task - degradation of pollutants, absorption of carbon, or nutrient cycle - has become more accessible with the incorporation of AI-based modeling (Renganathan et al., 2025). Compared to real environments, the complex and dynamic interplay between microbial species is not always predicted well by traditional approaches to community design (based on trial-and-error culturing and ecological intuition). In response, state-of-the-art computational systems implement graph neural networks (GNNs) to model inter-species interactions and forecast functional results based on environmental and genomic data (Yuan & Nault, 2025). Community-GNN builds weighted graphs by mixing metagenomic co-occurrence patterns with the annotations of functional genes, thus developing a system-level representation of microbial communities. It estimates the likelihood of species pairs when performing cooperative (mutualistic) or competitive actions- a vital parameter of creating stable and effective synthetic communities (Ruaud et al., 2024).

8 CASE STUDIES

8.1 AI-Predicted Biodegradative Enzymes

AI has also accelerated the identification of microbial enzymes for environmental remediation. At IISER Bhopal, researchers developed XenoBug, a web-based platform that predicts bacterial enzymes capable of degrading xenobiotic pollutants (Arora *et al.,* 2022). Using machine learning classifiers trained on over 3.3 million enzyme sequences from environmental and genomic datasets, XenoBug categorizes enzymes by reaction class and substrate specificity. It can accurately identify enzymes that break down plasticizers, pesticides, pharmaceuticals, and other persistent pollutants. By automating substrate–enzyme matching and reducing false positives, XenoBug streamlines bioremediation planning and synthetic ecology applications (Malwe et al., 2025).

These tools illustrate the growing maturity of AI-driven discovery in microbiology. Whether identifying antimicrobial candidates or biodegradative enzymes, machine learning models are pushing the boundaries of conventional biotechnology, enabling faster, more targeted, and functionally novel discoveries that have real-world impact across medicine and environmental science.

8.2 CRISPR and Advanced Genome Editing in Environmental Microbes

The Clustered Regularly Interspaced Short Palindromic Repeats (CRISPR) and CRISPR-associated (cas) proteins are a distinctive feature of the genome of most prokaryotes; specifically, bacteria and archaea. They form an adaptive immune system that is naturally present in prokaryote. This sophisticated immune system enables these microorganisms to gain sequence-specific memory of previous exposure to foreign mobile genetic elements (MGEs) in the form of short sequences (spacers) incorporated into the CRISPR array from viruses and plasmids and subsequently neutralize future invasions (Faure *et al.*, 2019). A complete CRISPR-Cas system essentially consists of the CRISPR array, the Cas proteins, and the guide

RNAs (gRNAs). The CRISPR array is a unique genomic locus characterized by short (two to several hundred), conserved often partially palindromic direct repeats interspaced by unique spacers. These spacers are short fragments of foreign DNA obtained from invading nucleic acids and serve as a genetic memory or storehouse of previous infections, and are complementary to the targeted foreign genetic material. The average size of these repeats is approximately base pairs 32 (bp); however, the size may vary from 21 to 47 bp. The spacers are nucleotide sequences with a fixed length, but they are highly variable in sequences. Spacers typically range from 20 to 72 bp (Karginov & Hannon, 2010). Adjacent to these are clusters of multiple Cas genes which are organized in one or more operons encoding both the adaptation and the effector modules, often along with accessory genes. These Cas genes code for Cas proteins which are a diverse family of enzymes crucial for the execution of the immune response. Over 93 Cas genes have been identified, and these genes have been classified into 35 families based on their sequence similarities (Table 2). Transcripts of the inserted spacers are processed into guide RNAs (gRNAs), which either comprise CRISPR RNA (crRNA) alone, or the crRNA and in some systems trans-activating crRNA (tracrRNA) (Makarova et al., 2025). These gRNAs are responsible for directing the Cas proteins to their cognate targets.

Table 2. Key Applications of CRISPR-Cas in Environmental Microbes with Specific Examples

Application Area	Specific Target/Pollutant/Compound/ Ecosystem Service	Engineered Microbe/Organism	Observed Outcome/ Benefit
Bioremediation	Cadmium (Cd)	*Mesorhizobium huakuii* subsp. *rengei* Strain B3	Increased Cd tolerance and accumulation
Bioremediation	Chromium(VI)	*Methylococcus capsulatus* (Bath)	Enhanced Cr(VI) reduction to less toxic Cr(III)
Bioremediation	Polycyclic Aromatic Hydrocarbons (PAHs), Lignin	*Aspergillus niger*, *Phanerochaete chrysosporium*	Enhanced PAH/ lignin degradation efficiency
Metabolic Engineering (Biofuels)	Lipids (for biodiesel)	Microalgae (*Haematococcus spp.*)	Doubled lipid production, enhanced carbon-to-lipid conversion
Metabolic Engineering (Biofuels)	Ethanol, Butanol	*E. coli*	Increased biofuel yield
Metabolic Engineering (Biofuels)	Biofuel tolerance, Thermotolerance	*Zymomonas mobilis*, *Saccharomyces cerevisiae*	Improved tolerance to biofuel toxicity and heat

continued on following page

Table 2. Continued

Application Area	Specific Target/Pollutant/Compound/ Ecosystem Service	Engineered Microbe/Organism	Observed Outcome/ Benefit
Metabolic Engineering (Bioplastics)	Polyhydroxyalkanoates (PHAs)	*E. coli*, *Pseudomonas putida*, *Yarrowia lipolytica*	Enhanced PHA production, controlled polymer composition
Metabolic Engineering (Industrial Chemicals)	Succinic Acid	Microorganisms	Increased succinic acid yield by disrupting competing pathways
Microbial Community Engineering	Nitrogen Fixation	Soil microbes	More efficient nitrogen cycling, reduced fertilizer need
Microbial Community Engineering	Phosphorus Solubilization	Soil microbes	Increased phosphorus availability to plants
Microbial Community Engineering	Plant Growth Promotion	Microbes	Production of plant growth-promoting hormones
Pathogen Control (Agricultural)	Rice Blast (fungus), other pathogens	Rice (engineered plant)	High yield, resistance to multiple pathogens
Pathogen Control (Waterborne)	*Burkholderia pseudomallei*	Water samples (detection method)	Improved detection sensitivity and specificity
Pathogen Control (Waterborne)	Antibiotic Resistance Genes	Phages, *E. coli* (via conjugation)	Targeted elimination of ARGs, modulation of bacterial function

The molecular mechanisms of CRISPR-Cas systems proceed through three distinct but often intertwined stages: (i) adaptation, (ii) expression, CRISPR RNA (crRNA) biogenesis, and crRNA processing, and (iii) interference.

i. Adaptation (Insertion or Spacer Acquisition): Upon the invasion of a foreign nucleic acid, specific Cas proteins, capture short fragments (protospacers) from the invader's genetic material. These protospacers are then integrated into the CRISPR array as new spacers, thereby establishing a genetic memory of the infection (figure 6). Integration of a new spacer is mediated by the heterohexameric protein complex ((Cas12–Cas2)2) to the leader sequence of the CRISPR array. There are two types of spacer acquisition system named type I and type II. Type I system utilizes integration host factor (IHF) bonded to the leader sequence, which induces DNA bending. This bending enables the (Cas12–Cas2)2 complex to perform an initial cleavage for insertion of the spacer. While in type

II system, the leader anchoring sequence (LAS) of the leader is recognized using the Cas1 protein of the (Cas12–Cas2)2 complex and, then, the polar spacer is inserted.

Figure 6.

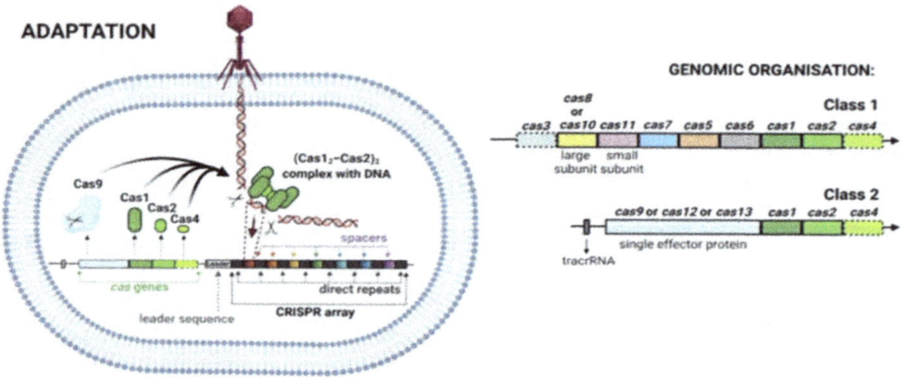

Source: (Nidhi et al., 2021)

ii. Expression or CRISPR RNA (crRNA) biogenesis or (crRNA Processing): The entire CRISPR array is transcribed into a long precursor CRISPR RNA (pre-crRNA). Further processing involves the cleavage of pre-crRNA within each direct repeat sequence to afford shorter, mature CRISPR RNAs (crRNAs). In certain systems, such as Type II, the crRNA further combines with tracrRNA to form a single guide RNA (sgRNA) complex, which is the functional unit for target recognition.

iii. Interference (Target Recognition and Cleavage): Interference involves the formation of a multiprotein effector complex or single effector protein. The multiprotein effector complex is typical for class 1 systems and is composed of multiple Cas proteins and crRNA. In class 2, the single effector protein is used and contains only one Cas protein with crRNA. The main purpose of both complexes is to recognize the same or very similar sequences in the genome of the invading virus or plasmid. Once recognised, the complex cleaves then inactivates the invasive genome. The interference step of some systems requires recognition of a protospacer adjacent motif (PAM) in the invading genome. The PAM is a short DNA sequence that is not present in the bacterial host genome.

Therefore, PAM is an essential targeting sequence which some Cas proteins need to bind and then be cleaved (Figure 7a,b) (Nidhi *et al.*, 2021).

Figure 7.

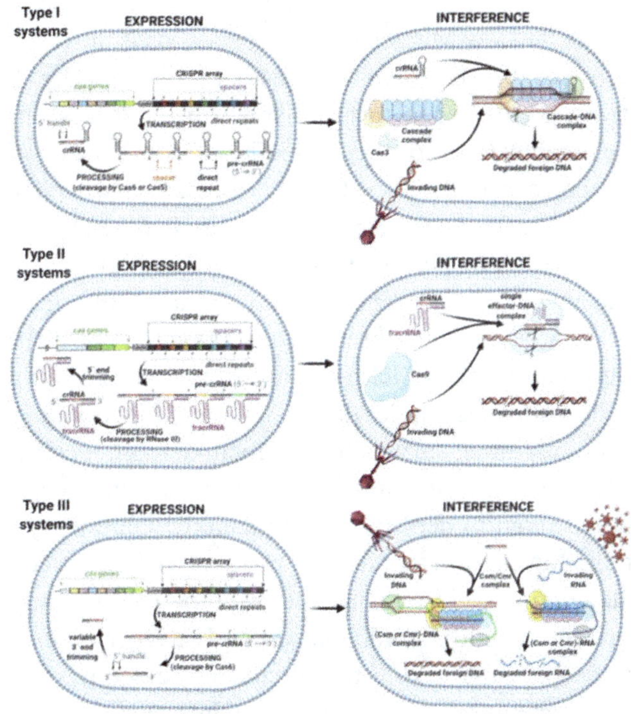

Source: (Nidhi et al., 2021)

CRISPR-Cas systems exhibit significant diversity, and their classification is based on differences in Cas protein composition and sequence divergence between the effector complexes. They are broadly classified into two main classes based on the architecture of their effector (Makarova et al., 2025).

a. **Class 1 Systems**: These systems are characterized by multi-subunit effector complexes, meaning that multiple Cas proteins work together to cleave the target (Type I, III, and IV). Each of these featuring effector complexes made up of multiple Cas proteins and crRNA during the interference stage (Liu & Doudna, 2020).
b. **Class 2 Systems:** The class 2 CRISPR-Cas systems include types II, V, and VI, each featuring a single, large multidomain Cas protein bound to crRNA. Type

II systems have three subtypes (II-A, II-B, II-C), with pre-crRNA processing involving RNase III, tracrRNA, and Cas9, which cleaves DNA using two nuclease domains (Shmakov et al., 2017).

The evolutionary origins and divergence of these effector modules are complex. The component of adaptation, catalyzing the insertion of spacers, has been proposed to be descended of casposons, a distinct kind of transposons, having Cas1 protein as a homolog of transposase. Much more heterogeneous is the origin of the effector modules (Makarova et al., 2025). Class 2 effectors, which include widely used Cas9 and Cas12 proteins, is believed to have evolved from nucleases, encoded by varied mobile genetic elements. One powerful hypothesis is that the evolution of Class 1 effector modules, and especially Type III systems, may have been related to ancient stress-response systems. These ancestral systems might have triggered programmed cell death upon activation by alarmones such as cyclic oligoA. This proposed evolutionary trajectory suggests a deeper, more integrated role of CRISPR systems in microbial survival beyond simple immunity (Faure et al., 2019). The intricate evolutionary interrelationship of CRISPR-Cas systems with mobile genetic elements and stress-response pathways implies that their natural function is far more dynamic and integrated into microbial survival than a mere defense mechanism (Koonin & Makarova, 2019b). This understanding is critical because engineering these systems for new applications requires deep insight of their co-evolutionary history. Without this, predicting the long-term stability and potential off-target ecological impacts when introducing engineered CRISPR systems into novel environments becomes challenging. Such interventions could inadvertently trigger or interfere with native cellular survival mechanisms, leading to unpredictable outcomes like unintended cell death, altered microbial community dynamics. They may also enhance horizontal gene transfer of the introduced genetic material, extending beyond simple off-target DNA cleavage to systemic ecological intrusions.

In response to these challenges, CRISPR-Cas toolkit has grown extensively to provide a comprehensive collection of programmable nucleases and editing capabilities since its discovery. Four most prominent of them are Cas9, Cas12, Cas13, Cas14 and more complex modifications such as Base and Prime Editors (Hu & Li, 2022).

8.3 CRISPR-Cas applications for Environmental Biotechnology

CRISPR-Cas systems have revolutionized the field of microbial engineering, offering a powerful collection of genetic editing tools that can be adopted to a diverse range of microorganisms. The technology also enables precise and efficient genetic editing, which promotes sustainability by providing solutions to global challenges, including climate change, environmental pollution, and food security Makarova et al., 2025). Bioremediation CRISPR-Cas9 greatly enhances microbial degradation potential of recalcitrant pollutants, such as plastics, heavy metals, pesticides, polychlorinated biphenyls (PCBs), polycyclic aromatic hydrocarbons (PAHs). This achieved by precisely dissecting and modifying specific genes involved in metabolic pathways, introducing novel degradation pathways, or increasing the expression of existing ones (Minhas et al., 2024).

8.4 Specific examples of CRISPR applications in environmental biotechnology include:

Heavy Metal Detoxification: *Mesorhizobium huakuii* subspecies. *rengei* Strain B3, modified with CRISPR-Cas9, has been shown to have increased tolerance and uptake of cadmium (Cd) due to upregulation of phytochelatin production. This modification makes the strain very effective in bioremediation of agricultural land causing contaminated agricultural zones such as rice fields. but with an organism (Singh et al, 2011). *Cupriavidus metallidurans CH34,* modified with CRISPR-Cas9, has been demonstrated to be genetically tractable via a single-plasmid CRISPR toolbox making it possible to delete or insert genes associated with metal resistance, biomineralization, or metal-handling pathways. This modification capacity makes *C. metallidurans* CH34 (or its derivatives) a promising chassis for bioremediation of metal-contaminated soils or waste streams, due to its innate broad resistance to heavy metals and ability to biomineralize or sequester toxic metals (Turco et al., 2022). Similarly, *Methylococcus capsulatus* (Bath) has been modified to enhance the reduction of highly toxic chromium (VI) to its less toxic Cr (III) form, exhibiting efficacy across a wide concentration range (1.4 to 1000 mg/L). This highlights its potential for use in diverse contaminated environments, including industrial waste sites and freshwater ecosystems (Singh et al., 2011).

Organic Pollutant Degradation: CRISPR-modified *Aspergillus niger* and *Phanerochaete chrysosporium* have demonstrated enhanced abilities to degrade PAHs and lignin, respectively, by targeting the cytochrome P450 pathway. In addition, microorganisms can be engineered to produce enzymes capable of degrading polymer chains in plastic waste or hazardous chemicals (Yuan et al., 2024). Notable plastic-degrading bacteria include *Pseudomonas* species (*P. aeruginosa*, *P. citronellolis*, *P. putida*) and Bacillus species. (*B. cereus*, *B. subtilis*) (Kotova et al., 2021).

Enhancing Nutrient Cycling: CRISPR-Cas can engineer soil microbes for more efficient nitrogen fixation, thereby reducing the reliance on synthetic fertilizers. It can also engineer microbes to solubilize phosphorus, making this essential nutrient more available to plants (García-Berumen et al., 2024).

Promoting Plant Growth and Resilience: CRISPR-Cas can engineer microbes to produce plant growth-promoting hormones, effectively acting as biofertilizers.

It can also be utilized to develop biocontrol agents against plant pathogens. Microbiome engineering aims to optimize ecosystem functions to promote plant growth and development, including enhancing nutrient uptake and stress tolerance (Singh & Ramakrishna, 2021).

Synthetic Microbial Consortia: CRISPR enables the precise design of specialized microbial consortia that work synergistically to degrade complex pollutants, such as hydrocarbons in oil spills or heavy metals in wastewater. These artificially created communities can be meticulously designed, focusing on selecting species with desired genetic and biological characteristics to achieve specific environmental outcomes (Zhou, 2024).

Microbial Community Engineering for Ecosystem Services: CRISPR-Cas technology has opened new avenues for understanding and manipulating microbial communities and their roles in ecosystems. The application of CRISPR to engineer microbial communities, rather than solely individual strains, signifies a paradigm shift towards addressing ecosystem-level challenges. This approach recognizes that many environmental processes are mediated by complex microbial interactions, and manipulating these interactions offers a more holistic strategy for enhancing ecosystem services like nutrient cycling and bioremediation (Dudeja et al., 2025). This represents a growing maturity in microbial biotechnology, moving from reductionist to more integrated and ecologically informed approaches Table 18.2.

8.5 CRISPR Biosensors for Environmental Monitoring

CRISPR biosensors are rapidly transforming environmental monitoring by enabling highly sensitive, specific, and rapid detection of a broad range of contaminants and pathogens, often in field-deployable formats. This advancement shifts environmental surveillance from laborious lab-based methods to real-time, on-site diagnostics, which is crucial for proactive environmental management and rapid response to emerging threats. The ability to detect both nucleic acid and non-nucleic acid targets further broadens their utility, moving from reactive monitoring to proactive surveillance (Wani et al., 2024).

The integration of CRISPR and advanced genome editing technologies into microbial environmental biotechnology represents a transformative frontier with profound implications for addressing global environmental challenges. The detailed examination of CRISPR-Cas systems, from their fundamental prokaryotic immune mechanisms to the diverse array of Cas enzymes and their refined functionalities, underscores the remarkable precision and versatility now available for microbial engineering (Verdezoto-Prado et al., 2025a). However, the path to widespread environmental deployment of CRISPR is fraught with significant technical, ecological, and ethical hurdles. The challenges of efficient and stable in situ delivery in complex natural environments, compounded by concerns about off-target effects, the long-term stability of engineered traits, and potential horizontal gene transfer, necessitate continued rigorous research and development (Karimi et al., 2025). The ecological risks, including the disruption of native microbial communities, the potential for engineered microbes to become invasive, and the irreversible nature of some genetic modifications, underscore the need for extreme caution. Ethically, questions surrounding human intervention in nature, equitable access to benefits, and biosecurity demand ongoing societal dialogue and robust governance. Current regulatory frameworks, while evolving, struggle to keep pace with the rapid advancements and the global nature of this technology (Caplan et al., 2015). Ultimately, realizing the full promise of CRISPR in microbial environmental biotechnology hinges on a concerted effort to overcome these challenges. This requires not only continued scientific innovation in developing safer and more precise tools and delivery methods but also the establishment of adaptive, internationally harmonized regulatory frameworks and transparent public engagement to build trust and ensure responsible deployment. The future of microbial environmental biotechnology, integrating AI, CRISPR, and systems biology, holds the key to sustainable solutions, provided that scientific ambition is balanced with a profound commitment to ecological integrity and ethical stewardship.

9 SYSTEMS BIOLOGY AND SYNTHETIC MICROBIAL COMMUNITIES

Microbial environmental biotechnology is an evolving field focused on solving environmental challenges such as the pressing demand of climate change action, pollution management and sustainability globally (Verdezoto-Prado *et al.*, 2025b). Among the most transformative tools in this field is systems biology, an interdisciplinary field that combines biology, computer science, mathematics, and engineering to comprehend complex biological systems in their entirety. Instead of studying single genes or proteins or their activities in isolation, systems biology seeks to reveal the interactions of various components of a system to create functions and behaviors. It aims to gain insight into biological systems at many different levels (genes to proteins, cells to community), through modelling of their dynamic interactions, feedback loops and network properties (Zupanic *et al.*, 2020). This field is particularly essential in environmental contexts, as microbial communities operate in a dynamic and heterogeneous environment (Voit, 2022). In environmental biotechnology, systems biology applies this system's approach to the study of environmental processes, particularly of microbial communities and their functions in the degradation of pollutants, bioremediation, biosensing and production of biofuels. It typically involves identification of the whole microbial communities, the context of the environment, and the connections between the microorganisms and the environment, instead of considering specific microbial species or particular metabolic pathways (Chakraborty *et al.*, 2012). According to Leggieri *et al.* (2021), systems biology represents a significant shift from traditional descriptive biology to a more predictive and quantitative science. By modelling how microbial communities (microbiomes) respond to environmental inputs, interact with other organisms, and regulate internal functions, researchers can design organisms and consortia that optimize microbial pathways for sustainable functions, and engineer new biological solutions to press global challenges such as pollution, climate change, and resource depletion.

Systems biology uses multi-omics integration; the combination of genomic, transcriptomic, proteomic, metabolomic and other datasets to create a layered model of biological state and functions (Pinu *et al.*, 2019). This approach, frequently facilitated by high-throughput (-omics) technologies, enables researchers to simultaneously determine which microbial genes are being transcribed, which proteins are being translated, and which metabolic processes are in action. Such combinations of data types enable the mapping of regulatory networks, the reconstruction of gene networks and the measurement of the fluxes in metabolic networks to allow a complete view of microbial communities' system dynamics (Wang *et al.*, 2024a). It is based on these technologies that systems biology creates computational models, that is, quantitative representations that are used to simulate the dynamics of a network and predict the response. These can be ordinary differential equations - describing how the concentrations of the various components vary with time. They can also be constraint-based models such as flux-balance analysis, which attempt to predict metabolic fluxes (Kim *et al.*, 2018), Boolean or agent-based networks which encode regulatory logic, or probabilistic models learned by fitting large datasets. These types of models can be accurately simulated in response to changes, e.g. gene knockdowns, environmental stressors and determine key control points and behaviors at a functional level of the system (Cruz & Kemp, 2022).

In environmental microbial biotechnology, such a systems-level approach is indispensable as these microorganisms play a central role in many ecological processes such as biogeochemical cycling, degradation of pollutants, carbon fixation and nitrogen turnover. Systems biology offers computational platforms to model these interactions, define community structure-function relationships and design synthetic microbial consortia (SynComs) that perform specific environmental functions substantially more effectively than single strains (Shayanthan *et al.*, 2022). Systems biology has been employed in the case of *Pseudomonas putida*, commonly applied in the case of bioremediation processes, to reconstruct its metabolic network, aiding in increasing the efficiency of its role in degrading toxic compounds (Nikel & de Lorenzo, 2018; Qattan, 2025).

Despite its transformative potential, there are difficulties with the implementation of the concept of system biology into environmental biotechnology. This is mainly due to the quality, completeness, and representativeness of the data, which are critical limitations. Most environmental microorganisms cannot be cultured, or are very rarely culturable, and their genomes are poorly annotated; therefore, it is not easy to describe their metabolic potential or roles in their ecosystem (Bodor *et al.*, 2020). Metagenomic and single-cell sequencing research are already very useful in that regard, but not without their challenges, which are uneven coverages, sequence chimaeras, contamination, assembly challenges, and annotation bottlenecks, particularly when dealing with low-abundance species (Xu & Zhao, 2018). However, there are efforts to address these challenges through initiatives that promote open data sharing, standardization of omics workflows, and the development of user-friendly modelling platforms. In addition, systems biology is becoming more widespread and detailed with the advent of metagenomics, single-cell technologies and real-time biosensing, allowing a growing number of natural microbial systems to be modelled (Zhang *et al.,* 2021). In retrospect, systems biology provides the intellectual and technological foundation for the next generation of microbial environmental biotechnology. It enables researchers to advance beyond trial-and-error experimentation in predictive modelling of microbial systems using a holistic, quantitative, and predictive understanding of microbial systems, which then provides a basis to design and optimize such systems.

10 OMICS TECHNOLOGIES: TRANSCRIPTOMICS, METABOLOMICS, PROTEOMICS

Omics technologies form the core of systems biology, providing a comprehensive and data-rich perspective of the understanding of the biological system at the molecular level. Unlike traditional research into individual components, omics methodology provides researchers with the ability to analyse all the components of gene expression, protein synthesis, and metabolic processes in organisms or microbial communities (Pinu *et al.*, 2019). They are based on high-throughput systems and bioinformatics resources capable of working with huge amounts of data and showing dynamic relationships between the components of the cell and the environment. Omics techniques are fundamental in the study of microbial ecosystems in learning about the regulatory, functional, and metabolic pathways that occur in interspecies interactions, community stability, and systems-level traits. These data can be integrated, helping build extensive, predictive models and assisting in ensuring more rational design of synthetic microbial communities (Wang *et al.*, 2024b).

a) Transcriptomics

Transcriptomics, especially RNA seq and metatranscriptomics, are important in microbial environmental biotechnology, especially in the context of system biology, as it is used to study gene expression, which occurs actively under certain environmental conditions. RNA seq enables the complete documentation of transcriptome in non-model or uncultured environmental microbes by enabling de novo assembly and annotation of the transcripts without utilising a reference genome (Hennessy *et al.*, 2017; Mante *et al.*, 2024; Kraus *et al.*, 2025). Whereas transcriptomics is the study of gene expression profile of a given strain at a specific growth phase, metatranscriptomics is the study of the gene expression profile of the microbial community by extraction of RNA sample in an environmental sample (Alidoosti *et al.*, 2024).

In environmental samples, metatranscriptomics identifies which genes are actively transcribed in situ, revealing ecologically relevant functional dynamics. Applying metatranscriptomics to the microbiome of a polluted soil or water sample can enable the researcher to discover which genes are actively being transcribed in situ, providing ecologically relevant functional dynamics of the microbiome. The use of transcriptomics in system biology has been demonstrated in a study conducted by Sun *et al.* (2024) to study the response of *Stenotrophomonas rhizophila* JC1 to the exposure to lead (Pb^{2+}) to unravel the kinetics of microbial adaptation to environmental stress. By application of comparative transcriptomics, the researchers traced the overall gene expression during lead exposure of *S. rhizophila* in tracking a complex of interrelated gene repertoires encompassing metal biosorption, transport and cellular defense. This shows how these genes interact as part of larger regulatory networks that confer environmental resilience. It was revealed that genes co-regulated in metal efflux, oxidative stress response, and membrane remodeling indicated network behaviour, not an independent response of different isolated genes. These findings can be used to construct the model of gene regulatory networks (GRNs) detailing how *S. rhizophila* coordinates its molecular resources in response to heavy metal stress, which gives us a systems-level understanding of its survival mechanism.

Also, Sharma *et al.* (2019) using meta-transcriptomics carried out a comparative analysis of the metatranscriptome of microbial communities from two distinct soil ecosystems referred to as M1 (agriculture soil) and O1 (organic soil), the results identified the significant role of archaea in the nitrification process in metal-contaminated and pesticide-contaminated soils compared to bacteria, which reveals their activity in polluted soils. It showed the strong expression of transcripts of the glyoxalase and/or bleomycin resistance dioxygenase, 4-hydroxyphenylpyruvate dioxygenase,

2-nitropropane dioxygenase, metapyrocatechase, ring hydroxylating dioxygenases and intradiol dioxygenase (belonging to *Novosphingobium* spp.) expressed in this domain, dedicated to aromatic hydrocarbon degradation in agricultural soil. By analyzing these transcriptomic profiles, researchers can reconstruct metabolic networks and identify key nodes and pathways responsible for environmental resilience by linking gene expression to environmental function, providing predictive insights into how microbial communities respond to emerging organic contaminants (EOCs). Singh *et al.* (2018) found 118 transcripts of 47 enzymes using metatranscriptomics responsible for xenobiotic degradation pathways were discovered across major bacterial phyla such as Proteobacteria, Actinobacteria, and Cyanobacteria, highlighting efficient pathways of pollutant detoxification. Such a dataset is invaluable to systems biology models that aim to predict microbial behavior and resilience under environmental stresses.

Nevertheless, transcriptomic data should be interpreted cautiously: mRNA abundance does not necessarily reflect protein levels or metabolic activity, and in microbial consortia, post-transcriptional regulation is frequent (Da Silva *et al.*, 2022). Maan *et al.* (2023) suggested that transcriptomics gives a primary layer of functional context; however, they have to be coupled with proteomics and metabolomics to establish adequate predictive models. In addition to this, environmental samples pose a challenge due to degradation of RNA, uneven coverage and even contamination with host or eukaryotic RNA (Wood *et al.*, 2019). The development of newer protocols, such as more efficient methods of RNA preservation and rRNA depletion, has improved most of these problems (Singh *et al.*, 2024). The transcriptomics data, integrated with other omics layers data, are invaluable resources that support the systems biology strategies to make mathematical models of gene regulatory networks and ecological responses to environmental stimuli.

b) Proteomics

Proteomics, or more precisely, metaproteomics, extends beyond transcriptomics, with the quantification of functional microbial proteins, enzyme activities, and interactions directly providing evidence of functional activity. In a recent experimental controlled consortium by Shrestha *et al.* (2021), they showed that protein biomass was dominated by *Pseudomonas* sp. GM17 across various growth media and mediated antagonistic interactions through siderophore and antibiotic production to influence community assembly and stability through metaproteomic analysis. These insights on the protein level identify competition, adaptation, and interaction or relationships between species cannot be resolved with transcriptomics. Metaproteomics has developed due to advances in High-resolution LC-MS/MS methods and optimized data workflows, enabling robust quantification of community-wide protein expression. This strategy has identified ecological functions-such as nitrogen cycling proteins in semiarid soils, carbon fixation enzymes in cyanobacteria and extracellular degradative enzymes central to litter decomposition and pollutant degradation (Arsène-Ploetze *et al.*, 2014).

Despite these benefits, proteomics still continues to face obstacles in the field of environmental biotechnology, as some proteins that are of low abundance, such as transcription factors and stress-response enzymes, are still difficult to detect. Moreover, identification of proteins relies critically on good reference genome catalogues that, in many cases, are not complete in environmental taxa. Bioinformatics tools have advanced to solve these problems with matched Metagenomic databases and more advanced algorithms that can infer proteins to identify and quantify proteins even in diverse biomes (Patel *et al.*, 2023). Importantly, the combination of metaproteomics with transcriptomics and metabolomics provides complementary evidence for function, where proteomics confirms which proteins are actually present and active in the community. Integrating proteomic data into systems biology models allows researchers to map enzyme abundance to metabolic fluxes, validate regulatory interactions inferred from transcriptomics, and quantify functional guilds within microbial consortia

c) Metabolomics

Metabolomics measures changes in metabolite concentrations, signaling pathway bottlenecks, redox imbalance or community-level metabolic flux through systems such as LC MS, GC MS and NMR. Metabolomics provides the closest phenotype-proximal readout of the omics levels by profiling of small molecules and metabolic intermediates, which reflect the real biochemical functioning of microbial systems (Hajnajafi & Iqbal, 2025). Metabolomics data forms part of a metabolic network reconstruction in systems biology modelling, which forms a basis of flux balance analysis and inverse modelling methods, such as MetConSIN. The molecular concentrations provide feedback in adjusting network stoichiometry and thermodynamic limitations and enhance the prediction. This enables them, in environmental applications, to model community-level metabolism, e.g. carbon or sulfur cycling, and characterise rate-limiting enzymes/ cross-feeding pathways (Palermo, 2022).

Moreira *et al.* (2022) highlighted how systems biology is shaping the future of environmental biotechnology, particularly in addressing the challenge of endocrine-disrupting compounds (EDCs) such as 17-β estradiol (E2). By using metabolomics tools, the researchers were able to identify the pathway in which E2 could be degraded by the bacterium *Rhodococcus* sp. ED55 in wastewater through their application of metabolomics tools. This whole-organism perspective was made possible by the ultra-performance liquid chromatography coupled with high-resolution mass spectrometry (UPLC/ESI/HRMS), which made it possible to detect and profile time-resolved 17 different metabolic intermediates. The prospect of metabolomics data incorporation within the analysis of microbial function through the systems biology approach allows for a complete picture of the biochemical conversions taking place in the process of bioremediation. Besides the potential to discover the key metabolites and pathways, this strategy also provides predictive understanding of the dynamics that occur in the course of degradation, the effectiveness of bioaugmentation and how treatment processes affect the ecological soundness, as reflected by the toxicity reduction and the estrogenic activity decline in the treated effluent.

In the study of Park *et al.* (2022) on *Chryseobacterium* sp. BSC2-3 utilizing a metabolomics technique based on LC-MS, the researchers managed to visualize not only the extracellular degradation of a pesticide, carbofuran, to 3-hydroxycarbofuran but also the resulting metabolic changes that occurred inside the cells. Such an analysis at the systems level showed derailment of pathways including aminobenzoate degradation, ubiquinone and terpenoid-quinone biosynthesis, arginine and proline metabolism, providing perspectives of how the organism defends against stress states and detoxification processes. The result of this omics approach can help inform the development of microbial agents both in pollutant degradation and in farming improvement, which is a new way of precision environmental biotechnology.

Nonetheless, metabolomics also has limitations, in that no single technology can access all metabolites, resulting in possible false negatives; the lack of comprehensive databases of metabolites prevents accurate annotation of compounds generated, and re-tracing individual metabolites to conventional pathways or taxa may be unclear. Such obstacles demand the inclusion of other layers of omics (Mahieu & Patti, 2017). As a matter of fact, the multi-omics studies reviewed favor the integration of metabolomics with transcriptomics and proteomics, a process through which a coherent representation of the microbial metabolism can be designed to validate gene expressions and protein functionalities against real metabolite production or consumption (Pinu *et al.*, 2019; Eicher *et al.*, 2020). Figure 8 showcases the different omics approaches involved in systems biology.

Figure 8.

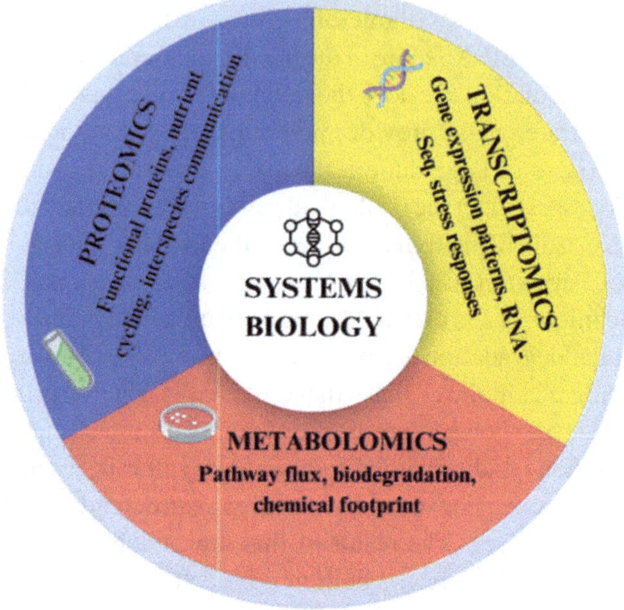

11 MODELLING MICROBIAL NETWORKS AND INTERACTIONS

Within the context of systems biology and environmental biotechnology, modelling microbial networks is essential for engineering microbial communities with predictable, robust functions (Altaf-Ul-Amin *et al.*, 2014). High-throughput omics such as genomics, transcriptomics and metabolomics, including mathematical modelling to capture biological systems as integrated networks, are unified through systems biology (Pinu *et al.*, 2019). Within environmental biotechnology, such frameworks are used to enable the logical design of engineered systems in activities such as the degradation of pollutants, cycling of nutrients, and bioenergy generation. Incorporating constraint-based metabolic analysis, dynamic modelling, spatial simulation, and statistical network inference, these methods can enable researchers to make predictions on inter-species interactions, nutrient availability, and determine spatial organization in biofilms or soil matrices and system stability in response to environmental fluctuations (Srinivasan *et al.*, 2024; Yadav *et al.*, 2022).

a) Genome-Scale Metabolic Models (GEMs)

Genome-Scale Metabolic Models (GEMs) are a comprehensive reconstruction of the metabolic network of an organism, constituting the whole set of enzyme catalytic reactions, encoded in a genome. In environmental microbial biotechnology, GEMs have been successfully applied to design synthetic consortia that degrade complex substrates and produce value-added products (Passi *et al.,* 2021). GEMs are typically combined with constraint-based modelling, more specifically Flux Balance Analysis (FBA), to attempt to predict growth rates, nutrient uptake and production fluxes under measured environmental conditions (Morrissey *et al.,* 2025). GEMs have been used to predict inter-species interactions, e.g., the existence of metabolic cross-feeding or syntrophic exchange, the determination of nutrient bottlenecks, optimal resource utilisation/ partitioning and the impact of induced environmental changes (e.g. oxygen availability, substrate abundance) on species communities and community performance in the context of synthetic biology and microbial consortia (Du *et al*., 2023). This model was used in the study of Schwalm *et al.* (2019), which used *Clostridium beijerinckii NCIMB 8052 and Yokenella regensburgei ATCC 43003* that were co-cultured in food-waste substrates. The authors hypothesised a cross-feeding interaction mediated by lactate: the *Y. regensburgei* would secrete lactate, which *C. beijerinckii* would use to increase biomass and hydrogen generation. The model predicted better biomass and hydrogen yields; in the experiment, cross-feeding on lactate was confirmed and resulted in higher butyrate or lactate production, but not higher yield of hydrogen as compared to cultures with only one species present, demonstrating the utility of the model to make inferences and optimize consortium composition efficiently. Additionally, tools such as the computation of Microbial Ecosystems in Time and Space (COMETS) extend GEM/FBA further with spatially structured dynamic models, which forecast diffusion of metabolites, growth of communities and spatial patterns in consortia and even community design, using frameworks such as FLYCOP, which optimise combinations, such as *Synechococcus elongatus* with *Pseudomonas putida* to make bioplastics (Du *et al.,* 2023).

b) Agent-Based Models (ABMs)

Agent-Based Model (ABMs, henceforth) functions to simulate the appearance of macroscopic properties of a colony because of microscopic interactions between agents, the properties of which can also have a large amount of variability. Their abilities to naturally incorporate heterogeneity in microbial colonies make ABMs the natural choice of model whenever such variability is known to play a significant role in the phenomenon of interest (Hellweger *et al*., 2016). ABMs are inherently spatial and time-resolved, and are capable of heterogeneous environmental profiles (gradients, such as oxygen, pH), and physical limitations. They are uniquely effective in simulating biofilm growth, the development of spatial niches, microbial patterning, and engineered SynCom incorporation in complicated systems such as soil pores/sediments, host tissue (Nagarajan *et al.*, 2022). A classic example is the Individual-based Dynamics of Microbial Communities Simulator) iDynoMiCS platform, which simulates biofilm growth in 3D and models nutrient diffusion, cell division, movement, quorum sensing, and pressure fields.

Helicobacter pylori biofilm formation have been modelled by exploiting the iDynoMiCS platform with classical versions to now model three-dimensional chemotaxis toward or away AI-2 quorum sensing molecules. The simulations that were generated recreated the observed shifts in experimental biofilm architecture, cooler, patchy in AI-2 overproducers, and warmer, homogeneous in sensing-incapacitated cells, and revealed how emergent spatial order was constructed based on spatially localized, individual chemotactic behavior. The result illustrated the importance of the local chemotactic responses to metabolic clues in the development of the global spatial patterns.

12 ECOLOGICAL NETWORK INFERENCE TOOLS

Network inference tools like SPIEC-EASI, SparCC, CoNet or FlashWeave use statistical and compositional data analysis to infer microbial interaction networks from high-throughput sequencing data. These techniques are based on co-occurrence patterns (positive associations) and mutual exclusivity (negative associations). They are useful in finding out ecological relations like mutualism, competition, keystone species as well as functional guilds, long before experiment validation commences (Matchado *et al.*, 2020; Lam & Ye, 2022).

13 DYNAMIC FLUX BALANCE ANALYSIS (DFBA)

Dynamic Flux Balance Analysis (dFBA) is an extension of conventional Flux Balance Analysis (FBA), incorporating time-dependent dynamics with metabolic modelling. Further than steady-state flux distributions, dFBA links genome-scale metabolic models to ordinary differential equations (ODEs), successively revising extracellular concentrations (i.e. substrates, products) and cell biomass with time (Kuriya & Araki, 2020). It enables simulation of batch or fed-batch fermentation, substrate exhaustion and formation of metabolites. dFBA is capable of modelling interactions between co-cultured species (e.g. cross-feeding or inhibition) by dynamically varying the constraints on fluxes or maintenance requirements according to the varying environmental conditions (Dehkordi *et al.*, 2023). Kuriya & Araki (2020) have used dynamic flux balance analysis (dFBA) in a model which evaluated a shikimic acid-producing *E. coli* strain. The experimental measurements of glucose consumption and growth were transformed into time-dependent constraints, and two-stage optimization was applied: optimizing first the biomass and then the flux of shikimic acid. The model suggested that the actual strain attained about 84 percent of the theoretical maximum yield under similar conditions. In the context of environmental microbiology, such engineered strains could be deployed in the bioprocessing of biomass waste, enabling sustainable chemical production. This highlights how systems biology tools bridge metabolic engineering and environmental biotechnology for resource-efficient biomanufacturing.

Table 3 outlines key computational tools and modelling approaches used to analyze microbial networks and interactions within the context of synthetic biology and environmental biotechnology while Table 4 shows the advantages and disadvantages different Modelling Approaches in Environmental Microbial Systems.

Table 3. Approaches for Modelling Microbial Networks and Interactions in Environmental Biotechnology

Model Type	Description	Applications in Environmental Biotechnology	Example Study / Case
Genome-Scale Metabolic Models (GEMs)	Reconstruction of an organism's entire metabolic network; often integrated with Flux Balance Analysis (FBA).	Design of microbial consortia for pollutant degradation, nutrient cycling, and bioproduct generation.	Schwalm et al. (2019): Predicted lactate cross-feeding between *Yokenella regensburgei* and *Clostridium beijerinckii*, improving biomass and butyrate production in food-waste media.
Agent-Based Models (ABMs)	Simulate individual microbial agents interacting in space and time; model heterogeneity and microenvironmental gradients.	Used for modelling biofilm formation, spatial niche development, and SynCom behaviour in soil, sediments, or host tissues.	Sweeney *et al.* (2019): Modeled *H. pylori* chemotaxis using iDynoMiCS to show how AI-2 gradients influence emergent biofilm structure and spatial order.
Ecological Network Inference Tools (e.g., SPIEC-EASI, SparCC, CoNet)	Use statistical inference on sequencing data to predict microbial interactions based on co-occurrence and exclusivity patterns.	Identify key ecological relationships (e.g. mutualism, competition, keystone taxa) for environmental community design.	Matchado *et al.* (2020); Lam & Ye (2022): Inferred functional networks in natural systems, guiding synthetic consortia design for nutrient fixation or pollutant degradation.
Dynamic Flux Balance Analysis (dFBA)	Extends FBA to model time-dependent changes in metabolic fluxes using ODEs; simulates fermentation and co-culture dynamics.	Models co-culture interactions (e.g., cross-feeding), simulates substrate depletion, metabolite formation, and strain optimization.	Kuriya & Araki (2020): Modeled *E. coli* producing shikimic acid from biomass; 84% theoretical yield predicted using dynamic flux modeling.

Table 4. Advantages and Disadvantages of different Modelling Approaches in Environmental Microbial Systems

Modelling Approach	Advantages	Disadvantages
Genome-Scale Metabolic Models (GEMs)	•Captures detailed metabolic pathways and enables strain/community design for pollutant degradation, nutrient cycling, and bioproduct synthesis (Passi *et al.*, 2021) • Efficient SteadyState simulation via FBA under constrained conditions, scalable to large models (Ranganathan et al., 2009)	• Dependent on accurate genome annotation and curation, uncertainty in reconstruction reduces predictive reliability (Bernstein *et al.*, 2021) • Steady-state assumption limits dynamic behaviours unless extended with time-varying methods
Agent-Based Models (ABMs)	• Captures individual-level heterogeneity (e.g. chemotaxis, quorum sensing), emergent spatial patterns in biofilms and SynComs (Nagarajan *et al.*, 2022) • Naturally spatial and time-resolved—suitable for environmental gradients and microenvironments	• Requires detailed rule sets, parameterisation, and often scarce experimental data • Computationally intensive; scaling to large-scale communities is challenging
Ecological Network Inference Tools (e.g. SPIEC-EASI, CoNet)	• Reconstruction of microbial interaction networks and identification of keystone taxa from sequencing data without culture (Faust & Raes, 2012) • Generates testable hypotheses for further modelling or experimentation	• Based on correlations may result in spurious or non-causal associations (Faust & Raes, 2012) • Sensitive to data compositionality, sequencing depth, and preprocessing biases
Dynamic Flux Balance Analysis (dFBA)	• Integrates time-varying extracellular metabolite concentrations and biomass growth dynamics; well-suited to simulate batch or fed-batch systems and co-cultures • Facilitates modelling of cross-feeding, substrate depletion, and community interactions	• Requires kinetic parameters and accurate environmental data not always available • More computational complexity compared to classic FBA, especially with system scaling

14 DESIGNING SYNTHETIC MICROBIAL CONSORTIA (SYNCOMS)

The design of synthetic microbial consortia (SynComs) leverages systems biology and synthetic biology to construct multi-species communities to fulfil specific tasks in environmental biotechnology, including the degradation of pollutants, cycling of nutrients, promotion of plant growth and conversion of bioenergy. There are two conventional approaches to designing SynComs: the top-down and bottom-up methods. The top-down approach aims at mimicking the natural microbial communities by introducing a great diversity of beneficial species, and thus has a low probability that useful functions will be excluded. This process requires robust species diversity measurements in the indigenous microbiome that inform dynamics, composition and service of the ensuing SynCom. Conversely, the bottom-up technique builds microbial communities using individual strains because some functional characteristics are chosen. These chosen microbes are then iteratively constructed and refined by successive metabolic engineering and testing until they perform as desired (Toju *et al.*, 2020).

In the top-down approach, one can start with a more complex natural microbial community exhibiting a phenotype or functionality desired. Environmental samples are collected, microbial cultures are isolated from the environment, and metagenomics and metabolomics methods are used to reveal the important microbial actors. Based on this, a simpler but functionally effective microbial consortium is constructed. Conversely, the bottom-up approach begins with known individual microbial isolates. Scientists put together and create SynComs using certain characteristics, test the usefulness of SynComs, and then enhance them using genomic editing. The top-down approach, although being data-guided and based on natural systems, enables the bottom approach to maximise precision and control over SynCom design (Tariq *et al.*, 2025).

Other than the assembly strategy, experimental methods are also critical as a SynCom has to be designed through culturing, genomic and functional profiling, and systematic composition in specific ratios of microbes. Designing effective SynComs is not just about picking the right microbes, it's about putting them together in the right amounts, roles, and places so they do not compete but collaborate efficiently (Johns *et al.*, 2016).

The outcomes of the design of SynComs are simulated and predicted by the use of computational models. The models combine AI and ML techniques to optimise SynComs with dynamic patterns of microbial population and metabolic cross-feeding networks. The computational methods are based upon genome and genome expression databases to find useful microbes having desirable functional characteristics (Finkel *et al.*, 2017). Ultimately, a well-structured SynCom depends on how effectively tasks are distributed across its members, hence the critical need for the division of labour in these engineered microbial communities.

Figure 9 visually summarizes the process of constructing SynComs for environmental and agricultural applications starting from identifying a functional goal to validating and deploying the SynCom

Figure 9.

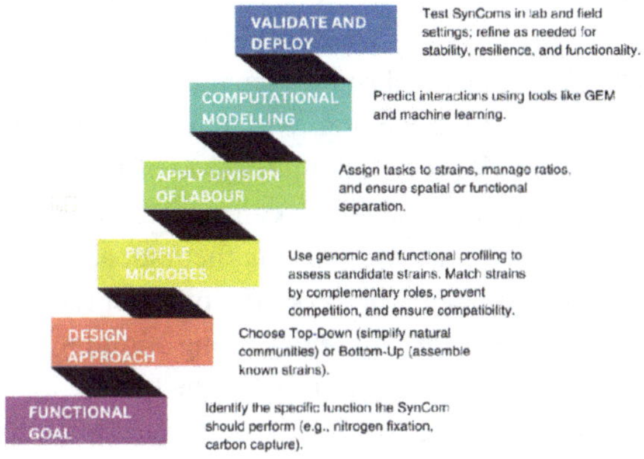

14.1 Division of Labour in Engineered Communities

Division of labour (DoL) is an essential strategy in the design of SynComs as it enables engineered consortia to complete a complex, multi-step task more efficiently than individual strains. DoL entails specialisation of metabolic or ecological functions among the members of the community, which reduces the metabolic burden on each community member and uses interspecies collaboration to increase overall functionality (Roell *et al.*, 2019). In principle, niche-specific strain specialisation, i.e. different enzymatic tasks being assigned to a specific member, and spatial separation methods, such as microencapsulation or isolation via hydrogel, can be used to implement DoL, whilst keeping the levels of functional complementarity and simultaneously avoiding interference between strains. Population control is achieved by engineered population control circuits (including quorum-sensing-based feedback), which maintain the strain ratios without expressing dominance or cheating. Computational tools, such as genome-scale metabolic models and dynamic flux balance analysis, support metabolic pathway division and prediction of metabolic pathway design prior to starting any bench work (Duncker *et al.*, 2021; Matuszyńska *et al.*, 2024).

DoL in designing SynComs has been well-demonstrated in agricultural and bioremediation contexts. Li *et al.*, (2021) studied, a synthetic pyrene-degrading microbial consortium where *Mycobacterium* strains are involved in the primary degradation of pyrene, and *Novosphingobium* and *Ochrobactrum* species are used in the degradation of the downstream intermediates, which provides an increase in efficiency over the monocultures. DoL also plays a role in lignocellulose-degrading microbial consortia that are designed for biofuels applications. Lin (2022) reported combining *Pseudomonas putida, Cellulomonas fimi,* and *Methylorubrum extorquens* in such a way that each strain specialized in lignin, cellulose or intermediate utilization, respectively. This functional and spatial separation allowed the decomposition of the biomass of plants by cross-feeding associations effectively. It is important to note that engineering SynComs based on division of labour includes intelligent selection of strains and specialization, which can be confirmed both experimentally (e.g. through binary association assays) and mathematically (through modelling). The ability to evenly allocate metabolic tasks-the nutrient recycling role, stress resistance, or the degradation of pollutants, etc., among complementary strains within a DoL-built SynCom leads to greater stability, less metabolic impact, and improved performance in complicated environments.

14.2 Host-Environment-Microbe Interactions

The functionality of synthetic microbial communities (SynComs) is heavily influenced by microbial characteristics alone, host biology and the environment, including the soil structure, including moisture, texture, structure and chemical gradients. The dynamics of microbial communities in the environment contribute to functionality, resilience and stability of synthetic microbial consortia in soil ecosystems. Major drivers are abiotic gradients of the soil moisture, pH, nutrient availability, texture and the organic matter. These contribute to the metabolism of the microbes, the community and the interaction among the species through mechanisms such as chemical signaling, resource competition and spatial structuring (De Roy *et al.*, 2013). Studies have shown the effect of these interactions in the effective design of SynComs. Hao *et al.* (2024) carried out a two-year field trial on desert regions, which is the host by utilizing SynCom (microbe), composed of indigenous probiotic bacteria, which significantly improved the soil's multifunctionality. In this context, the environment comprises the desert region's conditions (e.g., arid climate, low fertility, degraded structure). Their interactions led to an increase in the activity of enzymes and the abundance of microbial genes associated with carbon, nitrogen and phosphorus cycling. The SynCom played an immensely positive role in drastically improving soil function as compared to controls and chemical fertilizer treatment, and a very strong in situ local environment and microbiome context match.

Microbial interaction is also mediated by the spatial and physical structure of soil, the pore network and heterogeneity. Mathematical modelling has shown that geometry and diffusion pathway plays a significant role in the way the microbes interact in terms of metabolite transfer as well as the formation of the cooperative networks. Long pore channels in natural soil matrices might allow or prevent cooperation between strains by microbial spreading, a property which fundamentally changes cooperation among strains. As an example, environmental factors, like antibiotic stress and inoculation density, could determine the dynamics of mutualism, commensalism or extinction of populations in a synthetic micro-ecosystem (Peaudecerf *et al.*, 2017).

Design of good SynComs needs more than picking the microbial strains of interesting traits; one needs to incorporate the host and soil environment as a co-driver. Microbial colonisation, metabolic activity, and community stability are affected by soil moisture, texture, pollutants, compositions of the indigenous microbiome and environmental heterogeneity. Engineers can design microbial consortia that enrich the host systems and live adaptively in the desired ecological niche, and not only complementary to it but also dynamically enliven the ecosystems by reinstating soil functions, better resilience and higher ecosystem services in degraded or stressed landscapes, because of grounding the SynComs design in the realism of the environment.

15 CASE STUDIES: DEGRADATION, CARBON CAPTURE, NITROGEN FIXATION

a) Degradation: PET & Nylon Monomer Bioremediation

In the study of Qi *et al.* (2021), a synthetic microbial consortium (SynCom) capable of degrading polyethene terephthalate (PET), a commonly used plastic, was created. It was made up of two metabolically engineered *Bacillus subtilis*, engineered to express the proteins PET hydrolase (PETase) and mono(-2-hydroxyethyl) terephthalate hydrolase (MHETase), *Rhodococcus jostii* to ferment terephthalic acid and *Pseudomonas putida* to metabolise ethylene glycol. In this 4-species SynCom, 23.2 % loss of PET film was observed in 3 days, surpassing two-species or three-species consortia (~13-17 % loss of PET).

Edwards *et al.* (2021), using the multi-omic approaches, key genes having the role of degradation of PET were identified in the pangenome and the transcriptome of SynComs (consortium of five strains containing *Pseudomonas* and *Bacillus* species) previously utilised in the study of Roberts *et al.* (2020) for PET degradation. This resulted in the discovery of a new PETase enzyme, EstB, which is able to act on PET polymer and also its intermediate, BHET. Besides the PET-degrading genes identified in the study, several other genes belonging to a diversity of biodegradation pathways were retrieved. This highlighted a systems biology approach, where genome-scale models of the metabolic networks, as well as omics data, were used, in a systems biology perspective, to inform pathway partitioning, which strain would be best in dealing with which intermediate. In the complete consortium, RNA-seq identified upregulation of hydrolase genes, transport, and aromatic degradation and biopolymer (PHA/PHB) biosynthesis pathways, which revealed the strain-specific functions toward TPA and EG degradation pathways. Furthermore, pangenomic and transcriptomic analyses at a systems level have shown redundant and complementary enzyme functions to be present across strains, permitting the degradation of mixed plastics (e.g., LDPE, polyurethane) and plasticizers (another important status of functioning waste steams) (Edwards *et al.*, 2021; Qi *et al*, 2021; Fernández *et al.*, 2022). These studies illustrated how system biology combines multi-omic profiles and modelling of metabolic pathways to facilitate rational design of SynCom assembly- reducing the formation of toxic intermediates, maximising the enzymatic flux, and providing the biofilm-mediated contact to the substrate in the degradation of PET.

b) Carbon Capture & Bioproduction: CO_2-Fixed SynCom Systems

Zhang *et al.* (2020) utilized a synthetic microbial consortium allowing successive sequestration of CO_2 to form sucrose and then the platform chemical 3-hydroxypropionic acid (3-HP) in photoautotrophic growth conditions involving the fast-growing cyanobacterium *Synechococcus elongatus* UTEX 2973 and *Escherichia coli*. The cyanobacterium was engineered to efficiently secrete sucrose by overexpressing the gene *cscB* under a strong promoter, enabling it to convert CO_2 into sucrose under light-driven (photoautotrophic) conditions. Meanwhile, *E. coli* was modified to take in that sucrose and convert it into 3-HP by introducing two pathways: one to break down sucrose and another to produce 3-HP from malonyl-CoA. When the team optimised the growth temperature to 30 °C, they established a stable co-culture that produced up to 68.29 mg/L of 3-HP directly from CO_2. Interestingly, the growth of *S. elongatus* was also improved in the co-culture, likely because *E. coli* helped reduce harmful reactive oxygen species (ROS), thereby boosting photosynthesis in the cyanobacteria.

c) Nitrogen Fixation SynComs in Crop Systems

In agriculture, SynComs with free-living nitrogen-fixing bacteria- *Azospirillum* species, *Azotobacter* species, and *Rhizobium* species were used on the tomato system and the rice system. Using an open-field tomato trial, it was demonstrated that inoculation with the SynCom increased the microbial diversity, facilitated co-colonisation with mycorrhizal fungi, and increased the yield using low input of chemical nitrogen fertiliser (Novello *et al.*, 2024). Also, Chen et al. (2024) focusing on rice grown in greenhouses, six species SynCom (*Phenylobacterium, Rhizobium, Pleomorphomonas, Devosia, Sphingomonas,* and *Azospirillum*) in low-nitrogen soils greatly boosted shoot dry mass (1.34 x), showed shoot length (1.16 x), total plant nitrogen (1.48 x) and chlorophyll content quantified by SPAD (Soil-Plant Analyses Development) meters more than in untreated controls. From a systems biology perspective, diazotrophs species producing proteins of the nitrogenase complex (i.e. nifH, nifD, nifK) have been found to be predominant by metagenomic and proteomic profiling (multiomic approach). Genome reconstructions of metabolic networks and in silico analysis of flux variability contributed to the choice of non-nodulating strains with very high activities of nitrogen fixing and to community models to estimate carbon requirements and nitrogen productions (Fernandez *et al.*, 2020). These models contributed to the equilibrium in the cross-feeding and the carbon cost to host plants that maintained the ecology and the continuity of nitrogen supply without any negative growth effect. This demonstrated how systems biology enables predictive design and optimisation of agricultural SynComs: combining metabolic modelling, omics profiling, and empirical validation for robust, low-input enhancement of crop nitrogen nutrition

16 TECHNICAL BARRIERS AND LIMITATIONS

The current technical restrictions of integrating artificial intelligence (AI), CRISPR, and systems biology in microbial environmental biotechnology are a number of limitations to its full potential. The first difficulty is the accuracy of gene editing and off-target effects that may undermine the desired effects of microbial engineering. Although there are improvements in the technology of CRISPR, the off-target edits are still of concern, and more studies should be conducted to improve specificity and reduce these risks (Dara *et al.,* 2024). Contemporary technology in AI and CRISPR are already attempting to resolve some of these technical obstacles, but it can additionally create other complexities (Liv, 2024). AI has a tremendous impact on the performance of CRISPRs by enhancing downsides on target design and off-target effects to increase the level of precision and specificity when editing a gene (Chia *et al.,* 2025). In particular, deep learning models can be used to effectively model the activity and specificity of variant Cas9 and enable choosing the most appropriate enzyme and reducing off-target effects that play a critical role in safety and efficacy (Vibhuti Choubisa, 2024). There is also the use of AI and deep learning algorithms to automate other functions of CRISPR systems, including the prediction of guide RNA (gRNA) activity and the optimization of Cas9 variants, which considerably decreases the time and resources spent on CRISPR-based research and applications (Lee, 2023). In addition to CRISPR, it is hoped that AI will increase the reproducibility and predictability of biological experiments by transforming the large data sets and identifying unanticipated interactions between genes that may be overlooked by humans, thus allowing new hypothesis and experimental design (Abbasi, *et al.*, 2025). Nevertheless, certain AI techniques, such as constructing random forest trees, may be used as black boxes, in which outcomes have no apparent ways that humans can perceive, which is disturbing (Wackett, 2022).

17 VISION FOR THE NEXT DECADES

The decade ahead is set to produce major breakthroughs in microbial biotechnology, as further developments in AI, CRISPR, and systems biology lead to far-reaching effects in global sustainability. The various breakthroughs anticipated are the invention of new antimicrobial therapeutics, diagnostic tools, and advances in current microbial products like biofertilizers and biopesticides (Batista & Singh, 2021). It is believed that microbial biotechnology will transform the different industries, and it will provide new means in the field of bioremediation and carbon capture and renewable energy (Lee *et al.*, 2025). In the next few years, there will be more characterization of new microbes as well as the products of these microbes due to optimized microbiology and microbiome tools (Natarajan & Bhatt, 2020). The future of microbial biotechnology will rest on AI, CRISPR, and systems biology. In the near term, AI will play a central role in omics annotation, enabling rapid identification and characterization of new microbes and their metabolic products. This approach will accelerate the discovery of antimicrobial therapeutics, diagnostic tools, and microbial products such as biofertilizers and biopesticides, while providing detailed insights into protein structures, microbial interactions, and adaptation mechanisms (Dara et al., 2024; Robinson, 2022; Gong et al., 2024). In the mid-term, CRISPR technologies will allow precise engineering of microbial consortia, optimizing metabolic pathways and designing stable communities for applications in bioremediation, carbon capture, and renewable energy production (Lee et al., 2025; Mariam et al., 2025). The integration of CRISPR with AI will further enhance predictive modeling of microbial community dynamics, ensuring robust and effective microbial solutions. In the long term, ecosystem-scale synthetic biology will enable rational design of microbial systems to restore, enhance, or improve natural ecosystems. Such interventions are expected to mitigate pollution, promote clean water and land, advance renewable energy, and enable large-scale carbon capture, contributing to several Sustainable Development Goals (SDGs 6, 7, 13, 14, 15) (Behera et al., 2021; Hoffmann et al., 2023; Soberón-Chávez, 2023; Mishra et al., 2023; Maglione et al., 2024). Across all stages, careful regulatory oversight will be critical to ensure the ethical and safe deployment of these technologies, positioning microbial biotechnology as a transformative force for environmental sustainability and industrial innovation In order to prevent global warming, microorganisms are also important in supporting biogenic cycles to provide viable sources in sustainable energy production in renewable energy, and carbon capture (SDG 7, 13) (Behera *et al.*, 2021).

18 CONCLUSION: TOWARDS A SUSTAINABLE BIOENGINEERED FUTURE

Applying AI, CRISPR, and systems biology in microbial environmental biotechnology poses huge challenges as well as the invaluable potential of a sustainable future. The major problems are the technicality of off-target and gene editing, the complex character of microbial application transitioning between laboratory and industrial production and retaining efficiency and efficacy. Ecological hazards, including the proliferation of genetically engineered microorganisms out of control as well as accidental transfer of genes, are such issues that require strict biosafety measures and sound risk assessment systems. Moreover, existing regulatory mechanisms generally find it difficult to keep abreast with fast-paced innovations, creating uncertainties and inconsistencies that limit innovation and implementation. Although intellectual property rights are important in enhancing innovation incentives, they may hinder data sharing and cooperation, a development that prevents a rapid advance in research. The acceptance of society, which depends on the behavioral perception of society and the ethics, also stands as a major impediment, and this should be addressed through proper stakeholder involvement and sensitization of society. The solution to this dilemma is to actively participate in interaction with various stakeholders, such as policymakers, representatives of various industries, local communities, and advocacy groups, to make sure that their issues are addressed. Citizen science projects can also enhance the knowledge and engagement of the civil society as it would enable the population to take an active role in data gathering, testing and surveilling of biotechnological uses. Equity in the benefit-sharing approaches should also be undertaken in order to make sure that economic, environmental, and health advantages of microbial innovations are shared equitably among people, especially those who are directly affected by the technology. With its emerging capability to design, test and optimize microbial systems, AI and systems biology are poised to transform precision in gene editing, and develop robust data-driven synthetic microbes over the next decade with applications spanning from biopharmaceuticals to bioremediation. Also, through these technologies, microbial biotechnology is being recognized as having an immeasurable potential to solve some of the current world problems like environmental cleanup, carbon sequestration, sustainable farming, and health management in the world. The potential to design microbes to carry out superior bioremediation, nutrient cycling, the creation of biofuels, and bioplastics presents alternative, non-pollutive ways of carrying out otherwise pollutive industrial activities. In furtherance, it is imperative to have regulatory frameworks that are both adaptive and harmonized, based on objective-based science-grounded assessments and not based on a process-based differentiation as a means to bring clarity and predictability to innovators. International collaboration and harmony among the microbial laws will be essential in order to avoid the legislative fragmentation and the relaxation of the positive technology deployment throughout the world.

DECLARATION OF COMPETING INTEREST

The author declares that no competing interests exist.

ACKNOWLEDGEMENT

The author acknowledges all reviewers whose constructive criticisms had improved this manuscript.

REFERENCES

Abbasi, A. F., Asim, M. N., & Dengel, A. (2025). Transitioning from wet lab to artificial intelligence: A systematic review of AI predictors in CRISPR. *Journal of Translational Medicine*, *23*(1), 153. Advance online publication. DOI: 10.1186/s12967-024-06013-w PMID: 39905452

Ahmed, F. J. (2022). Microbial biotechnology and Bioremediation. *Futuristic Biotechnology*, *01*, 01. Advance online publication. DOI: 10.54393/fbt.v2i02.13

Al-Salihi, S. A., Ford, K. L., Mahyudin, M. M., & Bunawan, H. (2025). Synthetic biology: Pioneering the next bio revolution for a sustainable planet. *Journal of Applied Microbiology*, *136*(9), lxaf202. Advance online publication. DOI: 10.1093/jambio/lxaf202 PMID: 40833632

Alidoosti, F., Giyahchi, M., Moien, S., & Moghimi, H. (2024). Unlocking the potential of soil microbial communities for bioremediation of emerging organic contaminants: Omics-based approaches. *Microbial Cell Factories*, *23*(210), 1–21. DOI: 10.1186/s12934-024-02485-z PMID: 39054471

Altaf-Ul-Amin, M., Afendi, F. M., Kiboi, S. K., & Kanaya, S. (2014). Systems biology in the context of big data and networks. *BioMed Research International*, *428570*, 1–11. Advance online publication. DOI: 10.1155/2014/428570 PMID: 24982882

Arora, P. K., Kumar, A., Srivastava, A., Garg, S. K., & Singh, V. P. (2022). Current bioinformatics tools for biodegradation of xenobiotic compounds. *Frontiers in Environmental Science*, *10*, 980284. Advance online publication. DOI: 10.3389/fenvs.2022.980284

Arsène-Ploetze, F., Bertin, P. N., & Carapito, C. (2014). Proteomic tools to decipher microbial community structure and functioning. *Environmental Science and Pollution Research International*, *22*(18), 13599–13612. DOI: 10.1007/s11356-014-3898-0 PMID: 25475614

Ayilara, M. S., & Babalola, O. O. (2023). Bioremediation of environmental wastes: The role of microorganisms. *Frontiers in Agronomy*, *5*, 1183691. DOI: 10.3389/fagro.2023.1183691

Batista, B. D., & Singh, B. K. (2021). Realities and hopes in the application of microbial tools in agriculture. *Microbial Biotechnology, 14*(4), 1258–1268. DOI: 10.1111/1751-7915.13866 PMID: 34156754

Behera, H. T., Mojumdar, A., Das, S. R., Mohapatra, C., & Ray, L. (2021). Microbial Intervention in Sustainable Production of Biofuels and Other Bioenergy Products. In Mishra, B. B., Nayak, S. K., Mohapatra, S., & Samantaray, D. (Eds.), *Environmental and Agricultural Microbiology* (1st ed., pp. 361–381). Wiley., DOI: 10.1002/9781119525899.ch17

Bernstein, D. B., Sulheim, E., Almaas, E., & Segrè, D. (2021). Addressing uncertainty in genome-scale metabolic model reconstruction and analysis. *Genome Biology, 22*(1), 64. DOI: 10.1186/s13059-021-02289-z PMID: 33602294

Bizzotto, E., Fraulini, S., Zampieri, G., Orellana, E., Treu, L., & Campanaro, S. (2024). MICROPHERRET: MICRObial PHEnotypic tRait ClassifieR using machine learning techniques. *Environmental Microbiome, 19*(1), 58. DOI: 10.1186/s40793-024-00600-6 PMID: 39113074

Boadu, F., Cao, H., & Cheng, J. (2023). Combining protein sequences and structures with transformers and equivariant graph neural networks to predict protein function. *Bioinformatics (Oxford, England), 39*(1, Supplement_1), i318–i325. DOI: 10.1093/bioinformatics/btad208 PMID: 37387145

Bodor, A., Bounedjoum, N., Vincze, G. E., Kis, Á. E., Laczi, K., Bende, G., & Rákhely, G. (2020). Challenges of unculturable bacteria: Environmental perspectives. *Reviews in Environmental Science and Biotechnology, 19*(1), 1–22. DOI: 10.1007/s11157-020-09522-4

Brutesco, C., Prévéral, S., Escoffier, C., Descamps, E. C. T., Prudent, E., Cayron, J., Dumas, L., Ricquebourg, M., Adryanczyk-Perrier, G., De Groot, A., Garcia, D., Rodrigue, A., Pignol, D., & Ginet, N. (2017). Bacterial host and reporter gene optimization for genetically encoded whole cell biosensors. *Environmental Science and Pollution Research International, 24*(1), 52–65. DOI: 10.1007/s11356-016-6952-2 PMID: 27234828

Cameron, D. E., Bashor, C. J., & Collins, J. J. (2014). A brief history of synthetic biology. *Nature Reviews. Microbiology, 12*(5), 381–390. DOI: 10.1038/nrmicro3239 PMID: 24686414

Caplan, A. L., Parent, B., Shen, M., & Plunkett, C. (2015). No time to waste—the ethical challenges created by CRISPR: CRISPR/Cas, being an efficient, simple, and cheap technology to edit the genome of any organism, raises many ethical and regulatory issues beyond the use to manipulate human germ line cells. *EMBO Reports*, *16*(11), 1421–1426. DOI: 10.15252/embr.201541337 PMID: 26450575

Chakraborty, R., Wu, C. H., & Hazen, T. C. (2012). Systems biology approach to bioremediation. *Current Opinion in Biotechnology*, *23*(3), 483–490. DOI: 10.1016/j.copbio.2012.01.015 PMID: 22342400

Chaumeil, A., Mussig, A. J., Hugenholtz, P., & Parks, D. H. (2022). GTDB-Tk v2: Memory friendly classification with the genome taxonomy database. *Bioinformatics (Oxford, England)*, *38*(23), 5315–5316. DOI: 10.1093/bioinformatics/btac672 PMID: 36218463

Chen, M., Feng, S., Lv, H., Wang, Z., Zeng, Y., Shao, C., Lin, W., & Zhang, Z. (2024). OsCIPK2 mediated rice root microorganisms and metabolites to improve plant nitrogen uptake. *BMC Plant Biology*, *24*(285), 1–11. DOI: 10.1186/s12870-024-04982-0 PMID: 38627617

Chia, B. S., Seah, Y. F. S., Wang, B., Shen, K., Srivastava, D., & Chew, W. L. (2025) Engineering a New Generation of Gene Editors: Integrating Synthetic Biology and AI Innovations. *ACS Synthetic Biology*. 14 (3), 636–647. DOI: 10.1021/acssynbio.4c00686

Choubisa, V. (2024). CRISPR-Cas9 off-target predictions using CNN and double CNN: Comparative analysis. *International Journal of Science and Research Archive*, *12*(1), 1074–1080. DOI: 10.30574/ijsra.2024.12.1.0927

Cruz, D. A., & Kemp, M. L. (2022). Hybrid computational modeling methods for systems biology. *Progress in Biomedical Engineering (Bristol, England)*, *4*(1), 1–19. DOI: 10.1088/2516-1091/ac2cdf PMID: 41287231

Da Silva, L. C., Aoki, J. I., & Floeter-Winter, L. M. (2022). Finding correlations between mRNA and protein levels in Leishmania development: Is there a discrepancy? *Frontiers in Cellular and Infection Microbiology*, *12*, 852902. Advance online publication. DOI: 10.3389/fcimb.2022.852902 PMID: 35903202

Dada, E. G., Oyewola, D. O., Joseph, S. B., Emebo, O., & Oluwagbemi, O. O. (2022). Ensemble Machine Learning for Monkeypox Transmission Time Series Forecasting. *Applied Sciences (Basel, Switzerland)*, *12*(23), 12128. DOI: 10.3390/app122312128

Dara, M., Dianatpour, M., Azarpira, N., & Omidifar, N. (2024). Convergence of CRISPR and artificial intelligence: A paradigm shift in biotechnology. *Human Gene*, *41*, 201297. DOI: 10.1016/j.humgen.2024.201297

De Almeida, D. G., Soares Da Silva, C. F., Luna, J. M., Rufino, R. D., Santos, V. A., Banat, I. M., & Sarubbo, L. A. (2016). Biosurfactants: Promising Molecules for Petroleum Biotechnology Advances. *Frontiers in Microbiology*, *7*, 1718. DOI: 10.3389/fmicb.2016.01718 PMID: 27843439

De Roy, K., Marzorati, M., Van Den Abbeele, P., Van De Wiele, T., & Boon, N. (2013). Synthetic microbial ecosystems: An exciting tool to understand and apply microbial communities. *Environmental Microbiology*, *16*(6), 1472–1481. DOI: 10.1111/1462-2920.12343 PMID: 24274586

Dehkordi, M. B., Pishvaie, M. R., & Vafa, E. (2023). Model predictive control of a fermenter using dynamic flux balance analysis coupled with convolutional neural networks. *Computers & Chemical Engineering*, *179*, 108444. DOI: 10.1016/j.compchemeng.2023.108444

Du, H., Li, M., & Liu, Y. (2023). Towards applications of genome-scale metabolic model-based approaches in designing synthetic microbial communities. *Quantitative Biology*, *11*(1), 15–30. DOI: 10.15302/J-QB-022-0313 PMID: 41674535

Dudeja, C., Mishra, A., Ali, A., Singh, P. P., & Jaiswal, A. K. (2025). Microbial Genome Editing with CRISPR–Cas9: Recent Advances and Emerging Applications Across Sectors. *Fermentation (Basel, Switzerland)*, *11*(7), 410. DOI: 10.3390/fermentation11070410

Duncker, K. E., Holmes, Z. A., & You, L. (2021). Engineered microbial consortia: Strategies and applications. *Microbial Cell Factories*, *20*(1), 211. Advance online publication. DOI: 10.1186/s12934-021-01699-9 PMID: 34784924

Edwards, S., Ditter, R., Laster, H., Sheehan, G., Anderson, O., Beattie, T., & Mellies, J. L. (2021). Microbial Consortia and Mixed Plastic Waste: Pangenomic Analysis Reveals Potential for Degradation of Multiple Plastic Types via Previously Identified PET Degrading Bacteria. *International Journal of Molecular Sciences*, *23*(10), 1–22. DOI: 10.3390/ijms23105612 PMID: 35628419

Eicher, T., Kinnebrew, G., Patt, A., Spencer, K., Ying, K., Ma, Q., Machiraju, R., & Mathé, E. A. (2020). Metabolomics and Multi-Omics Integration: A survey of Computational methods and resources. *Metabolites*, *10*(5), 1–25. DOI: 10.3390/metabo10050202 PMID: 32429287

Faure, G., Shmakov, S. A., Yan, W. X., Cheng, D. R., Scott, D. A., Peters, J. E., Makarova, K. S., & Koonin, E. V. (2019). CRISPR–Cas in mobile genetic elements: Counter-defence and beyond. *Nature Reviews. Microbiology*, *17*(8), 513–525. DOI: 10.1038/s41579-019-0204-7 PMID: 31165781

Fernández, C. D. B., Castillo, M. P. G., Pérez, S. Q., & Rodríguez, L. V. C. (2022). Microbial degradation of polyethylene terephthalate: A systematic review. *SN Applied Sciences*, *4*(10), 1–20. DOI: 10.1007/s42452-022-05143-4

Fernandez, L., Peura, S., Eiler, A., Linz, A. M., McMahon, K. D., & Bertilsson, S. (2020). Diazotroph Genomes and Their Seasonal Dynamics in a Stratified Humic Bog Lake. *Frontiers in Microbiology*, *11*, 1500. DOI: 10.3389/fmicb.2020.01500 PMID: 32714313

Finkel, O. M., Castrillo, G., Herrera Paredes, S., Salas González, I., & Dangl, J. L. (2017). Understanding and exploiting plant beneficial microbes. *Current Opinion in Plant Biology*, *38*, 155–163. DOI: 10.1016/j.pbi.2017.04.018 PMID: 28622659

García-Berumen, J. A., Flores de la Torre, J. A., De los Santos-Villalobos, S., Espinoza-Canales, A., Echavarría-Cháirez, F. G., & Gutiérrez-Bañuelos, H. (2024). Phosphorus dynamics and sustainable agriculture: The role of microbial solubilization and innovations in nutrient management. *Current Research in Microbial Sciences*, *8*, 100326. DOI: 10.1016/j.crmicr.2024.100326 PMID: 39687549

Gong, X., Zhang, J., Gan, Q., Teng, Y., Hou, J., Lyu, Y., Liu, Z., Wu, Z., Dai, R., Zou, Y., Wang, X., Zhu, D., Zhu, H., Liu, T., & Yan, Y. (2024). Advancing microbial production through artificial intelligence-aided biology. *Biotechnology Advances*, *74*, 108399. DOI: 10.1016/j.biotechadv.2024.108399 PMID: 38925317

Hajnajafi, K., & Iqbal, M. A. (2025). Mass-spectrometry based metabolomics: An overview of workflows, strategies, data analysis and applications. *Proteome Science*, *23*(5), 1–15. DOI: 10.1186/s12953-025-00241-8 PMID: 40420110

Hao, X., Gu, Y., Zhang, H., Wang, X., Liu, X., Chen, C., Wang, C., Zhang, X., Liu, X., & Shen, X. (2024). Synthetic microbial community promotes bacterial communities leading to soil multifunctionality in desertified land. *Microorganisms*, *12*(6), 1–16. DOI: 10.3390/microorganisms12061117 PMID: 38930499

Haruta, S., & Yamamoto, K. (2018). Model Microbial Consortia as Tools for Understanding Complex Microbial Communities. *Current Genomics*, *19*(8), 723–733. DOI: 10.2174/1389202919666180911131206 PMID: 30532651

Hasibi, R., Michoel, T., & Oyarzún, D. A. (2024). Integration of graph neural networks and genome-scale metabolic models for predicting gene essentiality. *NPJ Systems Biology and Applications, 10*(1), 24. DOI: 10.1038/s41540-024-00348-2 PMID: 38448436

Hellweger, F. L., Clegg, R. J., Clark, J. R., Plugge, C. M., & Kreft, J. U. (2016). Advancing microbial sciences by individual-based modelling. *Nature Reviews. Microbiology, 14*(7), 461–471. DOI: 10.1038/nrmicro.2016.62 PMID: 27265769

Hennessy, R. C., Glaring, M. A., Olsson, S., & Stougaard, P. (2017). Transcriptomic profiling of microbe–microbe interactions reveals the specific response of the biocontrol strain P. Fluorescens In5 to the phytopathogen Rhizoctonia solani. *BMC Research Notes, 10*(1), 376. DOI: 10.1186/s13104-017-2704-8 PMID: 28807055

Hoffmann, S. A., Diggans, J., Densmore, D., Dai, J., Knight, T., Leproust, E., Boeke, J. D., Wheeler, N., & Cai, Y. (2023). Safety by design: Biosafety and biosecurity in the age of synthetic genomics. *iScience, 26*(3), 106165. DOI: 10.1016/j.isci.2023.106165 PMID: 36895643

Hu, Y., & Li, W. (2022). Development and Application of CRISPR-Cas Based Tools. *Frontiers in Cell and Developmental Biology, 10*, 834646. DOI: 10.3389/fcell.2022.834646 PMID: 35445018

Johns, N. I., Blazejewski, T., Gomes, A. L., & Wang, H. H. (2016). Principles for designing synthetic microbial communities. *Current Opinion in Microbiology, 31*, 146–153. DOI: 10.1016/j.mib.2016.03.010 PMID: 27084981

Karimi, M. A., Paryan, M., Fard, G. B., Sadeghian, H., Zarrinfar, H., & Bafghi, M. H. (2025). Challenges and Opportunities in the Application of CRISPR-Cas9: A Review on Genomic Editing and Therapeutic Potentials. *Medical Principles and Practice : International Journal of the Kuwait University, Health Science Centre*, 1–30. Advance online publication. DOI: 10.1159/000547334 PMID: 40675140

Katoh, K., & Standley, D. M. (2016). A simple method to control over-alignment in the MAFFT multiple sequence alignment program. *Bioinformatics (Oxford, England), 32*(13), 1933–1942. DOI: 10.1093/bioinformatics/btw108 PMID: 27153688

Kim, O. D., Rocha, M., & Maia, P. (2018). A Review of Dynamic Modeling Approaches and Their Application in Computational Strain Optimization for Metabolic Engineering. *Frontiers in Microbiology, 9*, 378885. DOI: 10.3389/fmicb.2018.01690 PMID: 30108559

Kotova, I. B., Taktarova, Y. V., Tsavkelova, E. A., Egorova, M. A., Bubnov, I. A., Malakhova, D. V., Shirinkina, L. I., Sokolova, T. G., & Bonch-Osmolovskaya, E. A. (2021). Microbial Degradation of Plastics and Approaches to Make it More Efficient. *Microbiology*, *90*(6), 671–701. DOI: 10.1134/S0026261721060084

Kraus, E. A., Prithiviraj, B., & Hernandez, M. (2025). Advancing transcriptomic profiling of airborne bacteria. *Applied and Environmental Microbiology*, *91*(5), 1–14. DOI: 10.1128/aem.00148-25 PMID: 40293243

Kulikova, A. V., Diaz, D. J., Chen, T., Cole, T. J., Ellington, A. D., & Wilke, C. O. (2023). Two sequence- and two structure-based ML models have learned different aspects of protein biochemistry. *BioRxiv*, 2023.03.20.533508. DOI: 10.1101/2023.03.20.533508

Kuriya, Y., & Araki, M. (2020). Dynamic Flux Balance Analysis to Evaluate the Strain Production Performance on Shikimic Acid Production in Escherichia coli. *Metabolites*, *10*(5), 198. DOI: 10.3390/metabo10050198 PMID: 32429049

Lam, T. J., & Ye, Y. (2022). Meta-analysis of microbiome association networks reveal patterns of dysbiosis in diseased microbiomes. *Scientific Reports*, *12*(1), 17482. DOI: 10.1038/s41598-022-22541-1 PMID: 36261472

Lee, D. H., Park, E. G., Lee, Y. J., Jeong, H., Roh, H., Jeong, G., Kim, S., Kim, H., Lee, D. H., Park, E. G., Lee, Y. J., Jeong, H., Roh, H., Jeong, G., Kim, S., & Kim, H. (2025). Advances in artificial intelligence (AI) models and generative algorithms represent a new paradigm for genomics research. *International Journal of Molecular Sciences*, *26*(22), 10925. DOI: 10.3390/ijms262210925 PMID: 41303412

Lee, M. (2023). Deep learning in CRISPR-Cas systems: A review of recent studies. *Frontiers in Bioengineering and Biotechnology*, *11*, 1226182. Advance online publication. DOI: 10.3389/fbioe.2023.1226182 PMID: 37469443

Lee, S. Y., Li, H., Weber, W., & Yang, Z. (2025). Turning Pollution Into Solutions Using Microbes. *Frontiers for Young Minds*, *13*, 1575505. DOI: 10.3389/frym.2025.1575505

Leggieri, P. A., Liu, Y., Hayes, M., Connors, B., Seppälä, S., & Venturelli, O. S. (2021). Integrating Systems and Synthetic Biology to Understand and Engineer Microbiomes. *Annual Review of Biomedical Engineering*, *23*(1), 169–201. DOI: 10.1146/annurev-bioeng-082120-022836 PMID: 33781078

Li, X., Wu, S., Dong, Y., Fan, H., Bai, Z., & Zhuang, X. (2021). Engineering Microbial Consortia towards Bioremediation. *Water (Basel)*, *13*(20), 1–11. DOI: 10.3390/w13202928

Lin, L. (2022). Bottom-up synthetic ecology study of microbial consortia to enhance lignocellulose bioconversion. *Biotechnology for Biofuels and Bioproducts, 15*(1), 14. Advance online publication. DOI: 10.1186/s13068-022-02113-1 PMID: 35418100

Liu, G., Catacutan, D. B., Rathod, K., Swanson, K., Jin, W., Mohammed, J. C., Chiappino-Pepe, A., Syed, S. A., Fragis, M., Rachwalski, K., Surette, M. G., Coombes, B. K., Magolan, J., Jaakkola, T., Barzilay, R., Collins, J. J., & Stokes, J. M. (2023). Deep learning-guided discovery of an antibiotic targeting *Acinetobacter baumannii. Nature Chemical Biology, 19*(11), [page numbers if known]. https://doi.org/DOI: 10.1038/s41589-023-01349-8

Liu, T. Y., & Doudna, J. A. (2020). Chemistry of Class 1 CRISPR-Cas effectors: Binding, editing, and regulation. *The Journal of Biological Chemistry, 295*(42), 14473–14487. DOI: 10.1074/jbc.REV120.007034 PMID: 32817336

Liv, B. (2024). Unlocking the Future of Biotechnology: The Transformative Potential of CRISPR and Emerging Gene-Editing Technologies. *Next Generation Journal for The Young Researchers, 8*(1). Advance online publication. DOI: 10.62802/8ewmr712

Maan, K., Baghel, R., Dhariwal, S., Sharma, A., Bakhshi, R., & Rana, P. (2023). Metabolomics and transcriptomics based multi-omics integration reveals radiation-induced altered pathway networking and underlying mechanism. *NPJ Systems Biology and Applications, 9*(1), 1–13. DOI: 10.1038/s41540-023-00305-5 PMID: 37689794

Maglione, G., Zinno, P., Tropea, A., Mussagy, C. U., Dufossé, L., Giuffrida, D., & Mondello, A. (2024). Microbes' role in environmental pollution and remediation: A bioeconomy focus approach. *AIMS Microbiology, 10*(3), 723–755. DOI: 10.3934/microbiol.2024033 PMID: 39219757

Mahieu, N. G., & Patti, G. J. (2017). Systems-level annotation of a metabolomics data set reduces 25,000 features to fewer than 1,000 unique metabolites. *Analytical Chemistry, 89*(19), 10397–10406. DOI: 10.1021/acs.analchem.7b02380 PMID: 28914531

Makarova, K. S., Shmakov, S. A., Wolf, Y. I., Mutz, P., Beisel, C. L., Brouns, S. J., Charpentier, E., Cheng, D., Doudna, J., Haft, D. H., Horvath, P., Moineau, S., Mojica, F. J., Pausch, P., Shah, S. A., Siksnys, V., Terns, M. P., Tordoff, J., Venclovas, Č., & Koonin, E. V. (2025). An updated evolutionary classification of CRISPR–Cas systems including rare variants. *Nature Microbiology, 10*(12), 3346–3361. DOI: 10.1038/s41564-025-02180-8 PMID: 41198952

Malwe, A. S., Longwani, U., & Sharma, V. K. (2025). XenoBug: Machine learning-based tool to predict pollutant-degrading enzymes from environmental metagenomes. *NAR Genomics and Bioinformatics*, *7*(2), lqaf037. Advance online publication. DOI: 10.1093/nargab/lqaf037 PMID: 40314024

Mante, J., Groover, K. E., & Pullen, R. M. (2024). Environmental community transcriptomics: Strategies and struggles. *Briefings in Functional Genomics*, *24*(1), 1–15. DOI: 10.1093/bfgp/elae033 PMID: 39183066

Mariam, I., Rova, U., Christakopoulos, P., Matsakas, L. & Patel, A. (2025) Data-driven synthetic microbes for sustainable future. *npj Systems Biology and Applications*. 11 (1), 74. .DOI: 10.1038/s41540-025-00556-4

Matchado, M. S., Lauber, M., Reitmeier, S., Kacprowski, T., Baumbach, J., Haller, D., & List, M. (2020). Network analysis methods for studying microbial communities: A mini review. *Computational and Structural Biotechnology Journal*, *19*(1), 2687–2698. DOI: 10.1016/j.csbj.2021.05.001 PMID: 34093985

Matuszyńska, A., Ebenhöh, O., Zurbriggen, M. D., Ducat, D. C., & Axmann, I. M. (2024). A new era of synthetic biology—Microbial community design. *Synthetic Biology*, *9*(1), 1–15. DOI: 10.1093/synbio/ysae011 PMID: 39086602

Minhas, W. R., Saleemi, H., Raheem, M. F., Khalid, S. S., Aslam, H., & Ahmad, W. (2024). CRISPR-Engineered Microbes for Enhanced biodegradation of recalcitrant pollutants a genomic approach to environmental remediation. *Frontiers in Health Informatics*, •••, 8281–8304. DOI: 10.63682/fhi787

Mishra, K., Siwal, S. S., Nayaka, S. C., Guan, Z., & Thakur, V. K. (2023). Waste-to-chemicals: Green solutions for bioeconomy markets. *The Science of the Total Environment*, *887*, 164006. DOI: 10.1016/j.scitotenv.2023.164006 PMID: 37172858

Mlost, J., Dawli, R., Liu, X., Costa, A. R., & Pollak Dorocic, I. (2025). Evaluation of unsupervised learning algorithms for the classification of behavior from pose estimation data. *Patterns (New York, N.Y.)*, *6*(5), 101237. DOI: 10.1016/j.patter.2025.101237 PMID: 40486967

Moreira, I. S., Murgolo, S., Mascolo, G., & Castro, P. M. L. (2022). Biodegradation and Metabolic Pathway of 17β-Estradiol by *Rhodococcus* sp. ED55. *International Journal of Molecular Sciences*, *23*(11), 6181. DOI: 10.3390/ijms23116181 PMID: 35682859

Morrissey, J., Barberi, G., Strain, B., Facco, P., & Kontoravdi, C. (2025). NEXT-FBA: A hybrid stoichiometric/data-driven approach to improve intracellular flux predictions. *Metabolic Engineering, 91*, 130–144. DOI: 10.1016/j.ymben.2025.03.010 PMID: 40118205

Nagarajan, K., Ni, C., & Lu, T. (2022). Agent-Based Modeling of Microbial Communities. *ACS Synthetic Biology, 11*(11), 3564–3574. DOI: 10.1021/acssynbio.2c00411 PMID: 36315012

Natarajan, A., & Bhatt, A. S. (2020). Microbes and microbiomes in 2020 and beyond. *Nature Communications, 11*(1), 4988. DOI: 10.1038/s41467-020-18850-6 PMID: 33020496

Nikel, P. I., & De Lorenzo, V. (2018). *Pseudomonas putida* as a functional chassis for industrial biocatalysis: From native biochemistry to trans-metabolism. *Metabolic Engineering, 50*, 142–155. DOI: 10.1016/j.ymben.2018.05.005 PMID: 29758287

Novello, G., Bona, E., Nasuelli, M., Massa, N., Sudiro, C., Campana, D. C., Gorrasi, S., Hochart, M. L., Altissimo, A., Vuolo, F., & Gamalero, E. (2024). The Impact of Nitrogen-Fixing Bacteria-Based Biostimulant Alone or in Combination with Commercial Inoculum on Tomato Native Rhizosphere Microbiota and Production: An Open-Field Trial. *Biology (Basel), 13*(6), 400. DOI: 10.3390/biology13060400 PMID: 38927280

Oluwagbemi, O., & Awe, O. (2018). A comparative computational genomics of Ebola Virus Disease strains: In-silico Insight for Ebola control, Informatics in Medicine Unlocked, Volume 12, 2018; Pages 106-119, ISSN 2352-9148; https://doi.org/DOI: 10.1016/j.imu.2018.07.004

Oluwagbemi, O. O., Oladipo, E. K., Kolawole, O. M., Oloke, J. K., Adelusi, T. I., Irewolede, B. A., Dairo, E. O., Ayeni, A. E., Kolapo, K. T., Akindiya, O. E., Oluwasegun, J. A., Oluwadara, B. F., & Fatumo, S. (2022). Bioinformatics, Computational Informatics, and Modeling Approaches to the Design of mRNA COVID-19 Vaccine Candidates. *Computation (Basel, Switzerland), 10*(7), 117. DOI: 10.3390/computation10070117

Oyewola, D. O., Dada, E. G., Emebo, O., & Oluwagbemi, O. O. (2022b). Using Deep 1D Convolutional Grated Recurrent Unit Neural Network to Optimize Quantum Molecular Properties and Predict Intramolecular Coupling Constants of Molecules of Potential Health Medications and Other Generic Molecules. *Applied Sciences (Basel, Switzerland), 12*(14), 7228. DOI: 10.3390/app12147228

Oyewola, D. O., Dada, E. G., Omotehinwa, T. O., Emebo, O., & Oluwagbemi, O. O. (2022a). Application of Deep Learning Techniques and Bayesian Optimization with Tree Parzen Estimator in the Classification of Supply Chain Pricing Datasets of Health Medications. *Applied Sciences (Basel, Switzerland)*, *12*(19), 10166. DOI: 10.3390/app121910166

Palermo, A. (2022). Metabolomics- and systems-biology-guided discovery of metabolite lead compounds and druggable targets. *Drug Discovery Today*, *28*(2), 1–10, 103460. DOI: 10.1016/j.drudis.2022.103460 PMID: 36427778

Park, H., Seo, S. I., Lim, J. H., Song, J., Seo, J. H., & Kim, P. I. (2022). Screening of Carbofuran-Degrading Bacteria *Chryseobacterium* sp. BSC2-3 and Unveiling the Change in Metabolome during Carbofuran Degradation. *Metabolites*, *12*(3), 1–16. DOI: 10.3390/metabo12030219 PMID: 35323662

Passi, A., Tibocha-Bonilla, J. D., Kumar, M., Tec-Campos, D., Zengler, K., & Zuniga, C. (2021). Genome-Scale Metabolic Modeling Enables In-Depth Understanding of Big Data. *Metabolites*, *12*(1), 14. DOI: 10.3390/metabo12010014 PMID: 35050136

Patel, A. I., Vyas, D. R., Vyas, A. J., Patel, A. B., Dudhrejiya, A. V., & Chotaliya, U. J. (2023). Proteomics: Technology, challenges and advances. *Asian Journal of Pharmaceutical Research*, *13*(4), 249–255. DOI: 10.52711/2231-5691.2023.00046

Peaudecerf, F. J., Bunbury, F., Bhardwaj, V., Bees, M. A., Smith, A. G., Goldstein, R. E., & Croze, O. A. (2017). *Microbial mutualism at a distance: The role of geometry in diffusive exchanges.* ArXiv., DOI: 10.1103/PhysRevE.97.022411

Pilalis, E., Zisis, D., Andrinopoulou, C., Karamanidou, T., Antonara, M., Stavropoulos, T. G., & Chatziioannou, A. (2025). Genome-wide functional annotation of variants: A systematic review of state-of-the-art tools, techniques and resources. *Frontiers in Pharmacology*, *16*, 1474026. DOI: 10.3389/fphar.2025.1474026 PMID: 40098614

Pinu, F. R., Beale, D. J., Paten, A. M., Kouremenos, K., Swarup, S., Schirra, H. J., & Wishart, D. (2019). Systems Biology and Multi-Omics Integration: Viewpoints from the Metabolomics Research Community. *Metabolites*, *9*(4), 76. DOI: 10.3390/metabo9040076 PMID: 31003499

Przymus, P., Rykaczewski, K., Truu, J., Kolev, M., Naskinova, I., Gruca, A., Sampri, A., Frohme, M., & Nechyporenko, A. (2025). Deep learning in microbiome analysis: A comprehensive review of neural network models. *Frontiers in Microbiology*, *15*, 1516667. DOI: 10.3389/fmicb.2024.1516667 PMID: 39911715

Qattan, S. Y. A. (2025). Harnessing bacterial consortia for effective bioremediation: Targeted removal of heavy metals, hydrocarbons, and persistent pollutants. *Environmental Sciences Europe*, *37*(1), 85. Advance online publication. DOI: 10.1186/s12302-025-01103-y

Qi, X., Ma, Y., Chang, H., Li, B., Ding, M., & Yuan, Y. (2021). Evaluation of PET Degradation Using Artificial Microbial Consortia. *Frontiers in Microbiology*, *12*, 778828. DOI: 10.3389/fmicb.2021.778828 PMID: 35003008

Rafeeq, H., Afsheen, N., Rafique, S., Arshad, A., Intisar, M., Hussain, A., Bilal, M., & Iqbal, H. M. (2022). Genetically engineered microorganisms for environmental remediation. *Chemosphere*, *310*, 136751. DOI: 10.1016/j.chemosphere.2022.136751 PMID: 36209847

Ranganathan, S., Suthers, P. F., & Maranas, C. D. (2009). Flux balance analysis of biological systems. *Briefings in Bioinformatics*, *10*(4), 435–449. DOI: 10.1093/bib/bbp011 PMID: 19287049

Reiman, D., Layden, B. T., & Dai, Y. (2021). MiMeNet: Exploring microbiome–metabolome relationships using neural networks. *PLoS Computational Biology*, *17*(5), e1009021. DOI: 10.1371/journal.pcbi.1009021 PMID: 33999922

Renganathan, P., Gaysina, L. A., Gutiérrez, C. G., Rueda Puente, E. O., & Sainz-Hernández, J. C. (2025). Harnessing Engineered Microbial Consortia for Xenobiotic Bioremediation: Integrating Multi-Omics and AI for Next-Generation Wastewater Treatment. *Journal of Xenobiotics*, *15*(4), 133. DOI: 10.3390/jox15040133 PMID: 40863340

Rengasamy, D., Mase, J. M., Kumar, A., Rothwell, B., Torres, M. T., Alexander, M. R., Winkler, D. A., & Figueredo, G. P. (2022). Feature importance in machine learning models: A fuzzy information fusion approach. *Neurocomputing*, *511*(1), 163–174. DOI: 10.1016/j.neucom.2022.09.053

Roberts, C., Edwards, S., Vague, M., León-Zayas, R., Scheffer, H., Chan, G., Swartz, N. A., & Mellies, J. L. (2020). Environmental consortium containing pseudomonas and bacillus species synergistically degrades polyethylene terephthalate plastic. *MSphere*, *5*(6), 1–20. DOI: 10.1128/mSphere.01151-20 PMID: 33361127

Robinson, S. L. (2022). Artificial intelligence for microbial biotechnology: Beyond the hype. *Microbial Biotechnology*, *15*(1), 65–69. DOI: 10.1111/1751-7915.13943 PMID: 34606686

Roell, G. W., Zha, J., Carr, R. R., Koffas, M. A., Fong, S. S., & Tang, Y. J. (2019). Engineering microbial consortia by division of labour. *Microbial Cell Factories*, *18*(35), 1–11. DOI: 10.1186/s12934-019-1083-3 PMID: 30736778

Ruaud, A., Sancaktar, C., Bagatella, M., Ratzke, C., & Martius, G. (2024, July 8). *Modelling Microbial Communities with Graph Neural Networks*. PMLR. https://proceedings.mlr.press/v235/ruaud24a.html

Satam, H., Joshi, K., Mangrolia, U., Waghoo, S., Zaidi, G., Rawool, S., Thakare, R. P., Banday, S., Mishra, A. K., Das, G., & Malonia, S. K. (2023). Next-Generation Sequencing Technology: Current Trends and Advancements. *Biology (Basel)*, *12*(7), 997. DOI: 10.3390/biology12070997 PMID: 37508427

Schwalm, N. D.III, Mojadedi, W., Gerlach, E. S., Benyamin, M., Perisin, M. A., & Akingbade, K. L. (2019). Developing a Microbial Consortium for Enhanced Metabolite Production from Simulated Food Waste. *Fermentation (Basel, Switzerland)*, *5*(4), 98–105. DOI: 10.3390/fermentation5040098

Shajil, U. K., Sujith, A., & Nair, B. G. (2025). Comparison of human intelligence with artificial intelligence. In *Artificial Intelligence and Biological Sciences* (pp. 68–92). CRC Press. DOI: 10.1201/9781003492726-4

Sharma, P. K., Sharma, V., Sharma, S., Bhatia, G., Singh, K., & Sharma, R. (2019). Comparative metatranscriptome analysis revealed broad response of microbial communities in two soil types, agriculture versus organic soil. *Journal of Genetic Engineering and Biotechnology*, *17*(1), 6. DOI: 10.1186/s43141-019-0006-3 PMID: 31659568

Shayanthan, A., Ordoñez, P. C., & Oresnik, I. J. (2022). The role of Synthetic Microbial Communities (SYNCOM) in sustainable agriculture. *Frontiers in Agronomy*, *4*, 896307. Advance online publication. DOI: 10.3389/fagro.2022.896307

Sheng, J., Amankwah-Amoah, J., & Wang, X. (2019). Technology in the 21st century: New challenges and opportunities. *Technological Forecasting and Social Change*, *143*, 321–335. DOI: 10.1016/j.techfore.2018.06.009

Shi, Y., Zhang, L., Peterson, C. B., Do, A., & Jenq, R. R. (2022). Performance determinants of unsupervised clustering methods for microbiome data. *Microbiome*, *10*(1), 25. DOI: 10.1186/s40168-021-01199-3 PMID: 35120564

Shmakov, S., Smargon, A., Scott, D., Cox, D., Pyzocha, N., Yan, W., Abudayyeh, O. O., Gootenberg, J. S., Makarova, K. S., Wolf, Y. I., Severinov, K., Zhang, F., & Koonin, E. V. (2017). Diversity and evolution of class 2 CRISPR–Cas systems. *Nature Reviews. Microbiology, 15*(3), 169–182. DOI: 10.1038/nrmicro.2016.184 PMID: 28111461

Shrestha, H. K., Appidi, M. R., Villalobos Solis, M. I., Wang, J., Carper, D. L., Burdick, L., Pelletier, D. A., Doktycz, M. J., Hettich, R. L., & Abraham, P. E. (2021). Metaproteomics reveals insights into microbial structure, interactions, and dynamic regulation in defined communities as they respond to environmental disturbance. *BMC Microbiology, 21*(308), 1–17. DOI: 10.1186/s12866-021-02370-4 PMID: 34749649

Singh, A., Xue, A., Tai, J., Mbadugha, F., Obi, P., Mascarenhas, R., Tyagi, A., Siena, A., & Chen, Y. G. (2024). A scalable and cost-efficient rRNA depletion approach to enrich RNAs for molecular biology investigations. *RNA (New York, N.Y.), 30*(6), 728–738. DOI: 10.1261/rna.079761.123 PMID: 38485192

Singh, D. P., Prabha, R., Gupta, V. K., & Verma, M. K. (2018). Metatranscriptome Analysis Deciphers Multifunctional Genes and Enzymes Linked With the Degradation of Aromatic Compounds and Pesticides in the Wheat Rhizosphere. *Frontiers in Microbiology, 9*(1331), 1–15. DOI: 10.3389/fmicb.2018.01331 PMID: 30034370

Singh, J. S., Abhilash, P., Singh, H., Singh, R. P., & Singh, D. (2011). Genetically engineered bacteria: An emerging tool for environmental remediation and future research perspectives. *Gene, 480*(1-2), 1–9. DOI: 10.1016/j.gene.2011.03.001 PMID: 21402131

Singh, S., & Ramakrishna, W. (2021). Application of CRISPR–Cas9 in plant–plant growth-promoting rhizobacteria interactions for next Green Revolution. *3 Biotech, 11*(12), 492. https://doi.org/DOI: 10.1007/s13205-021-03041-x

Soberón-Chávez, G. (2023). Some insights on traditional and novel approaches in microbial biotechnology that contribute to the United Nations Sustainable Development Goals. *Microbial Biotechnology, 16*(11), 2015–2018. DOI: 10.1111/1751-7915.14318 PMID: 37452713

Song, Y., Yuan, Q., Chen, S., Zeng, Y., Zhao, H., & Yang, Y. (2024). Accurately predicting enzyme functions through geometric graph learning on ESMFold-predicted structures. *Nature Communications, 15*(1), 8180. DOI: 10.1038/s41467-024-52533-w PMID: 39294165

Srinivasan, S., Jnana, A., & Murali, T. S. (2024). Modeling Microbial Community Networks: Methods and Tools for Studying Microbial Interactions. *Microbial Ecology, 87*(1), 56. DOI: 10.1007/s00248-024-02370-7 PMID: 38587642

Sun, S., Wang, Y., He, B., Chen, J., Leng, F., & Luo, W. (2024). Comparative transcriptomics revealed the mechanism of Stenotrophomonas rhizophila JC1 response and biosorption to Pb2+. *Environmental Geochemistry and Health*, *46*(7), 1–18. DOI: 10.1007/s10653-024-02019-x PMID: 38849682

Tariq, A., Guo, S., Farhat, F., & Shen, X. (2025). Engineering Synthetic Microbial Communities: Diversity and applications in soil for plant resilience. *Agronomy (Basel)*, *15*(3), 513. DOI: 10.3390/agronomy15030513

Tawfiq, R., Niu, K., Hoehndorf, R., & Kulmanov, M. (2024). DeepGOMeta for functional insights into microbial communities using deep learning-based protein function prediction. *Scientific Reports*, *14*(1), 31813. DOI: 10.1038/s41598-024-82956-w PMID: 39738309

Toju, H., Abe, M. S., Ishii, C., Hori, Y., Fujita, H., & Fukuda, S. (2020). Scoring Species for Synthetic Community Design: Network Analyses of Functional Core Microbiomes. *Frontiers in Microbiology*, *11*(543143), 11–17. DOI: 10.3389/fmicb.2020.01361 PMID: 32676061

Turco, F., Garavaglia, M., Houdt, R. V., Hill, P., Rawson, F. J., & Kovacs, K. (2022). Synthetic Biology Toolbox, Including a Single-Plasmid CRISPR-Cas9 System to Biologically Engineer the Electrogenic, Metal-Resistant Bacterium Cupriavidus metallidurans CH34. *ACS Synthetic Biology*, *11*(11), 3617–3628. DOI: 10.1021/acssynbio.2c00130 PMID: 36278822

Verdezoto-Prado, J., Chicaiza-Ortiz, C., Mejía-Pérez, A. B., Freire-Torres, C., Viteri-Yánez, M., Deng, L., Barba-Ostria, C., & Guamán, L. P. (2025a). Advances in environmental biotechnology with CRISPR/Cas9: Bibliometric review and cutting-edge applications. *Discover Applied Sciences*, *7*(3), 167. Advance online publication. DOI: 10.1007/s42452-025-06609-x

Verdezoto-Prado, J., Chicaiza-Ortiz, C., Mejía-Pérez, A. B., Freire-Torres, C., Viteri-Yánez, M., Deng, L., Barba-Ostria, C., & Guamán, L. P. (2025b). Advances in environmental biotechnology with CRISPR/Cas9: Bibliometric review and cutting-edge applications. *Discover Applied Sciences*, *7*(3), 167. Advance online publication. DOI: 10.1007/s42452-025-06609-x

Voit, E. O. (2022). Perspective: Systems biology beyond biology. *Frontiers in Systems Biology*, *2*, 987135. DOI: 10.3389/fsysb.2022.987135

Wackett, L. P. (2022). The future of *Microbial Biotechnology*. *Microbial Biotechnology*, *15*(1), 79–80. DOI: 10.1111/1751-7915.13920 PMID: 34612578

Wang, F., Xiang, L., Sze-Yin Leung, K., Elsner, M., Zhang, Y., Guo, Y., Pan, B., Sun, H., An, T., Ying, G., Brooks, B. W., Hou, D., Helbling, D. E., Sun, J., Qiu, H., Vogel, T. M., Zhang, W., Gao, Y., Simpson, M. J., & Tiedje, J. M. (2024a). Emerging contaminants: A One Health perspective. *Innovation (Cambridge (Mass.))*, *5*(4), 100612. DOI: 10.1016/j.xinn.2024.100612 PMID: 38756954

Wang, X.-W., Wang, T., & Liu, Y.-Y. (2024b). Artificial intelligence for microbiology and microbiome research. *Comprehensive Reviews in Microbiology*, *18*(4), 601–629.

Wang, Z., Wang, S., He, Q., Yang, X., Zhao, B., Zhang, H., & Deng, Y. (2025). Ecological design of high-performance synthetic microbial communities: From theoretical foundations to functional optimization. *ISME Communications*, *5*(1), ycaf133. Advance online publication. DOI: 10.1093/ismeco/ycaf133 PMID: 40860568

Wani, A. K., Akhtar, N., Mir, T. U. G., Chopra, C., Singh, R., Hong, J. C., & Kadam, U. S. (2024). CRISPR/Cas12a-based biosensors for environmental monitoring and diagnostics. *Environmental Technology & Innovation*, *34*, 103625. DOI: 10.1016/j.eti.2024.103625

Weissman, J. L., Hou, S., & Fuhrman, J. A. (2021). Estimating maximal microbial growth rates from cultures, metagenomes, and single cells via codon usage patterns. *Proceedings of the National Academy of Sciences of the United States of America*, *118*(12), e2016810118. DOI: 10.1073/pnas.2016810118 PMID: 33723043

Wood, S., Biessy, L., Latchford, J. L., von Ammon, U., Laroche, O., Zaiko, A., & Pochon, X. (2019). Release and degradation of environmental DNA and RNA in a marine system. *The Science of the Total Environment*, *704*, 135314. DOI: 10.1016/j.scitotenv.2019.135314 PMID: 31780169

Xu, Y., & Zhao, F. (2018). Single-cell metagenomics: Challenges and applications. *Protein & Cell*, *9*(5), 501–510. DOI: 10.1007/s13238-018-0544-5 PMID: 29696589

Yadav, A. K., Banerjee, S. K., Das, B., & Chaudhary, K. (2022). Editorial: Systems Biology and Omics Approaches for Understanding Complex Disease Biology. *Frontiers in Genetics*, *13*, 896818. DOI: 10.3389/fgene.2022.896818 PMID: 35495146

Yuan, G., Deng, S., Czajka, J. J., Dai, Z., Hofstad, B. A., Kim, J., & Pomraning, K. R. (2024). CRISPR-Cas9/Cas12a systems for efficient genome editing and large genomic fragment deletions in Aspergillus niger. *Frontiers in Bioengineering and Biotechnology*, *12*, 1452496. DOI: 10.3389/fbioe.2024.1452496 PMID: 39479294

Yuan, K., & Nault, R. (2025). Application of a metabolic network-based graph neural network for the identification of toxicant-induced perturbations. *Toxicological Sciences : An Official Journal of the Society of Toxicology*, *206*(1), 19–29. DOI: 10.1093/toxsci/kfaf065 PMID: 40445920

Zhang, L., Chen, F., Zeng, Z., Xu, M., Sun, F., Yang, L., Bi, X., Lin, Y., Gao, Y., Hao, H., Yi, W., Li, M., & Xie, Y. (2021). Advances in Metagenomics and Its Application in Environmental Microorganisms. *Frontiers in Microbiology*, *12*, 766364. DOI: 10.3389/fmicb.2021.766364 PMID: 34975791

Zhang, L., Chen, L., Diao, J., Song, X., Shi, M., & Zhang, W. (2020). Construction and analysis of an artificial consortium based on the fast-growing cyanobacterium Synechococcus elongatus UTEX 2973 to produce the platform chemical 3-hydroxypropionic acid from CO2. *Biotechnology for Biofuels*, *13*(1), 1–12. DOI: 10.1186/s13068-020-01720-0 PMID: 32391082

Zhou, Y., & Zeeshan Ul Haq, M. (2024). Engineering of Synthetic Microbial Consortia for Sustainable Management of Wastewater and Polyethylene Terephthalate: A Comprehensive Review. *International Journal of Molecular Sciences*, *26*(23), 11623. DOI: 10.3390/ijms262311623 PMID: 41373776

Zupanic, A., Bernstein, H. C., & Heiland, I. (2020). Systems biology: Current status and challenges. *Cellular and Molecular Life Sciences : CMLS*, *77*(3), 379–380. DOI: 10.1007/s00018-019-03410-z PMID: 31932855

Chapter 10
Policy and Governance in Microbial Bioremediation: Regulatory Framework, Risk Assessment, and Global Standards

Chioma Blaise Chikere
http://orcid.org/0000-0003-3004-9780
University of Port Harcourt, Nigeria

Chioma Bertha Ehis-Eriakha
http://orcid.org/0000-0002-3284-2021
Faculty of Science, Admiralty University of Nigeria, Ibusa, Nigeria

Rosemary Agbeko
http://orcid.org/0009-0004-9813-3676
World Bank Africa Centre of Excellence, Centre for Oilfield Chemicals Research (ACE-CEFOR), University of Port Harcourt, Nigeria

Peace Omoikhudu Oleghe
Faculty of Science, Edo State University, Iyamho, Nigeria

ABSTRACT

Microbial bioremediation uses microbes to degrade pollutants in water, soil, and air and offers a cost-effective and sustainable alternative to conventional remediation methods. However, its widespread application is limited by the absence of global safety standards, weak international collaboration, limited technology transfer, and public awareness. Existing regulatory frameworks are often fragmented and narrowly focused on specific contaminants, ignoring the broader range of pollutants that can be treated biologically. These inconsistencies are further influenced by differing national priorities and policies. This review highlights key gaps in environmental

DOI: 10.4018/979-8-2600-2998-5.ch010

risk assessment, biosafety guidelines, legal frameworks, and global cooperation. It identifies lack of standardization and poor public acceptance as major constraints. To address these challenges, it recommends adaptive, risk-informed policies, improved public awareness and communication, increased funding for research, stronger stakeholder engagement, and development of an international treaty to harmonize regulations.

1 INTRODUCTION

The increasing pollution of the environment resulting from human activities such as oil spills, urban expansion, industrial, and agricultural activities pose significant challenges (Ghosal et al., 2016; Sharma, 2020; Tripathi et al., 2021). To address these challenges, microbial bioremediation has emerged as a promising solution. This biological process utilizes naturally occurring or genetically modified organisms (GMOs), such as bacteria, fungi, and algae, to break down harmful environmental pollutants, including hydrocarbons, heavy metals, pesticides, and industrial solvents, found in water, soil, and air matrices (Abatenh et al., 2017; Vaksmaa et al., 2023; Nordin et al., 2025).

This approach has proven to be a green, cost-efficient, and eco-friendly alternative to conventional remediation methods like physical excavation, neutralization, incineration, and chemical oxidation (Ajona and Vasanthi, 2021; Bala et al., 2022). Researchers have advocated for the use of diverse microbes to enhance the effectiveness of microbial bioremediation, considering factors like temperature, nutrient availability, and aerobic or anaerobic conditions (Krzmarzick et al., 2018; Alaira et al., 2021; Masindi et al., 2021; Chikere et al., 2021). By adopting this socially acceptable and sustainable approach, we can restore ecosystems to their original state while mitigating further anthropogenic pollution in environments prone to pollution challenges (Arregui et al., 2019; Bharath et al., 2019; Berillo et al., 2021).

Several postulated protocols aimed at confronting this problem, have not yielded much. These challenges require effective and efficient governance and pollution containment policy frameworks that centers on the deployment and utilization of microbiological technologies as a critical bio-sustainable alternative since microorganisms are ubiquitous in the biosphere. Furthermore, regulatory frameworks and global standards including those from International Organization for Standardization (ISO), Organization for Economic Co-operation and Development (OECD), and the Cartagena Protocol ensure responsible implementation of bioremediation practices since it dictate intervention and target values for remediation efforts. This is essential because the absence of a well-articulated and distinct regulatory policy framework could jeopardize the health of living organisms, creates imbalance within the ecological space, leading to ethical debates around the economy of bio-systems (Bohua et al., 2023). Ultimately, bioremediation stands as a pivotal technology for managing environmental contamination while achieving ecological restoration.

Despite these laudable potentials, microbial bioremediation still possesses multifaceted challenges that transcends the frontiers of microbiological techniques. The need for a stringent and comprehensive governance policy legal framework in microbial bioremediation that robustly regulates, evaluates the risk involvement and globally standardize the operational protocols cannot be overstated. Also, an effective balance between technological benefits and attendant risks to the entire ecosystem in the use of GMOs is imperative.

This paper examines the policy and governance dimensions of microbial bioremediation, with a focus on regulatory frameworks, risk assessment, biosafety, and global standardization. It evaluates existing frameworks, identifies critical gaps, and proposes improvements to ensure the safe and effective application of microbial technologies, while promoting stakeholder engagement and alignment with environmental protection and climate change strategies. The chapter further provides a critical analysis of the environmental and health risks associated with microbial bioremediation, alongside the precautionary measures required to mitigate them. It also explores global biosafety standards and regulatory guidelines governing the deployment of microbial agents, highlighting both their strengths and limitations. In addition, comparative case studies are presented to demonstrate how different countries have implemented policy and governance frameworks to ensure responsible bioremediation practices.

2 CONCEPTS

The microbial bioremediation concept relies on using the innate capability of microbes and their adaptability to diverse environmental conditions to break down environmental pollutants, exploiting and metabolizing their various components as “substrates or energy sources” under natural environmental conditions (Chikere et al., 2019, Ehis-Eriakha et al., 2025).

2.1 Mechanisms

Using mechanisms like biodegradation (microbial degradation of pollutants into simpler products), mineralization (comprehensive alteration of pollutants into inorganic, non-toxic end products), biotransformation (partial alteration of pollutants into less toxic forms), biosorption (inactive binding of metal and dye pollutants onto microbial surfaces), bioaccumulation (ingestion and intracellular storage of pollutants) and cometabolism (supplementary breakdown of pollutants while metabolizing another substrate); these processes make use of microbial enzymes to facilitate rapid decomposition of these pollutants through aerobic and anaerobic pathways, producing non-toxic end products (Ali *et al.,* 2017).

2.2 Methods

Bioremediation can be carried out using *ex situ* (off-site treatment) or *in situ* (on-site treatment) methods, based on considerations like microbial community, cost implication, environmental situation, nutrient availability, bioavailability, nature and properties of pollutants (Romero et al., 2019; Bala et al., 2022).

Depending on oxygen availability, microbial diversity significantly affects the success of bioremediation efforts. Key microbial genera including *Pseudomonas, Bacillus, Mycobacterium, and Flavobacterium* have demonstrated unique ability to remediate pollutants utilizing natural or enriched procedures to enhance microbial activity like biostimulation (adding nutrients that boost original microbes), bioaugmentation (adding specified microbes), biosurfactant (surface active, amphiphilic compounds with hydrophobic and hydrophilic moieties produced by microorganisms to enhance hydrocarbon degradation), bioventing (introducing controlled air and oxygen to stimulate original microbes in unsaturated areas), bioattenuation (reliance of natural biochemical processes, no external interference), bioreactors (using fermenting vessels to cultivate microbes), biopiles (using a constructed engineered system to access contaminated and excavated soil), biosparging (injecting air or gas under pressure to stimulate bioactivity in saturated areas), composting and land farming are among the *ex situ* and *in situ* engineered bioremediation systems designed to treat contaminated water and soil environments (Che and Men, 2019; Jaiswal and Shukla, 2020; Sanghvi et al., 2020; Giri et al., 2021; Phale et al., 2022). Incorporating microbial metabolic metagenomic indices in all bioremediation schemes, with global best practice alignments, would guarantee their efficacy and safety at all times (Techtmann and Hazen, 2016; Nikel and de Lorenzo, 2021; Sharma et al., 2022).

2.3 Applications

Bioremediation application has swiftly developed over the years, tackling diverse environmental issues across the biosphere from contamination (oil spillage, explosives, metropolitan wastewater, Industrial effluents, agricultural waste, heavy metals, chlorinated solvents, etc) to forensic applications and oil spill liability cases (Sanghvi et al., 2020; Phale et al., 2022; Meena et al., 2025). This environmental biotechnological approach utilizes the natural metabolic potentials of microbes and their enzymes to degrade xenobiotic and recalcitrant compounds using a cascade of oxidation–reduction processes (Phale *et al.*, 2022). Through modernized and advanced knowledge and skill-set in genetic engineering, biotechnology and nanotechnology. The frontiers of microbial bioremediation are constantly expanding, overcoming previously existing challenges associated with environmental dependence, recalcitrant pollutants, slow kinesis and kinetics (Dangi et al., 2019; Dar and Naseer, 2022).

3 MICROBIAL BIOREMEDIATION GLOBAL POLICY AND LEGAL FRAMEWORKS

Global policy and governance are essential for sustainable microbial bioremediation, as pollution transcends national borders. Consequently, there is an urgent need for dynamic protocols that provide a robust framework for responsible implementation, ensuring that safety, effectiveness, and ethical considerations are adequately addressed in environmental cleanup operations. This framework includes regulations, ethical guidelines, and monitoring systems to manage possible risks associated with the introduction of microbes into each environmental setting (Manisha *et al.*, 2021). Some developing nations like Brazil and India, have successfully introduced sustainable cost-effective areas of innovation and adaptation in bioremediation framework as alternative treatment protocols (to the existing traditional physico-chemical remediation protocols) geared towards tackling soil, groundwater, urban and industrial pollutions. These innovations concentrate on leveraging cutting-edge biotechnological tools to local ecological situations, indigenous microbial consortia, adapting to the use of low-cost biosorbents technologies, phytoremediation incorporation, increased bioavailability and solubility by utilizing bacterial biosurfactant on hydrophobic hydrocarbons from oil spills, as well as providing funding and expertise towards scalability of these processes into pilot-scale levels (Rezaei et al., 2025).

3.1 Importance of Policy and Governance Legal Frameworks

- **Safety:** Bioremediation involves introducing microorganisms into the environment, and policies are needed to ensure that these microorganisms are safe for humans, animals, and the broader ecosystem. This includes assessing the potential for these microorganisms to become invasive, mutate, or cause harm.
- **Effectiveness:** Policies and regulations can help ensure that bioremediation strategies are well-designed and implemented effectively. This includes setting standards for the selection of appropriate microbial strains, optimizing environmental conditions, and monitoring progress.

- **Ethical Considerations:** Bioremediation can raise ethical questions, particularly regarding the use of bioengineered microorganisms and its potential impact on ecosystems. Policies can help address these ethical considerations by establishing clear guidelines and promoting transparency.
- **Public Awareness and Acceptance:** Public understanding and acceptance of bioremediation are essential for its successful implementation. Policies can support public education and outreach efforts, ensuring that the public is informed about the potential benefits and risks of bioremediation.
- **Long-term Monitoring and Management:** Bioremediation may require long-term monitoring to assess its effectiveness and identify any potential unintended consequences. Policies can ensure that there are mechanisms in place for ongoing monitoring, evaluation, and adaptive management.
- **Regulatory Oversight:** Regulatory bodies play a critical role in overseeing bioremediation projects, ensuring compliance with environmental regulations and standard.

3.2 National Regulations, Permit and Licensing Requirements

Microbial bioremediation epitomizes a transformative environmental management policy, yet its operations are reliant upon healthy supervisory oversight. To effectively monitor and manage practices, different countries across regions have designed and established indigenous regulatory framework systems specific to their ecosystems. The aim is to guarantee the effectiveness, safety, and compliance of microbial bioremediation technologies with community and ecological health standards.

Most countries nationally operate permit-based frameworks to alleviate community and ecological health related risks linked to microbial interruption. Here are some examples;

3.2.1 United States of America

In the United States, microbial bioremediation with GMOs is controlled by the Environmental Protection Agency (EPA) using the Toxic Substances Control Act (TSCA), while the Office of Pollution Prevention and Toxics (OPPT) handles risk assessments. The United States Congress in 1976, mandated the EPA to regulate pre-manufacture notice, all-inclusive risk assessment, and continuing post-release monitoring (United States Congress, 1976). Key legislation includes the Comprehensive Environmental Response, Compensation, and Liability Act (CERCLA), commonly known as the Superfund program. The EPA oversees site-specific bioremediation activities, ensuring they meet cleanup goals without posing additional risks. These processes are supplemented by the Coordinated Framework for Biotechnology, which integrates risk-based regulatory principles across Federal agencies, mandating a scientifically defensible approval mechanism (Wozniak et al., 2012; U.S EPA, 2022).

3.2.2 European Union

In the European Union, the European Commission (EC) governs and enforces microbial bioremediation activities under the Environmental Liability Directive (ELD) which emphasizes oversight of the 'polluter pays' principle. They also govern the deliberate release of GMOs through Directive 2001/18/EC, and regulates their contained use through Directive 2009/41/EC. These directives mandate comprehensive environmental risk assessments, permit approvals, public consultation, and licensing approval before environmental deployment. This structured regulatory system has been lauded for balancing innovation with precautionary principles, although critics highlight bureaucratic delays that may hinder timely remediation efforts (Chemla et al., 2025). Additionally, the REACH regulation (Registration, Evaluation, Authorization and Restriction of Chemicals) mandates that bioremediation substances and microbial formulations meet safety standards before use. Member states are required to follow the precautionary principle, ensuring high levels of environmental protection (European Commission, 2020).

3.2.3 India

In India, microbial bioremediation is regulated through the Ministry of Environment, Forest and Climate Change (MoEFCC). Key instruments include the National Environmental Policy and the Hazardous and Other Wastes (Management and Transboundary Movement) Rules, 2016. The Central Pollution Control Board (CPCB) also plays a significant role in approving and monitoring bioremediation projects, especially those dealing with industrial waste and sewage treatment (MoEFCC, 2016).

3.2.4. Nigeria

In Nigeria, regulatory oversight for enforcing compliance with National Biosafety Laws is shared between the National Environmental Standards and Regulations Enforcement Agency (NESREA) and the National Biosafety Management Agency (NBMA). In addition, the Environmental Guidelines and Standards for the Petroleum Industry in Nigeria (EGASPIN), issued by the Department of Petroleum Resources (now part of the Nigerian Upstream Petroleum Regulatory Commission), serves as a critical regulatory framework for petroleum operations. EGASPIN provides comprehensive environmental standards and guidelines to ensure that oil and gas activities are conducted in a safe and environmentally sustainable manner. Among its key functions is oil spill management, which encompasses spill detection, reporting, remediation, and cleanup, thereby playing a vital role in mitigating environmental degradation and supporting effective environmental governance in the petroleum sector (Omotuyi 2023; Olawuyi and Tubodenyefa, 2018). The core requirements for microbial bioremediation for these institutions, typically include submission of technical dossiers; detailing microbial strains, anticipated environmental impacts assessments reports, biosafety obligations, and post-implementation monitoring strategies (Springer, 1993; Azubuike et al., 2016). Unlike the U.S. or EU frameworks, Nigeria's regulatory system is still developing in terms of institutional capacity and enforcement rigor, highlighting the challenges faced by many developing nations.

3.2.5 Other Outstanding Examples

In Canada, the Canadian Environmental Protection Act (CEPA) mandates that new biotechnology products, including microbial agents, undergo pre-market assessments (Environment and Climate Change Canada, 2022). In China, the Ministry of Ecology and Environment regulates biotechnology applications through its Biosafety Management Regulations, promoting ecological risk management (Ministry of Ecology and Environment of China, 2020). Australia applies the Gene Technology Act 2000 to oversee genetically modified organisms in bioremediation contexts (Commonwealth of Australia, 2024).

The comparative scrutiny of the mandates from the agencies of various nations, shows that their regulatory legal frameworks differ in strictness, bureaucratic intricacy, and dimensions of enforcement. For example, while the U.S. and EU have been able to provide formalized guidelines that is legally-binding with evidently well-defined institutional responsibilities, Nigeria and other African countries frequently rely on technical assistance from international donor-driven organizations/governments to build capacity and they are still in the process of operationalizing their biosafety laws and by-laws (Biomed Central, 2024). These significant differences, has globally impacted the pace, possibilities, and supposed legitimacy of microbial bioremediation projects.

3.3 Environmental Protection Enforcement Agencies

The effectiveness of the microbial bioremediation governance framework hinges on stringent and competent oversight by specialized enforcement agencies, with mechanisms detailing responsibility for enforcing compliance with microbial bioremediation protocols, a robust enforcement requires both legal authority and institutional capacity. These institutions ensure that microbial applications are conducted in alignment with national laws and, where relevant, international treaties. However, weak institutional capacity continues to hinder enforcement in many jurisdictions (Azubuike et al., 2016). Some of these agencies are United States, the EPA; European Union, European Environment Agency (EEA) and Nigeria NESREA and NBMA (Table 20.1).

Table 1. Environmental Protection Enforcement Agencies Across Different African Countries

Name of Agency	Abbreviation	Country / Region	Year Established	Mandate (Brief Description)	Website
Ministry of Environment and Renewable Energy	ONM	Algeria	1970	Manages natural resources and oversees renewable energy initiatives	https://www.me.gov.dz
Ministry of Environment and Tourism (Department of Environment and Conservation)	MET	Botswana	1985	Regulates pollution, environmental, tourism policy, waste management, and bioconservation enforcement.	https://www.gov.bw
Ministry of Environment, Protection of Nature and Sustainable Development	MINEPDED	Cameroon	2004	Coordinates environmental protection, regulation, and sustainable development enforcement.	https://www.minepded.gov.cm
National Agency of Meteorology and Teledetection by Satellite	METTELSAT	Democratic Republic of the Congo	1991	Handles meteorology and satellite-based environmental monitoring and protection.	https://www.meteordcongo.cd
Ministry of Environment Egyptian Environmental Affairs Agency	EEAA	Egypt	1994	Regulates and enforces environmental policies at national level.	https://www.eeaa.gov.eg/

continued on following page

Table 1. Continued

Name of Agency	Abbreviation	Country / Region	Year Established	Mandate (Brief Description)	Website
Ethiopian Environmental Protection Authority	EPA	Ethiopia	1995	Enforces environmental laws and standards nationwide.	https://www.epa.gov.et
Ministry for the Protection of the Environment and Natural Resources, Forest and the Sea of Gabon	DGEPN	Gabon	2014	Enforces climate and land use plan, protection of national parks and protected areas.	https://www.eaux-forets.gouv.ga/
Ghana Environmental Protection Agency	EPA	Ghana	1994	Implements policy, regulates, and enforces environmental compliance for sustainable development.	https://www.epa.gov.gh
National Environment Management Authority	NEMA	Kenya	1999	Regulates, supervise and coordinates all matters relating to sustainable environmental management	https://nema.go.ke
Ministry of Tourism, Environment and Culture (Department of Environment)	DoE	Lesotho	2008	Coordinates environmental protection and enforcement activities.	https://www.environment.gov.ls
Liberia Environmental Protection Agency	EPA	Liberia	2002	Co-ordinate, Monitor, Supervise and Consult on all activities of the environment and sustainable use of natural resources	https://epa.gov.lr
Ministry of Environment and Sustainable Development	MEDD	Madagascar	1990	Developing and implementing environmental policies	https://www.ecologie.gov.mg/

continued on following page

Table 1. Continued

Name of Agency	Abbreviation	Country / Region	Year Established	Mandate (Brief Description)	Website
Federal Ministry of Environment	FMEnv	Nigeria	1999	Formulates policies, coordinates protection and conservation enforcement.	https://environment.gov.ng/
National Environmental Standards and Regulations Enforcement Agency	NESREA	Nigeria	2007	Protects environment, enforces laws, standards, and agreements on pollution, biodiversity, climate.	https://www.nesrea.gov.ng
National Oil Spill Detection and Response Agency	NOSDRA	Nigeria	2006	Detects, manages, and enforces responses to oil spills.	https://www.nosdra.gov.ng
National Biosafety Management Agency	NBMA	Nigeria	2015	Regulates biotechnology, ensures biosafety for people and environment, enforces GMO and biotech protocols.	https://www.nbma.gov.ng
Rwanda Environment Management Authority	REMA	Rwanda	2003	Coordinates environmental protection, set policies and implement strategies for sustainable use of natural resources.	https://www.rema.gov.rw/
Department of Forestry Fisheries and the Environment	DFFE	South Africa	1996	Oversees agriculture and environmental enforcement.	https://www.dffe.gov.za
Environmental Management Inspectorate	EMI	South Africa	2005	Coordinates enforcement across biodiversity, pollution, waste and water sub-sectors	https://www.dffe.gov.za
Environmental Management Agency	EMA	Zimbabwe	2003	Enforces laws, issues permits, promotes sustainable development.	https://www.ema.co.zw/

Global cooperation has produced treaties that provide the legal basis for biosafety governance for microbial bioremediation operations, providing critical guidance for enforceable stewardship in the environmental space, across territories. Saddled with this responsibility are multilateral organizations such as United Nations Environment Programme (UNEP), Organization for Economic Co-operation and Development (OECD), International Maritime Organization (IMO). World Health Organization (WHO) and Food and Agriculture Organization (FAO). UNEP provides important regulatory support for sound environmental global best practices in biotechnology management, offer strategic technical guidelines as well as organize programs to develop and build capacity for stakeholders in developing nations (UNEP, 2019). OECD has produced global standards and guidance for testing procedures, streamlining regulatory guidelines for assessing microbial products' safety and performance. They also have a system called Mutual Acceptance of Data (MAD) which ensures that safety data generated from one affiliate nation are accepted by others (OECD, 2020).

Similarly, IMO controls maritime pollution activities using the developed treaty of the International Convention for the Prevention of Pollution from Ships (MARPOL). They support the use of bioremediation agents in oil spill responses highlighting the need for scientific validation and environmental risk assessments for any microbial applications within the marine ecosystems (IMO, 2018). Both WHO and FAO are both agencies under the United Nations Organizations (UNO). They both collaborate together frequently in areas bothering on nutrition, health and food safety. Their contribution to the evaluation agents of bioremediation that may affect public health and food chains are notable (WHO/FAO, 2017).

3.4 Biosafety Laws and Protocol

There is a need for mutual cooperation among countries across regions for global microbial bioremediation governance to be effective and efficient. Since pollution does not respect boundaries, race or biospheres, there is need for standard protocol-setting bodies, and multilateral agreements. All global biosafety laws and protocols are "warehoused" in the Cartagena Protocol on Biosafety, under the Convention on Biological Diversity. This was adopted in 2000 in Montreal, Canada but became enforced in 2003. It provides the cornerstone of global legal microbial bioremediation framework governing the harmless transfer, handling, and use of all GMOs transboundary. It sets-up the procedures requiring obligatory risk assessment conducts, advanced notification agreement measures, accountability, and emergency actions required towards inadvertent release of GMOs during transboundary movement, with a view to upholding transparency. This led to the establishment of an information-sharing Biosafety Clearing-House among signatory countries (Convention on Biological Diversity, 2000). Each country is expected to be politically accountable by committedly harmonizing, then ingraining it into their domestic biosafety legislation, operationalizing and coherently fostering the embedded legal responsibilities to align with international best practice.

In spite of this laudable microbial bioremediation roadmap however, there still exist significant implementation lacuna across global regions. According to Bharagava, (2020), there are apparent global jurisdictions disparities as some Developed nations have comprehensively implemented all aspects of the biosafety laws, while most developing nations either often face limited resource, erratic enforcement, and disjointed or no official legal frameworks. All these, compromises both ecological protection and erodes public confidence (Chemla et al., 2025).

Therefore, this clarion call is an advocacy for an urgent political willingness to harmonize and standardized biosafety laws as well as organize programs to build the capacity of all stakeholders in developing countries in order to bridge these gaps.

4 GOVERNANCE AND STAKEHOLDER CONTRIBUTION

Governance in microbial bioremediation environmental cleanup operations refers to the policies, structures for enforcement and the procedural processes aimed at solving the issues, with a view to alignment with international treaties, biosafety standards and environmentally protectionist regulations (UNEP, 2019; U.S. EPA, 2022)

4.1 Importance of Policy and Governance in Environmental Cleanup Operations

Trustworthy governing policy operations are the pivot of all environmental cleanup operations, being that they provide the enforcement and ethical structure required for successful microbial bioremediation processes that would lead to a safe environment and healthy populace. The components of an efficient governance microbial remediation policy are based on comprehensive data-driven scientific proofs, ethical responsibility and legally justifiable (Azubuike et al., 2016).

This policy framework should have clearly defined operational roles for all stakeholders that ensures accountability, responsibilities, communal oversight, eliminates uncertainty, and proper risk evaluation procedure of the entire ecosystem (Chemla et al., 2025). It should also provide the legitimate national microbial bioremediation technology suitable to be adopted in line with international best quality standards that would guarantee public and institutional confidence in the process (Bharagava, 2020). Chemla et al., (2025) observed that transparent policy guided projects have higher acceptance communal ratings, this is crucial factor towards a successful microbial bioremediation environmental cleanup operation.

4.2. Ethical, Social, and Cultural Considerations

Microbial bioremediation environmental cleanup operation is not only a technical process, but includes both ethical and sociocultural intersections. Ethical concerns arise from safeguarding the responsibility component bothering around prolong negative environmental effects, safety protocols and probability of inadvertent hazardous horizontal genetic transmissions. These can be addressed by adopting the free, prior, and informed consent (FPIC) principle especially among defenseless aboriginal societies who are at risk of exposure to hazardous microorganisms' release.

Also, sociocultural acceptance of microbes is determined by perspectives, insights and idiosyncrasies of each community (Azubuike et al., 2016). These forms the basis of social intervention acceptance of environmental cleanup operation using microorganisms. Therefore, in order to carry out a sustainable cleanup of the environment, government must transparently engage an all-inclusive coalition of stakeholder during consultation, prior to commencement and communicate lucidly the mechanisms involved (Azubuike et al., 2016). Incorporating these strategies during consultation would not only ensure legitimacy and compliance, but promote lasting stability and public trust before and after cleanup.

4.3. Challenges in Regulation and Governance

As with other aspects of life, microbial bioremediation policy governance regulatory framework for risk assessment, in order to attain global standards is not without its numerous multifaceted in-country challenges. Some of these according to Azubuike et al., (2016) and Bharagava (2020), are itemized below:

(i). **Regulatory Overlap:** The governing directive for supervising and enforcing microbial bioremediation policies lies at intersections between environmentalist, biotechnologist and health experts. This fragmentation in functionality is creating regulatory overlap among these sector stakeholders leading to confusion and sometimes institutional ineptitude

(ii). **Personnel and Infrastructural Capacity Shortfalls:** In most cases especially within the developing nations threshold, regulatory supervision and enforcement are seriously hampered owing to shortfalls in technical personnel expertise, equipment and infrastructural capacity.

(iii). **Analytical unpredictability:** Microbial bioremediation is a data-driven science that steadily relies on results obtained from the interactions of microorganisms in an intricate natural ecological system. This could be very tricky as regulatory evaluation capabilities towards a comprehensive risk assessment becomes varied, limited and subjective.

(iv). **Universal Divergence:** In today's world, there are significant disparities between the developed and developing countries leading to prejudiced socio-political agreement implementations with regards to enforcing environmental and health issues.

The challenges of bioremediation regulations and governance are profoundly interrelated, establishing an intricate scenario where ethical uncertainties, technical restrictions, socioeconomic factors, and policy gaps feed into one another. All these hinder the formation of vibrant, supportive strategies, which in turn asphyxiates investments and application. This intricate relationship frequently creates a disjointed, lethargic, or excessively cautious regulatory frameworks that cannot keep pace with current innovation, (such as the use of GMOs), struggling with other site-specific and biological complexities of bioremediation. Disjointed governance also encumbers the standardization required for widespread implementation.

These identified challenges can be adequately addressed with the right political-willingness to implement past and future agreements, using multilateral governmental convergence to discuss current issues, intercontinental alliances, continuous research and development, capacity-building trainings, a combination of ethical science-based data precision and committed social responsibility.

5 ENVIRONMENTAL RISK ASSESSMENT

The fundamental role of risk assessment in microbial bioremediation is to detect, measure, and control potential risks to ecosystems and human health. The process mainly focuses on hazard identification, modeling exposure routes, and risk characterization, as shown in Fig.1 (Meena et al., 2025).

• Hazard Identification

Hazard identification, the first stage of environmental risk assessment, focuses on possible risks associated with microbial bioremediation. It involves evaluating the properties of microorganisms introduced or stimulated through bioaugmentation or biostimulation, the nature of pollutants targeted, and any metabolites formed during biodegradation (Ehis-Eriakha et al., 2025). Characterizing microbial traits based on spore formation, survival capabilities, pathogenicity, and antagonistic interactions is crucial. Also, the type of contaminants such as hydrocarbons, heavy metals, and chlorinated compounds and their degradation pathways deeply influence hazard potential (Ayilara et al., 2023; Smith et al., 2023).

• Exposure Assessment

The second phase, exposure assessment, quantifies the extent, frequency, and duration of exposure of environmental receptors. These receptors include soil, aquatic organisms, plants, animals, and humans to the identified hazards. It assesses how microbes and chemicals move through exposure routes such as soil or water contact, inhalation of bioaerosols, and food chain. Techniques used include environmental sampling, microbial plate counts, chemical analyses, and predictive modeling. Accurate exposure assessment helps prevent unintended microbial spread and buildup of hazardous intermediates in ecosystems (Liu et al., 2020). Together, hazard identification and exposure assessment provide the foundation for risk characterization.

• Risk Characterization

Risk Characterization step combines hazard and exposure data to estimate the likelihood and severity of adverse effects associated with microbial bioremediation. This step assesses microbial pathogenicity, contaminant toxicity and persistence, and receptor sensitivity (Ayilara et al., 2023). Both qualitative and quantitative tools such as risk matrices or probabilistic models are employed to categorize risks as low, moderate, or high (Skaar et al., 2016).

These three interconnected stages of environmental risk assessment require continuous monitoring to track progress and detect negative effects of microbial bioremediation. The real-time data generated supports flexible management and allow adjustments to maintain a balance between pollutant breakdown and environmental protection (Fig. 20.1).

Figure 1. Environmental risk assessment process of microbial Bioremediation

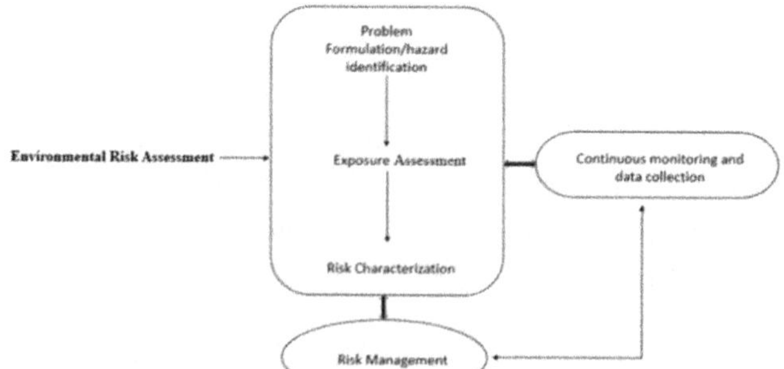

(Meena et al., 2025).

6 BIOCONTAINMENT STRATEGIES AND MONITORING

The application of microbial entities in microbial bioremediation requires reliable strategies for containment and monitoring. There is an inherent risk that microbes may spread beyond their target areas, impact native ecosystems, or transfer genetic material. Biocontainment strategies address these concerns by ensuring that microbes remain confined and their activities are controlled during the remediation process (Kim & Lee, 2020). Methods used in biocontainment have been discussed below;

6.1 Genetic Modifications

Biocontainment can be achieved through genetic modifications that enable microbes to survive only in environments supplying specific essential nutrients or metabolic precursors (Tosi et al., 2022). Another genetic approach is through engineered gene circuits that trigger cell death or inhibit growth when exposed to specific conditions. (Martínez-García et al., 2023). These genetic approaches act as a biocontainment mechanism by restricting microbial growth to designated areas

6.2 Physical Containment

Another method is by physical containment such as spatial isolation or controlled application techniques. Spatial isolation features the integration of gene regulatory systems that activate microbial functions only in the presence of specific environmental cues or pollutants. This ensures that microbial degradation is spatially and temporally limited to contaminated zones, reducing off-target effects. Also, microbial encapsulation within matrices like gels, beads, or biofilms physically restricts their movement and dispersal. This approach protects microbes from environmental stressors and enhances degradation efficiency while providing containment (Jaiswal & Shukla, 2020; Ayilara & Babalola, 2023).

6.3 Molecular and Sensor Technologies Monitoring

This biocontainment strategy involves monitoring with molecular and sensor technologies to provide detailed tracking of microbial communities and pollutant degradation. Molecular-based techniques such as 16S rRNA gene sequencing, metagenomics, metatranscriptomics, and proteomics can be used to analyze microbial diversity, gene expression profiles, and metabolic activity *in situ*. These approaches give comprehensive insights into microbial functions and their responses to environmental changes during remediation. Moreover, biosensors, electrochemical and optical devices integrated with Internet of Things (IoT) platforms, enable real-time data collection on pollutants, microbial activity, pH, temperature, and oxygen levels. This monitoring infrastructure, combined with artificial intelligence algorithms, optimize bioremediation processes by predicting outcomes and adapting conditions dynamically (Ayilara, & Babalola, 2023; Akintola, 2024).

7 RISK AND BENEFIT ANALYSIS

An essential consideration in applying microbial bioremediation within environmental management is ensuring that, microbes used can effectively breakdown complex pollutants, generally at lower cost (Table 20.2). Also, the technology should be capable of addressing a wide array of contaminants to achieve complete transformation.

Table 2. Comparative summary of benefits and risk of microbial bioremediation technology

Benefits	Risks	Reference(s)
Reduces hazardous pollutants in soil and water	May promote growth of harmful pathogens	Ayilara et al., 2023; Maglione, et al., 2024
More affordable than traditional cleanup methods	May leave toxic residues or create more toxic byproducts	Chinthala, 2020
Uses natural processes, reducing chemical and energy use	Can disrupt existing microbial ecosystems	Ayilara et al., 2023
Enhances soil pH and supports organic matter decomposition	Formation of biofilms that hinder remediation	Liu et al., 2020
Converts toxins into less harmful forms	Bioaccumulation of pollutants in microbes or transfer to the food chain	Kuppan et al., 2024; Kakde, & Sharma, 2024

Despite these benefits, significant scientific and practical gaps remain that constrain effective microbial bioremediation. One major challenge is incomplete comprehension of microbial interactions and pollutant degradation pathways in complex environments (Chinthala, 2020). Environmental conditions such as pH, temperature, nutrient levels, and oxygen availability strongly affect microbial activity and pollutant breakdown (Lahel et al., 2016; Maletić et al., 2021; Li et al., 2022). This can slow remediation and make outcomes difficult to predict. Also, limited data on stability of specific pollutants in different matrices that hinders effective field operations of laboratory successes must be considered (Maglione et al., 2024).

Moreover, the risk of incomplete pollutant degradation is significant as intermediate products formed during biotransformation might be toxic, sometimes more so than the original contaminants (Kuppan et al., 2024). Another critical factor is the potential for these microbes to extend beyond the targeted remediation zones and transfer genetic material to native organisms, which might have unforeseen ecological repercussions (Datta et al., 2025). More studies are needed to assess short- and medium-term risks of pollutants in different media or matrices to serve as a basis of decision making focused on environmental protection and risk management. This integrated assessment of benefits and risks improves regulatory compliance and public trust in bioremediation (Ayilara and Babalola, 2023; Akintola, 2024).

8 HEALTH RISK AND PRECAUTIONARY PROTOCOL

Health Risk to Workers

Health risks associated with microbial bioremediation must be carefully identified and managed to protect workers. Personnel involved in bioremediation activities may be exposed to microbial agents that are pathogenic or toxigenic, posing respiratory, allergic, or infectious hazards. Such exposure can occur through inhalation of aerosols, direct skin contact, or accidental ingestion (Skaar et al., 2016).

Studies have shown that certain bioremediation sites contain airborne bacterial and fungal species capable of causing infections, underscoring the critical need for thorough exposure assessment (Ma et al., 2022; Nombamba et al., 2024). In addition, exposure to microbial spores or metabolic by-products may trigger allergic responses, including asthma or hypersensitivity, particularly among workers lacking proper protective equipment.

Additionally, occupational environments such as soil treatment sites or waste sorting facilities where microbial aerosols are frequently generated have reported cases of respiratory irritation and allergen sensitization associated with bioremediation activities (Viegas and Gonçalves, 2021). These findings emphasize the necessity of implementing effective respiratory protection and environmental control measures to minimize health risks.

Risk to the Environment

Incomplete degradation of pollutants can generate toxic metabolites that pose additional health hazards when released into the environment or inhaled. This risk necessitates thorough monitoring of pollutant transformation products and adjustment of remediation protocols to minimize toxic exposure (Skaar et al., 2016). Another significant concern is the potential for horizontal gene transfer between introduced microbial strains and native microbiota. Such gene exchange can facilitate the spread of antibiotic resistance or virulence factors, thereby complicating the treatment of infections and disrupting the balance of microbial ecosystems (Kakde and Sharma, 2024).

Precautionary Protocols

To address these health risks, precautionary protocols emphasize the use of personal protective equipment (PPE) to limit direct contact and inhalation. This involves wearing gloves, masks or respirators, appropriate protective clothing, and eye protection to minimize contact with microbial agents or toxic degradation products. In addition, implementing strict hygiene measures, such as regular handwashing and proper sanitation of equipment, play a crucial role in preventing microbial transmission during bioremediation activities. Furthermore, regular health surveillance of workers, including clinical assessments and exposure evaluations, helps in the early detection of potential health issues. Implementing access control measures around remediation sites also reduces unnecessary exposure, collectively enhancing worker safety (Kuppan et al., 2024).

Also, proper labeling and documentation of microbial products according to regulatory standards allow for transparent risk communication and informed usage. Regulations such as the Norwegian Pollution Control Act, European Union biocontrol directives, and U.S. EPA microbial risk guidelines establish mandatory procedures for microbial product registration, environmental release permissions, and workplace safety standards. Adhering to these legal requirements guarantees that bioremediation efforts conform to public health and environmental protection objectives (Skaar et al., 2016). Environmental safeguards, such as ventilation and site isolation, help protect surrounding communities from unintended exposure (Ayilara and Babalola, 2023; Kakde and Sharma, 2024).

Finally, training programs and awareness campaigns are indispensable for equipping personnel and stakeholders with knowledge on microbial hazards, proper PPE usage, and emergency response measures. Ongoing education fosters safe operational behavior, encourages early hazard recognition, and promotes transparent communication among all parties involved (Central Pollution Control Board, 2022). Collectively, these measures form the foundation for safe microbial bioremediation practices that balance remediation efficacy with health and environmental protection.

9 GLOBAL BIOSAFETY STANDARDS AND GUIDELINES

Biosafety protocols for microbial bioremediation focus on policies and frameworks established to guarantee the safe application of microorganisms in cleaning up the environment. Risk assessments based on microbial traits like disease-causing ability, survival in the environment, genetic behavior, and potential to transfer genes are key components informing biosafety standards. Different countries employ classification schemes to assess the biosafety levels needed when working with microbes, especially genetically engineered ones. For instance, Thailand's BIOTEC illustrates a system dividing microbial work into four risk classes, from microbes suitable for large-scale industrial use under strict practice protocols to those needing stringent containment due to higher hazards. Also, globally recognized guidelines such as the United States' Biosafety in Microbiological and Biomedical Laboratories (BMBL) manual and the World Health Organization's (WHO) Laboratory Biosafety Manual provide foundational principles for safely handling microorganisms in research and industrial contexts, including environmental remediation (Richmond, & McKinney, 2009; WHO, 2020).

These international biosafety regulations for microbial bioremediation highlight the need for standardized rules to ensure safety and global uniformity. Although individual countries regulate genetically modified microorganisms and other microbial agents through their own frameworks, significant gaps and variations persist in risk assessment, containment measures, and emergency response strategies (Ivanov et al., 2019). These inconsistencies can hinder effective oversight and create disparities in safety practices. Public concerns about the unintended spread of engineered microbes and their genetic material further emphasize the importance of transparency, robust oversight, and clear communication. At the same time, climate-driven pollution is inherently transboundary, meaning that pollutants can cross national borders. This makes inconsistent biosafety frameworks a major bottleneck to effective environmental management.

Currently, the diversity of regulatory protocols across countries can slow innovation and limit international collaboration. These differences often create obstacles for the approval, transfer, and implementation of bioremediation technologies across borders. To address these challenges, establishing common international guidelines is essential. Such guidelines would provide consistent benchmarks for evaluating environmental and health risks, as well as for containment procedures, waste management, and accident response. Harmonizing biosafety regulations would simplify compliance, ensure uniform safety standards, and promote the safe application of microbial technologies worldwide. Ultimately, this alignment is crucial for fostering global cooperation and effectively addressing pollution that transcends national boundaries (Maglione et al., 2024).

10 COMPARATIVE CASE STUDIES

An analysis of various case studies reveals how governmental regulations and policies have been instrumental in facilitating microbial bioremediation application and are summarized in (Table 20.3). In Florida, indigenous microbial populations were used to degrade polycyclic aromatic hydrocarbons (PAHs) at Brown wood preserving site. This approach reduced PAH concentrations drastically from initial levels of 100-280 mg/kg to 3-23 mg/kg for over 18 months. Central to this achievement were guidelines set forth by the United States Environmental Protection Agency's (USEPA) Federal Remediation Technologies Roundtable. These guidelines provided frameworks for site evaluation, technology application, and monitoring, assuring environmental protection and high remediation standards through rigorous oversight (Federal Remediation Technologies Roundtable FRTR, 1995).

Similarly, in Pakistan, an oil production facility implemented *in situ* bio-piling complemented by "ultra-microbes". A 70% reduction in petroleum hydrocarbons was achieved within two weeks. Although Pakistan's regulatory system is comparatively less codified, growing alignment with global environmental protocols has encouraged environmentally sound remediation practices. Local regulatory bodies supported these efforts through approvals and periodic environmental assessments, thus enabling controlled microbial deployment with minimized ecological impacts (Mukhtar, 2023). In addition, consortia of microbes including *Saccharomyces cerevisiae*, *Pseudomonas*, and fungi such as *Aspergillus* species have also successfully been utilized to bioaccumulate and adsorb metals like lead, mercury, and cadmium at a multiple heavy metal-contaminated sites. National hazardous waste policies demanded strict characterization, containment, and monitoring of bioremediation agents, mitigating risks of secondary contamination and horizontal gene transfer, thereby ensuring the safety and acceptance of bioremediation strategies (Maglione et al., 2024).

In Alaska, native aerobic microbes were stimulated to remediate sub-arctic soils contaminated with JP-4 jet fuel hydrocarbons under extreme environmental conditions. EPA protocols for environmental remediation, coupled with state-level oversight, provided frameworks mandating environmental impact assessments and ongoing surveillance (Song and Bartha, 1990). Various petroleum-contaminated soils benefited from aerobic microbial degradation facilitated by genera such as *Pseudomonas* and *Rhodococcus*, augmented with nutrient amendments. Compliance with soil cleanup regulations that stipulate microbial profiling, toxicity evaluations, and remediation benchmarks was essential to protect ecosystems and human health (Liu et al., 2025). Lastly, the biodegradation of textile dyes using fungal species like *Cladosporium* was driven by stringent wastewater treatment regulations and industrial effluent standards compelling the implementation of bioremediation.

Regulatory frameworks mandate regular monitoring to prevent microbial treatment from introducing new environmental hazards (Maglione et al., 2024).

Across these cases, regulatory oversight, whether rigorous (U.S.A, EU) or emerging (Pakistan), proved central to safe and effective microbial bioremediation. Where clear frameworks existed, risks of microbial spread and gene transfer were minimized; where frameworks were evolving, international collaboration filled the gap. Government interventions through legislation, policies, and guidelines have decisively influenced these successful outcomes. Thorough microbial evaluation, effective containment methods, and continuous environmental monitoring prescribed by policies have minimized risks related to microbial spread, contaminant migration, and gene transfer, bolstering public confidence. Moreover, required documentation and periodic performance reviews generate critical data that inform and improve bioremediation technologies (Ehis-Eriakha et al., 2024). In countries with emerging or evolving regulatory frameworks, the promotion of voluntary environmental standards, international collaboration, and capacity building have further encouraged the integration of microbial bioremediation into industrial cleanup operations, enhancing sustainability.

Table 3. Summary of Case Studies on Microbial Bioremediation

Location/Source	Microbial Strains used	Target Contaminants	Remediation Method	Key outcomes and Remarks	Risk Assessment Measure	Regulatory Frameworks	Reference
Brown Wood Preserving Site, Florida	Indigenous microbes	Polycyclic Aromatic Hydrocarbons (PAHs)	Landfarming with aeration and soil layering	PAHs reduced from ~100-280 mg/kg to 3-23 mg/kg in 18 months successful cleanup	Hazard identification, exposure monitoring of PAHs and microbes; dose response analysis	U. S. EPA remediation standards; state environmental regulations	Federal Remediation Technologies Roundtable, (FRTR), 1995
Oil Production Facility, Pakistan	Ultra-Microbes	Petroleum hydrocarbons	*In situ* biopiling with aeration and nutrient addition	70% hydrocarbon reduction in two weeks	Microbial population surveys; contaminant concentration tracking	Local environmental protection compliance	Mukhtar, 2023
Heavy metal sites	*Saccharomyces cerevisiae, Pseudomonas,* fungi	Lead, Mercury, Cadmium, Cobalt	Bioaccumulation and biosorption with fungi and bacteria	Efficient removal of metals via microbial metabolism and sorption	Metal bioavailability assays; microbial toxicity tests	National hazardous waste management policies	Maglione et al., 2024
Subarctic soil, Alaska	Native aerobic microbes enhanced by warming	JP-4 jet fuel hydrocarbons	Bioventing with temperature control	Successful hydrocarbon degradation under cold conditions	Temperature-dependent biodegradation rate measurement; microbial viability assessment	State-level pollution control acts	EPA, 2015

continued on following page

Table 3. Continued

Location/Source	Microbial Strains used	Target Contaminants	Remediation Method	Key outcomes and Remarks	Risk Assessment Measure	Regulatory Frameworks	Reference
Petroleum contaminated soils	*Pseudomonas, Burkholderia, Rhodococcus*	Hydrocarbon chains C_5- C_{16}	Aerobic microbial degradation with nutrient amendment	Key enzymatic pathways identified; enhanced degradation rates	Assessment of enzymatic activity; microbial community profiling	Soil contamination and cleanup regulations	EPA, 2015
Textile dye contamination	Fungi (*Cladosporium, Paecilomyces*)	Industrial dyes	Mycoremediation with enzymatic degradation	Effective breakdown of complex dyes by fungal enzymes	Monitoring dye degradation products; fungal spore and metabolite safety evaluation	Waste water discharge standards	Maglione et al., 2024

11 CONCLUSIONS

This chapter highlights the critical importance of robust policies and governance frameworks for microbial bioremediation. By evaluating existing regulatory frameworks, identifying gaps, and recommending improvements, the safety and effective use of this technology can be ensured. Also, the findings from this review emphasize the need for harmonized global standards, enhanced stakeholder engagement, and integration with broader environmental protection and climate change mitigation strategies. The full potential of microbial bioremediation to address pressing environmental challenges while minimizing risks to human health and the environment could be unlocked by working together. Ultimately, this review contributes to the development of more effective and sustainable bioremediation practices, supporting a healthier, more sustainable and resilient planet.

REFERENCES

Abatenh, E., Gizaw, B., Tsegaye, Z., & Wassie, M. (2017). The role of microorganisms in bioremendiation- a review. *Open Journal of Environmental Biology*, *2*(1), 38–46. DOI: 10.17352/ojeb.000007

Ajona, M., & Vasanthi, P. (2021). Bioremediation of Petroleum contaminated soil- A review. *Materials Today: Proceedings*, *45*, 7117–7122. DOI: 10.1016/j.matpr.2021.01.949

Akintola, A. A.Akeem Akinkunmi Akintola. (2024). AI-driven monitoring systems for bioremediation: Real-time data analysis and predictive modelling. *World Journal of Advanced Research and Reviews*, *24*(1), 788–803. DOI: 10.30574/wjarr.2024.24.1.3099

Alaira, S., Padilla, C., Alcantara, E., & Aggangan, N. (2021). Social Acceptability of the Bioremediation Technology for the Rehabilitation of an Abandoned Mined-Out Area in Mogpog, Marinduque, Philippines. *Journal of Environmental Science and Management*, *24*(1), 77. Advance online publication. DOI: 10.47125/jesam/2021_1/08

Ali, A., Guo, O. D., Mahar, A., Wang, P., Shen, F., Li, R., & Zhang, Z. (2017). Mycoremediation of potentially toxic trace elements—A biological tool for soil cleanup: A review. *Pedosphere*, *27*(2), 205–222. DOI: 10.1016/S1002-0160(17)60311-4

Arregui, L., Ayala, M., Gomez-Gil, ´. X., Gutiérrez-Soto, G., Hernández-Luna, C. E., de los Santos, M. H., Levin, L., Domínguez, A. R., Martínez, D. R., Saparrat, M. C. N., Trujillo-Roldan, ´. M. A., & Valdez-Cruz, N. A. (2019). Laccases: Structure, function, and potential application in water bioremediation. *Microbial Cell Factories*, *18*(1), 200. DOI: 10.1186/s12934-019-1248-0 PMID: 31727078

Ayilara, M. S., & Babalola, O. O. (2023). Bioremediation of environmental wastes: The role of microorganisms. *Frontiers in Agronomy*, *5*, 1183691. DOI: 10.3389/fagro.2023.1183691

Azubuike, C. C., Chikere, C. B., & Okpokwasili, G. C. (2016). Bioremediation techniques—classification based on site of application: Principles, advantages, limitations and prospects. *World Journal of Microbiology & Biotechnology*, *32*(11), 180. DOI: 10.1007/s11274-016-2137-x PMID: 27638318

Bala, S., Garg, D., Thirumalesh, B. V., Sharma, M., Sridhar, K., Inbaraj, B. S., & Tripathi, M. (2022). Recent Strategies for Bioremediation of Emerging Pollutants: A Review for a Green and Sustainable Environment. *Toxics*, *10*(8), 484. DOI: 10.3390/toxics10080484 PMID: 36006163

Berillo, D., Al-Jwaid, A., & Caplin, J. (2021). Polymeric materials used for immobilisation of bacteria for the bioremediation of contaminants in water. *Polymers*, *13*(7), 1073. DOI: 10.3390/polym13071073 PMID: 33805360

Bharagava, R. N. (2020). Genetically engineered microorganisms for environmental remediation. *The Science of the Total Environment*, *717*, 137222. DOI: 10.1016/j.scitotenv.2020.137222

Bharath, Y., Singh, S. N., Keerthiga, G., & Prabhakar, R. (2019). Mycoremediation of contaminated soil in MSW sites. In Ghosh, S. K. (Ed.), *Waste Management and Resource Efficiency* (pp. 321–329). Springer Nature., DOI: 10.1007/978-981-10-7290-1_28

Bohua, L., Yuexin, W., Yakun, O., Kunlan, Z., Huan, L., & Ruipeng, L. (2023). Ethical framework on risk governance of synthetic biology. *Journal of Biosafety and Biosecurity*, *5*(2), 45–56. DOI: 10.1016/j.jobb.2023.03.002

Central, B. (2024). Microbial remediation of polluted environment by using recombinant organisms. *Biotechnology for the Environment*, *1*(1), 8. Advance online publication. DOI: 10.1186/s44314-024-00008-z

Central Pollution Control Board (CPCB). (2022). Annual Report on Solid Waste Management Rules (2020–21). Central pollution control board. Government of India. https://cpcb.nic.in/uploads/MSW/MSW_AnnualReport_2020-21.pdf

Che, S., & Men, Y. (2019). Synthetic microbial consortia for biosynthesis and biodegradation: Promises and challenges. *Journal of Industrial Microbiology & Biotechnology*, *46*(9-10), 1343–1358. DOI: 10.1007/s10295-019-02211-4 PMID: 31278525

Chemla, Y., Sweeney, C. J., Wozniak, C. A., & Voigt, C. A. (2025). Design and regulation of engineered bacteria for environmental release. *Nature Microbiology*, *10*(2), 281–300. DOI: 10.1038/s41564-024-01918-0 PMID: 39905169

Chikere, C. B., Tekere, M., & Adeleke, R. (2019). Enhanced microbial hydrocarbon biodegradation as stimulated during field-scale landfarming of crude oil-impacted soil. *Sustainable Chemistry and Pharmacy*, *14*, 100177. DOI: 10.1016/j.scp.2019.100177

Chikere, C. B., Tekere, M., & Adeleke, R. (2021). Microbial communities in field-scale oil-polluted soil remediation using 16S rRNA amplicon sequencing. *The International Journal of Environmental Studies*, *78*(3), 410–426. DOI: 10.1080/00207233.2020.1817276

Chinthala, L. K. (2020). Microbial applications in bioremediation and waste treatment. *Environmental Remediation and Resource Recovery*. 1-10. DOI: 10.2139/ssrn.5232417

Commonwealth of Australia. (2024). Gene Technology Act 2000. No. 169, 2000. Compilation No.14, Authorised Version C2024C00451. Federal Register of Legislation. https://www.legislation.gov.au/C2004A00762/latest

Convention on Biological Diversity. (2000). Cartagena Protocol on Biosafety to the Convention on Biological Diversity: Text and annexes. ISBN: 92-807-1924-6 Montreal: Secretariat of the CBD. Retrieved from https://www.cbd.int/doc/legal/cartagena-protocol-en.pdf

Dangi, A. K., Sharma, B., Hill, R. T., & Shukla, P. (2019). Bioremediation through microbes: Systems biology and metabolic engineering approach. *Critical Reviews in Biotechnology*, *39*(1), 79–98. DOI: 10.1080/07388551.2018.1500997 PMID: 30198342

Dar, A., & Naseer, A. (2022). *Hazardous Waste Management. IntechOpen*. Recent Applications of Bioremediation and Its Impact., DOI: 10.5772/intechopen.104959

Datta, S., Manna, S., & Roy, D. (2025). Attachment Mechanism of In-Situ EPS of *Bacillus megaterium* and *Lysinibacillus* sp. on Sand Surface from Soil Improvement Perspective. *Geomicrobiology Journal*, *42*(5), 406–419. DOI: 10.1080/01490451.2025.2470213

Ehis-Eriakha, C. B., Ajuzieogu, C. A., Orogun, J. O., & Akemu, S. E. (2024). Overview of petroleum hydrocarbon pollution and bioremediation technologies –. *Bioremediation Journal*, •••, 1–23. DOI: 10.1080/10889868.2024.234901

Ehis-Eriakha, C. B., Chikere, C. B., & Akemu, S. E. (2025). A comparative assessment of biostimulation potential in microbiome-based ecorestoration of polycyclic aromatic hydrocarbon polluted soil. *Brazilian Journal of Microbiology : Publication of the Brazilian Society for Microbiology*, *56*(1), 203–224. DOI: 10.1007/s42770-024-01556-y PMID: 39602070

Environment and Climate Change Canada. (2022). Canadian Environmental Protection Act, 1999: Annual Report to Parliament for April 2020 to March 2021. Environment and Climate Change Canada, Cat. No.: En81-3E-PDF ISBN: 1492-0212 EC21270. Retrieved from https://www.canada.ca/en/environment-climate-change/services/canadian-environmental-protection-act-registry/publications/annual-report-2020-2021.html

European Commission. (2020). Improving Financial Security in the Context of the Environmental Liability Directive (ELD). Brussels: European Commission, Directorate-General for Environment. Retrieved from https://www.impel.eu/contents/files/eu.pdf

Federal Remediation Technologies Roundtable (US). (1995). *Abstracts of remediation case studies*. Federal Remediation Technologies Roundtable.

Food and Agriculture Organization of the United Nations (FAO). (2017). FAO/WHO food safety assessment tools for trainers [ToolKit component]. Food and Agriculture Organization of the United Nations. Retrieved from https://www.fao.org/food-safety/scientific-advice/crosscutting-and-emerging-issues/biotechnology/en/

Ghosal, D., Ghosh, S., Dutta, T. K., & Ahn, Y. (2016). Current state of knowledge in microbial degradation of polycyclic aromatic hydrocarbons (PAHs): A Review. *Frontiers in Microbiology*, *7*, 1369. DOI: 10.3389/fmicb.2016.01369 PMID: 27630626

Giri, B. S., Geed, S., Vikrant, K., Lee, S. S., Kim, K. H., Kailasa, S. K., Vithanage, M., Chaturvedi, P., Rai, B. N., & Singh, R. S. (2021). Progress in bioremediation of pesticide residues in the environment. *Environmental Engineering Research*, *26*(6), 200446. DOI: 10.4491/eer.2020.446

Internal Maritime Organization (IMO). (2018). Initial IMO strategy on reduction of GHG emissions from ships. Retrieved from https://www.imo.org/en/ourwork/environment/pages/imo-strategy-on-reduction-of-ghg-emmissions-from-ships.aspx

Ivanov, V., Stabnikov, V., Stabnikova, O., & Kawasaki, S. (2019). Environmental safety and biosafety in construction biotechnology. *World Journal of Microbiology & Biotechnology*, *35*(2), 26. DOI: 10.1007/s11274-019-2598-9 PMID: 30666430

Jaiswal, S., & Shukla, P. (2020). Alternative strategies for microbial remediation of pollutants via synthetic biology. *Frontiers in Microbiology*, *11*, 808. DOI: 10.3389/fmicb.2020.00808 PMID: 32508759

Kakde, P., & Sharma, J. (2024). Microbial Bioremediation of Petroleum Contaminated Soil: Structural Complexity, Degradation Dynamics and Advanced Remediation Techniques. *Journal of Pure & Applied Microbiology*, *18*(4), 2244–2261. DOI: 10.22207/JPAM.18.4.28

Kim, D., & Lee, J. W. (2020). Genetic biocontainment systems for the safe use of engineered microorganisms. *Biotechnology and Bioprocess Engineering; BBE*, *25*(6), 974–984. DOI: 10.1007/s12257-020-0070-1 PMID: 33437151

Krzmarzick, M. J., Taylor, D. K., Fu, X., & McCutchan, A. L. (2018). Diversity and niche of archaea in bioremediation. *Archaea : an International Microbiological Journal, 319*, 1–17. Advance online publication. DOI: 10.1155/2018/3194108 PMID: 30254509

Kuppan, N., Padman, M., Mahadeva, M., Srinivasan, S., & Devarajan, R. (2024). A comprehensive review of sustainable bioremediation techniques: Eco friendly solutions for waste and pollution management. *Waste Management Bulletin, 2*(3), 154–171. DOI: 10.1016/j.wmb.2024.07.005

Lahel, A., Fanta, A. B., Sergienko, N., Shakya, M., López, M. E., Behera, S. K., & Park, H. S. (2016). Effect of process parameters on the bioremediation of diesel contaminated soil by mixed microbial consortia. *International Biodeterioration & Biodegradation, 113*, 375–385. DOI: 10.1016/j.ibiod.2016.05.005

Li, L., Dong, Y., Chen, Y., Jiao, J., & Zou, X. (2022). A new method for environmental risk assessment of pollutants based on multi-dimensional risk factors. *Toxics, 10*(11), 659. DOI: 10.3390/toxics10110659 PMID: 36355950

Liu, C. Y., Tian, A. M., Sun, F., & Yuan, P. (2020). Progress of ecological risk assessment methods and applications. *Keji Guanli Yanjiu, 40*(02), 79–83.

Liu, X., Ma, Y., Jiang, Y., Guo, Y., Deng, Z., & He, X. (2025). Microbial Community Succession During Bioremediation of Petroleum-Contaminated Soils Using *Rhodococcus* sp. OS62-1 and Pseudomonas sp. P35. *Microorganisms, 14*(1), 7. DOI: 10.3390/microorganisms14010007 PMID: 41597527

Ma, K., Wang, Y., Yang, X., Wang, C., Han, Y., Huang, X., & Pan, J. (2022). Analysis of the composition of culturable airborne microorganisms in the archaeological excavation protection site of the Nanhai No. 1 ancient shipwreck. *Frontiers in Microbiology, 13*, 958914. DOI: 10.3389/fmicb.2022.958914 PMID: 36090103

Maglione, G., Zinno, P., Tropea, A., Mussagy, C. U., Dufossé, L., Giuffrida, D., & Mondello, A. (2024). Microbes' role in environmental pollution and remediation: A bioeconomy focus approach. *AIMS Microbiology, 10*(3), 723–755. DOI: 10.3934/microbiol.2024033 PMID: 39219757

Maletić, S., Beljin, J., Tamindžija, D., Grgić, M., Jazić, J. M., Isakovski, M. K., & Rončević, S. (2021). Bioremediation perspective of historically contaminated sediment with polycyclic aromatic hydrocarbons. *International Journal of Sediment Research, 36*(4), 479–488. DOI: 10.1016/j.ijsrc.2020.11.004

Martínez-García, E., Fraile, S., Algar, E., Aparicio, T., Velázquez, E., Calles, B., Tas, H., Blázquez, B., Martín, B., Prieto, C., Sánchez-Sampedro, L., Nørholm, M. H. H., Volke, D. C., Wirth, N. T., Dvořák, P., Alejaldre, L., Grozinger, L., Crowther, M., Goñi-Moreno, A., & de Lorenzo, V. (2023). SEVA 4.0: An update of the Standard European Vector Architecture database for advanced analysis and programming of bacterial phenotypes. *Nucleic Acids Research*, *51*(D1), D1558–D1567. DOI: 10.1093/nar/gkac1059 PMID: 36420904

Masindi, V., Osman, M. S., & Tekere, M. (2021). *Water Pollution and Remediation: Heavy Metals. Springer.* Mechanisms and Approaches for the Removal of Heavy Metals from Acid Mine Drainage and Other Industrial Effluents., DOI: 10.1007/978-3-030-52421-0_15

Meena, P. M., Aggarwal, R. K., Meena, R., & Rathore, M. S. (2025). Soil Contamination, Risk Assessment, and Remediation. In Ivascu, L., Pislaru, M., & Alexa, L. (Eds.), *The Future of Risk Management* (pp. 234–264). IntechOpen., DOI: 10.5772/intechopen.1005045

Ministry of Ecology and Environment of China. (2020). Biosafety management regulations [Press release/Regulation summary]. Retrieved from http://english.mee.gov.cn

Ministry of Environment. Forest and Climate Change (MoEFCC). (2016). The Harzadous and other wastes (Management and Transboundary Movement) Rules, 2016. Government of India. Retrieved from https://upload.indiacode.nic.in/showfile?actid=AC_RJ_83-1096-00001_00001_1563872109827&filename=harzadous_and_other_wastes_%28management_and_transboundary_movement%29_rules%2C_2016.pdf&type=rule

Mukhtar, Z. (2023). Environmental pollution and regulatory and non-regulatory environmental responsibility (reviewing Pakistan environmental protection act). *American Journal of Industrial and Business Management*, *13*(6), 443–456. DOI: 10.4236/ajibm.2023.136028

Nikel, P. I., & de Lorenzo, V. (2021). Metabolic engineering for large-scale environmental bioremediation. *Metabolic Engineering*, *13*, 859–890. DOI: 10.1002/9783527823468.ch22

Nombamba, A., Fadiji, A. E., & Babalola, O. O. (2024). Exploring the role of endophytic fungi in the amelioration of drought stress in plants. *Plant Protection Science*, *60*(3), 213–228. DOI: 10.17221/25/2023-PPS

Nordin, A. R. R., Navarro, A. R., Reyes, J. C., Maragathavalli, S., Kristanti, R. A., Wulandari, R., & Bunrith, S. (2025). Microbial Bioremediation of Petroleum Contaminated Soil: A Sustainable Approach. *Topical Aquatic and Soil Pollution*, *5*(1), 71–87. DOI: 10.53623/tasp.v5i1.683

Olawuyi, D. S., & Tubodenyefa, Z. (2018). *Review of the environmental guidelines and standards for the petroleum industry in Nigeria (EGASPIN)*. OGEES Institute.

Omotuyi, O. Y. (2023). A Critical Assessment of the Regulatory Framework for Oil and Gas Decommissioning in Nigeria. *Journal of Sustainable Developmental Law & Policy*, *14*(1), 140–163. DOI: 10.4314/jsdlp.v14i1.7s

Organisation for Economic Co-operation and Development (OECD). (2020). The Mutual Acceptance of Data (MAD) system-harmonised test methods and data quality requirements. Retrieved from https://www.oecd.org/en/topics/sub-issues/testing-of-chemicals/mutual-acceptance-of-data-system.html

Phale, P. S., Mohapatra, B., Malhotra, H., & Shah, B. A. (2022). Eco-physiological portrait of a novel *Pseudomonas* sp. CSV86: An ideal host/candidate for metabolic engineering and bioremediation. *Environmental Microbiology*, *24*(6), 2797–2816. DOI: 10.1111/1462-2920.15694 PMID: 34347343

Rezaei, Z., Amoozegar, M. A., & Moghimi, H. (2025). Innovative approaches in bioremediation: The role of halophilic microorganisms in mitigating hydrocarbons, toxic metals, and microplastics in hypersaline environments. *Microbial Cell Factories*, *24*(1), 184. DOI: 10.1186/s12934-025-02817-7 PMID: 40804729

Richmond, J. Y., & McKinney, R. W. (2009). *Biosafety in microbiological and biomedical laboratories*. US Government Printing Office.

Romero, M., Gallego, D., Blaz, J., Lechuga, A., Martínez, J. F., Barajas, H. R., Hayano-Kanashiro, C., Peimbert, M., Cruz-Ortega, R., & Molina-Freaner, F. E. (2019). Rhizosphere metagenomics of mine tailings colonizing plants: Assembling and selecting synthetic bacterial communities to enhance in situ bioremediation. *bioRxiv*., 664805. DOI: 10.1101/664805

Sanghvi, G., Thanki, A., Pandey, S., & Singh, N. K. (2020). Bioremediation of Pollutants. Elsevier, Amsterdam, The Netherlands: *Engineered bacteria for bioremediation*, pp. 359–374. DOI: 10.1016/B978-0-12-819025-8.00017-X

Sharma, I. (2020). Bioremediation Techniques for Polluted Environment: Concept, Advantages, Limitations, and Prospects. In Trace Metals in the Environment-New Approaches and Recent Advances. *IntechOpen*. DOI: DOI: 10.5772/intechopen.90453

Sharma, M., Sharma, S., Mazumder, S., & Negi, R. K. (2022). Application of "OMICs" Approaches in Bioremediation. In Tripathi, M., & Singh, D. S. (Eds.), *Bioremediation: Challenges and Advancements* (Vol. 69, pp. 191–223). Bentham Science Publisher. DOI: 10.2174/9789815036039122010010

Skaar, I., Asare, N. Y. O., Klein, J., Okoli, A. S., Ruus, A., Gjøen, T., & Ytrehus, B. (2016). *Health and environmental risk evaluation of microorganisms used in bioremediation. Scientific Opinion of the Panel on Microbial Ecology of the Norwegian Scientific Committee for Food Safety. (VKM Report 2016: 42)*. Norwegian Scientific Committee for Food Safety.

Smith, K., Fearnley, C. J., Dixon, D., Bird, D. K., & Kelman, I. (2023). *Environmental Hazards: Assessing Risk and Reducing Disaster* (7th ed.). Routledge., DOI: 10.4324/9781351261647

Song, H. G., & Bartha, R. (1990). Effects of jet fuel spills on the microbial community of soil. *Applied and Environmental Microbiology, 56*(3), 646–651. DOI: 10.1128/aem.56.3.646-651.1990 PMID: 16348138

Springer, A. (1993). Impact of present and future regulations on bioremediation. *Bioremediation Journal, 1*(1), 1–12. DOI: 10.1007/BF01575586

Techtmann, S. M., & Hazen, T. C. (2016). Metagenomics applications in environmental monitoring and bioremediation. *Journal of Industrial Microbiology & Biotechnology, 43*(10), 1345–1354. DOI: 10.1007/s10295-016-1809-8 PMID: 27558781

Tosi, S., Sfeir, C., Carnesecchi, E., & Chauzat, M. P. (2022). Lethal, sublethal, and combined effects of pesticides on bees: A meta-analysis and new risk assessment tools. *The Science of the Total Environment, 844*, 156857. DOI: 10.1016/j.scitotenv.2022.156857 PMID: 35760183

Tripathi, M., Singh, D. N., Prasad, N., & Gaur, R. (2021). Advanced Bioremediation Strategies for Mitigation of Chromium and Organics Pollution in Tannery. In Kumar, V., Prasad, R., & Kumar, M. (Eds.), *Rhizobiont in Bioremediation of Hazardous Waste* (pp. 195–215). Springer., DOI: 10.1007/978-981-16-0602-1_10

Tyagi, M., da Fonseca, M. M. R., & de Carvalho, C. C. (2011). Bioaugmentation and biostimulation strategies to improve the effectiveness of bioremediation processes. *Biodegradation, 22*(2), 231–241. DOI: 10.1007/s10532-010-9394-4 PMID: 20680666

United Nations Environment Programme (UNEP). (2019). Environmental Governance Update-December 2019. Retrieved from https://www.unep.org/resources/report/environmental-governance-update-december-2019

United Nations Environment Programme (UNEP). (2024). Why dirty air costs us trillions every year. Retrieved from https;//www.unep.org

United States Congress. (1976). Toxic Substances Control Act (TSCA), 15 U.S.C. § 2601–2629. U.S. Government Publishing Office. https://www.govinfo.gov/content/pkg/USCODE-2011-title15/pdf/USCODE-2011-title15-chap53.pdf

U.S. Environmental Protection Agency (EPA). (2022). Overview of Biotechnology under the Toxic Substances Control Act (TSCA): Role of the office of pollution Prevention and Toxics (OPPT). U.S. EPA, office of Chemical Safety and Pollution Prevention. Retrieved from: https//www.epa.gov/regulation-biotechnology-under-tsca-and-fifra/overview-biotechnology-under-tsca

Vaksmaa, A., Guerrero-Cruz, S., Ghosh, P., Zeghal, E., Hernando-Morales, V., & Niemann, H. (2023). Role of fungi in bioremediation of emerging pollutants. *Frontiers in Marine Science, 10*, 1070905. DOI: 10.3389/fmars.2023.1070905

Viegas, C., & Gonçalves, M. (2021). Integrated treatment of pig production wastewaters using pre-treatment with biomass ash and bioremediation by microalgae. In *International Conference on Water Energy Food and Sustainability* (pp. 245-257). Cham: Springer International Publishing. DOI: 10.1007/978-3-030-75315-3_29

World Bank. (2025). Accelerating access to clean air for a livable planet (English). Washington, D. C., World Bank Group. https://documents.worldbank.org/curated/en/099032625132535486

World Health Organization. (2020). *Laboratory biosafety manual* (4th ed.). WHO Press. https://www.who.int/publications/i/item/9789240011311

World Health Organization (WHO). (2017). Internal Code of Conduct on Pesticide Mangement; Guidelines for the registration of microbial, botanical and semiochemical pest control agents for plant protection and public health uses. Retrieved from https://www.who.int/publications/i/item/WHO-HTM-NTD-WHOPES-2017.05

Wozniak, C. A., McClung, G., Gagliardi, J., & Matthews, K. (2012). Regulation of genetically engineered microorganisms under FIFRA, FFDCA and TSCA. In C. A. Wozniak, G. McClung, J. Gagliardi, and K. Matthews (Eds.), Regulation of agricultural biotechnology: The United States and Canada (pp. 57–94). Springer. DOI: 10.1007/978-94-007-2156-2_4

Compilation of References

Abatenh, E., Gizaw, B., Tsegaye, Z., & Wassie, M. (2017). The role of microorganisms in bioremendiation- a review. *Open Journal of Environmental Biology*, *2*(1), 38–46. DOI: 10.17352/ojeb.000007

Abbasi, A. F., Asim, M. N., & Dengel, A. (2025). Transitioning from wet lab to artificial intelligence: A systematic review of AI predictors in CRISPR. *Journal of Translational Medicine*, *23*(1), 153. Advance online publication. DOI: 10.1186/s12967-024-06013-w PMID: 39905452

Abbas, M. M., Husien, K. S., & Zainal, I. G. (2024). Assessment of Lead and Cadmium in Some Industrial Wastewater in Kirkuk City. *Journal of Pioneering Medical Sciences*, *13*(1), 13. DOI: 10.61091/jpms202413101

Abdulsalam, S., & Omale, A. B. (2009). Comparison of biostimulation and bioaugmentation techniques for the remediation of used motor oil contaminated soil. *Brazilian Archives of Biology and Technology*, *52*(3), 747–754. DOI: 10.1590/S1516-89132009000300027

Abraham, A. P., & Schopf, S. (2026). Bioleaching as a biotechnological tool for metal recovery: From sewage to space mining. *Frontiers in Bioengineering and Biotechnology*, *13*, 1712157. DOI: 10.3389/fbioe.2025.1712157 PMID: 41625076

Adams, G. O., Fufeyin, P. T., Okoro, S. E., & Ehinomen, I. (2015). Bioremediation, biostimulation and bioaugmention: A review. *International Journal of Environmental Bioremediation & Biodegradation*, *3*(1), 28–39. DOI: 10.12691/ijebb-3-1-5

Adams, M., Xie, J., Xie, J., Chang, Y., Guo, M., Chen, C., & Zhang, T. C. (2020). The effect of carrier addition on Anammox start-up and microbial community: A review. *Reviews in Environmental Science and Biotechnology*, *19*(2), 355–368. DOI: 10.1007/s11157-020-09530-4

Adebiyi, F. M. (2022). Air quality and management in petroleum refining industry: A review. *Environmental Toxicology and Chemistry*, *4*, 89–96.

Adekanmbi, A. O., Oyeladun, W. O., & Olaposi, A. V. (2018). Degradation of surfactant and metal-removal by bacteria from a Nigerian laundry environment. *European Journal of Biological Research, 8*(4), 243.

Adelodun, A. A., Olajire, T., Afolabi, N. O., Akinwumiju, A. S., Akinbobola, E., & Hassan, U. O. (2021). Phytoremediation potentials of *Eichhornia crassipes* for nutrients and organic pollutants from textile wastewater. *International Journal of Phytoremediation, 23*(13), 1333–1341. DOI: 10.1080/15226514.2021.1895719 PMID: 33788648

Adetunji, C. O., & Anani, O. A. (2021). *Bioaugmentation: a powerful biotechnological techniques for sustainable Ecorestoration of soil and groundwater contaminants. Microbial Rejuvenation of Polluted Environment* (Vol. 1). Springer.

Adli, M. (2018). The CRISPR tool kit for genome editing and beyond. *Nature Communications, 9*(1), 1911. DOI: 10.1038/s41467-018-04252-2 PMID: 29765029

Ahmadnezhad, Z., Vaezihir, A., Schüth, C., & Zarrini, G. (2021). Combination of zeolite barrier and bio sparging techniques to enhance efficiency of organic hydrocarbon remediation in a model of shallow groundwater. *Chemosphere, 273*, 128555. DOI: 10.1016/j.chemosphere.2020.128555 PMID: 33087257

Ahmed, J., Thakur, A., and Goyal, A. (2021). Industrial wastewater and its toxic effects. https://doi.org/DOI: 10.1039/97818391165399-00001

Ahmed, R., Obaidi, A. & Bahaa, M. 2024. Influence of Environmental Conditions on Bacterial Remediation of Oil-Contaminated Soil: A Case Study at Baiji Refinery. *Al-Rafidain Journal of Engineering Sciences*, 383-391.

Ahmed, F. J. (2022). Microbial biotechnology and Bioremediation. *Futuristic Biotechnology, 01*, 01. Advance online publication. DOI: 10.54393/fbt.v2i02.13

Ahmed, M. A. M. (2023). Evaluation of Industrial Waste in the Paints Industry Sector: Khartoum North Industrial Area. *Journal of Nature. Life and Applied Sciences, 7*(4), 56–64. DOI: 10.26389/AJSRP.R080622

Ahmed, S. F., Mofijur, M., Nuzhat, S., Chowdhury, A. T., Rafa, N., Uddin, M. A., Inayat, A., Mahlia, T. M. I., Ong, H. C., Chia, W. Y., & Show, P. L. (2021). Recent developments in physical, biological, chemical, and hybrid treatment techniques for removing emerging contaminants from wastewater. *Journal of Hazardous Materials, 416*, 125912. DOI: 10.1016/j.jhazmat.2021.125912 PMID: 34492846

Ahmed, S., Sanni, K. A., Azeez, O. S., Saka, A. A., & Oluwafemi, S. (2018). Production of 100 kg/day of zeolite a as a builder for powdered detergent from Nigerian Ahoko Kaolin using locally fabricated mini zeolite plant. [IOP Publishing.]. *IOP Conference Series. Materials Science and Engineering, 413*, 012064. DOI: 10.1088/1757-899X/413/1/012064

Ajaz, M., Shakeel, S., & Rehman, A. (2020). Microbial use for azo dye degradation—A strategy for dye bioremediation. *International Microbiology, 23*(2), 149–159. DOI: 10.1007/s10123-019-00103-2 PMID: 31741129

Ajona, M., & Vasanthi, P. (2021). Bioremediation of Petroleum contaminated soil- A review. *Materials Today: Proceedings, 45*, 7117–7122. DOI: 10.1016/j.matpr.2021.01.949

Akhtar, N., & Mannan, M. A. U. (2020). Mycoremediation: Expunging environmental pollutants. *Biotechnology Reports (Amsterdam, Netherlands), 26*, e00452. DOI: 10.1016/j.btre.2020.e00452 PMID: 32617263

Akintola, A. A.Akeem Akinkunmi Akintola. (2024). AI-driven monitoring systems for bioremediation: Real-time data analysis and predictive modelling. *World Journal of Advanced Research and Reviews, 24*(1), 788–803. DOI: 10.30574/wjarr.2024.24.1.3099

Akpasi, S. O., Anekwe, I. M. S., Tetteh, E. K., Amune, U. O., Shoyiga, H. O., Mahlangu, T. P., & Kiambi, S. L. (2023). Mycoremediation as a potentially promising technology: current status and prospects: a review. *Applied Sciences (Basel, Switzerland), 13*(8), 4978. DOI: 10.3390/app13084978

Alaira, S., Padilla, C., Alcantara, E., & Aggangan, N. (2021). Social Acceptability of the Bioremediation Technology for the Rehabilitation of an Abandoned Mined-Out Area in Mogpog, Marinduque, Philippines. *Journal of Environmental Science and Management, 24*(1), 77. Advance online publication. DOI: 10.47125/jesam/2021_1/08

Alavian, F., & Khodabakhshi, F. (2025). Integrating artificial intelligence with microbial biotechnology for sustainable environmental remediation. *Environmental Monitoring and Assessment, 197*(11), 1183. DOI: 10.1007/s10661-025-14666-3 PMID: 41055779

Al-Dhabi, N. A., & Arasu, M. V. (2022). Biosorption of hazardous waste from the municipal wastewater by marine algal biomass. *Environmental Research, 204*, 112115. DOI: 10.1016/j.envres.2021.112115 PMID: 34563525

Aleid, G. M., Alshammari, A. S., Ahmad, A. R. D., Hussain, F., Oh, S. E., Ahmad, A., & Umar, K. (2023). Advancement in microbial fuel cells technology by using waste extract as an organic substrate to produce energy with metal removal. *Processes (Basel, Switzerland)*, *11*(8), 2434. DOI: 10.3390/pr11082434

Al-Hussieny, A. A., Imran, S. G., & Jabur, Z. A. (2020). The use of local blue-green algae in the bioremediation of hydrocarbon pollutants in wastewater from oil refineries. *Plant Archives*, *20*, 797–802.

Ali, A., Guo, O. D., Mahar, A., Wang, P., Shen, F., Li, R., & Zhang, Z. (2017). Mycoremediation of potentially toxic trace elements—A biological tool for soil cleanup: A review. *Pedosphere*, *27*(2), 205–222. DOI: 10.1016/S1002-0160(17)60311-4

Alidoosti, F., Giyahchi, M., Moien, S., & Moghimi, H. (2024). Unlocking the potential of soil microbial communities for bioremediation of emerging organic contaminants: Omics-based approaches. *Microbial Cell Factories*, *23*(210), 1–21. DOI: 10.1186/s12934-024-02485-z PMID: 39054471

Ali, H. (2010). Biodegradation of Synthetic Dyes—A Review. *Water, Air, and Soil Pollution*, *213*(1–4), 251–273. DOI: 10.1007/s11270-010-0382-4

Ali, N., Hameed, A., & Ahmed, S. (2009). Physicochemical characterization and Bioremediation perspective of textile effluent, dyes and metals by indigenous Bacteria. *Journal of Hazardous Materials*, *164*(1), 322–328. DOI: 10.1016/j.jhazmat.2008.08.006 PMID: 18818017

Ali, S. S., Al-Tohamy, R., & Sun, J. (2025). *Unmasking microplastics in anaerobic digestion: Hidden threats, synergistic pollutants, and biodegradation Frontiers—A comprehensive hotspot review*. Environmental Chemistry and Ecotoxicology.

Alisi, C., Musella, R., Tasso, F., Ubaldi, C., Manzo, S., Cremisini, C., & Sprocati, A. R. (2009). Bioremediation of diesel oil in a co-contaminated soil by bioaugmentation with a microbial formula tailored with native strains selected for heavy metals resistance. *The Science of the Total Environment*, *407*(8), 3024–3032. DOI: 10.1016/j.scitotenv.2009.01.011 PMID: 19201450

Allard, S. M., Costa, M. T., Bulseco, A. N., Helfer, V., Wilkins, L. G., Hassenrück, C., & Bowman, J. (2020). Introducing the mangrove microbiome initiative: Identifying microbial research priorities and approaches to better understand, protect, and rehabilitate mangrove ecosystems. *mSystems*, *5*(5), e00658–e20. DOI: 10.1128/mSystems.00658-20 PMID: 33082281

Allison, S. D., & Martiny, J. B. H. (2008). Resistance, resilience, and redundancy in microbial communities. *Proceedings of the National Academy of Sciences of the United States of America, 105*(Suppl. 1), 11512–11519. DOI: 10.1073/pnas.0801925105 PMID: 18695234

Al-Mailem, D. M., Al-Deieg, M., Eliyas, M., & Radwan, S. S. (2017). Biostimulation of indigenous microorganisms for bioremediation of oily hypersaline microcosms from the Arabian Gulf Kuwaiti coasts. *Journal of Environmental Management, 193*, 576–583. DOI: 10.1016/j.jenvman.2017.02.054 PMID: 28262419

Al-Najar, J. A., Lutfee, T., & Alwan, N. F. (2021). The action of yeast as an adsorbent in wastewater treatment: A Brief Review. *IOP Conference Series. Earth and Environmental Science, 779*(1), 012054. Advance online publication. DOI: 10.1088/1755-1315/779/1/012054

Alori, E. T., Gabasawa, A. I., Elenwo, C. E., & Agbeyegbe, O. O. (2022). Bioremediation techniques as affected by limiting factors in soil environment. *Frontiers in Soil Science, 2*, 937186. DOI: 10.3389/fsoil.2022.937186

Al-Salihi, S. A., Ford, K. L., Mahyudin, M. M., & Bunawan, H. (2025). Synthetic biology: Pioneering the next bio revolution for a sustainable planet. *Journal of Applied Microbiology, 136*(9), lxaf202. Advance online publication. DOI: 10.1093/jambio/lxaf202 PMID: 40833632

Alsukaibi, A. K. (2022). Various approaches for the detoxification of toxic dyes in wastewater. *Processes (Basel, Switzerland), 10*(10), 1968. DOI: 10.3390/pr10101968

Altaf-Ul-Amin, M., Afendi, F. M., Kiboi, S. K., & Kanaya, S. (2014). Systems biology in the context of big data and networks. *BioMed Research International, 428570*, 1–11. Advance online publication. DOI: 10.1155/2014/428570 PMID: 24982882

Al-Tohamy, R., Ali, S. S., Li, F., Okasha, K. M., Mahmoud, Y. A. G., Elsamahy, T., Jiao, H., Fu, Y., & Sun, J. (2022). A critical review on the treatment of dye-containing wastewater: Ecotoxicological and health concerns of textile dyes and possible remediation approaches for environmental safety. *Ecotoxicology and Environmental Safety, 231*, 113160. DOI: 10.1016/j.ecoenv.2021.113160 PMID: 35026583

Al-tohamy, R., Kenawy, E., Sun, J., Ali, S. S., & Di Gregorio, S. (2020). Performance of a Newly Isolated Salt-Tolerant Yeast Strain *Sterigmatomyces halophilus* SSA-1575 for Azo Dye Decolorization and Detoxification. *Frontiers in Microbiology, 11*(June), 1–19. DOI: 10.3389/fmicb.2020.01163 PMID: 32595618

Alzain, H., Kalimugogo, V., & Hussein, K. (2023). A Review of Environmental Impact of Azo Dyes. *International Journal of Research and Review, 10*(6), 64–689. DOI: 10.52403/ijrr.20230682

Ameen, G. (2023). Industrial Chemistry: Chemicals And Technology In Everyday Life. *Shengming De Huaxue*, •••, 213.

Amobonye, A., Bhagwat, P., Singh, S., & Pillai, S. (2021). Plastic biodegradation: Frontline microbes and their enzymes. *The Science of the Total Environment, 759*, 143536. DOI: 10.1016/j.scitotenv.2020.143536 PMID: 33190901

Amran, M., Onaizi, A. M., Fediuk, R., Vatin, N. I., Muhammad Rashid, R. S., Abdelgader, H., & Ozbakkaloglu, T. (2022). Self-healing concrete as a prospective construction material: A Review. *Materials (Basel), 15*(9), 3214. DOI: 10.3390/ma15093214 PMID: 35591554

Anastasi, A., Parato, B., Spina, F., Tigini, V., Prigione, V., & Varese, G. C. (2011). Decolourisation and detoxification in the fungal treatment of textile wastewaters from dyeing processes. *New Biotechnology, 29*(1), 38–45. DOI: 10.1016/j.nbt.2011.08.006 PMID: 21911089

Andrews, K., Landeryou, T., & Nale, J. Y. N. (2024). Fleas of fleas: The potential role of bacteriophages in *Salmonella* diversity and pathogenicity. In *Microbiology Society Annual Conference 2024*.

Angeles-De Paz, G., Cubero-Cardoso, J., Pozo, C., Calvo, C., Aranda, E., & Robledo-Mahón, T. (2025). Optimizing bioaugmentation for pharmaceutical stabilization of sewage sludge: A study on short-term composting under real conditions. *Journal of Fungi (Basel, Switzerland), 11*(1), 67. DOI: 10.3390/jof11010067 PMID: 39852486

Aniyikaiye, T. E., Oluseyi, T., Odiyo, J. O., & Edokpayi, J. N. (2019). Physico-chemical analysis of wastewater discharge from selected paint industries in Lagos, Nigeria. *International Journal of Environmental Research and Public Health, 16*(7), 1235. DOI: 10.3390/ijerph16071235 PMID: 30959965

Ansori, A. N., Antonius, Y., Susilo, R. J., Hayaza, S., Kharisma, V. D., Parikesit, A. A., Zainul, R., Jakhmola, V., Saklani, T., Rebezov, M., Ullah, M. E., Maksimiuk, N., Derkho, M., & Burkov, P. (2023). Application of CRISPR-Cas9 genome editing technology in various fields: A review. *Narra J, 3*(2), e184. DOI: 10.52225/narra.v3i2.184 PMID: 38450259

Anyanwu, C. C. (2023). Establishment of Anammox wastewater treatment process in the Expanded Bed Biofilm Reactor (EBBR) with ABDite. Available at: https://core.ac.uk/download/568419276.pdf

Anyanwu, C., Akhidime, D., Dempsey, M., & Melling, L. (2024). Establishment of the Anammox wastewater treatment process in a lab-scale expanded bed biofilm reactor with ABDite®. In: Ekpo, S.C. (ed.) *The Second International Adaptive and Sustainable Science, Engineering and Technology Conference (ASSET 2023)*. Signals and Communication Technology. Springer, Cham. https://doi.org/DOI: 10.1007/978-3-031-53935-0_4

Anzalone, A. V., Randolph, P. B., Davis, J. R., Sousa, A. A., Koblan, L. W., Levy, J. M., Chen, P. J., Wilson, C., Newby, G. A., Raguram, A., & Liu, D. R. (2019). Search-and-replace genome editing without double-strand breaks or donor DNA. *Nature*, *576*(7785), 149–157. DOI: 10.1038/s41586-019-1711-4 PMID: 31634902

Aparicio, J. D., Raimondo, E. E., Gil, R. A., Benimeli, C. S., & Polti, M. A. (2018). Actinobacteria consortium as an efficient biotechnological tool for mixed polluted soil reclamation: Experimental factorial design for bioremediation process optimization. *Journal of Hazardous Materials*, *342*, 408–417. DOI: 10.1016/j.jhazmat.2017.08.041 PMID: 28854393

Appleton, E., Madsen, C., Roehner, N., & Densmore, D. (2017). Design automation in synthetic biology. *Cold Spring Harbor Perspectives in Biology*, *9*(4), a023978. DOI: 10.1101/cshperspect.a023978 PMID: 28246188

Aragaw, T. (2021). Functions of various bacteria for specific pollutants degradation and their application in wastewater treatment: A review. *International Journal of Environmental Science and Technology*, *18*(7), 2063–2076. DOI: 10.1007/s13762-020-03022-2

Aragaw, T. A. (2024). A review of dye biodegradation in textile wastewater, challenges due to wastewater characteristics, and the potential of alkaliphiles. *Journal of Hazardous Materials Advances*, *16*(August), 100493. DOI: 10.1016/j.hazadv.2024.100493

Armanu, E. G., Bertoldi, S., Chrzanowski, Ł., Volf, I., Heipieper, H. J., & Eberlein, C. (2025). Benefits of Immobilized Bacteria in Bioremediation of Sites Contaminated with Toxic Organic Compounds. *Microorganisms*, *13*(1), 155. DOI: 10.3390/microorganisms13010155 PMID: 39858923

Arora, J., Ranjan, A., Chauhan, A., Biswas, R., Rajput, V. D., Sushkova, S., Mandzhieva, S., Minkina, T., & Jindal, T. (2022). Surfactant pollution, an emerging threat to ecosystem: Approaches for effective bacterial degradation. *Journal of Applied Microbiology*, *133*(3), 1229–1244. DOI: 10.1111/jam.15631 PMID: 35598183

Arora, P. K., Kumar, A., Srivastava, A., Garg, S. K., & Singh, V. P. (2022). Current bioinformatics tools for biodegradation of xenobiotic compounds. *Frontiers in Environmental Science, 10*, 980284. Advance online publication. DOI: 10.3389/fenvs.2022.980284

Arora, U., Khuntia, H. K., Chanakya, H. N., & Kapley, A. (2023). Surfactants: Combating the fate, impact, and aftermath of their release in the environment. *International Journal of Environmental Science and Technology, 20*(10), 11551–11574. DOI: 10.1007/s13762-022-04549-2

Arregui, L., Ayala, M., Gomez-Gil, ´. X., Gutiérrez-Soto, G., Hernández-Luna, C. E., de los Santos, M. H., Levin, L., Domínguez, A. R., Martínez, D. R., Saparrat, M. C. N., Trujillo-Roldan, ´. M. A., & Valdez-Cruz, N. A. (2019). Laccases: Structure, function, and potential application in water bioremediation. *Microbial Cell Factories, 18*(1), 200. DOI: 10.1186/s12934-019-1248-0 PMID: 31727078

Arsène-Ploetze, F., Bertin, P. N., & Carapito, C. (2014). Proteomic tools to decipher microbial community structure and functioning. *Environmental Science and Pollution Research International, 22*(18), 13599–13612. DOI: 10.1007/s11356-014-3898-0 PMID: 25475614

Arslan Younas, A. S., Maqbool, H., Qamar, M. S., Parveen, A., Afzal, R., & Saleem, A., and Ullah10, Q. (2024). Nanoparticle Innovations in Wastewater Treatment: A Comparative study of titanium dioxide, gold, and nano zero-valent iron nanoparticles for lead, chromium, and mercury removal. *Global Scientific and Academic Research Journal of Multidisciplinary Studies, 2*(8), 15–31.

Arya, S., & Kumar, S. (2020). Bioleaching: Urban mining option to curb the menace of E-waste challenge. *Bioengineered, 11*(1), 640–660. DOI: 10.1080/21655979.2020.1775988 PMID: 32538256

Asgher, M., Yasmeen, Q., Muhammad, H., & Iqbal, N. (2013). Enhanced decolorization of Solar brilliant red 80 textile dye by an indigenous white rot fungus *Schizophyllum commune* IBL-06. *Saudi Journal of Biological Sciences, 20*(4), 347–352. DOI: 10.1016/j.sjbs.2013.03.004 PMID: 24235871

Asim, N., Badiei, M., Torkashvand, M., Mohammad, M., Alghoul, M. A., Gasaymeh, S. S., & Sopian, K. (2021). Wastes from the petroleum industries as sustainable resource materials in construction sectors: Opportunities, limitations, and directions. *Journal of Cleaner Production, 284*, 125459. DOI: 10.1016/j.jclepro.2020.125459

Atlas, R. M., & Bartha, R. (1998). *Microbial ecology: Fundamentals and applications* (4th ed.). Benjamin/Cummings.

Austin, H. P., Allen, M. D., Donohoe, B. S., Rorrer, N. A., Kearns, F. L., Silveira, R. L., Pollard, B. C., Dominick, G., Duman, R., El Omari, K., Mykhaylyk, V., Wagner, A., Michener, W. E., Amore, A., Skaf, M. S., Crowley, M. F., Thorne, A. W., Johnson, C. W., Woodcock, H. L., & Beckham, G. T. (2018). Characterization and engineering of a plastic-degrading aromatic polyesterase. *Proceedings of the National Academy of Sciences of the United States of America, 115*(19), E4350–E4357. DOI: 10.1073/pnas.1718804115 PMID: 29666242

Ayangbenro, A. S., & Babalola, O. O. (2018). Metal(loid) bioremediation: Strategies employed by microbial polymers. *Sustainability (Basel), 10*(9), 3028. DOI: 10.3390/su10093028

Aydemir, C., & Özsoy, S. A. (2020). Environmental impact of printing inks and printing process. *Journal of Graphic Engineering and Design, 11*(2), 11–18. DOI: 10.24867/JGED-2020-2-011

Ayele, A., Suresh, A., & Benor, S. (2021). Phycoremediation of Heavy Metals, Factors Involved and Mechanisms Related to Functional Groups in the Algae Cell Surface – A Review. In Aravind, J., Kamaraj, M., Prashanthi Devi, M., & Rajakumar, S. (Eds.), *Strategies and Tools for Pollutant Mitigation* (pp. 269–289). Springer., DOI: 10.1007/978-3-030-63575-6_13

Ayilara, M. S., Adeleke, B. S., Adebajo, M. T., Akinola, S. A., Fayose, C. A., Adeyemi, U. T., Gbadegesin, L. A., Omole, R. K., Johnson, R. M., Edhemuino, M., Ogundolie, F. A., & Babalola, O. O. (2023). Remediation by enhanced natural attenuation; an environment-friendly remediation approach. *Frontiers in Environmental Science, 11*, 1182586. DOI: 10.3389/fenvs.2023.1182586

Ayilara, M. S., & Babalola, O. O. (2023). Bioremediation of environmental wastes: The role of microorganisms. *Frontiers in Agronomy, 5*, 1183691. DOI: 10.3389/fagro.2023.1183691

Ayub, A., Wani, A. K., Malik, S. M., Ayub, M., Chopra, C., Singh, R., & Malik, T. (2025). Harnessing microbes and plants for bioremediation of heavy metal contaminants: Current paradigms and future perspectives. *Environmental Challenges, 20*, 101220. DOI: 10.1016/j.envc.2025.101220

Ayub, A., Wani, A. K., Malik, S. M., Ayub, M., Singh, R., Chopra, C., & Malik, T. (2025). Green nanoscience for healthcare: Advancing biomedical innovation through eco-synthesized nanoparticle. *Biotechnology Reports (Amsterdam, Netherlands), 00913*, e00913. Advance online publication. DOI: 10.1016/j.btre.2025.e00913 PMID: 40895589

Azanaw, A., Birlie, B., Teshome, B., & Jemberie, M. (2022). Textile effluent treatment methods and eco-friendly resolution of textile wastewater. *Case Studies in Chemical and Environmental Engineering*, *6*, 100230. DOI: 10.1016/j.cscee.2022.100230

Aziz, Z. S., Jazza, S. H., Dageem, H. N., Banoon, S. R., Balboul, B. A., & Abdelzaher, M. (2024). Bacterial biodegradation of oil-contaminated soil for pollutant abatement contributing to achieve sustainable development goals: A comprehensive review. *Results in Engineering*, *22*, 102083. DOI: 10.1016/j.rineng.2024.102083

Aznar-Sánchez, J. A., García-Gómez, J. J., Velasco-Muñoz, J. F., & Carretero-Gómez, A. (2018). Mining Waste and Its Sustainable Management: Advances in Worldwide Research. *Minerals (Basel)*, *8*(7), 284. DOI: 10.3390/min8070284

Azubuike, C. C., Chikere, C. B., & Okpokwasili, G. C. (2016). Bioremediation techniques—classification based on site of application: Principles, advantages, limitations and prospects. *World Journal of Microbiology & Biotechnology*, *32*(11), 180. DOI: 10.1007/s11274-016-2137-x PMID: 27638318

Baij, L., Hermans, J., Ormsby, B., Noble, P., Iedema, P., & Keune, K. (2020). A review of solvent action on oil paint. *Heritage Science*, *8*(1), 1–23. DOI: 10.1186/s40494-020-00388-x

Bajracharya, S., ter Heijne, A., Dominguez Benetton, X., Vanbroekhoven, K., Buisman, C. J. N., Strik, D. P. B. T. B., & Pant, D. (2017). Carbon dioxide reduction by mixed and pure cultures in microbial electrosynthesis using an assembly of graphite felt and stainless steel as a cathode. *Bioresource Technology*, *195*, 14–24. DOI: 10.1016/j.biortech.2015.05.081 PMID: 26066971

Balaji, S., Gopi, K., & Muthuvelan, B. (2013). A review on production of poly β hydroxybutyrates from cyanobacteria for the production of bio plastics. *Algal Research*, *2*(3), 278–285. DOI: 10.1016/j.algal.2013.03.002

Bala, S., Garg, D., Thirumalesh, B. V., Sharma, M., Sridhar, K., Inbaraj, B. S., & Tripathi, M. (2022). Recent strategies for bioremediation of emerging pollutants: A review for a green and sustainable environment. *Toxics*, *10*(8), 484. DOI: 10.3390/toxics10080484 PMID: 36006163

Balson, T., & Felix, M. S. B. (1995). Biodegradability of non-ionic surfactants. In Karsa, D. R., & Porter, M. R. (Eds.), *Biodegradability of Surfactants* (pp. 204–230). Springer Netherlands. DOI: 10.1007/978-94-011-1348-9_7

Barrangou, R., Fremaux, C., Deveau, H., Richards, M., Boyaval, P., Moineau, S., Romero, D. A., & Horvath, P. (2007). CRISPR provides acquired resistance against viruses in prokaryotes. *Science, 315*(5819), 1709–1712. DOI: 10.1126/science.1138140 PMID: 17379808

Batista, B. D., & Singh, B. K. (2021). Realities and hopes in the application of microbial tools in agriculture. *Microbial Biotechnology, 14*(4), 1258–1268. DOI: 10.1111/1751-7915.13866 PMID: 34156754

Batstone, D. J., Keller, J., Angelidaki, I., Kalyuzhnyi, S. V., Pavlostathis, S. G., Rozzi, A., Sanders, W. T. M., Siegrist, H., & Vavilin, V. A. (2002). The IWA Anaerobic Digestion Model No 1 (ADM1). *Water Science and Technology, 45*(10), 65–73. DOI: 10.2166/wst.2002.0292 PMID: 12188579

Batstone, D. J., Puyol, D., Flores-Alsina, X., & Rodríguez, J. (2015). Mathematical modelling of anaerobic digestion processes: Applications and future needs. *Reviews in Environmental Science and Biotechnology, 14*(4), 595–613. DOI: 10.1007/s11157-015-9376-4

Bayatian, M., Pourbabaee, A. A., & Amoozegar, M. A. (2025). Revealing the composition of bacterial communities in various oil-contaminated soils and investigating their intrinsic traits in hydrocarbon degradation. *Scientific Reports, 15*(1), 22016. Advance online publication. DOI: 10.1038/s41598-025-05519-7 PMID: 40596159

Behera, H. T., Mojumdar, A., Das, S. R., Mohapatra, C., & Ray, L. (2021). Microbial Intervention in Sustainable Production of Biofuels and Other Bioenergy Products. In Mishra, B. B., Nayak, S. K., Mohapatra, S., & Samantaray, D. (Eds.), *Environmental and Agricultural Microbiology* (1st ed., pp. 361–381). Wiley., DOI: 10.1002/9781119525899.ch17

Behera, S., & Das, S. (2023). Potential and prospects of Actinobacteria in the bioremediation of environmental pollutants: Cellular mechanisms and genetic regulations. *Microbiological Research, 273*, 127399. DOI: 10.1016/j.micres.2023.127399 PMID: 37150049

Bell, E. L., Smithson, R., Kilbride, S., Foster, J., Hardy, F. J., Ramachandran, S., Tedstone, A. A., Haigh, S. J., Garforth, A. A., Day, P. J. R., Levy, C., Shaver, M. P., & Green, A. P. (2022). Directed evolution of an efficient and thermostable PET depolymerase. *Nature Catalysis, 5*(8), 673–681. DOI: 10.1038/s41929-022-00821-3

Bener, S., Atalay, S., & Ersöz, G. (2020). The hybrid process with eco-friendly materials for the treatment of the real textile industry wastewater. *Ecological Engineering, 148*(December 2019). DOI: 10.1016/j.ecoleng.2020.105789

Benner, S. A. (2005). Synthetic biology: Act natural. *Nature*, *421*(6919), 118. DOI: 10.1038/421118a PMID: 12520282

Benner, S. A., & Sismour, A. M. (2005). Synthetic biology. *Nature Reviews. Genetics*, *6*(7), 533–543. DOI: 10.1038/nrg1637 PMID: 15995697

Bereza-Malcolm, L. T., Mann, G., & Franks, A. E. (2015). Environmental sensing of heavy metals through whole cell microbial biosensors: A synthetic biology approach. *ACS Synthetic Biology*, *4*(5), 535–546. DOI: 10.1021/sb500286r PMID: 25299321

Berillo, D., Al-Jwaid, A., & Caplin, J. (2021). Polymeric materials used for immobilisation of bacteria for the bioremediation of contaminants in water. *Polymers*, *13*(7), 1073. DOI: 10.3390/polym13071073 PMID: 33805360

Bering, S., Mazur, J., Tarnowski, K., Dąbkowska, N., Janus, M., Mozia, S., & Morawski, A. W. (2018). Removal of organic pollutants and surfactants from laundry wastewater in membrane bioreactor (MBR). *Desalination and Water Treatment*, *134*, 281–288. DOI: 10.5004/dwt.2018.23207

Bernal-Jácome, L. A., Izar-Landeta, J. M., Flores-Ramírez, R., González-Salazar, L. F., & Vargas-Berrones, K. X. (2024). Nonylphenol ethoxylate degradation in detergents during shelf time, a new exposure pathway, and a perspective on their substitution. *Environmental Science and Pollution Research International*, *31*(21), 30497–30508. DOI: 10.1007/s11356-024-33260-7 PMID: 38607492

Bernstein, D. B., Sulheim, E., Almaas, E., & Segrè, D. (2021). Addressing uncertainty in genome-scale metabolic model reconstruction and analysis. *Genome Biology*, *22*(1), 64. DOI: 10.1186/s13059-021-02289-z PMID: 33602294

Bertels, F., & Rainey, P. B. (2011). Within-genome evolution of REPINs: A new family of miniature mobile DNA in bacteria. *PLOS Genetics*, *7*(6), e1002132. DOI: 10.1371/journal.pgen.1002132 PMID: 21698139

Bhandari, S., Poudel, D. K., Marahatha, R., Dawadi, S., Khadayat, K., Phuyal, S., & Parajuli, N. (2021). Microbial enzymes used in bioremediation. *Journal of Chemistry*, *2021*(1), 8849512.

Bharagava, R. N. (2020). Genetically engineered microorganisms for environmental remediation. *The Science of the Total Environment*, *717*, 137222. DOI: 10.1016/j.scitotenv.2020.137222

Bharathiraja, B., Jayamuthunagai, J., Praveenkumar, R., & Iyyappan, J. (2018). Phytoremediation Techniques for the Removal of Dye in Wastewater. *Energy. Environmental Sustainability*, •••, 243–252. DOI: 10.1007/978-981-10-7485-1_12

Bharath, Y., Singh, S. N., Keerthiga, G., & Prabhakar, R. (2019). Mycoremediation of contaminated soil in MSW sites. In Ghosh, S. K. (Ed.), *Waste Management and Resource Efficiency* (pp. 321–329). Springer Nature., DOI: 10.1007/978-981-10-7290-1_28

Bhatia, D., Sharma, N. R., Singh, J., & Kanwar, R. S. (2017). Biological methods for textile dye removal from wastewater: A review. *Critical Reviews in Environmental Science and Technology, 47*(19), 1836–1876. DOI: 10.1080/10643389.2017.1393263

Bhatt, P., Gangola, S., Bhandari, G., Zhang, W., Maithani, D., Mishra, S., & Chen, S. (2021). New insights into the degradation of synthetic pollutants in contaminated environments. *Chemosphere, 268*, 128827. DOI: 10.1016/j.chemosphere.2020.128827 PMID: 33162154

Bhavsar, R. A., & Sardesai, A. (2022). Investigation of effect of type of pigment/ extender on the stability of high pigment volume concentration water-based architectural paint. *Journal of Coatings Technology and Research, 19*(3), 919–930. DOI: 10.1007/s11998-021-00568-9

Bidu, J. M. (2025). Integrated anaerobic reactor and intermittently aerated constructed wetland for enhanced textile wastewater treatment. *Cleaner Water, 4*(November 2024). DOI: 10.1016/j.clwat.2025.100093

Bidu, J. M., Bruggen, B. Van der, Rwiza, M. J., & Njau, N. K. (2021). *Current status of textile wastewater management practices and ef fl uent characteristics in Tanzania J. M. Bidu, B. Van der Bruggen, M. J. Rwiza and K. N. Njau*. 2363–2376. DOI: 10.2166/wst.2021.133

Bidu, J. M. (2024). Sequential anaerobic–intermittently aerated treatment of textile and domestic wastewater. *Water Practice & Technology, 19*(12), 4875–4888. DOI: 10.2166/wpt.2024.283

Bilal, M., Ashraf, S. S., Barceló, D., & Iqbal, H. M. (2019). Biocatalytic degradation/ redefining "removal" fate of pharmaceutically active compounds and antibiotics in the aquatic environment. *The Science of the Total Environment, 691*, 1190–1211. DOI: 10.1016/j.scitotenv.2019.07.224 PMID: 31466201

Bizzotto, E., Fraulini, S., Zampieri, G., Orellana, E., Treu, L., & Campanaro, S. (2024). MICROPHERRET: MICRObial PHEnotypic tRait ClassifieR using machine learning techniques. *Environmental Microbiome, 19*(1), 58. DOI: 10.1186/s40793-024-00600-6 PMID: 39113074

Blaettle, K. G. (2018). The Effects of Sewage on Aquatic Ecosystems. Sciencing. Available at: https://sciencing.com/effects-sewage-aquatic-ecosystems-21773.html (Accessed 09 A. Nov. 2025).

Blánquez, P., Sarrà, M., & Vicent, T. (2008). Development of a continuous process to adapt the textile wastewater treatment by fungi to industrial conditions. *Process Biochemistry (Barking, London, England)*, *43*(1), 1–7. DOI: 10.1016/j.procbio.2007.10.002

Boadu, F., Cao, H., & Cheng, J. (2023). Combining protein sequences and structures with transformers and equivariant graph neural networks to predict protein function. *Bioinformatics (Oxford, England)*, *39*(1, Supplement_1), i318–i325. DOI: 10.1093/bioinformatics/btad208 PMID: 37387145

Bobate, S., Mahalle, S., Dafale, N. A., & Bajaj, A. (2023). Emergence of environmental antibiotic resistance: Mechanism, monitoring and management. *Environmental Advances*, *13*, 100409. DOI: 10.1016/j.envadv.2023.100409

Bodor, A., Bounedjoum, N., Vincze, G. E., Kis, Á. E., Laczi, K., Bende, G., & Rákhely, G. (2020). Challenges of unculturable bacteria: Environmental perspectives. *Reviews in Environmental Science and Biotechnology*, *19*(1), 1–22. DOI: 10.1007/s11157-020-09522-4

Bohua, L., Yuexin, W., Yakun, O., Kunlan, Z., Huan, L., & Ruipeng, L. (2023). Ethical framework on risk governance of synthetic biology. *Journal of Biosafety and Biosecurity*, *5*(2), 45–56. DOI: 10.1016/j.jobb.2023.03.002

Botchkova, E. A., Plakunov, V. K., & Nozhevnikova, A. N. (2015). Dynamics of biofilm formation on microscopic slides submerged in an Anammox bioreactor. *Microbiology*, *84*(3), 456–460. DOI: 10.1134/S0026261715030029

BOUCHEZ, PATUREAU, DABERT, JURETSCHKO, DORÉ, DELGENÈS, MOLETTA & WAGNER (2000). Ecological study of a bioaugmentation failure. *Environmental microbiology,* 2, 179-190.

Bouchez, T., Blieux, A. L., Dequiedt, S., Domaizon, I., Dufresne, A., Ferreira, S., Godon, J. J., Hellal, J., Joulian, C., Quaiser, A., Martin-Laurent, F., Mauffret, A., Monier, J. M., Peyret, P., Schmitt-Koplin, P., Sibourg, O., D'oiron, E., Bispo, A., Deportes, I., & Ranjard, L. (2016). Molecular microbiology methods for environmental diagnosis. *Environmental Chemistry Letters*, *14*(4), 423–441. DOI: 10.1007/s10311-016-0581-3

Boyd, C. E. (2019). Nitrogen. In *Water Quality: An Introduction* (pp. 269–290). Springer International Publishing.

Brazil, G., Kenefick, L., Callanan, M., Haro, A., De Lorenzo, V., Dowling, D., & O'Gara, F. (1995). Construction of a rhizosphere pseudomonad with potential to degrade polychlorinated biphenyls and detection of bph gene expression in the rhizosphere. *Applied and Environmental Microbiology*, *61*(5), 1946–1952. DOI: 10.1128/aem.61.5.1946-1952.1995 PMID: 7646029

Brenner, K., You, L., & Arnold, F. H. (2008). Engineering microbial consortia: A new frontier in synthetic biology. *Trends in Biotechnology*, *26*(9), 483–489. DOI: 10.1016/j.tibtech.2008.05.004 PMID: 18675483

Brewer, A. M., & George, D. R. (2025). Meeting the moment: Opportunities and limitations for genetically engineered microbes for bioremediation. *Journal of Science Policy & Governance*, *26*(2), 1–13. DOI: 10.38126/JSPG260201

Brouns, S. J., Jore, M. M., Lundgren, M., Westra, E. R., Slijkhuis, R. J., Snijders, A. P., Dickman, M. J., Makarova, K. S., Koonin, E. V., & Van Der Oost, J. (2008). Small CRISPR RNAs guide antiviral defense in prokaryotes. *Science*, *321*(5891), 960–964. DOI: 10.1126/science.1159689 PMID: 18703739

Brutesco, C., Prévéral, S., Escoffier, C., Descamps, E. C. T., Prudent, E., Cayron, J., Dumas, L., Ricquebourg, M., Adryanczyk-Perrier, G., De Groot, A., Garcia, D., Rodrigue, A., Pignol, D., & Ginet, N. (2017). Bacterial host and reporter gene optimization for genetically encoded whole cell biosensors. *Environmental Science and Pollution Research International*, *24*(1), 52–65. DOI: 10.1007/s11356-016-6952-2 PMID: 27234828

Budania, R., & Dangayach, S. (2023). A comprehensive review on permeable reactive barrier for the remediation of groundwater contamination. *Journal of Environmental Management*, *332*, 117343. DOI: 10.1016/j.jenvman.2023.117343 PMID: 36758361

Burns, R. G., Deforest, J. L., Marxsen, J., Sinsabaugh, R. L., Stromberger, M. E., Wallenstein, M. D., Weintraub, M. N., & Zoppini, A. (2013). Soil enzymes in a changing environment: Current knowledge and future directions. *Soil Biology & Biochemistry*, *58*, 216–234. DOI: 10.1016/j.soilbio.2012.11.009

Bu, T., Yang, R., Zhang, Y., Cai, Y., Tang, Z., Li, C., Wu, Q., & Chen, H. (2020). Improving decolorization of dyes by laccase from Bacillus licheniformis by random and site-directed mutagenesis. *PeerJ*, *8*, e10267. DOI: 10.7717/peerj.10267 PMID: 33240620

Cameron, D. E., Bashor, C. J., & Collins, J. J. (2014). A brief history of synthetic biology. *Nature Reviews. Microbiology*, *12*(5), 381–390. DOI: 10.1038/nrmicro3239 PMID: 24686414

Canton, B., Labno, A., & Endy, D. (2008). Refinement and standardization of synthetic biological parts and devices. *Nature Biotechnology*, *26*(7), 787–793. DOI: 10.1038/nbt1413 PMID: 18612302

Caplan, A. L., Parent, B., Shen, M., & Plunkett, C. (2015). No time to waste—the ethical challenges created by CRISPR: CRISPR/Cas, being an efficient, simple, and cheap technology to edit the genome of any organism, raises many ethical and regulatory issues beyond the use to manipulate human germ line cells. *EMBO Reports*, *16*(11), 1421–1426. DOI: 10.15252/embr.201541337 PMID: 26450575

Carlson, R. (2016). Estimating the biotech sector's contribution to the US economy. *Nature Biotechnology*, *34*(3), 247–255. DOI: 10.1038/nbt.3491 PMID: 26963545

Carmen, Z., & Daniela, S. (2012). *Textile organic dyes-characteristics, polluting effects and separation/elimination procedures from industrial effluents-a critical overview* (Vol. 3). IntechOpen. DOI: 10.5772/32373

Carney, A. B., Cartine, J., Jönander, C., Karlsson, M., Langlois, J., Lindström, M., Lundin, J., Melander, N., Pesqueda, A., Rahmqvist, I., Renaux, J., Roos, J., Spilsbury, F., Svalin, J., Vestlund, H., Zhao, L., Asker, N., Ašmonaitė, G., Birgersson, L., … Sturve, J. (2021). Assessing the effects of textile leachates in fish using multiple testing methods: From gene expression to behavior. *Ecotoxicology and Environmental Safety, 207*(July 2020). DOI: 10.1016/j.ecoenv.2020.111523

Carrascal-Hernández, D. C., Orozco-Beltrán, E. J., Insuasty, D., Márquez, E., & Grande-Tovar, C. D. (2025). Systematic Evaluation of Biodegradation of Azo Dyes by Microorganisms: Efficient Species, Physicochemical Factors, and Enzymatic Systems. *International Journal of Molecular Sciences*, *26*(16), 1–37. DOI: 10.3390/ijms26167973 PMID: 40869295

Carvalho, G., Fouchet, D., Danesh, G., Godeux, A.-S., Laaberki, M.-H., Pontier, D., Charpentier, X., & Venner, S. (2020). Bacterial transformation buffers environmental fluctuations through the reversible integration of mobile genetic elements. *mBio*, *11*(2), e02443-19. DOI: 10.1128/mBio.02443-19 PMID: 32127449

Casini, A., Storch, M., Baldwin, G. S., & Ellis, T. (2015). Bricks and blueprints: Methods and standards for DNA assembly. *Nature Reviews. Molecular Cell Biology*, *16*(9), 568–576. DOI: 10.1038/nrm4014 PMID: 26081612

Cavazzoli, S., Selonen, V., Rantalainen, A.-L., Sinkkonen, A., Romantschuk, M., & Squartini, A. (2022). Natural additives contribute to hydrocarbon and heavy metal co-contaminated soil remediation. *Environmental Pollution*, *307*, 119569. DOI: 10.1016/j.envpol.2022.119569 PMID: 35680061

Cavazzoli, S., Squartini, A., Sinkkonen, A., Romantschuk, M., Rantalainen, A.-L., Selonen, V., & Roslund, M. I. (2023). Nutritional additives dominance in driving the bacterial communities succession and bioremediation of hydrocarbon and heavy metal contaminated soil microcosms. *Microbiological Research*, *270*, 127343. DOI: 10.1016/j.micres.2023.127343 PMID: 36841130

Cebada, J. D. P. (2016). Mining corporations and air pollution science before the Age of Ecology. *Ecological Economics*, *123*, 77–83. DOI: 10.1016/j.ecolecon.2015.12.001

Ceballos Rodriguez-Conde, F., Zhu, S., & Dikicioglu, D. (2025). Harnessing microbial division of labor for biomanufacturing: A review of laboratory and formal modeling approaches. *Critical Reviews in Biotechnology*, *45*(6), 1–19. DOI: 10.1080/07388551.2025.2455607 PMID: 39972973

Central Pollution Control Board (CPCB). (2022). Annual Report on Solid Waste Management Rules (2020–21). Central pollution control board. Government of India. https://cpcb.nic.in/uploads/MSW/MSW_AnnualReport_2020-21.pdf

Central, B. (2024). Microbial remediation of polluted environment by using recombinant organisms. *Biotechnology for the Environment*, *1*(1), 8. Advance online publication. DOI: 10.1186/s44314-024-00008-z

Centurion, V. B., Moura, A. G. L., Delforno, T. P., Okada, D. Y., Dos Santos, V. P., Varesche, M. B. A., & Oliveira, V. M. (2018). Anaerobic co-digestion of commercial laundry wastewater and domestic sewage in a pilot-scale EGSB reactor: The influence of surfactant concentration on microbial diversity. *International Biodeterioration & Biodegradation*, *127*, 77–86. DOI: 10.1016/j.ibiod.2017.11.017

Ceretta, M. B., Nercessian, D., & Wolski, E. A. (2021). Current Trends on Role of Biological Treatment in Integrated Treatment Technologies of Textile Wastewater. *Frontiers in Microbiology*, *12*(March), 1–7. DOI: 10.3389/fmicb.2021.651025 PMID: 33841377

Chae, T. U., Choi, S. Y., Kim, J. W., Ko, Y. S., & Lee, S. Y. (2017). Recent advances in systems metabolic engineering tools and strategies. *Current Opinion in Biotechnology*, *47*, 67–82. DOI: 10.1016/j.copbio.2017.06.007 PMID: 28675826

Chakraborty, R., Wu, C. H., & Hazen, T. C. (2012). Systems biology approach to bioremediation. *Current Opinion in Biotechnology*, *23*(3), 483–490. DOI: 10.1016/j.copbio.2012.01.015 PMID: 22342400

Chakraborty, S., Ghosh, S., Banerjee, S., Kumar, S., & Bhattacharyya, P. (2024). Elucidating the synergistic effect of acidity and metalloid poisoning on the microbiome through metagenomics and machine learning approaches. *Environmental Research, 243*, 117885. DOI: 10.1016/j.envres.2023.117885 PMID: 38072100

Chakraborty, S., Talukdar, A., Dey, S., & Bhattacharya, S. (2025). Role of fungi, bacteria and microalgae in bioremediation of emerging pollutants with special reference to pesticides, heavy metals and pharmaceuticals. *Discover Environment, 3*(1), 91. DOI: 10.1007/s44274-025-00217-7

Chan, C. T. Y., Lee, J. W., Cameron, D. E., Bashor, C. J., & Collins, J. J. (2016). 'Deadman' and 'Passcode' microbial kill switches for bacterial containment. *Nature Chemical Biology, 12*(2), 82–86. DOI: 10.1038/nchembio.1979 PMID: 26641934

Chandanshive, V. V., Kadam, S. K., Khandare, R. V., Kurade, M. B., Jeon, B. H., Jadhav, J. P., & Govindwar, S. P. (2018). In situ phytoremediation of dyes from textile wastewater using garden ornamental plants, effect on soil quality and plant growth. *Chemosphere, 210*, 968–976. DOI: 10.1016/j.chemosphere.2018.07.064 PMID: 30208557

Chandra, S., Sharma, R., Singh, K., & Sharma, A. (2013). Application of bioremediation technology in the environment contaminated with petroleum hydrocarbon. *Annals of Microbiology, 63*(2), 417–431. DOI: 10.1007/s13213-012-0543-3

Chang, C. Y., Vila, J. C., Bender, M., Li, R., Mankowski, M. C., Bassette, M., Borden, J., Golfier, S., Sanchez, P. G. L., Waymack, R., Zhu, X., Diaz-Colunga, J., Estrela, S., Rebolleda-Gomez, M., & Sanchez, A. (2021a). Engineering complex communities by directed evolution. *Nature Ecology & Evolution, 5*(7), 1011–1023. DOI: 10.1038/s41559-021-01457-5 PMID: 33986540

Chang, Y., Guo, X., Chen, Y., & Wang, J. (2021b). Machine learning approaches for microbial community analysis in environmental biotechnology. *Biotechnology Advances, 47*, 107696.

Chao, H., Cai, A., Heimbuger, B., Wu, Y., Zhao, D., Sun, M., & Hu, F. (2024). Keystone taxa enhance the stability of soil bacterial communities and multifunctionality under steelworks disturbance. *Journal of Environmental Management, 356*, 120664. DOI: 10.1016/j.jenvman.2024.120664 PMID: 38508006

Chaudhari, A. U., Tapase, S. R., Markad, V. L., & Kodam, K. M. (2013). Simultaneous decolorization of reactive Orange M2R dye and reduction of chromate by Lysinibacillus sp. KMK-A. *Journal of Hazardous Materials, 262*, 580–588. https://doi.org/dx.doi.org/10.1016/j.jhazmat.2013.09.006. DOI: 10.1016/j.jhazmat.2013.09.006 PMID: 24095998

Chaudhary, D. K., Bajagain, R., Jeong, S.-W., & Kim, J. (2021). Effect of consortium bioaugmentation and biostimulation on remediation efficiency and bacterial diversity of diesel-contaminated aged soil. *World Journal of Microbiology & Biotechnology, 37*(3), 46. DOI: 10.1007/s11274-021-02999-3 PMID: 33554294

Chaumeil, A., Mussig, A. J., Hugenholtz, P., & Parks, D. H. (2022). GTDB-Tk v2: Memory friendly classification with the genome taxonomy database. *Bioinformatics (Oxford, England), 38*(23), 5315–5316. DOI: 10.1093/bioinformatics/btac672 PMID: 36218463

Chavan, A. R., and Kaur, G. (2024). Keeping Food Safe: Tackling The Health Risks Of Phthalates And Azo Dyes.

Chemla, Y., Sweeney, C. J., Wozniak, C. A., & Voigt, C. A. (2025). Design and regulation of engineered bacteria for environmental release. *Nature Microbiology, 10*(2), 281–300. DOI: 10.1038/s41564-024-01918-0 PMID: 39905169

Chen, C., Li, K., Lin, X., Li, W., & Xu, X. (2025). Engineering natural microbial communities: Harnessing synthetic communities for bioremediation. *Current Opinion in Microbiology, 87*, 102652. DOI: 10.1016/j.mib.2025.102652 PMID: 40768900

Chen, J., Liu, Y., & Mahadevan, R. (2023). Genetic engineering of *Acidithiobacillus ferridurans* using CRISPR systems to mitigate toxic release in biomining. *Environmental Science & Technology, 57*(33), 12315–12324. DOI: 10.1021/acs.est.3c02492 PMID: 37556825

Chen, M., Feng, S., Lv, H., Wang, Z., Zeng, Y., Shao, C., Lin, W., & Zhang, Z. (2024). OsCIPK2 mediated rice root microorganisms and metabolites to improve plant nitrogen uptake. *BMC Plant Biology, 24*(285), 1–11. DOI: 10.1186/s12870-024-04982-0 PMID: 38627617

Chen, W., Sun, J., Ji, R., Min, J., Wang, L., Zhang, J., Qiao, H., & Cheng, S. (2024). Crude Oil Biodegradation by a Biosurfactant-Producing Bacterial Consortium in High-Salinity Soil. *Journal of Marine Science and Engineering, 12*(11), 2033. DOI: 10.3390/jmse12112033

Chen, W., Zuo, Y., Hou, Z., Wang, B., Xiong, S., Ding, X., Peng, B., Zhou, K., Li, J., Liu, R., Ding, G., Wei, Y., & Xu, T. (2023). Effect of Rhodococcus bioaugmentation and biostimulation on dibenzothiophene biodegradation and bacterial community interaction in petroleum-contaminated soils. *Frontiers in Environmental Science, 11*, 1270599. DOI: 10.3389/fenvs.2023.1270599

Chen, Y.-M., Lin, T.-F., Huang, C., Lin, J.-C., & Hsieh, F.-M. (2007). Degradation of phenol and TCE using suspended and chitosan-bead immobilized Pseudomonas putida. *Journal of Hazardous Materials, 148*(3), 660–670. DOI: 10.1016/j.jhazmat.2007.03.030 PMID: 17434262

Chen, Z., Pan, X., Chen, H., Guan, X., & Lin, Z. (2016). Biomineralization of Pb (II) into Pb-hydroxyapatite induced by *Bacillus cereus* 12-2 isolated from Lead–Zinc mine tailings. *Journal of Hazardous Materials, 301*, 531–537. DOI: 10.1016/j.jhazmat.2015.09.023 PMID: 26468754

Cherubini, F. (2010). The biorefinery concept: Using biomass instead of oil for producing energy and chemicals. *Energy Conversion and Management, 51*(7), 1412–1421. DOI: 10.1016/j.enconman.2010.01.015

Che, S., & Men, Y. (2019). Synthetic microbial consortia for biosynthesis and biodegradation: Promises and challenges. *Journal of Industrial Microbiology & Biotechnology, 46*(9-10), 1343–1358. DOI: 10.1007/s10295-019-02211-4 PMID: 31278525

Chettri, D., Verma, A. K., & Verma, A. K. (2024). Bioaugmentation: An approach to biological treatment of pollutants. *Biodegradation, 35*(2), 117–135. DOI: 10.1007/s10532-023-10050-5 PMID: 37684525

Chia, B. S., Seah, Y. F. S., Wang, B., Shen, K., Srivastava, D., & Chew, W. L. (2025) Engineering a New Generation of Gene Editors: Integrating Synthetic Biology and AI Innovations. *ACS Synthetic Biology*. 14 (3), 636–647. DOI: 10.1021/acssynbio.4c00686

Chikere, C. B., Tekere, M., & Adeleke, R. (2019). Enhanced microbial hydrocarbon biodegradation as stimulated during field-scale landfarming of crude oil-impacted soil. *Sustainable Chemistry and Pharmacy, 14*, 100177. DOI: 10.1016/j.scp.2019.100177

Chikere, C. B., Tekere, M., & Adeleke, R. (2021). Microbial communities in field-scale oil-polluted soil remediation using 16S rRNA amplicon sequencing. *The International Journal of Environmental Studies, 78*(3), 410–426. DOI: 10.1080/00207233.2020.1817276

Chinthala, L. K. (2020). Microbial applications in bioremediation and waste treatment. *Environmental Remediation and Resource Recovery*. 1-10. DOI: 10.2139/ssrn.5232417

Choppala, G., Saifullah, , Bolan, N., Bibi, S., Iqbal, M., Rengel, Z., Kunhikrishnan, A., Ashwath, N., & Ok, Y. S. (2014). Cellular mechanisms in higher plants governing tolerance to cadmium toxicity. *Critical Reviews in Plant Sciences, 33*(5), 374–391. DOI: 10.1080/07352689.2014.903747

Choubisa, V. (2024). CRISPR-Cas9 off-target predictions using CNN and double CNN: Comparative analysis. *International Journal of Science and Research Archive, 12*(1), 1074–1080. DOI: 10.30574/ijsra.2024.12.1.0927

Chowdhury, U. K., Rahman, M. M., Samanta, G., Biswas, B. K., Basu, G. K., Chanda, C. R., & Chakraborti, D. (2003). In Naidu, R., Gupta, V. V. S. R., Rogers, S., Kookana, R. S., Bolan, N. S., & Adriano, D. (Eds.), *Groundwater arsenic contamination in West Bengal-India and Bangladesh: case study on bioavailibility of geogenic arsenic. Bioavailibility, Toxicity and Risk Relationships in Ecosystems* (pp. 291–329). Science Publishers.

Christie, R., & Abel, A. (2021). Organic pigments: General principles. *Physical Sciences Reviews, 6*(12), 807–834. DOI: 10.1515/psr-2020-0187

Church, G. M., Elowitz, M. B., Smolke, C. D., Voigt, C. A., & Weiss, R. (2014). Realizing the potential of synthetic biology. *Nature Reviews. Molecular Cell Biology, 15*(4), 289–294. DOI: 10.1038/nrm3767 PMID: 24622617

Commonwealth of Australia. (2024). Gene Technology Act 2000. No. 169, 2000. Compilation No.14, Authorised Version C2024C00451. Federal Register of Legislation. https://www.legislation.gov.au/C2004A00762/latest

Compernolle, T., Van Passel, S., & Lebbe, L. (2013). Bioremediation: How to deal with removal efficiency uncertainty? An economic application. *Journal of Environmental Management, 127*, 77–85. DOI: 10.1016/j.jenvman.2013.04.016 PMID: 23685120

Contreras-Salgado, E. A., Sánchez-Morán, A. G., Rodríguez-Preciado, S. Y., Sifuentes-Franco, S., Rodríguez-Rodríguez, R., Macías-Barragán, J., & Díaz-Zaragoza, M. (2024). Multifaceted applications of synthetic microbial communities: Advances in biomedicine, bioremediation, and industry. *Microbiology Research, 15*(3), 1709–1727. DOI: 10.3390/microbiolres15030113

Convention on Biological Diversity. (2000). Cartagena Protocol on Biosafety to the Convention on Biological Diversity: Text and annexes. ISBN: 92-807-1924-6 Montreal: Secretariat of the CBD. Retrieved from https://www.cbd.int/doc/legal/cartagena-protocol-en.pdf

Corominas, L., Garrido-Baserba, M., Villez, K., Olsson, G., Cortés, U., & Poch, M. (2018). Transforming data into knowledge for improved wastewater treatment operation: A critical review of techniques. *Environmental Modelling & Software, 106*, 89–103. DOI: 10.1016/j.envsoft.2017.11.023

Corrochano, N., de Tuesta, J. D., de Mora, A., Pariente, M. I., Segura, Y., Molina, R., & Martínez, F. (2025). Valorization of refinery sludges by hydrothermal carbonization towards circular economy. *Waste Management, 204*, 114949. DOI: 10.1016/j.wasman.2025.114949 PMID: 40527174

Cosgrove, L., Mcgeechan, P., Handley, P., & Robson, G. (2010). Effect of biostimulation and bioaugmentation on degradation of polyurethane buried in soil. *Applied and Environmental Microbiology, 76*(3), 810–819. DOI: 10.1128/AEM.00534-09 PMID: 19948849

Costa, C. S. D., Bertagnolli, C., Boos, A., da Silva, M. G. C., & Vieira, M. G. A. (2020). Application of a dealginated seaweed derivative for the simultaneous metal ions removal from real and synthetic effluents. *Journal of Water Process Engineering, 37*, 101546. DOI: 10.1016/j.jwpe.2020.101546

Coulon, F., Pelletier, E., Gourhant, L., & Delille, D. (2005). Effects of nutrient and temperature on degradation of petroleum hydrocarbons in contaminated sub-Antarctic soil. *Chemosphere, 58*(10), 1439–1448. DOI: 10.1016/j.chemosphere.2004.10.007 PMID: 15686763

Crouzet, M., Le Senechal, C., Brozel, V. S., Costaglioli, P., Barthe, C., Bonneu, M., Garbay, B., & Vilain, S. (2014). Exploring early steps in biofilm formation: Set-up of an experimental system for molecular studies. *BMC Microbiology, 14*(1), 253. DOI: 10.1186/s12866-014-0253-z PMID: 25266973

Cruz, D. A., & Kemp, M. L. (2022). Hybrid computational modeling methods for systems biology. *Progress in Biomedical Engineering (Bristol, England), 4*(1), 1–19. DOI: 10.1088/2516-1091/ac2cdf PMID: 41287231

Cuadros-Pedraza, A., Blanco-Zuñiga, C. R., & Rojas-Arias, N. (2021). Efficiency of biostimulation and bioaugmentation processes in the treatment of soils contaminated with hydrocarbons. *Journal of Applied Research and Technology, 19*(5), 540–547. DOI: 10.22201/icat.24486736e.2021.19.5.1589

Culp, E. J., & Goodman, A. L. (2023). Cross-feeding in the gut microbiome: Ecology and mechanisms. [External Link]. *Cell Host & Microbe, 31*(4), 485–499. DOI: 10.1016/j.chom.2023.03.016 PMID: 37054671

Cuperlovic-Culf, M. (2018). Machine learning methods for analysis of metabolic data and metabolic pathway modeling. *Metabolites*, *8*(1), 4. DOI: 10.3390/metabo8010004 PMID: 29324649

Curiel-Alegre, S., Velasco-Arroyo, B., Rumbo, C., Khan, A. H. A., Tamayo-Ramos, J. A., Rad, C., Gallego, J. L. R., & Barros, R. (2022). Evaluation of biostimulation, bioaugmentation, and organic amendments application on the bioremediation of recalcitrant hydrocarbons of soil. *Chemosphere*, *307*, 135638. DOI: 10.1016/j.chemosphere.2022.135638 PMID: 35817192

Cycoń, M., Mrozik, A., & Piotrowska-Seget, Z. (2019). Antibiotics in the soil environment—Degradation and their impact on microbial activity and diversity. *Frontiers in Microbiology*, *10*, 338. DOI: 10.3389/fmicb.2019.00338 PMID: 30906284

D'Souza, G., Shitut, S., Preussger, D., Yousif, G., Waschina, S., & Kost, C. (2018). Ecology and evolution of metabolic cross-feeding interactions in bacteria. *Natural Product Reports*, *35*(5), 455–488. DOI: 10.1039/C8NP00009C PMID: 29799048

Da Silva, L. C., Aoki, J. I., & Floeter-Winter, L. M. (2022). Finding correlations between mRNA and protein levels in Leishmania development: Is there a discrepancy? *Frontiers in Cellular and Infection Microbiology*, *12*, 852902. Advance online publication. DOI: 10.3389/fcimb.2022.852902 PMID: 35903202

Dada, E. G., Oyewola, D. O., Joseph, S. B., Emebo, O., & Oluwagbemi, O. O. (2022). Ensemble Machine Learning for Monkeypox Transmission Time Series Forecasting. *Applied Sciences (Basel, Switzerland)*, *12*(23), 12128. DOI: 10.3390/app122312128

Dahabieh, M. S., Thevelein, J. M., & Gibson, B. (2020). Multimodal microorganism development: Integrating top-down biological engineering with bottom-up rational design. *Trends in Biotechnology*, *38*(3), 241–253. DOI: 10.1016/j.tibtech.2019.09.006 PMID: 31653446

Dai, F., De Prá, M. C., Vanotti, M. B., Gilmore, K. R., & Cumbie, W. E. (2019). Microbial characteristics of nitrifiers, denitrifiers, and anammox bacteria on different support media to treat space mission wastewater. *Journal of Environmental Management*, *232*, 943–951. DOI: 10.1016/j.jenvman.2018.12.008 PMID: 33395762

Daims, H., Lebedeva, E. V., Pjevac, P., Han, P., Herbold, C., Albertsen, M., Jehmlich, N., Palatinszky, M., Vierheilig, J., Bulaev, A., Kirkegaard, R. H., von Bergen, M., Rattei, T., Bendinger, B., Nielsen, P. H., & Wagner, M. (2015). Complete nitrification by *Nitrospira* bacteria. *Nature*, *528*(7583), 504–509. DOI: 10.1038/nature16461 PMID: 26610024

Dang, H., & Chen, C. A. (2017). Ecological Energetic Perspectives on Responses of Nitrogen-Transforming Chemolithoautotrophic Microbiota to Changes in the Marine Environment. *Frontiers in Microbiology*, *8*, 1246. DOI: 10.3389/fmicb.2017.01246 PMID: 28769878

Dangi, A. K., Sharma, B., Hill, R. T., & Shukla, P. (2019). Bioremediation through microbes: Systems biology and metabolic engineering approach. *Critical Reviews in Biotechnology*, *39*(1), 79–98. DOI: 10.1080/07388551.2018.1500997 PMID: 30198342

Danouche, M., El Arroussi, H., Bahafid, W., & El Ghachtouli, N. (2021). An overview of the biosorption mechanism for the bioremediation of synthetic dyes using yeast cells. *Environmental Technology Reviews*, *10*(1), 58–76. DOI: 10.1080/21622515.2020.1869839

Dar, A., & Naseer, A. (2022). *Hazardous Waste Management. IntechOpen.* Recent Applications of Bioremediation and Its Impact., DOI: 10.5772/intechopen.104959

Dara, M., Dianatpour, M., Azarpira, N., & Omidifar, N. (2024). Convergence of CRISPR and artificial intelligence: A paradigm shift in biotechnology. *Human Gene*, *41*, 201297. DOI: 10.1016/j.humgen.2024.201297

Dash, D. M., & Osborne, W. J. (2023). A systematic review on the implementation of advanced and evolutionary biotechnological tools for efficient bioremediation of organophosphorus pesticides. *Chemosphere*, *313*, 137506. DOI: 10.1016/j.chemosphere.2022.137506 PMID: 36526134

Das, S., Cherwoo, L., & Singh, R. (2023). Decoding dye degradation: Microbial remediation of textile industry effluents. *Biotechnology Notes*, *4*(August), 64–76. DOI: 10.1016/j.biotno.2023.10.001 PMID: 39416919

Datta, S., Manna, S., & Roy, D. (2025). Attachment Mechanism of In-Situ EPS of *Bacillus megaterium* and *Lysinibacillus* sp. on Sand Surface from Soil Improvement Perspective. *Geomicrobiology Journal*, *42*(5), 406–419. DOI: 10.1080/01490451.2025.2470213

Daughton, C. G., & Ternes, T. A. (1999). Pharmaceuticals and personal care products in the environment: Agents of subtle change? *Environmental Health Perspectives*, *107*(Suppl 6), 907–938. DOI: 10.1289/ehp.99107s6907 PMID: 10592150

Dave, S., & Das, J. (2021). Role of microbial enzymes for biodegradation and bioremediation of environmental pollutants: challenges and future prospects. Bioremediation for environmental sustainability, 325-346.

Davoodi, S. M., Miri, S., Taheran, M., Brar, S. K., Galvez-Cloutier, R., & Martel, R. (2020). Bioremediation of unconventional oil contaminated ecosystems under natural and assisted conditions: A review. *Environmental Science & Technology, 54*(4), 2054–2067. DOI: 10.1021/acs.est.9b00906 PMID: 31904944

Dawkar, V. V., Jadhav, Æ. U. U., Ghodake, G. S., & Govindwar, Æ. S. P. (2009). Effect of inducers on the decolorization and biodegradation of textile azo dye Navy blue 2GL by *Bacillus sp.* VUS. *Biodegradation, 20*(6), 777–787. DOI: 10.1007/s10532-009-9266-y PMID: 19468842

De Almeida, D. G., Soares Da Silva, C. F., Luna, J. M., Rufino, R. D., Santos, V. A., Banat, I. M., & Sarubbo, L. A. (2016). Biosurfactants: Promising Molecules for Petroleum Biotechnology Advances. *Frontiers in Microbiology, 7*, 1718. DOI: 10.3389/fmicb.2016.01718 PMID: 27843439

De Pourcq, K., Ayora, C., García-Gutiérrez, M., Missana, T., & Carrera, J. (2015). A clay permeable reactive barrier to remove Cs-137 from groundwater: Column experiments. *Journal of Environmental Radioactivity, 149*, 36–42. DOI: 10.1016/j.jenvrad.2015.06.029 PMID: 26197347

De Roy, K., Marzorati, M., Van Den Abbeele, P., Van De Wiele, T., & Boon, N. (2013). Synthetic microbial ecosystems: An exciting tool to understand and apply microbial communities. *Environmental Microbiology, 16*(6), 1472–1481. DOI: 10.1111/1462-2920.12343 PMID: 24274586

de Santana da Silva, J., Heck, M. C., Buzo, M. G., Almeida, I. V., & Vicentini, V. E. P. (2018). Evaluation of textile laundry effluents and their cytotoxic effects on *Allium cepa. Environmental Science and Pollution Research International, 25*(28), 27890–27898. DOI: 10.1007/s11356-018-2813-5 PMID: 30056543

De Sousa, C. S., Hassan, S. S., Pinto, A. C., Silva, W. M., De Almeida, S. S., Soares, S. D. C., Azevedo, M. S., Rocha, C. S., Barh, D., & Azevedo, V. (2018). Microbial omics: applications in biotechnology. In Debmalya, B., & Vasco, A. (Eds.), *Omics technologies and bio-engineering* (Vol. 2, pp. 3–20). Elsevier. DOI: 10.1016/B978-0-12-815870-8.00001-2

Dehkordi, M. B., Pishvaie, M. R., & Vafa, E. (2023). Model predictive control of a fermenter using dynamic flux balance analysis coupled with convolutional neural networks. *Computers & Chemical Engineering, 179*, 108444. DOI: 10.1016/j.compchemeng.2023.108444

Dehnavi, S., & Ebrahimipour, G. (2022). Comparative remediation rate of biostimulation, bioaugmentation, and phytoremediation in hydrocarbon contaminants. *International Journal of Environmental Science and Technology*, *19*(11), 11561–11586. DOI: 10.1007/s13762-022-04343-0

Dejonghe, W., Goris, J., El Fantroussi, S., Höfte, M., De Vos, P., Verstraete, W., & Top, E. M. (2000). Effect of dissemination of 2, 4-dichlorophenoxyacetic acid (2, 4-D) degradation plasmids on 2, 4-D degradation and on bacterial community structure in two different soil horizons. *Applied and Environmental Microbiology*, *66*(8), 3297–3304. DOI: 10.1128/AEM.66.8.3297-3304.2000 PMID: 10919784

Delgado-Baquerizo, M., Reich, P. B., Trivedi, C., Eldridge, D. J., Abades, S., Alfaro, F. D., Bastida, F., Berhe, A. A., Cutler, N. A., Gallardo, A., García-Velázquez, L., Hart, S. C., Hayes, P. E., He, J.-Z., Hseu, Z.-Y., Hu, H.-W., Kirchmair, M., Neuhauser, S., Pérez, C. A., & Singh, B. K. (2020). Multiple elements of soil biodiversity drive ecosystem functions across biomes. *Nature Ecology & Evolution*, *4*(2), 210–220. DOI: 10.1038/s41559-019-1084-y PMID: 32015427

Delkash, M., Al-Faraj, F. A., & Scholz, M. (2018). Impacts of anthropogenic land use changes on nutrient concentrations in surface waterbodies: A review. *Clean (Weinheim)*, *46*(5), 1800051. DOI: 10.1002/clen.201800051

Dell'Anno, F., Rastelli, E., Sansone, C., Brunet, C., Ianora, A., & Dell'Anno, A. (2021). Bacteria, fungi and microalgae for the bioremediation of marine sediments contaminated by petroleum hydrocarbons in the omics era. *Microorganisms*, *9*(8), 1695. DOI: 10.3390/microorganisms9081695 PMID: 34442774

Dexter, H. S., & Lu, T. (2022). Engineering microbial consortia with rationally designed cellular interactions. *Current Opinion in Biotechnology*, *76*, 102730. DOI: 10.1016/j.copbio.2022.102730 PMID: 35609504

Diao, J., Tian, Y., Park, S., Cho, S.-M., & Moon, T. S. (2025). Engineering microbial consortia for mixed plastic upcycling. *Nature Communications*, *17*(1), 637. DOI: 10.1038/s41467-025-67409-w PMID: 41381587

Díaz-García, L., Huang, S., Spröer, C., Sierra-Ramírez, R., Bunk, B., Overmann, J. & Overmann D. J. (2021). Dilution-to-stimulation/extinction method: a combination enrichment strategy to develop a minimal and versatile lignocellulolytic bacterial consortium. *Applied Environmental Microbiology*, (2) 2021, 87:e02427-20, https://doi.org/10. 1128/AEM.02427-20

Dimitrova, I., Dabrowska, A., & Ekström, S. (2020). Start-up of a full-scale partial nitritation-anammox MBBR without inoculum at Klagshamn WWTP. *Water Science and Technology*, *81*(9), 2033–2042. DOI: 10.2166/wst.2020.271 PMID: 32666956

Dinh, C. T., Tran, N. H., Nguyen, M. N., Hong, C. S., Bao, W., Zomaya, A. Y., & Gramoli, V. (2020). Federated learning over wireless networks: Convergence analysis and resource allocation. *IEEE/ACM Transactions on Networking*, *29*(1), 398–409. DOI: 10.1109/TNET.2020.3035770

Diversity, C. P. B. C. B. (2000). *Cartagena Protocol on Biosafety to the Convention on Biological Diversity: text and annexes*. Secretariat of the Convention of Biological Diversity.

Dixit, S., & Garg, S. (2018). Biodegradation of Environmentally Hazardous Azo Dyes and Aromatic Amines Using *Klebsiella pneumoniae*. *Journal of Environmental Engineering*, *144*(6), 04018035. Advance online publication. DOI: 10.1061/(ASCE)EE.1943-7870.0001353

Dolganova, I., Dolganov, I., Ivanchina, E., & Ivashkina, E. (2018). Alkylaromatics in detergents manufacture: Modeling and optimizing linear alkylbenzene sulfonation. *Journal of Surfactants and Detergents*, *21*(1), 175–184. DOI: 10.1002/jsde.12009

Donkadokula, N. Y., Kola, A. K., Naz, I., & Saroj, D. (2020). A review on advanced physico-chemical and biological textile dye wastewater treatment techniques. *Reviews in Environmental Science and Biotechnology*, *19*(3), 543–560. DOI: 10.1007/s11157-020-09543-z

Doroszkowski, A. (2020). Paints. In *Technological Applications of Dispersions* (pp. 1–68). CRC Press. DOI: 10.1201/9781003067252-1

Dreher, J., Scheiber, J., Stiefl, N., & Baumann, K. (2018). xMaP An interpretable alignment-free four-dimensional quantitative structure–activity relationship technique based on molecular surface properties and conformer ensembles. *Journal of Chemical Information and Modeling*, *58*(1), 165–181.

Dudeja, C., Mishra, A., Ali, A., Singh, P. P., & Jaiswal, A. K. (2025). Microbial Genome Editing with CRISPR–Cas9: Recent Advances and Emerging Applications Across Sectors. *Fermentation (Basel, Switzerland)*, *11*(7), 410. DOI: 10.3390/fermentation11070410

Du, H., Li, M., & Liu, Y. (2023). Towards applications of genome-scale metabolic model-based approaches in designing synthetic microbial communities. *Quantitative Biology*, *11*(1), 15–30. DOI: 10.15302/J-QB-022-0313 PMID: 41674535

Du, L., & Liu, W. (2012). Occurrence, fate, and ecotoxicity of antibiotics in agro-ecosystems. A review. *Agronomy for Sustainable Development*, *32*(2), 309–327. DOI: 10.1007/s13593-011-0062-9

Duncker, K. E., Holmes, Z. A., & You, L. (2021). Engineered microbial consortia: Strategies and applications. *Microbial Cell Factories*, *20*(1), 211. Advance online publication. DOI: 10.1186/s12934-021-01699-9 PMID: 34784924

Dvořák, P., Nikel, P. I., Damborský, J., & de Lorenzo, V. (2017). Bioremediation 3.0: Engineering pollutant-removing bacteria in the times of systemic biology. *Biotechnology Advances*, *35*(7), 845–866. DOI: 10.1016/j.biotechadv.2017.08.001 PMID: 28789939

Edwards, S., Ditter, R., Laster, H., Sheehan, G., Anderson, O., Beattie, T., & Mellies, J. L. (2021). Microbial Consortia and Mixed Plastic Waste: Pangenomic Analysis Reveals Potential for Degradation of Multiple Plastic Types via Previously Identified PET Degrading Bacteria. *International Journal of Molecular Sciences*, *23*(10), 1–22. DOI: 10.3390/ijms23105612 PMID: 35628419

Ehis-Eriakha, C. B., Ajuzieogu, C. A., Orogun, J. O., & Akemu, S. E. (2024). Overview of petroleum hydrocarbon pollution and bioremediation technologies –. *Bioremediation Journal*, •••, 1–23. DOI: 10.1080/10889868.2024.234901

Ehis-Eriakha, C. B., Chikere, C. B., & Akemu, S. E. (2025). A comparative assessment of biostimulation potential in microbiome-based ecorestoration of polycyclic aromatic hydrocarbon polluted soil. *Brazilian Journal of Microbiology : Publication of the Brazilian Society for Microbiology*, *56*(1), 203–224. DOI: 10.1007/s42770-024-01556-y PMID: 39602070

Eicher, T., Kinnebrew, G., Patt, A., Spencer, K., Ying, K., Ma, Q., Machiraju, R., & Mathé, E. A. (2020). Metabolomics and Multi-Omics Integration: A survey of Computational methods and resources. *Metabolites*, *10*(5), 1–25. DOI: 10.3390/metabo10050202 PMID: 32429287

El Awady, M. E., El-Shall, F. N., Mohamed, G. E., Abd-Elaziz, A. M., Abdel-Monem, M. O., & Hassan, M. G. (2024). Exploring the decolorization efficiency and biodegradation mechanisms of different functional textile azo dyes by *Streptomyces albidoflavus* 3MGH. *BMC Microbiology*, *24*(1), 1–22. DOI: 10.1186/s12866-024-03347-9 PMID: 38877404

El Fantroussi, S., & Agathos, S. N. (2005). Is bioaugmentation a feasible strategy for pollutant removal and site remediation? *Current Opinion in Microbiology*, *8*(3), 268–275. DOI: 10.1016/j.mib.2005.04.011 PMID: 15939349

Ellen MacArthur Foundation. (2015). *Towards a circular economy: Business rationale for an accelerated transition.* Ellen MacArthur Foundation. https://www.ellenmacarthurfoundation.org/towards-a-circular-economy-business-rationale-for-an-accelerated-transition

Elmore, J. R., Dexter, G. N., Salvachúa, D., Martinez-Baird, J., Hatmaker, E. A., Huenemann, J. D., Klingeman, D. M., Peabody, G. L., Peterson, D. J., Singer, C., Beckham, G. T., & Guss, A. M. (2023). Production of itaconic acid from alkali pretreated lignin by dynamic two stage bioconversion. *Nature Communications*, *14*, 2494. DOI: 10.1038/s41467-023-38012-5 PMID: 33859194

Elshafei, A. M., & Mansour, R. (2024). Microbial bioremediation of soils contaminated with petroleum hydrocarbons. *Discover Soil*, *1*(1), 9. DOI: 10.1007/s44378-024-00004-5

Endy, D. (2005). Foundations for engineering biology. *Nature*, *438*(7067), 449–453. DOI: 10.1038/nature04342 PMID: 16306983

Environment and Climate Change Canada. (2022). Canadian Environmental Protection Act, 1999: Annual Report to Parliament for April 2020 to March 2021. Environment and Climate Change Canada, Cat. No.: En81-3E-PDF ISBN: 1492-0212 EC21270. Retrieved from https://www.canada.ca/en/environment-climate-change/services/canadian-environmental-protection-act-registry/publications/annual-report-2020-2021.html

Ertu rul, S., San, N. O., & Dönmez, G. (2009). Treatment of dye (Remazol Blue) and heavy metals using yeast cells with the purpose of managing polluted textile wastewaters. *Ecological Engineering*, *35*(1), 128–134. DOI: 10.1016/j.ecoleng.2008.09.015

European Commission. (2020). Improving Financial Security in the Context of the Environmental Liability Directive (ELD). Brussels: European Commission, Directorate-General for Environment. Retrieved from https://www.impel.eu/contents/files/eu.pdf

Fabre, E., Dias, M., Costa, M., Henriques, B., Vale, C., Lopes, C. B., & Pereira, E. (2020). Negligible effect of potentially toxic elements and rare earth elements on mercury removal from contaminated waters by green, brown and red living marine macroalgae. *The Science of the Total Environment*, *724*, 138133. DOI: 10.1016/j.scitotenv.2020.138133 PMID: 32268287

Fan, Y., Song, Z., Wu, Y., Ren, X., Bi, C., Ye, W., & Xu, Y. (2024). Chemicals of emerging concern in water-based paint products. *Environmental Science & Technology Letters*, *11*(5), 445–452. DOI: 10.1021/acs.estlett.4c00052

Farias, C. B. B., Rita de Cássia, F., Almeida, F. C. G., Santos, V. A., & Sarubbo, L. A. (2021). Removal of heavy oil from contaminated surfaces with a detergent formulation containing biosurfactants produced by *Pseudomonas* spp. *PeerJ*, *9*, e12518. DOI: 10.7717/peerj.12518 PMID: 34900433

Fashola, M. O., Ngole-Jeme, V. M., & Babalola, O. O. (2015). Diversity of acidophilic bacteria and archaea and their roles in bioremediation of acid mine drainage. *British Microbiology Research Journal, 8*(3), 443–456. DOI: 10.9734/BMRJ/2015/14365

Fashola, M. O., Ngole-Jeme, V. M., & Babalola, O. O. (2016). Heavy metal pollution from gold mines: Environmental effects and bacterial strategies for resistance. *International Journal of Environmental Research and Public Health, 13*(11), 1047. DOI: 10.3390/ijerph13111047 PMID: 27792205

Fashola, M. O., Ngole-Jeme, V. M., & Babalola, O. O. (2020a). Physicochemical properties, heavy metals, and metal-tolerant bacteria profiles of abandoned gold mine tailings in Krugersdorp, South Africa. *Canadian Journal of Soil Science, 100*(3), 217–233. DOI: 10.1139/cjss-2018-0161

Fashola, M. O., Ngole-Jeme, V. M., & Babalola, O. O. (2020b). Heavy metal immobilization potential of indigenous bacteria isolated from gold mine tailings. *International Journal of Environmental Research, 14*(1), 71–86. DOI: 10.1007/s41742-019-00240-6

Faure, G., Shmakov, S. A., Yan, W. X., Cheng, D. R., Scott, D. A., Peters, J. E., Makarova, K. S., & Koonin, E. V. (2019). CRISPR–Cas in mobile genetic elements: Counter-defence and beyond. *Nature Reviews. Microbiology, 17*(8), 513–525. DOI: 10.1038/s41579-019-0204-7 PMID: 31165781

Federal Remediation Technologies Roundtable (US). (1995). *Abstracts of remediation case studies*. Federal Remediation Technologies Roundtable.

Fedeson, D. T., Saake, P., Calero, P., Nikel, P. I., & Ducat, D. C. (2020). Biotransformation of 2,4-dinitrotoluene in a phototrophic co-culture of engineered *Synechococcus elongatus* and *Pseudomonas putida. Microbial Biotechnology, 13*(4), 997–1011. DOI: 10.1111/1751-7915.13544 PMID: 32064751

Fensham, E., Ling, N., and Tempero, G. W. (2022). Toxicological effects of aluminium in relation to diel pH changes on fish and kōura.

Fernández, C. D. B., Castillo, M. P. G., Pérez, S. Q., & Rodríguez, L. V. C. (2022). Microbial degradation of polyethylene terephthalate: A systematic review. *SN Applied Sciences, 4*(10), 1–20. DOI: 10.1007/s42452-022-05143-4

Fernández, J. D., Fajardo, C., Campos, J. L., Mosquera-Corral, A., & Méndez, R. (2012). Short- and long-term effects of ammonium and nitrite on the Anammox process. *Journal of Environmental Management, 95*, S170–S174. DOI: 10.1016/j.jenvman.2010.10.044 PMID: 21074312

Fernandez, L., Peura, S., Eiler, A., Linz, A. M., McMahon, K. D., & Bertilsson, S. (2020). Diazotroph Genomes and Their Seasonal Dynamics in a Stratified Humic Bog Lake. *Frontiers in Microbiology, 11*, 1500. DOI: 10.3389/fmicb.2020.01500 PMID: 32714313

Fernández, P. M., Vinarta, S. C., Bernal, A. R., Cruz, E. L., & Figueroa, L. I. (2018). Bioremediation strategies for chromium removal: Current research, scale-up approach and future perspectives. *Chemosphere, 208*, 139–148. DOI: 10.1016/j.chemosphere.2018.05.166 PMID: 29864705

Ferreira-Filipe, D. A., Duarte, A. C., Hursthouse, A. S., Rocha-Santos, T., & Silva, A. L. P. (2025). Biobased Strategies for E-Waste Metal Recovery: A Critical Overview of Recent Advances. *Environments (Basel, Switzerland), 12*(1), 26. DOI: 10.3390/environments12010026

Finkel, O. M., Castrillo, G., Herrera Paredes, S., Salas González, I., & Dangl, J. L. (2017). Understanding and exploiting plant beneficial microbes. *Current Opinion in Plant Biology, 38*, 155–163. DOI: 10.1016/j.pbi.2017.04.018 PMID: 28622659

Fisher, C. (2024). What are the 5 R's of Waste Management — Refuse, Reduce, Reuse, Repurpose, Recycle. https://www.rts.com/blog/the-5-rs-refuse-reduce-reuse-repurpose-recycle/. Published on March 10, 2024. Downloaded 12 October 2025

Flemming, H. C., van Hullebusch, E. D., Neu, T. R., Nielsen, P. H., Seviour, T., Stoodley, P., Wingender, J., & Wuertz, S. (2023a). The biofilm matrix: Multitasking in a shared space. *Nature Reviews. Microbiology, 21*(2), 70–86. DOI: 10.1038/s41579-022-00791-0 PMID: 36127518

Floros, M., Hojabri, L., Abraham, E., Jose, J., Thomas, S., Pothan, L., Leao, A. L., & Narine, S. (2012). Enhancement of thermal stability, strength and extensibility of lipid-based polyurethanes with cellulose-based nanofibers. *Polymer Degradation & Stability, 97*(10), 1970–1978. DOI: 10.1016/j.polymdegradstab.2012.02.016

Food and Agriculture Organization of the United Nations (FAO). (2017). FAO/WHO food safety assessment tools for trainers [ToolKit component]. Food and Agriculture Organization of the United Nations. Retrieved from https://www.fao.org/food-safety/scientific-advice/crosscutting-and-emerging-issues/biotechnology/en/

Fredens, J., Wang, K., de la Torre, D., Funke, L. F. H., Robertson, W. E., Christova, Y., Chia, T., Schmied, W. H., Dunkelmann, D. L., Beránek, V., Uttamapinant, C., Llamazares, A. G., Elliott, T. S., & Chin, J. W. (2019). Total synthesis of *Escherichia coli* with a recoded genome. *Nature, 569*(7757), 514–518. DOI: 10.1038/s41586-019-1192-5 PMID: 31092918

Fu, F., & Wang, Q. (2011). Removal of heavy metal ions from wastewaters: A review. *Journal of Environmental Management*, *92*(3), 407–418. DOI: 10.1016/j.jenvman.2010.11.011 PMID: 21138785

Fujita, H., Ushio, M., Suzuki, K., Abe, M. S., Yamamichi, M., Iwayama, K., Canarini, A., Hayashi, I., Fukushima, K., Fukuda, S., Kiers, E. T., & Toju, H. (2023). Alternative stable states, nonlinear behavior, and predictability of microbiome dynamics. *Microbiome*, *11*(1), 63. DOI: 10.1186/s40168-023-01474-5 PMID: 36978146

Fu, P., & Secundo, F. (2016). Algae and their bacterial consortia for soil bioremediation. *Chemical Engineering Transactions*, *49*, 427–432.

Fu, R., Wang, Y., Liu, Y., Liu, H., Zhao, Q., Zhang, Y., Wang, C., Li, Z., Jiao, B., & He, Y. (2022). CRISPR-Cas12a based fluorescence assay for organophosphorus pesticides in agricultural products. *Food Chemistry*, *387*, 132919. DOI: 10.1016/j.foodchem.2022.132919 PMID: 35421656

Furuya, E. Y., & Lowy, F. D. (2006). Antimicrobial-resistant bacteria in the community setting. *Nature Reviews. Microbiology*, *4*(1), 36–45. DOI: 10.1038/nrmicro1325 PMID: 16357859

Gadd, G. M. (2010). Metals, minerals and microbes: Geomicrobiology and bioremediation. *Microbiology (Reading, England)*, *156*(3), 609–643. DOI: 10.1099/mic.0.037143-0 PMID: 20019082

Gao, W., Wu, Z., Zuo, K., Xiang, Q., Chen, X., Zhang, L., & Liu, H. (2025). Biosafety concept: Origins, evolution, and prospects. *Biosafety and Health*, *7*(4), 209–217. DOI: 10.1016/j.bsheal.2025.07.003 PMID: 40918204

Garbini, G. L., Barra Caracciolo, A., & Grenni, P. (2023). Electroactive bacteria in natural ecosystems and their applications in microbial fuel cells for bioremediation: A review. *Microorganisms*, *11*(5), 1255. DOI: 10.3390/microorganisms11051255 PMID: 37317229

Garbisu, C., Garaiyurrebaso, O., Epelde, L., Grohmann, E., & Alkorta, I. (2017). Plasmid-mediated bioaugmentation for the bioremediation of contaminated soils. *Frontiers in Microbiology*, *8*, 1966. DOI: 10.3389/fmicb.2017.01966 PMID: 29062312

García-Berumen, J. A., Flores de la Torre, J. A., De los Santos-Villalobos, S., Espinoza-Canales, A., Echavarría-Cháirez, F. G., & Gutiérrez-Bañuelos, H. (2024). Phosphorus dynamics and sustainable agriculture: The role of microbial solubilization and innovations in nutrient management. *Current Research in Microbial Sciences*, *8*, 100326. DOI: 10.1016/j.crmicr.2024.100326 PMID: 39687549

Garousin, H., Pourbabaee, A. A., Alikhani, H. A., & Yazdanfar, N. (2021). A combinational strategy mitigated old-aged petroleum contaminants: Ineffectiveness of biostimulation as a bioremediation technique. *Frontiers in Microbiology, 12*, 642215. DOI: 10.3389/fmicb.2021.642215 PMID: 33717040

Gasiunas, G., Barrangou, R., Horvath, P., & Siksnys, V. (2012). Cas9–crRNA ribonucleoprotein complex mediates specific DNA cleavage for adaptive immunity in bacteria. *Proceedings of the National Academy of Sciences of the United States of America, 109*(39), E2579–E2586. DOI: 10.1073/pnas.1208507109 PMID: 22949671

Gaylarde, C. C. & Da Fonseca, E. M. Biofertilization and Bioremediation—How Can Microbiological Technology Assist the Ecological Crisis in Developing Countries? Micro, 2025. MDPI, 18.

Geissdoerfer, M., Savaget, P., Bocken, N. M. P., & Hultink, E. J. (2017). The circular economy: A new sustainability paradigm? *Journal of Cleaner Production, 143*, 757–768. DOI: 10.1016/j.jclepro.2016.12.048

Gentry, T., Rensing, C., & Pepper, I. (2004). New approaches for bioaugmentation as a remediation technology. *Critical Reviews in Environmental Science and Technology, 34*(5), 447–494. DOI: 10.1080/10643380490452362

Geyer, R., Jambeck, J. R., & Law, K. L. (2017). Production, use, and fate of all plastics ever made. *Science Advances, 3*(7), e1700782. DOI: 10.1126/sciadv.1700782 PMID: 28776036

Ghisellini, P., Cialani, C., & Ulgiati, S. (2016). A review on circular economy: The expected transition to a balanced interplay of environmental and economic systems. *Journal of Cleaner Production, 114*, 11–32. DOI: 10.1016/j.jclepro.2015.09.007

Ghosal, D., Ghosh, S., Dutta, T. K., & Ahn, Y. (2016). Current state of knowledge in microbial degradation of polycyclic aromatic hydrocarbons (PAHs): A Review. *Frontiers in Microbiology, 7*, 1369. DOI: 10.3389/fmicb.2016.01369 PMID: 27630626

Giri, B. S., Geed, S., Vikrant, K., Lee, S. S., Kim, K. H., Kailasa, S. K., Vithanage, M., Chaturvedi, P., Rai, B. N., & Singh, R. S. (2021). Progress in bioremediation of pesticide residues in the environment. *Environmental Engineering Research, 26*(6), 200446. DOI: 10.4491/eer.2020.446

Glibert, P. M. (2017). Eutrophication, harmful algae, and biodiversity—Challenging paradigms in a world of complex nutrient changes. *Marine Pollution Bulletin, 124*(2), 591–606. DOI: 10.1016/j.marpolbul.2017.04.027 PMID: 28434665

Gomes, H. I., Dias-Ferreira, C., & Ribeiro, A. B. (2013). Overview of in-situ and ex-situ remediation technologies for PCB-contaminated soils and sediments and obstacles for full-scale application. *The Science of the Total Environment*, *445*, 237–260. DOI: 10.1016/j.scitotenv.2012.11.098 PMID: 23334318

Gomez, F., & Sartaj, M. (2013). Field scale ex-situ bioremediation of petroleum contaminated soil under cold climate conditions. *International Biodeterioration & Biodegradation*, *85*, 375–382. DOI: 10.1016/j.ibiod.2013.08.003

Gonçalves, I. M. C., Gomes, A., Brás, R., Ferra, M. I. A., Amorim, M. T. P., & Porter, R. S. (2000). Biological treatment of effluent containing textile dyes. *Coloration Technology*, *116*(12), 393–397. DOI: 10.1111/j.1478-4408.2000.tb00016.x

Gong, H., Gao, J., & Jin, Z. (2022). Biodegradation of pharmaceutical compounds in wastewater: Mechanisms and applications. In Varjani, S., Pandey, A., Gnansounou, E., Khanal, S. K., & Raveendran, S. (Eds.), *Bioremediation for environmental sustainability* (pp. 393–421). Elsevier., DOI: 10.1016/B978-0-323-89874-4.00010-4

Gong, T., Liu, R., Zuo, Z., Che, Y., Yu, H., Song, C., & Yang, C. (2016). Metabolic engineering of *Pseudomonas putida* KT2440 for complete mineralization of methyl parathion and γ-hexachlorocyclohexane. *ACS Synthetic Biology*, *5*(5), 434–442. DOI: 10.1021/acssynbio.6b00025 PMID: 26854500

Gong, T., Xu, X., Dang, Y., Kong, A., Wu, Y., Liang, P., Wang, S., Yu, H., Xu, P., & Yang, C. (2018). An engineered *Pseudomonas putida* can simultaneously degrade organophosphates, pyrethroids and carbamates. *The Science of the Total Environment*, *628-629*, 1258–1265. DOI: 10.1016/j.scitotenv.2018.02.143 PMID: 30045547

Gong, X., Zhang, J., Gan, Q., Teng, Y., Hou, J., Lyu, Y., Liu, Z., Wu, Z., Dai, R., Zou, Y., Wang, X., Zhu, D., Zhu, H., Liu, T., & Yan, Y. (2024). Advancing microbial production through artificial intelligence-aided biology. *Biotechnology Advances*, *74*, 108399. DOI: 10.1016/j.biotechadv.2024.108399 PMID: 38925317

González-Martínez, A., Muñoz-Palazon, B., Rodríguez-Sánchez, A., & González-López, J. (2018). New concepts in anammox processes for wastewater nitrogen removal: Recent advances and future prospects. *FEMS Microbiology Letters*, *365*(6), fny031. DOI: 10.1093/femsle/fny031 PMID: 29438563

Goswami, M., Chakraborty, P., Mukherjee, K., Mitra, G., Bhattacharyya, P., Dey, S., & Tribedi, P. (2018). Bioaugmentation and biostimulation: A potential strategy for environmental remediation. *Journal of Microbiology & Experimentation*, *6*, 223–231.

Grandel, N. E., Alexander, A. M., Peng, X., Palamountain, C., Alnahhas, R. N., Hirning, A. J., Josić, K., & Bennett, M. R. (2025). Long-term homeostasis in microbial consortia via auxotrophic cross-feeding. *Nature Communications*, *16*(1), 8573. DOI: 10.1038/s41467-025-63575-z PMID: 41022726

Greated, A., Lambertsen, L., Williams, P. A., & Thomas, C. M. (2002). Complete sequence of the IncP-9 TOL plasmid pWW0 from *Pseudomonas putida. Environmental Microbiology*, *4*(12), 856–871. DOI: 10.1046/j.1462-2920.2002.00305.x PMID: 12534468

Gricajeva, A., Nadda, A. K., & Gudiukaite, R. (2022). Insights into polyester plastic biodegradation by carboxyl ester hydrolases. *Journal of Chemical Technology and Biotechnology*, *97*(2), 359–380. DOI: 10.1002/jctb.6745

Guadie, A., Gessesse, A., & Xia, S. (2018). Halomonas sp. strain A55, a novel dye decolorizing bacterium from dye-uncontaminated Rift Valley Soda lake. *Chemosphere*, *206*, 59–69. DOI: 10.1016/j.chemosphere.2018.04.134 PMID: 29730566

Gui, Q., Lawson, T., Shan, S., Yan, L., & Liu, Y. (2017). The application of whole cell-based biosensors for use in environmental analysis and in medical diagnostics. *Sensors (Basel)*, *17*(7), 1623. DOI: 10.3390/s17071623 PMID: 28703749

Guo, R., Li, G., Jiang, T., Schuchardt, F., Chen, T., Zhao, Y., & Shen, Y. (2012). Effect of aeration rate, C/N ratio and moisture content on the stability and maturity of compost. *Bioresource Technology*, *112*, 171–178. DOI: 10.1016/j.biortech.2012.02.099 PMID: 22437050

Guo, Y., Gao, M., Jiang, L., Huang, H., Kang, G., & Yu, H. (2025). Synthetic microbial consortia based on quorum-sensing for disease therapy. *Bioresources and Bioprocessing*, *12*(1), 125. DOI: 10.1186/s40643-025-00962-w PMID: 41182556

Haider, H. U., Arshad, E., Amir, M. A., Zafar, N., Akhtar, I., Ahmad, M. T., and Ali, M. (2024). Integrating Nanotechnology and Bacteria-Based Remediation: A Synergistic Approach to Mitigating Climate Change and Soil Degradation in Agriculture. *Global Scientific and Academic Research Journal of Multidisciplinary Studies. 1-13*

Hajnajafi, K., & Iqbal, M. A. (2025). Mass-spectrometry based metabolomics: An overview of workflows, strategies, data analysis and applications. *Proteome Science*, *23*(5), 1–15. DOI: 10.1186/s12953-025-00241-8 PMID: 40420110

Halden, R. U., Tepp, S. M., Halden, B. G., & Dwyer, D. F. (1999). Degradation of 3-phenoxybenzoic acid in soil by Pseudomonas pseudoalcaligenes POB310 (pPOB) and two modified Pseudomonas strains. *Applied and Environmental Microbiology*, *65*(8), 3354–3359. DOI: 10.1128/AEM.65.8.3354-3359.1999 PMID: 10427019

Hamdi, H., Benzarti, S., Manusadžianas, L., Aoyama, I., & Jedidi, N. (2007). Bioaugmentation and biostimulation effects on PAH dissipation and soil ecotoxicity under controlled conditions. *Soil Biology & Biochemistry, 39*(8), 1926–1935. DOI: 10.1016/j.soilbio.2007.02.008

Handore, A. V., Jagtap, V. S., Ghangale, S. S., Khandelwal, S. R., Bholay, A. D., Karmakar, R., & Handore, D. V. (2023). *Bioaugmentation: an emerging ecofriendly technology for wastewater treatment. Integrative Strategies for Bioremediation of Environmental Contaminants, Volume Two*. Elsevier. DOI: 10.1016/B978-0-443-14013-6.00004-4

Hanlon, P., & Sewalt, V. (2021). GEMs: Genetically engineered microorganisms and the regulatory oversight of their uses in modern food production. *Critical Reviews in Food Science and Nutrition, 61*(6), 959–970. DOI: 10.1080/10408398.2020.1749026 PMID: 32274948

Hao, L., Zhang, D., Zhou, H., Wang, Z., Zhang, J., Zhao, Z., & Li, M. (2025). Machine learning applications in prediction and optimization of heavy metal bioremediation: A review. *Journal of Environmental Chemical Engineering, 13*(5), 118845. DOI: 10.1016/j.jece.2025.118845

Hao, X., Gu, Y., Zhang, H., Wang, X., Liu, X., Chen, C., Wang, C., Zhang, X., Liu, X., & Shen, X. (2024). Synthetic microbial community promotes bacterial communities leading to soil multifunctionality in desertified land. *Microorganisms, 12*(6), 1–16. DOI: 10.3390/microorganisms12061117 PMID: 38930499

Haque, M. M., Mosharaf, M. K., Haque, M. A., Tanvir, M. Z. H., & Alam, M. K. (2021). Biofilm formation, production of matrix compounds and biosorption of copper, nickel and lead by different bacterial strains. *Frontiers in Microbiology, 12*, 615113. DOI: 10.3389/fmicb.2021.615113 PMID: 34177820

Harms, H., Schlosser, D., & Wick, L. Y. (2011). Untapped potential: Exploiting fungi in bioremediation of hazardous chemicals. *Nature Reviews. Microbiology, 9*(3), 177–192. DOI: 10.1038/nrmicro2519 PMID: 21297669

Haruta, S., & Yamamoto, K. (2018). Model Microbial Consortia as Tools for Understanding Complex Microbial Communities. *Current Genomics, 19*(8), 723–733. DOI: 10.2174/1389202919666180911131206 PMID: 30532651

Hasibi, R., Michoel, T., & Oyarzún, D. A. (2024). Integration of graph neural networks and genome-scale metabolic models for predicting gene essentiality. *NPJ Systems Biology and Applications, 10*(1), 24. DOI: 10.1038/s41540-024-00348-2 PMID: 38448436

Hassan, Z. U., Ali, S., Rizwan, M., Ibrahim, M., Nafees, M., & Waseem, M. 2017. Role of bioremediation agents (bacteria, fungi, and algae) in alleviating heavy metal toxicity. Probiotics in agroecosystem, pp.517-537.

Hassan, M., Zhu, G., Yang, Z., & Lu, Y. (2021). Simultaneous removal of sulfamethoxazole and enhanced denitrification process from simulated municipal wastewater by a novel 3D-BER system. *Journal of Environmental Health Science & Engineering*, *19*(1), 23–38. DOI: 10.1007/s40201-020-00562-0 PMID: 34150216

Hassan, S., & Ganai, B. A. (2023). Deciphering the recent trends in pesticide bioremediation using genome editing and multi-omics approaches: A review. *World Journal of Microbiology and Biotechnology*, *39*(6), 151. DOI: 10.1007/s11274-023-03603-6 PMID: 37029313

Hawley, E. R., Piao, H., Scott, N. M., Malfatti, S., Pagani, I., Huntemann, M., Chen, A., Glavina del Rio, T., Foster, B., Copeland, A., Jansson, J., Pati, A., Tringe, S., Gilbert, J. A., Lorenson, T. D., & Hess, M. (2014). Metagenomic analysis of microbial consortium from natural crude oil that seeps into the marine ecosystem offshore Southern California. *Standards in Genomic Sciences*, *9*(3), 1259–1274. DOI: 10.4056/sigs.5029016 PMID: 25197496

Haynes, V. (2020). The Interactive Effects of Titanium Dioxide Nanoparticles and Light on Coastal Planktonic Organisms. Doctoral dissertation

Hazen, T. C., Dubinsky, E. A., DeSantis, T. Z., Andersen, G. L., Piceno, Y. M., Singh, N., Jansson, J. K., Probst, A., Borglin, S. E., Fortney, J. L., Stringfellow, W. T., Bill, M., Conrad, M. E., Tom, L. M., Chavarria, K. L., Alusi, T. R., Lamendella, R., Joyner, D. C., Spier, C., & Mason, O. U. (2010). Deep-sea oil plume enriches indigenous oil-degrading bacteria. *Science*, *330*(6001), 204–208. DOI: 10.1126/science.1195979 PMID: 20736401

Hazra, C., Kundu, D., & Chaudhari, A. (2012). Biosurfactant-assisted bioaugmentation in bioremediation. Microorganisms in Environmental Management, 631-664.

Hellweger, F. L., Clegg, R. J., Clark, J. R., Plugge, C. M., & Kreft, J. U. (2016). Advancing microbial sciences by individual-based modelling. *Nature Reviews. Microbiology*, *14*(7), 461–471. DOI: 10.1038/nrmicro.2016.62 PMID: 27265769

Hendrickson, D., Kelley, D. R., Tenen, D., Bernstein, B., & Rinn, J. L. (2016). Widespread RNA binding by chromatin-associated proteins. *Genome Biology*, *17*(1), 28. DOI: 10.1186/s13059-016-0878-3 PMID: 26883116

Hendrickx, T. L. G., Kampman, C., Zeeman, G., Temmink, B. G., Hu, Z., Kartal, B., & Buisman, C. J. N. (2014). High specific activity for Anammox bacteria enriched from activated sludge at 10 °C. *Bioresource Technology*, *163*, 214–222. DOI: 10.1016/j.biortech.2014.04.025 PMID: 24814247

Hennebel, T., Boon, N., Maes, S., & Lenz, M. (2015). Biotechnologies for critical raw material recovery from primary and secondary sources: R&D priorities and future perspectives. *New Biotechnology*, *32*(1), 121–127. DOI: 10.1016/j.nbt.2013.08.004 PMID: 23994422

Hennessy, R. C., Glaring, M. A., Olsson, S., & Stougaard, P. (2017). Transcriptomic profiling of microbe–microbe interactions reveals the specific response of the biocontrol strain P. Fluorescens In5 to the phytopathogen Rhizoctonia solani. *BMC Research Notes*, *10*(1), 376. DOI: 10.1186/s13104-017-2704-8 PMID: 28807055

Henze, M., Gujer, W., Mino, T., & van Loosdrecht, M. (2000). *Activated sludge models ASM1, ASM2, ASM2d and ASM3*. IWA Publishing.

Heuer, H., & Smalla, K. (2012). Plasmids foster diversification and adaptation of bacterial populations in soil. *FEMS Microbiology Reviews*, *36*(6), 1083–1104. DOI: 10.1111/j.1574-6976.2012.00337.x PMID: 22393901

Hirschi, S., Ward, T. R., Meier, W. P., Müller, D. J., & Fotiadis, D. (2022). Synthetic biology: Bottom-up assembly of molecular systems. *Chemical Reviews*, *122*(21), 16294–16328. DOI: 10.1021/acs.chemrev.2c00339 PMID: 36179355

Hobom, B. (1980). Gene surgery: On the threshold of synthetic biology. *Medizinische Klinik*, *75*(24), 834–841. PMID: 6160378

Hoffmann, S. A., Diggans, J., Densmore, D., Dai, J., Knight, T., Leproust, E., Boeke, J. D., Wheeler, N., & Cai, Y. (2023). Safety by design: Biosafety and biosecurity in the age of synthetic genomics. *iScience*, *26*(3), 106165. DOI: 10.1016/j.isci.2023.106165 PMID: 36895643

Hoffmann, T. D., Reeksting, B. J., & Gebhard, S. (2021). Bacteria-induced mineral precipitation: A mechanistic review. *Microbiology (Reading, England)*, *167*(4), 001049. DOI: 10.1099/mic.0.001049 PMID: 33881981

Holkar, C. R., Jadhav, A. J., Pinjari, D. V., Mahamuni, N. M., & Pandit, A. B. (2016). A critical review on textile wastewater treatments: Possible approaches. *Journal of Environmental Management*, *182*, 351–366. DOI: 10.1016/j.jenvman.2016.07.090 PMID: 27497312

Horozal, A., & Aksoy, Ö. (2020). Evaluation of the genotoxicity of some standard and eco-friendly detergents with Vicia faba. *Caryologia, 73*(4), 129–139. DOI: 10.13128/caryologia-856

Hossain, A., Lopez, E., Halper, S. M., Cetnar, D. P., Reis, A. C., Strickland, D., Klavins, E., & Tabor, J. J. (2020). Automated design of thousands of nonrepetitive parts for engineering stable genetic systems. *Nature Biotechnology, 38*(12), 1466–1475. DOI: 10.1038/s41587-020-0584-2 PMID: 32661437

Hossain, L., Sarker, S. K., & Khan, M. S. (2018). Evaluation of present and future wastewater impacts of textile dyeing industries in Bangladesh. *Environmental Development, 26*, 23–33. DOI: 10.1016/j.envdev.2018.03.005

Hossain, M. S., Sarker, P., Rahaman, M. S., Ahmed, F. F., Shaibur, M. R., & Khabir Uddin, M. (2022). Biological Treatment of Textile Wastewater by Total Aerobic Mixed Bacteria and Comparison with Chemical Fenton Process. *Pollution, 8*(4), 1418–1433. DOI: 10.22059/POLL.2022.340753.1408

Hossen, Z., Akhter, S., Rahman, T., & Khatun, M. (2020). Bacterial Degradation of Synthetic Dye by *Pseudomonas sp*. Obtained from Dyeing Mill Effluent. *American Journal of Pure and Applied Biosciences, 2(6), 192-199, 2020 Publisher, 2*(6), 192–199. https://doi.org/doi.org/10.34104/ajpab.020.01920199

Hoxha, D., Belayachi, N., Brunetaud, X., & Rémond, S. (2019). A Critical Review on Durability of Sustainable Materials and Structures. In *International Conference on Critical Thinking in Sustainable Rehabilitation and Risk Management of the Built Environment* (181-199). Cham: Springer International Publishing.

Hrouzková, S., & Szarka, A. (2021). Development of a modified QuEChERS procedure for the isolation of pesticide residues from textile samples, followed by GC–MS determination. *Separations, 8*(8), 106. DOI: 10.3390/separations8080106

Huang, J., Fu, Y., & Liu, Y. (2014). Comparison of Alkali-Tolerant Fungus *Myrothecium Sp*. IMER1 and White-Rot Fungi for Decolorization of Textile Dyes and Dye Effluents Bioremediation & Biodegradation. *Journal of Bioremediation & Biodegradation, 5*(3). Advance online publication. DOI: 10.4172/2155-6199.1000221

Huntington, V. E., Coulon, F., & Wagland, S. T. (2022). Innovative resource recovery from industrial sites: A critical review. *Sustainability (Basel), 15*(1), 489. DOI: 10.3390/su15010489

Huo, J., Song, B., Lin, X., Riaz, M., Zhao, X., Liu, S., & She, Q. (2024). Ecological characteristics of sugar beet plant and rhizosphere soil in response to high boron stress: A study of the remediation potential. *Journal of Environmental Management, 356*, 120655. DOI: 10.1016/j.jenvman.2024.120655 PMID: 38513589

Hussain, A., Rehman, F., Rafeeq, H., Waqas, M., Asghar, A., Afsheen, N., Rahdar, A., Bilal, M., & Iqbal, H. M. (2022). In-situ, Ex-situ, and nano-remediation strategies to treat polluted soil, water, and air–A review. *Chemosphere, 289*, 133252. DOI: 10.1016/j.chemosphere.2021.133252 PMID: 34902385

Hu, Y., & Li, W. (2022). Development and Application of CRISPR-Cas Based Tools. *Frontiers in Cell and Developmental Biology, 10*, 834646. DOI: 10.3389/fcell.2022.834646 PMID: 35445018

Iark, D., dos Reis Buzzo, A. J., Garcia, J. A. A., Côrrea, V. G., Helm, C. V., Corrêa, R. C. G., & Peralta, R. M. (2019). Enzymatic degradation and detoxification of azo dye Congo red by a new laccase from Oudemansiella canarii. *Bioresource Technology, 289*, 121655. DOI: 10.1016/j.biortech.2019.121655 PMID: 31247524

Ibo, E. M., Adeogun, A. O., Orji, M. U., & Umeh, O. R. (2024). Harnessing microbial synergy: A comprehensive evaluation of consortia-mediated bioremediation strategies for petroleum refinery wastewater treatment. *Cleaner Water, 2*, 100055. DOI: 10.1016/j.clwat.2024.100055

Ibrahim, A., Oginga, B., Zhang, Y., Ling, W., Tang, L., Elatafi, E., Abady, M., & Gao, Y. (2025). Bioremediation of soils with emerging organic contaminants using immobilized microorganisms. *Environmental Technology and Innovation, 40*, 104345. DOI: 10.1016/j.eti.2025.104345

Ibrahim, M., Raajaraam, L., & Raman, K. (2021). Modelling microbial communities: Harnessing consortia for biotechnological applications. *Computational and Structural Biotechnology Journal, 19*, 3892–3907. DOI: 10.1016/j.csbj.2021.06.048 PMID: 34584635

iGEM Foundation. (2021). *Registry of standard biological parts*. http://parts.igem.org

Ilyas, S., Anwar, M. A., Niazi, S. B., & Ghauri, M. A. (2007). Bioleaching of metals from electronic scrap by moderately thermophilic acidophilic bacteria. *Hydrometallurgy, 88*(1-4), 180–188. DOI: 10.1016/j.hydromet.2007.04.007

Ilyas, S., & Lee, J. C. (2014a). Biometallurgical recovery of metals from waste electrical and electronic equipment: A review. *ChemBioEng Reviews, 1*(4), 148–169. DOI: 10.1002/cben.201400001

Ilyas, S., & Lee, J. C. (2014b). Bioleaching of metals from electronic scrap in a stirred tank reactor. *Hydrometallurgy, 149*, 50–62. DOI: 10.1016/j.hydromet.2014.07.004

Imron, M. F., Kurniawan, S. B., Ismail, N. I., & Abdullah, S. R. S. (2020). Future challenges in diesel biodegradation by bacteria isolates: A review. *Journal of Cleaner Production, 251*, 119716. DOI: 10.1016/j.jclepro.2019.119716

Ingram, D., & Stan, G.-B. (2023). Modelling genetic stability in engineered cell populations. *Nature Communications, 14*(1), 3471. DOI: 10.1038/s41467-023-38850-6 PMID: 37308512

Internal Maritime Organization (IMO). (2018). Initial IMO strategy on reduction of GHG emissions from ships. Retrieved from https://www.imo.org/en/ourwork/environment/pages/imo-strategy-on-reduction-of-ghg-emmissions-from-ships.aspx

Işıldar, A., Rene, E. R., van Hullebusch, E. D., & Lens, P. N. L. (2016). Electronic waste as a secondary source of critical metals: Management and recovery technologies. *Resources, Conservation and Recycling, 135*, 296–312. DOI: 10.1016/j.resconrec.2017.07.031

Islam, M. R., & Mostafa, M. G. (2018). Textile Dyeing Effluents and Environment Concerns -. *RE:view, 11*, 131–144.

Islam, M., Aidid, A. R., Mohshin, J. N., Mondal, H., Ganguli, S., & Chakraborty, A. K. (2025). A critical review on textile dye-containing wastewater: Ecotoxicity, health risks, and remediation strategies for environmental safety. *Cleaner Chemical Engineering, 11*(January), 100165. DOI: 10.1016/j.clce.2025.100165

Ivanov, V., Stabnikov, V., Stabnikova, O., & Kawasaki, S. (2019). Environmental safety and biosafety in construction biotechnology. *World Journal of Microbiology & Biotechnology, 35*(2), 26. DOI: 10.1007/s11274-019-2598-9 PMID: 30666430

Jahan, N., Tahmid, M., Shoronika, A. Z., Fariha, A., Roy, H., Pervez, M., Cai, Y., Naddeo, V., & Islam, M. S. (2022). A Comprehensive Review on the Sustainable Treatment of Textile Wastewater: Zero Liquid Discharge and Resource Recovery Perspectives. *Sustainability (Basel), 14*(22), 15398. DOI: 10.3390/su142215398

Jain, M., Olsen, H. E., Paten, B., & Akeson, M. (2016). The Oxford Nanopore MinION: Delivery of nanopore sequencing to the genomics community. *Genome Biology, 17*(1), 239. DOI: 10.1186/s13059-016-1103-0 PMID: 27887629

Jaishankar, M., Tseten, T., Anbalagan, N., Mathew, B. B., & Beeregowda, K. N. (2014). Toxicity, mechanism and health effects of some heavy metals. *Interdisciplinary Toxicology, 7*(2), 60–72. DOI: 10.2478/intox-2014-0009 PMID: 26109881

Jaiswal, S., & Shukla, P. (2020). Alternative strategies for microbial remediation of pollutants via synthetic biology. *Frontiers in Microbiology*, *11*, 808. DOI: 10.3389/fmicb.2020.00808 PMID: 32508759

Jallouli, S., Buonerba, A., Borea, L., Hasan, S. W., Belgiorno, V., Ksibi, M., & Naddeo, V. (2023). Living membrane bioreactor for highly effective and eco-friendly treatment of textile wastewater. *The Science of the Total Environment*, *871*(January), 161963. DOI: 10.1016/j.scitotenv.2023.161963 PMID: 36737022

Jamee, R., & Siddique, R. (2019). Biodegradation of synthetic dyes of textile effluent by microorganisms: An environmentally and economically sustainable approach. *European Journal of Microbiology & Immunology*, *9*(4), 114–118. DOI: 10.1556/1886.2019.00018 PMID: 31934362

Janusz, G., Pawlik, A., Sulej, J., Świderska-Burek, U., Jarosz-Wilkołazka, A., & Paszczyński, A. (2017). Lignin degradation: Microorganisms, enzymes involved, genomes analysis and evolution. *FEMS Microbiology Reviews*, *41*(6), 941–962. DOI: 10.1093/femsre/fux049 PMID: 29088355

Jayakumar, V., Govindaradjane, S., Rajamohan, N., & Rajasimman, M. (2022). Biosorption potential of brown algae, Sargassum polycystum, for the removal of toxic metals, cadmium and zinc. *Environmental Science and Pollution Research International*, *29*(28), 41909–41922. DOI: 10.1007/s11356-021-15185-7 PMID: 34275071

Jayakumar, V., Govindaradjane, S., & Rajasimman, M. (2021). Efficient adsorptive removal of zinc by Green marine macro alga Caulerpa scalpelliformis–characterization, optimization, modeling, isotherm, kinetic, thermodynamic, desorption and regeneration studies. *Surfaces and Interfaces*, *22*, 100798. DOI: 10.1016/j.surfin.2020.100798

Jayasekara, S. K., Joni, H. D., Jayantha, B., Dissanayake, L., Mandrell, C., Sinharage, M. M. S., Molitor, R., Jayasekara, T., Sivakumar, P., & Jayakody, L. N. (2023). Trends in in-silico guided engineering of efficient polyethylene terephthalate (PET) hydrolyzing enzymes to enable bio-recycling and upcycling of PET. *Computational and Structural Biotechnology Journal*, *21*, 3513–3521. DOI: 10.1016/j.csbj.2023.06.004 PMID: 37484494

Jeong, S. H., Lee, H. J., & Lee, S. J. (2023). Recent advances in CRISPR-Cas technologies for synthetic biology. *The Journal of Microbiology*, *61*(1), 13–36. DOI: 10.1007/s12275-022-00005-5 PMID: 36723794

Jerez, C. A. (2017). Biomining of metals: How to access and exploit natural resource sustainably. *Microbial Biotechnology, 10*(5), 1191–1193. DOI: 10.1111/1751-7915.12792 PMID: 28771998

Jia, Y., Ruan, R., Qu, J., Tan, Q., Sun, H., & Niu, X. (2024). Multi-scale and trans-disciplinary research and technology developments of heap bioleaching. *Minerals (Basel), 14*(8), 808. DOI: 10.3390/min14080808

Johns, G., Samuel, V., Freemantle, L., Lewis, J., & Waddington, L. (2022). The global prevalence of depression and anxiety among doctors during the covid-19 pandemic: Systematic review and meta-analysis. *Journal of Affective Disorders, 298*, 431–441. DOI: 10.1016/j.jad.2021.11.026 PMID: 34785264

Johns, N. I., Blazejewski, T., Gomes, A. L. C., & Wang, H. H. (2016). Principles for designing synthetic microbial communities. *Current Opinion in Microbiology, 31*, 146–153. DOI: 10.1016/j.mib.2016.03.010 PMID: 27084981

Jones, T. S., Oliveira, S. M., Myers, C. J., Voigt, C. A., & Densmore, D. (2022). Genetic circuit design automation with Cello 2.0. *Nature Protocols, 17*(4), 1097–1113. DOI: 10.1038/s41596-021-00675-2 PMID: 35197606

Jorge, A. M. S., Athira, K. K., Alves, M. B., Gardas, R. L., & Pereira, J. F. B. (2023). Textile dyes effluents: A current scenario and the use of aqueous biphasic systems for the recovery of dyes. *Journal of Water Process Engineering, 55*, 104125. DOI: 10.1016/j.jwpe.2023.104125

Joshi, H. K., Vishwakarma, M. C., Kumar, R., Sharma, H., Bhandari, N. S., & Joshi, S. K. (2023). The biosorption of Zn2+ by various biomasses from wastewater: A review. *Journal of Water Process Engineering, 56*, 104389. DOI: 10.1016/j.jwpe.2023.104389

Juárez-Maldonado, A., Tortella, G., Rubilar, O., Fincheira, P., & Benavides-Mendoza, A. (2021). Biostimulation and toxicity: The magnitude of the impact of nanomaterials in microorganisms and plants. *Journal of Advanced Research, 31*, 113–126. DOI: 10.1016/j.jare.2020.12.011 PMID: 34194836

Kacser, H. (1973). The control of flux. Symposia of the Society for Experimental Biology, Kadier, A., Ilyas, R., Huzaifah, M., Harihastuti, N., Sapuan, S., Harussani, M., Azlin, M., Yuliasni, R., Ibrahim, R., & Atikah, M. (2021). Use of industrial wastes as sustainable nutrient sources for bacterial cellulose (BC) production: Mechanism, advances, and future perspectives. *Polymers, 13*(19), 3365.

Kaczorek, E., Pacholak, A., Zdarta, A., & Smułek, W. (2018). The Impact of Biosurfactants on Microbial Cell Properties Leading to Hydrocarbon Bioavailability Increase. *Colloids and Interfaces*, *2*(3), 35. DOI: 10.3390/colloids2030035

Kadier, A., Simayi, Y., Abdeshahian, P., Azman, N. F., Chandrasekhar, K., & Kalil, M. S. (2016). A comprehensive review of microbial electrolysis cells (MEC) reactor designs and configurations for sustainable hydrogen gas production. *Alexandria Engineering Journal*, *55*(1), 427–443. DOI: 10.1016/j.aej.2015.10.008

Kaida, H., Syed, M. A., Shukor, M. Y., & Othman, A. R.da. (2021). Biodegradation of linear alkylbenzene sulfonates (LAS): A mini review. *Bioremediation Science and Technology Research*, *9*(1), 1–6. DOI: 10.54987/bstr.v9i1.590

Kakde, P., & Sharma, J. (2024). Microbial Bioremediation of Petroleum Contaminated Soil: Structural Complexity, Degradation Dynamics and Advanced Remediation Techniques. *Journal of Pure & Applied Microbiology*, *18*(4), 2244–2261. DOI: 10.22207/JPAM.18.4.28

Kakoi, B., Kaluli, J. W., Ndiba, P., & Thiong'o, G. (2017). Optimization of Maerua Decumbent bio-coagulant in paint industry wastewater treatment with response surface methodology. *Journal of Cleaner Production*, *164*, 1124–1134. DOI: 10.1016/j.jclepro.2017.06.240

Kalhor, A. X., Movafeghi, A., Mohammadi-Nassab, A. D., Abedi, E., & Bahrami, A. (2017). Potential of the green alga Chlorella vulgaris for biodegradation of crude oil hydrocarbons. *Marine Pollution Bulletin*, *123*(1-2), 286–290. DOI: 10.1016/j.marpolbul.2017.08.045 PMID: 28844453

Kallawar, G. A., & Bhanvase, B. A. (2024). A review on existing and emerging approaches for textile wastewater treatments: Challenges and future perspectives. *Environmental Science and Pollution Research International*, *31*(2), 1748–1789. DOI: 10.1007/s11356-023-31175-3 PMID: 38055170

Kamm, B., & Kamm, M. (2004). Principles of biorefineries. *Applied Microbiology and Biotechnology*, *64*(2), 137–145. DOI: 10.1007/s00253-003-1537-7 PMID: 14749903

Kapoor, R. T., Salvadori, M. R., Rafatullah, M., Siddiqui, M. R., Khan, M. A., & Alshareef, S. A. (2021). Exploration of Microbial Factories for Synthesis of Nanoparticles – A Sustainable Approach for Bioremediation of Environmental Contaminants. *Frontiers in Microbiology*, *12*, 658294. DOI: 10.3389/fmicb.2021.658294 PMID: 34149647

Karigar, C. S., & Rao, S. S. (2011). Role of microbial enzymes in the bioremediation of pollutants: A review. *Enzyme Research, 2011*, 2011. DOI: 10.4061/2011/805187 PMID: 21912739

Karimi, M. A., Paryan, M., Fard, G. B., Sadeghian, H., Zarrinfar, H., & Bafghi, M. H. (2025). Challenges and Opportunities in the Application of CRISPR-Cas9: A Review on Genomic Editing and Therapeutic Potentials. *Medical Principles and Practice : International Journal of the Kuwait University, Health Science Centre*, 1–30. Advance online publication. DOI: 10.1159/000547334 PMID: 40675140

Karthika, A., Seenivasagan, R., Kasimani, R., Sudhakara Rao, J., & Poonkuzhali, K. (2024). Microbial technologies for sustainable textile effluent treatment: A review. *Journal of Environmental Chemical Engineering, 12*(5), 113275. DOI: 10.1016/j.jece.2024.113275

Katoh, K., & Standley, D. M. (2016). A simple method to control over-alignment in the MAFFT multiple sequence alignment program. *Bioinformatics (Oxford, England), 32*(13), 1933–1942. DOI: 10.1093/bioinformatics/btw108 PMID: 27153688

Kauppi, S., Sinkkonen, A., & Romantschuk, M. (2011). Enhancing bioremediation of diesel-fuel-contaminated soil in a boreal climate: Comparison of biostimulation and bioaugmentation. *International Biodeterioration & Biodegradation, 65*(2), 359–368. DOI: 10.1016/j.ibiod.2010.10.011

Kavitha, G., Govindhan, M., & Premkumar, S. (2025). Dye pollution and its implications for human health, aquatic ecosystems, and sustainable wastewater treatment: A comprehensive review. *Journal of Water Process Engineering, 80*, 109071. DOI: 10.1016/j.jwpe.2025.109071

Kaya, M. (2016). Recovery of metals and nonmetals from electronic waste by physical and chemical recycling processes. *Waste Management, 57*, 64–90. DOI: 10.1016/j.wasman.2016.08.004 PMID: 27543174

Keawboonchu, J., Thepanondh, S., Kultan, V., Pinthong, N., Malakan, W., & Robson, M. G. (2023). Integrated Sustainable Management of Petrochemical Industrial Air Pollution. *International Journal of Environmental Research and Public Health, 20*(3), 2280. DOI: 10.3390/ijerph20032280 PMID: 36767648

Kebede, G., Tafese, T., Abda, E. M., Kamaraj, M., & Assefa, F. (2021). Factors influencing the bacterial bioremediation of hydrocarbon contaminants in the soil: Mechanisms and impacts. *Journal of Chemistry, 2021*, 9823362. DOI: 10.1155/2021/9823362

Khalil, A. S., & Collins, J. J. (2010). Synthetic biology: Applications come of age. *Nature Reviews. Genetics, 11*(5), 367–379. DOI: 10.1038/nrg2775 PMID: 20395970

Khan, N. (2023). *Analysis of Market Preference in Paint Industry* (Doctoral dissertation, BIITM).

Khan, A. A., Gul, J., Naqvi, S. R., Ali, I., Farooq, W., Liaqat, R., AlMohamadi, H., Štěpanec, L., & Juchelková, D. (2022). Recent progress in microalgae-derived biochar for the treatment of textile industry wastewater. *Chemosphere, 306*(July), 135565. Advance online publication. DOI: 10.1016/j.chemosphere.2022.135565 PMID: 35793745

Khan, A., Nayarisseri, A., & Singh, S. K. (2025). Characterization and optimization of azo dyes degrading microbes isolated from textile effluent. *Scientific Reports, 15*(1), 1–23. DOI: 10.1038/s41598-025-95359-2 PMID: 40175441

Khan, M. D., Abdulateif, H., Ismail, I. M., Sabir, S., & Khan, M. Z. (2015). Bioelectricity Generation and Bioremediation of an Azo-Dye in a Microbial Fuel Cell Coupled Activated Sludge Process. *PLoS One, 10*(10), e0138448. DOI: 10.1371/journal.pone.0138448 PMID: 26496083

Khan, M. F. (2025). Recent Advances in Microbial Enzyme Applications for Sustainable Textile Processing and Waste Management. *Sci, 7*(2), 46. Advance online publication. DOI: 10.3390/sci7020046

Khan, R. A. (2022). Detergents. In *Environmental micropollutants* (pp. 117–130). Elsevier., DOI: 10.1016/B978-0-323-90555-8.00014-3

Khan, S., & Malik, A. (2013). Environmental and health effects of textile industry wastewater. In *Environmental deterioration and human health: Natural and anthropogenic determinants* (pp. 55–71). Springer Netherlands.

Khattab, T. A., Abdelrahman, M. S., & Rehan, M. (2020). Textile dyeing industry: Environmental impacts and remediation. *Environmental Science and Pollution Research International, 27*(4), 3803–3818. DOI: 10.1007/s11356-019-07137-z PMID: 31838699

Khodak, Y. A., Shaifutdinov, R. R., Khasanov, D. S., Orlova, N. A., & Vorobiev, I. I. (2023). Location and orientation of the genetic toxin–antitoxin element hok/sok in the plasmid affect expression of pharmaceutically significant proteins in bacterial cells. *Biochemistry. Biokhimiia, 88*(9), 1326–1337. DOI: 10.1134/S0006297923090122 PMID: 37770399

Khurshid, C. A., Mahdi, K., Ahmed, O. I., Osman, R., Rahman, M., & Ritsema, C. (2022). Assessment of potentially toxic elements in the urban soil and plants of Kirkuk City in Iraq. *Sustainability (Basel)*, *14*(9), 5655. DOI: 10.3390/su14095655

Kim, D., & Lee, J. W. (2020). Genetic biocontainment systems for the safe use of engineered microorganisms. *Biotechnology and Bioprocess Engineering; BBE*, *25*(6), 974–984. DOI: 10.1007/s12257-020-0070-1 PMID: 33437151

Kim, G. B., Choi, S. Y., Cho, I. J., Ahn, D.-H., & Lee, S. Y. (2023). Metabolic engineering for sustainability and health. *Trends in Biotechnology*, *41*(3), 425–451. DOI: 10.1016/j.tibtech.2022.12.014 PMID: 36635195

Kim, O. D., Rocha, M., & Maia, P. (2018). A Review of Dynamic Modeling Approaches and Their Application in Computational Strain Optimization for Metabolic Engineering. *Frontiers in Microbiology*, *9*, 378885. DOI: 10.3389/fmicb.2018.01690 PMID: 30108559

Kinejara, A. (2023). The 4 Rs of Recycling: Reduce, Reuse, Recycle, Repurpose. https://www.reelpaper.com/blogs/reel-talk/4rs-of-recycling. Published January 11, 2023. Downloaded 12 October 2025

Kirchherr, J., Reike, D., & Hekkert, M. (2017). Conceptualizing the circular economy: An analysis of 114 definitions. *Resources, Conservation and Recycling*, *127*, 221–232. DOI: 10.1016/j.resconrec.2017.09.005

Knight, R., Vrbanac, A., Taylor, B. C., Aksenov, A., Callewaert, C., Debelius, J., Gonzalez, A., Kosciolek, T., McCall, L. I., McDonald, D., Melnik, A. V., Morton, J. T., Navas, J., Quinn, R. A., Sanders, J. G., Swafford, A. D., Thompson, L. R., Tripathi, A., Xu, Z. Z., & Dorrestein, P. C. (2018). Best practices for analysing microbiomes. *Nature Reviews. Microbiology*, *16*(7), 410–422. DOI: 10.1038/s41579-018-0029-9 PMID: 29795328

Kocaman, S. (2025). Impacts of dye-contaminated effluents on the environment. In *Engineered Biocomposites for Dye Adsorption* (Number 2022, pp. 11–22). INC. DOI: 10.1016/B978-0-443-29877-6.00002-0

Koch, H., Van Kessel, M. A. H. J., & Lucker, S. (2019). Complete nitrification: Insights into the ecophysiology of comammox *Nitrospira*. *Applied Microbiology and Biotechnology*, *103*(1), 177–189. DOI: 10.1007/s00253-018-9486-3 PMID: 30415428

Kochhar, N., Shrivastava, S., Ghosh, A., Rawat, V. S., Sodhi, K. K., & Kumar, M. (2022). Perspectives on the microorganism of extreme environments and their applications. *Current Research in Microbial Sciences*, *3*, 100134. DOI: 10.1016/j.crmicr.2022.100134 PMID: 35909612

Kolya, H., & Kang, C. W. (2024). Toxicity of metal oxides, dyes, and dissolved organic matter in water: Implications for the environment and human health. *Toxics*, *12*(2), 111. DOI: 10.3390/toxics12020111 PMID: 38393206

Komor, A. C., Kim, Y. B., Packer, M. S., Zuris, J. A., & Liu, D. R. (2016). Programmable editing of a target base in genomic DNA without double-stranded DNA cleavage. *Nature*, *533*(7603), 420–424. DOI: 10.1038/nature17946 PMID: 27096365

Koohsaryan, E., Anbia, M., Sepehrian, M., & Maghsoodlu, M. (2021). Facile hydrothermal synthesis of hierarchical sodium P zeolite as a non-phosphate detergent builder. *Journal of Surfactants and Detergents*, *24*(1), 85–97. DOI: 10.1002/jsde.12452

Kosgey, K., Awolusi, O. O., Kiambi, S. L., Allam, M., Ismail, A., Bux, F., Chandran, K., & Kumari, S. (2021). Process performance and microbial community structures in three ANAMMOX-mediated systems with different mixing conditions. *Journal of Environmental Chemical Engineering*, *9*(6), 106466. DOI: 10.1016/j.jece.2021.106466

Kosgey, K., Zungu, P. V., Bux, F., & Kumari, S. (2022). Biological nitrogen removal from low carbon wastewater. *Frontiers in Microbiology*, *13*, 968812. DOI: 10.3389/fmicb.2022.968812 PMID: 36466689

Kotova, I. B., Taktarova, Y. V., Tsavkelova, E. A., Egorova, M. A., Bubnov, I. A., Malakhova, D. V., Shirinkina, L. I., Sokolova, T. G., & Bonch-Osmolovskaya, E. A. (2021). Microbial Degradation of Plastics and Approaches to Make it More Efficient. *Microbiology*, *90*(6), 671–701. DOI: 10.1134/S0026261721060084

Koul, B., & Taak, P. (2018). *Biotechnological strategies for effective remediation of polluted soils*. Springer. DOI: 10.1007/978-981-13-2420-8

Kour, D., Khan, S. S., Kour, H., Kaur, T., Devi, R., Rai, P. K., Judy, C., McQuestion, C., Bianchi, A., Spells, S., & Mohan, R. (2022). Microbe-mediated bioremediation: Current research and future challenges. *Journal of Applied Biology and Biotechnology*, *10*(2), 6–24. DOI: 10.7324/JABB.2022.10s202

Kraus, E. A., Prithiviraj, B., & Hernandez, M. (2025). Advancing transcriptomic profiling of airborne bacteria. *Applied and Environmental Microbiology*, *91*(5), 1–14. DOI: 10.1128/aem.00148-25 PMID: 40293243

Krieger, A. G., Zhang, J., & Lin, X. N. (2021). Temperature regulation as a tool to program synthetic microbial community composition. *Biotechnology and Bioengineering, 118*(3), 1381–1392. DOI: 10.1002/bit.27662 PMID: 33399224

Krishnan, J., Arvind Kishore, A., Suresh, A., Madhumeetha, B., & Gnana Prakash, D. (2017). Effect of pH, inoculum dose and initial dye concentration on the removal of azo dye mixture under aerobic conditions. *International Biodeterioration & Biodegradation, 119*, 16–27. DOI: 10.1016/j.ibiod.2016.11.024

Kroll, J., Klinter, S., Schneider, C., Voß, I., & Steinbüchel, A. (2010). Plasmid addiction systems: Perspectives and applications in biotechnology. *Microbial Biotechnology, 3*(6), 634–657. DOI: 10.1111/j.1751-7915.2010.00170.x PMID: 21255361

Krzmarzick, M. J., Taylor, D. K., Fu, X., & McCutchan, A. L. (2018). Diversity and niche of archaea in bioremediation. *Archaea : an International Microbiological Journal, 319*, 1–17. Advance online publication. DOI: 10.1155/2018/3194108 PMID: 30254509

Kubota, Y., Shiono, T., Kusumoto, B., & Fujinuma, J. (2020). Multiple drivers of the COVID-19 spread: The roles of climate, international mobility, and region-specific conditions. *PLoS One, 15*(9), e0239385. DOI: 10.1371/journal.pone.0239385 PMID: 32966315

Kuenen, J. G. (2020). Anammox and beyond. *Environmental Microbiology, 22*(2), 525–536. DOI: 10.1111/1462-2920.14904 PMID: 31867834

Kugarajah, V., Nisha, K. N., Jayakumar, R., Sahabudeen, S., Ramakrishnan, P., & Mohamed, S. (2023). Significance of microbial genome in environmental remediation. *Microbiological Research, 271*, 127360. DOI: 10.1016/j.micres.2023.127360 PMID: 36931127

Kulikova, A. V., Diaz, D. J., Chen, T., Cole, T. J., Ellington, A. D., & Wilke, C. O. (2023). Two sequence- and two structure-based ML models have learned different aspects of protein biochemistry. *BioRxiv*, 2023.03.20.533508. DOI: 10.1101/2023.03.20.533508

Kulikova, N. A., & Perminova, I. V. (2021). Interactions between humic substances and microorganisms and their implications for nature-like bioremediation technologies. *Molecules (Basel, Switzerland), 26*(9), 2706. DOI: 10.3390/molecules26092706 PMID: 34063010

Kulkarni, A. N., Watharkar, A. D., Rane, N. R., & Jeon, B. (2018). Ecotoxicology and Environmental Safety Decolorization and detoxification of dye mixture and textile effluent by lichen *Dermatocarpon vellereceum* in fixed-bed upflow bioreactor with subsequent oxidative stress study. *Ecotoxicology and Environmental Safety, 148*(September 2017), 17–25. DOI: 10.1016/j.ecoenv.2017.10.001

Kumar, A. Utkarsh, D., Kaman, S., Satya Prakash, G., and Mirza S. Jamal Beg. (2021). "Structure and properties of dyes and pigments." In *Dyes and pigments-novel applications and waste treatment*. IntechOpen, Available at DOI: 10.5772/intechopen.97104

Kumar, P., Chitara, D., Sengupta, S., Banerjee, P., & Rai, S. N. (2025). Microbial consortia in biotechnology: Applications and challenges in industrial processes. *3 Biotech, 15*(386). DOI: 10.1007/s13205-025-04558-1

Kumar, A., Thakur, A. K., Gaurav, G. K., Klemeš, J. J., Sandhwar, V. K., Pant, K. K., & Kumar, R. (2023). A critical review on sustainable hazardous waste management strategies: A step towards a circular economy. *Environmental Science and Pollution Research International, 30*(48), 105030–105055. DOI: 10.1007/s11356-023-29511-8 PMID: 37725301

Kumari, R., Singha, L. P., & Shukla, P. (2023). Biotechnological potential of microbial bio-surfactants, their significance, and diverse applications. *FEMS Microbes, 4*, xtad015. Advance online publication. DOI: 10.1093/femsmc/xtad015 PMID: 37614639

Kumari, U. (2024). Textile dyes and their impact on the natural environment. In *Dye Pollution from Textile Industry: Challenges and Opportunities for Sustainable Development* (pp. 17–30). Springer Nature Singapore. DOI: 10.1007/978-981-97-5341-3_2

Kumar, M., Mishra, A., Patel, S. K., Kushwaha, J., Singh, S., Mishra, V., Singh, D., Singh, V., Giri, B. S., Singhania, R. R., & Singh, D. (2025). Environmental Impacts and Strategies for Bioremediation of Dye-Containing Wastewater. *Bioengineering (Basel, Switzerland), 12*(10), 1043. DOI: 10.3390/bioengineering12101043 PMID: 41155042

Kumar, M., Rohilla, A., Singh, D., Panwar, A., & Manna, S. (2025). Microbial enzyme-based water treatment in smart cities: A synoptic review. *Current Opinion in Environmental Science and Health, 46*, 100640. DOI: 10.1016/j.coesh.2025.100640

Kumar, P., & Agrawal, K. (2025). *Ecological Restorations of Wetlands by Geomicrobes. Mineral Transformation and Bioremediation by Geo-Microbes*. Springer.

Kumar, R., Farda, B., Mignini, A., Djebaili, R., Koolman, L., Paul, A., Mondal, S., Joel, J. M., Pandit, A., Panneerselvam, P., Pellegrini, M., & Mitra, D. (2025). Microbial Solutions in Agriculture: Enhancing Soil Health and Resilience through Bio-Inoculants and Bioremediation. *Bacteria*, *4*(3), 28. DOI: 10.3390/bacteria4030028

Kumar, R., Singh, L., & Zularisam, A. W. (2017). Exoelectrogens: Recent advances in molecular drivers involved in extracellular electron transfer and strategies used to improve it for microbial fuel cell applications. *Renewable & Sustainable Energy Reviews*, *56*, 1322–1336. DOI: 10.1016/j.rser.2015.12.029

Kumar, S., Stecher, G., Li, M., Knyaz, C., & Tamura, K. (2018). MEGA X: Molecular evolutionary genetics analysis across computing platforms. *Molecular Biology and Evolution*, *35*(6), 1547–1549. DOI: 10.1093/molbev/msy096 PMID: 29722887

Kümmerer, K. (2009). The presence of pharmaceuticals in the environment due to human use: Present knowledge and future challenges. *Journal of Environmental Management*, *90*(8), 2354–2366. DOI: 10.1016/j.jenvman.2009.01.023 PMID: 19261375

Kuppan, N., Padman, M., Mahadeva, M., Srinivasan, S., & Devarajan, R. (2024). A comprehensive review of sustainable bioremediation techniques: Eco friendly solutions for waste and pollution management. *Waste Management Bulletin*, *2*(3), 154–171. DOI: 10.1016/j.wmb.2024.07.005

Kurade, M. B., Waghmode, T. R., Patil, S. M., Jeon, B. H., & Govindwar, S. P. (2017). Monitoring the gradual biodegradation of dyes in a simulated textile effluent and development of a novel triple layered fixed bed reactor using a bacterium-yeast consortium. *Chemical Engineering Journal*, *307*, 1026–1036. DOI: 10.1016/j.cej.2016.09.028

Kuriya, Y., & Araki, M. (2020). Dynamic Flux Balance Analysis to Evaluate the Strain Production Performance on Shikimic Acid Production in Escherichia coli. *Metabolites*, *10*(5), 198. DOI: 10.3390/metabo10050198 PMID: 32429049

Kusumlata, A., Ambade, B., Kumar, A., & Gautam, S. (2024). Sustainable Solutions: Reviewing the Future of Textile Dye Contaminant Removal with Emerging Biological Treatments. *Limnological Review*, *24*(2), 126–149. DOI: 10.3390/limnolrev24020007

Kuypers, M. M. M., Marchant, H. K., & Kartal, B. (2018). The microbial nitrogen-cycling network. *Nature Reviews. Microbiology*, *16*(5), 263–276. DOI: 10.1038/nrmicro.2018.9 PMID: 29398704

Kylilis, N., Tuza, Z. A., Stan, G. B., & Polizzi, K. M. (2018). Tools for engineering coordinated system behaviour in synthetic microbial consortia. *Nature Communications*, *9*(1), 2677. DOI: 10.1038/s41467-018-05046-2 PMID: 29992956

Lackner, S., Glibert, E. M., Vlaeminck, S. E., Joss, A., Horn, H., & Van Loosdrecht, M. C. M. (2014). Full-scale partial nitritation/anammox experiences – An application survey. *Water Research*, *55*, 292–303. DOI: 10.1016/j.watres.2014.02.032 PMID: 24631878

Lacy, P., & Rutqvist, J. (2015). *Waste to wealth: The circular economy advantage*. Palgrave Macmillan.

Lade, H. S., Waghmode, T. R., Kadam, A. A., & Govindwar, S. P. (2012). Enhanced biodegradation and detoxification of disperse azo dye Rubine GFL and textile industry effluent by defined fungal-bacterial consortium. *International Biodeterioration & Biodegradation*, *72*, 94–107. DOI: 10.1016/j.ibiod.2012.06.001

Lahel, A., Fanta, A. B., Sergienko, N., Shakya, M., López, M. E., Behera, S. K., & Park, H. S. (2016). Effect of process parameters on the bioremediation of diesel contaminated soil by mixed microbial consortia. *International Biodeterioration & Biodegradation*, *113*, 375–385. DOI: 10.1016/j.ibiod.2016.05.005

Lam, T. J., & Ye, Y. (2022). Meta-analysis of microbiome association networks reveal patterns of dysbiosis in diseased microbiomes. *Scientific Reports*, *12*(1), 17482. DOI: 10.1038/s41598-022-22541-1 PMID: 36261472

Lanfranconi, I., Ceretta, M. B., Bertola, N., Wolski, E. A., & Durruty, I. (2022). Textile dyeing wastewater treatment by *Penicillium chrysogenum*: Design of a sustainable process. *Water Science and Technology*, *86*(2), 292–301. DOI: 10.2166/wst.2022.204 PMID: 35906908

Langone, M., Yan, J., Haaijer, S. C. M., Op den Camp, H. J. M., Jetten, M. S. M., & Andreottola, G. (2014). Coexistence of nitrifying, anammox and denitrifying bacteria in a sequencing batch reactor. *Frontiers in Microbiology*, *5*, 28. DOI: 10.3389/fmicb.2014.00028 PMID: 24550899

Larson, R. J., & Woltering, D. M. (2020). Linear alkylbenzene sulfonate (LAS). In *Fundamentals of Aquatic Toxicology* (pp. 859–882). CRC Press. DOI: 10.1201/9781003075363-35

Larsson, D. G. J., Andremont, A., Bengtsson-Palme, J., Brandt, K. K., de Roda Husman, A. M., Fagerstedt, P., Fick, J., Flach, C. F., Gaze, W. H., Kuroda, M., Kvint, K., Laxminarayan, R., Manaia, C. M., Nielsen, K. M., Plant, L., Ploy, M. C., Segovia, C., Simonet, P., Smalla, K., & Wernersson, A. S. (2018). Environmental factors influencing the development and spread of antibiotic resistance. *FEMS Microbiology Reviews*, *42*(1), 68–80. DOI: 10.1093/femsre/fux053 PMID: 29069382

Larsson, D. G. J., de Pedro, C., & Paxeus, N. (2007). Effluent from drug manufactures contains extremely high levels of pharmaceuticals. *Journal of Hazardous Materials*, *148*(3), 751–755. DOI: 10.1016/j.jhazmat.2007.07.008 PMID: 17706342

Lassoued, R., Macall, D. M., Smyth, S. J., Phillips, P. W., & Hesseln, H. (2020). How should we regulate products of new breeding techniques? Opinion of surveyed experts in plant biotechnology. *Biotechnology Reports (Amsterdam, Netherlands)*, *26*, e00460. DOI: 10.1016/j.btre.2020.e00460 PMID: 32617264

Laureni, M., Weissbrodt, D. G., Szivak, I., Robin, O., Nielsen, J. L., Morgenroth, E., & Joss, A. (2015). Activity and growth of anammox biomass on aerobically pre-treated municipal wastewater. *Water Research*, *80*, 325–336. DOI: 10.1016/j.watres.2015.04.026 PMID: 26024830

Laurent, A., Bakas, I., Clavreul, J., Bernstad, A., Niero, M., Gentil, E., Hauschild, M. Z., & Christensen, T. H. (2014). Review of LCA studies of solid waste management systems: Part I—Lessons learned and perspectives. *Waste Management*, *34*(3), 573–588. DOI: 10.1016/j.wasman.2013.10.045 PMID: 24369845

Lawson, C. E., Harcombe, W. R., Hatzenpichler, R., Lindemann, S. R., Löffler, F. E., O'Malley, M. A., García Martín, H., Pfleger, B. F., Raskin, L., Venturelli, O. S., Weissbrodt, D. G., Noguera, D. R., & McMahon, K. D. (2019). Common principles and best practices for engineering microbiomes. *Nature Reviews. Microbiology*, *17*(12), 725–741. DOI: 10.1038/s41579-019-0255-9 PMID: 31548653

Le Pape, P., Battaglia-Brunet, F., Parmentier, M., Joulian, C., Gassaud, C., Fernandez-Rojo, L., Guigner, J. M., Ikogou, M., Stetten, L., Olivi, L., Casiot, C., & Morin, G. (2017). Complete removal of arsenic and zinc from a heavily contaminated acid mine drainage via an indigenous SRB consortium. *Journal of Hazardous Materials*, *321*, 764–772. DOI: 10.1016/j.jhazmat.2016.09.060 PMID: 27720469

Lea-Smith, D. J., Hassard, F., Coulon, F., Partridge, N., Horsfall, L., Parker, K. D. J., Smith, R. D. J., McCarthy, R. R., McKew, B., Gutierrez, T., Kumar, V., Dotro, G., Yang, Z., Curtis, T. P., Golyshin, P., Heaven, S., Jefferson, B., Jeffrey, P., Jones, D. L., & Krasnogor, N. (2025). Engineering biology applications for environmental solutions: Potential and challenges. *Nature Communications*, *16*(1), 3538. DOI: 10.1038/s41467-025-58492-0 PMID: 40229265

Lebeau, T. (2011). *Bioaugmentation for in situ soil remediation: how to ensure the success of such a process. Bioaugmentation, biostimulation and biocontrol*. Springer.

Lee, D. H., Park, E. G., Lee, Y. J., Jeong, H., Roh, H., Jeong, G., Kim, S., Kim, H., Lee, D. H., Park, E. G., Lee, Y. J., Jeong, H., Roh, H., Jeong, G., Kim, S., & Kim, H. (2025). Advances in artificial intelligence (AI) models and generative algorithms represent a new paradigm for genomics research. *International Journal of Molecular Sciences*, *26*(22), 10925. DOI: 10.3390/ijms262210925 PMID: 41303412

Lee, M. (2023). Deep learning in CRISPR-Cas systems: A review of recent studies. *Frontiers in Bioengineering and Biotechnology*, *11*, 1226182. Advance online publication. DOI: 10.3389/fbioe.2023.1226182 PMID: 37469443

Lee, S. Y., Li, H., Weber, W., & Yang, Z. (2025). Turning Pollution Into Solutions Using Microbes. *Frontiers for Young Minds*, *13*, 1575505. DOI: 10.3389/frym.2025.1575505

Lee, Y., Jeong, S. E., Hur, M., Ko, S., & Jeon, C. O. (2018). Construction and evaluation of a Korean native microbial consortium for the bioremediation of diesel fuel-contaminated soil in Korea. *Frontiers in Microbiology*, *9*, 2594. DOI: 10.3389/fmicb.2018.02594 PMID: 30425703

Leggieri, P. A., Liu, Y., Hayes, M., Connors, B., Seppälä, S., & Venturelli, O. S. (2021). Integrating Systems and Synthetic Biology to Understand and Engineer Microbiomes. *Annual Review of Biomedical Engineering*, *23*(1), 169–201. DOI: 10.1146/annurev-bioeng-082120-022836 PMID: 33781078

León, D. S., & Nogales, J. (2022). Toward merging bottom–up and top–down model-based designing of synthetic microbial communities. *Current Opinion in Microbiology*, *69*, 102169. DOI: 10.1016/j.mib.2022.102169 PMID: 35763963

Levin, L., Grassi, E., & Carballo, R. (2012). International Biodeterioration & Biodegradation Efficient azoic dye degradation by *Trametes trogii* and a novel strategy to evaluate products released. *International Biodeterioration & Biodegradation*, *75*, 214–222. DOI: 10.1016/j.ibiod.2012.10.005

Liang, Y., Ma, A., & Zhuang, G. (2022). Construction of environmental synthetic microbial consortia: Based on engineering and ecological principles. *Frontiers in Microbiology, 13*, 829717. DOI: 10.3389/fmicb.2022.829717 PMID: 35283862

Liao, C., Wang, T., Maslov, S., & Xavier, J. B. (2020). Modeling microbial cross-feeding at intermediate scale portrays community dynamics and species coexistence. *PLoS Computational Biology, 16*(8), e1008135. DOI: 10.1371/journal.pcbi.1008135 PMID: 32810127

Liao, X., Ye, M., Li, S., Liang, J., Zhou, S., Fang, X., Gan, Q., & Sun, S. (2021). Simultaneous recovery of valuable metal ions and tailings toxicity reduction using a mixed culture bioleaching process. *Journal of Cleaner Production, 316*, 128319. DOI: 10.1016/j.jclepro.2021.128319

Li, B., Deng, A., Li, K., Hu, Y., Li, Z., Shi, Y., & Lu, J. (2022). Viral infection and transmission in a large, well-traced outbreak caused by the SARS-CoV-2 Delta variant. *Nature Communications, 13*(1), 460. DOI: 10.1038/s41467-022-28089-y PMID: 35075154

Li, D., Ji, S., Guo, Y., & Sang, N. (2021). Ambient NO_2 exposure sex-specifically impairs myelin and contributes to anxiety and depression-like behaviors of C57BL/6J mice. *Journal of Hazardous Materials, 416*, 125836. DOI: 10.1016/j.jhazmat.2021.125836 PMID: 34492793

Li, J. F., Wang, J., Liu, Z., Liu, Q., Guo, C., Liao, C., Li, W., Guo, C., Zhang, T., & Liu, J. (2023). Comparative study on reproductive biology of two co-invasion tilapia species in the Shanmei reservoir. *Shui Sheng Sheng Wu Hsueh Bao, 47*, 657–665.

Li, J., Wang, X., Zou, J., Yang, K., Wang, X., Wang, Y., Zhang, H., Huang, H., Su, X., Yao, B., Luo, H., & Qin, X. (2023). Identification and characterization of the determinants of copper resistance in the acidophilic fungus *Acidomyces richmondensis* MEY-1 using the CRISPR/Cas9 System. *Applied and Environmental Microbiology, 89*(3), e02107–e02122. DOI: 10.1128/aem.02107-22 PMID: 36912653

Li, L., Dong, Y., Chen, Y., Jiao, J., & Zou, X. (2022). A new method for environmental risk assessment of pollutants based on multi-dimensional risk factors. *Toxics, 10*(11), 659. DOI: 10.3390/toxics10110659 PMID: 36355950

Li, L.-G., & Zhang, T. (2023). Plasmid-mediated antibiotic resistance gene transfer under environmental stresses: Insights from laboratory-based studies. *The Science of the Total Environment, 887*, 163870. DOI: 10.1016/j.scitotenv.2023.163870 PMID: 37149187

Lin, L. (2022). Bottom-up synthetic ecology study of microbial consortia to enhance lignocellulose bioconversion. *Biotechnology for Biofuels and Bioproducts, 15*(1), 14. Advance online publication. DOI: 10.1186/s13068-022-02113-1 PMID: 35418100

Liu, G., Catacutan, D. B., Rathod, K., Swanson, K., Jin, W., Mohammed, J. C., Chiappino-Pepe, A., Syed, S. A., Fragis, M., Rachwalski, K., Surette, M. G., Coombes, B. K., Magolan, J., Jaakkola, T., Barzilay, R., Collins, J. J., & Stokes, J. M. (2023). Deep learning-guided discovery of an antibiotic targeting *Acinetobacter baumannii. Nature Chemical Biology, 19*(11), [page numbers if known]. https://doi.org/DOI: 10.1038/s41589-023-01349-8

Liu, B., Dong, X., Cheng, H., Zheng, C., Chen, Z., Rodríguez, T. C., Liang, S.-Q., Xue, W., & Sontheimer, E. J. (2022). A split prime editor with untethered reverse transcriptase and circular RNA template. *Nature Biotechnology, 40*(9), 1388–1393. DOI: 10.1038/s41587-022-01255-9 PMID: 35379962

Liu, C. Y., Tian, A. M., Sun, F., & Yuan, P. (2020). Progress of ecological risk assessment methods and applications. *Keji Guanli Yanjiu, 40*(02), 79–83.

Liu, Q., Sun, S., Chen, S., Su, Y., Wang, Y., Tang, F., Zhao, C., & Li, L. (2023). A novel dehydrocoenzyme activator combined with a composite microbial agent TY for enhanced bioremediation of crude oil-contaminated soil. *Journal of Environmental Management, 331*, 117246. DOI: 10.1016/j.jenvman.2023.117246 PMID: 36642048

Liu, T. Y., & Doudna, J. A. (2020). Chemistry of Class 1 CRISPR-Cas effectors: Binding, editing, and regulation. *The Journal of Biological Chemistry, 295*(42), 14473–14487. DOI: 10.1074/jbc.REV120.007034 PMID: 32817336

Liu, X., Ma, Y., Jiang, Y., Guo, Y., Deng, Z., & He, X. (2025). Microbial Community Succession During Bioremediation of Petroleum-Contaminated Soils Using *Rhodococcus* sp. OS62-1 and Pseudomonas sp. P35. *Microorganisms, 14*(1), 7. DOI: 10.3390/microorganisms14010007 PMID: 41597527

Liu, Y. J., Nikolausz, M., & Wang, X. C. (2009). Biodegradation and detoxication of phenol by using free and immobilized cells of Acinetobacter sp. XA05 and Sphingomonas sp. FG03. *Journal of Environmental Science and Health. Part A, Toxic/Hazardous Substances & Environmental Engineering, 44*(2), 130–136. DOI: 10.1080/10934520802539673 PMID: 19123092

Liu, Z., Liu, Y., Zeng, G., Shao, B., Chen, M., Li, Z., Jiang, Y., Liu, Y., Zhang, Y., & Zhong, H. (2018). Application of molecular docking for the degradation of organic pollutants in the environmental remediation: A review. *Chemosphere, 203*, 139–150. DOI: 10.1016/j.chemosphere.2018.03.179 PMID: 29614407

Liv, B. (2024). Unlocking the Future of Biotechnology: The Transformative Potential of CRISPR and Emerging Gene-Editing Technologies. *Next Generation Journal for The Young Researchers, 8*(1). Advance online publication. DOI: 10.62802/8ewmr712

Liwarska-Bizukojc, E., Chojnacki, J., Klink, M., & Olejnik, D. (2018). Effect of the type of external carbon source on denitrification kinetics of wastewater. *Desalination and Water Treatment, 101*, 143–150. DOI: 10.5004/dwt.2018.21758

Li, X., Shen, X., Jiang, W., Xi, Y., & Li, S. (2024). Comprehensive review of emerging contaminants: Detection technologies, environmental impact, and management strategies. *Ecotoxicology and Environmental Safety, 278*, 116420. DOI: 10.1016/j.ecoenv.2024.116420 PMID: 38701654

Li, X., & Sung, S. (2015). Development of the combined nitritation-anammox process in an upflow anaerobic sludge blanket (UASB) reactor with anammox granules. *Chemical Engineering Journal, 281*, 837–843. DOI: 10.1016/j.cej.2015.07.016

Li, X., Wang, A., Wan, W., Luo, X., Zheng, L., He, G., Huang, D., Chen, W., & Huang, Q. (2021). High salinity inhibits soil bacterial community mediating nitrogen cycling. *Applied and Environmental Microbiology, 87*(21), e01366–e21. DOI: 10.1128/AEM.01366-21 PMID: 34406835

Li, X., Wu, S., Dong, Y., Fan, H., Bai, Z., & Zhuang, X. (2021). Engineering Microbial Consortia towards Bioremediation. *Water (Basel), 13*(20), 1–11. DOI: 10.3390/w13202928

Logan, B. E., Call, D., Cheng, S., Hamelers, H. V. M., Sleutels, T. H. J. A., Jeremiasse, A. W., & Rozendal, R. A. (2008). Microbial electrolysis cells for high yield hydrogen gas production from organic matter. *Environmental Science & Technology, 42*(23), 8630–8640. DOI: 10.1021/es801553z PMID: 19192774

Logan, B. E., & Rabaey, K. (2012). Conversion of wastes into bioelectricity and chemicals by using microbial electrochemical technologies. *Science, 337*(6095), 686–690. DOI: 10.1126/science.1217412 PMID: 22879507

Losoi, P. S., Santala, V. P., & Santala, S. M. (2019). Enhanced population control in a synthetic bacterial consortium by interconnected carbon cross-feeding. *ACS Synthetic Biology, 8*(12), 2642–2650. DOI: 10.1021/acssynbio.9b00316 PMID: 31751122

Lovley, D. R. (2003). Cleaning up with genomics: Applying molecular biology to bioremediation. *Nature Reviews. Microbiology, 1*(1), 35–44. DOI: 10.1038/nrmicro731 PMID: 15040178

Lovley, D. R. (2012). Electromicrobiology. *Annual Review of Microbiology, 66*(1), 391–409. DOI: 10.1146/annurev-micro-092611-150104 PMID: 22746334

Lovley, D. R. (2017). Syntrophy Goes Electric: Direct Interspecies Electron Transfer. *Annual Review of Microbiology, 71*(1), 643–664. DOI: 10.1146/annurev-micro-030117-020420 PMID: 28697668

Lovley, D. R., & Nevin, K. P. (2011). A shift in the current paradigm for electron exchange within microbial communities. *Energy & Environmental Science, 4*(12), 4896–4906. DOI: 10.1039/c1ee02229f

Lovley, D. R., & Nevin, K. P. (2013). Electrobiocommodities: Powering microbial production of fuels and commodity chemicals from carbon dioxide with electricity. *Current Opinion in Biotechnology, 24*(3), 385–390. DOI: 10.1016/j.copbio.2013.02.012 PMID: 23465755

Lúcia, A., Carissimi, E., Dotto, G. L., & Feris, L. A. (2018). *Biosorption of rhodamine B dye from dyeing stones effluents using the green microalgae Chlorella pyrenoidosa*. Cleaner Production., DOI: 10.1016/j.jclepro.2018.07.128

Lu, H., Diaz, D. J., Czarnecki, N. J., Zhu, C., Kim, W., Shroff, R., Acosta, D. J., Alexander, B. R., Cole, H. O., Zhang, Y., Lynd, N. A., Ellington, A. D., & Alper, H. S. (2022). Machine learning-aided engineering of hydrolases for PET depolymerization. *Nature, 604*(7907), 662–667. DOI: 10.1038/s41586-022-04599-z PMID: 35478237

Lü, H., Wei, J. L., Tang, G. X., Chen, Y. S., Huang, Y. H., Hu, R., & Li, Q. X. (2024). Microbial consortium degrading of organic pollutants: Source, degradation efficiency, pathway, mechanism and application. *Journal of Cleaner Production, 451*, 141913. DOI: 10.1016/j.jclepro.2024.141913

Luo, L., Chen, X., Liu, B., Nie, Y., & Wu, X. L. (2025). Strengthen or Weaken: Evolutionary Directions of Cross-Feeding After Formation. *Environmental Microbiology Reports, 17*(4), e70175. DOI: 10.1111/1758-2229.70175 PMID: 40784676

Luo, Y., Jin, X., Xie, H., Ji, X., Liu, Y., Guo, C., & Xu, J. (2023). Linear alkylbenzene sulfonate threats to surface waters at the national scale: A neglected traditional pollutant. *Journal of Environmental Management, 342*, 118344. DOI: 10.1016/j.jenvman.2023.118344 PMID: 37320921

Lu, Y. F., Ma, L. J., Ma, L., Shan, B., & Chang, J. J. (2018). Improvement of start-up and nitrogen removal of the anammox process in reactors inoculated with conventional activated sludge using biofilm carrier materials. *Environmental Technology, 39*(1), 59–67. DOI: 10.1080/09593330.2017.1294624 PMID: 28278777

Lyu, X., Nuhu, M., Candry, P., Wolfanger, J., Betenbaugh, M., Saldivar, A., & Shrestha, S. (2024). Top-down and bottom-up microbiome engineering approaches to enable biomanufacturing from waste biomass. *Journal of Industrial Microbiology & Biotechnology*, *51*, kuae025. Advance online publication. DOI: 10.1093/jimb/kuae025 PMID: 39003244

Maan, K., Baghel, R., Dhariwal, S., Sharma, A., Bakhshi, R., & Rana, P. (2023). Metabolomics and transcriptomics based multi-omics integration reveals radiation-induced altered pathway networking and underlying mechanism. *NPJ Systems Biology and Applications*, *9*(1), 1–13. DOI: 10.1038/s41540-023-00305-5 PMID: 37689794

MacArthur, E. (2013). Towards the circular economy. *Journal of Industrial Ecology*, *2*, 23–44.

Macchi, M., Festa, S., Nieto, E., Irazoqui, J. M., Vega-Vela, N. E., Junca, H., Valacco, M. P., Amadio, A. F., Morelli, I. S., & Coppotelli, B. M. (2021). Design and evaluation of synthetic bacterial consortia for optimized phenanthrene degradation through the integration of genomics and shotgun proteomics. *Biotechnology Reports (Amsterdam, Netherlands)*, *29*, e00588. DOI: 10.1016/j.btre.2021.e00588 PMID: 33489789

Madhav, S., Ahamad, A., Singh, P., & Mishra, P. K. (2018). A review of textile industry: Wet processing, environmental impacts, and effluent treatment methods. *Environmental Quality Management*, *27*(3), 31–41. DOI: 10.1002/tqem.21538

Madigan, M. T., Martinko, J. M., Stahl, D. A., & Clark, D. P. (2012). *Brock Biology of Microorganisms* (13th ed.). Pearson Education, Inc.

Maglione, G., Zinno, P., Tropea, A., Mussagy, C. U., Dufossé, L., Giuffrida, D., & Mondello, A. (2024). Microbes' role in environmental pollution and remediation: A bioeconomy focus approach. *AIMS Microbiology*, *10*(3), 723–755. DOI: 10.3934/microbiol.2024033 PMID: 39219757

Magnin, A., Pollet, E., Phalip, V., & Avérous, L. (2019). Evaluation of biological degradation of polyurethanes. *Biotechnology Advances*, *39*, 107457. DOI: 10.1016/j.biotechadv.2019.107457 PMID: 31689471

Magrí, A., & Lotti, T. (2023). Special issue on "Anammox-based processes for wastewater treatment". *Processes (Basel, Switzerland)*, *11*(5), 1422. DOI: 10.3390/pr11051422

Ma, H., Zhao, Y., Yang, K., Wang, Y., Zhang, C., & Ji, M. (2022). Application oriented bioaugmentation processes: Mechanism, performance improvement and scale-up. *Bioresource Technology, 344*, 126192. DOI: 10.1016/j.biortech.2021.126192 PMID: 34710609

Mahajan, P., & Kaushal, J. (2020). Phytoremediation of azo dye methyl red by macroalgae *Chara vulgaris L:* kinetic and equilibrium studies. *Environmental Science and Pollution Research.*

Mahajan, M., & Prakash, A. (2025). Bacterial consortia as potential bioremediation agents for wastewater treatment: A comprehensive review. *International Journal of Advancement in Life Sciences Research, 8*(1), 16–33. DOI: 10.31632/ijalsr.2025.v08i01.002

Mahartin, T. L. (2023). Waste management plan with reduce, reuse, recycle (3r) method. *Journal of Sustainability, Society, and eco-welfare*, 1 (1).

Mahieu, N. G., & Patti, G. J. (2017). Systems-level annotation of a metabolomics data set reduces 25,000 features to fewer than 1,000 unique metabolites. *Analytical Chemistry, 89*(19), 10397–10406. DOI: 10.1021/acs.analchem.7b02380 PMID: 28914531

Mahto, K. U., & Das, S. (2025). Electroactive biofilm communities in microbial fuel cells for the synergistic treatment of wastewater and bioelectricity generation. *Critical Reviews in Biotechnology, 45*(2), 434–453. DOI: 10.1080/07388551.2024.2372070 PMID: 39009474

Maile, F. J. (2021). Colorants in coatings. *Physical Sciences Reviews, 6*(11), 707–789. DOI: 10.1515/psr-2020-0160

Maitra, S. (2018). *In situ bioremediation: An overview.* Responsibility of Life Science Informatics Publications.

Ma, K., Wang, Y., Yang, X., Wang, C., Han, Y., Huang, X., & Pan, J. (2022). Analysis of the composition of culturable airborne microorganisms in the archaeological excavation protection site of the Nanhai No. 1 ancient shipwreck. *Frontiers in Microbiology, 13*, 958914. DOI: 10.3389/fmicb.2022.958914 PMID: 36090103

Makarova, K. S., Shmakov, S. A., Wolf, Y. I., Mutz, P., Beisel, C. L., Brouns, S. J., Charpentier, E., Cheng, D., Doudna, J., Haft, D. H., Horvath, P., Moineau, S., Mojica, F. J., Pausch, P., Shah, S. A., Siksnys, V., Terns, M. P., Tordoff, J., Venclovas, Č., & Koonin, E. V. (2025). An updated evolutionary classification of CRISPR–Cas systems including rare variants. *Nature Microbiology, 10*(12), 3346–3361. DOI: 10.1038/s41564-025-02180-8 PMID: 41198952

Makia, A. A., Al-Taeea, A. M., & Atwanb, Z. W. (2024). Enhancing the performance of *Methylorubrum extorquens* AAZ-using CRISPR/Cas9 for the degradation of hydrocarbon compounds. *Asia-Pacific Journal of Molecular Biology and Biotechnology, 32*, 39–48. DOI: 10.35118/apjmbb.2024.032.3.05

Malakar, N., Mitra, S., Toppo, P., & Mathur, P. (2020). Understanding the functional attributes of different microbial enzymes in bioremediation.

Maletić, S., Beljin, J., Tamindžija, D., Grgić, M., Jazić, J. M., Isakovski, M. K., & Rončević, S. (2021). Bioremediation perspective of historically contaminated sediment with polycyclic aromatic hydrocarbons. *International Journal of Sediment Research, 36*(4), 479–488. DOI: 10.1016/j.ijsrc.2020.11.004

Malla, M. A., Dubey, A., Yadav, S., Kumar, A., Hashem, A., & Abd Allah, E. F. (2018). Understanding and designing the strategies for the microbe-mediated remediation of environmental contaminants using omics approaches. *Frontiers in Microbiology, 9*, 1132. DOI: 10.3389/fmicb.2018.01132 PMID: 29915565

Malwe, A. S., Longwani, U., & Sharma, V. K. (2025). XenoBug: Machine learning-based tool to predict pollutant-degrading enzymes from environmental metagenomes. *NAR Genomics and Bioinformatics, 7*(2), lqaf037. Advance online publication. DOI: 10.1093/nargab/lqaf037 PMID: 40314024

Mandeep, L., Liu, H., & Shukla, P. (2021). Synthetic biology and biocomputational approaches for improving microbial endoglucanases toward their innovative applications. *ACS Omega, 6*(9), 6055–6063. DOI: 10.1021/acsomega.0c05744 PMID: 33718696

Mandell, D. J., Lajoie, M. J., Mee, M. T., Takeuchi, R., Kuznetsov, G., Norville, J. E., Gregg, C. J., Stoddard, B. L., & Church, G. M. (2015). Biocontainment of genetically modified organisms by synthetic protein design. *Nature, 518*(7537), 55–60. DOI: 10.1038/nature14121 PMID: 25607366

Mandpe, A., Bombaywala, S., Paliya, S., & Kumar, S. (2022). Efficacy of microbes in removal of pesticides from watershed system. In Kapoor, R. T., & Shah, M. P. (Eds.), *Synergistic Approaches for Bioremediation of Environmental Pollutants: Recent Advances and Challenges* (pp. 27–51). Academic Press. DOI: 10.1016/B978-0-323-91860-2.00010-5

Mani, M., Pachaiappan, R., Mahendra Gowda, R. V., & Aroulmoji, V. (2023). Environmental and Health Effects of Emerging Contaminants – A Critical Review. *International Journal of Advanced Science and Engineering, 10*(2), 3449–3470. DOI: 10.29294/IJASE.10.2.2023.3449-3470

Mante, J., Groover, K. E., & Pullen, R. M. (2024). Environmental community transcriptomics: Strategies and struggles. *Briefings in Functional Genomics, 24*(1), 1–15. DOI: 10.1093/bfgp/elae033 PMID: 39183066

Marchant, G. E., & Stevens, Y. A. (2015). A new window of opportunity to reject process-based biotechnology regulation. *GM Crops & Food*, *6*(4), 233–242. DOI: 10.1080/21645698.2015.1134406 PMID: 26930116

Margesin, R., & Schinner, F. (2001). Bioremediation (natural attenuation and biostimulation) of diesel-oil-contaminated soil in an alpine glacier skiing area. *Applied and Environmental Microbiology*, *67*(7), 3127–3133. DOI: 10.1128/AEM.67.7.3127-3133.2001 PMID: 11425732

Mariam, I., Rova, U., Christakopoulos, P., Matsakas, L. & Patel, A. (2025) Data-driven synthetic microbes for sustainable future. *npj Systems Biology and Applications*. 11 (1), 74. .DOI: 10.1038/s41540-025-00556-4

Maroušek, J., & Maroušková, A. (2021). Economic considerations on nutrient utilization in wastewater management. *Energies*, *14*(12), 3468. DOI: 10.3390/en14123468

Martínez-García, E., Fraile, S., Algar, E., Aparicio, T., Velázquez, E., Calles, B., Tas, H., Blázquez, B., Martín, B., Prieto, C., Sánchez-Sampedro, L., Nørholm, M. H. H., Volke, D. C., Wirth, N. T., Dvořák, P., Alejaldre, L., Grozinger, L., Crowther, M., Goñi-Moreno, A., & de Lorenzo, V. (2023). SEVA 4.0: An update of the Standard European Vector Architecture database for advanced analysis and programming of bacterial phenotypes. *Nucleic Acids Research, 51*(D1), D1558–D1567. DOI: 10.1093/nar/gkac1059 PMID: 36420904

Martinez, J. L., Sánchez, M. B., Martínez-Solano, L., Hernandez, A., Garmendia, L., Fajardo, A., & Alvarez-Ortega, C. (2009). Functional role of bacterial multidrug efflux pumps in microbial natural ecosystems. *FEMS Microbiology Reviews*, *33*(2), 430–449. DOI: 10.1111/j.1574-6976.2008.00157.x PMID: 19207745

Mashreghi, M. & Prosser, J. I. 2006. Survival and activity of lux-marked phenanthrene-degrading Pseudomonas stutzeri P16 under different conditions.

Masindi, V., Osman, M. S., & Tekere, M. (2021). *Water Pollution and Remediation: Heavy Metals. Springer*. Mechanisms and Approaches for the Removal of Heavy Metals from Acid Mine Drainage and Other Industrial Effluents., DOI: 10.1007/978-3-030-52421-0_15

Massa, V., Infantino, A., Radice, F., Orlandi, V., Tavecchio, F., Giudici, R., Conti, F., Urbini, G., Di Guardo, A., & Barbieri, P. (2009). Efficiency of natural and engineered bacterial strains in the degradation of 4-chlorobenzoic acid in soil slurry. *International Biodeterioration and Biodegradation*, *63*(1), 112–115. DOI: 10.1016/j.ibiod.2008.07.006

Massot, F., Bernard, N., Alvarez, L. M. M., Martorell, M. M., Mac Cormack, W. P., & Ruberto, L. A. (2022). Microbial associations for bioremediation. What does "microbial consortia" mean? *Applied Microbiology and Biotechnology*, *106*(7), 2283–2297. DOI: 10.1007/s00253-022-11864-8 PMID: 35294589

Mataigne, V., Vannier, N., Vandenkoornhuyse, P., & Hacquard, S. (2021). Microbial Systems Ecology to Understand Cross-Feeding in Microbiomes. *Frontiers in Microbiology*, *12*, 780469. DOI: 10.3389/fmicb.2021.780469 PMID: 34987488

Matchado, M. S., Lauber, M., Reitmeier, S., Kacprowski, T., Baumbach, J., Haller, D., & List, M. (2020). Network analysis methods for studying microbial communities: A mini review. *Computational and Structural Biotechnology Journal*, *19*(1), 2687–2698. DOI: 10.1016/j.csbj.2021.05.001 PMID: 34093985

Matuszyńska, A., Ebenhöh, O., Zurbriggen, M. D., Ducat, D. C., & Axmann, I. M. (2024). A new era of synthetic biology—Microbial community design. *Synthetic Biology*, *9*(1), 1–15. DOI: 10.1093/synbio/ysae011 PMID: 39086602

Ma, W., Tang, S., Dengzeng, Z., Zhang, D., Zhang, T., & Ma, X. (2022). Root exudates contribute to belowground ecosystem hotspots: A review. *Frontiers in Microbiology*, *13*, 937–940. DOI: 10.3389/fmicb.2022.937940 PMID: 36274740

Mawang, C.-I., Azman, A.-S., Fuad, A.-S. M., & Ahamad, M. (2021). Actinobacteria: An eco-friendly and promising technology for the bioaugmentation of contaminants. *Biotechnology Reports (Amsterdam, Netherlands)*, *32*, e00679. DOI: 10.1016/j.btre.2021.e00679 PMID: 34660214

Maymo-Gatell, X., Chien, Y. T., Gossett, J. M., & Zinder, S. H. (1997). Isolation of a bacterium that reductively dechlorinates tetrachloroethene to ethene. *Science*, *276*(5318), 1568–1571. DOI: 10.1126/science.276.5318.1568 PMID: 9171062

McCarty, N. S., & Ledesma-Amaro, R. (2019). Synthetic biology tools to engineer microbial communities for biotechnology. *Trends in Biotechnology*, *37*(2), 181–197. DOI: 10.1016/j.tibtech.2018.11.002 PMID: 30497870

Medfu Tarekegn, M., Zewdu Salilih, F., & Ishetu, A. I. (2020). Microbes used as a tool for bioremediation of heavy metal from the environment. *Cogent Food & Agriculture*, *6*(1), 1783174. DOI: 10.1080/23311932.2020.1783174

Meena, P. M., Aggarwal, R. K., Meena, R., & Rathore, M. S. (2025). Soil Contamination, Risk Assessment, and Remediation. In Ivascu, L., Pislaru, M., & Alexa, L. (Eds.), *The Future of Risk Management* (pp. 234–264). IntechOpen., DOI: 10.5772/intechopen.1005045

Megharaj, M., Ramakrishnan, B., Venkateswarlu, K., Sethunathan, N., & Naidu, R. (2011). Bioremediation approaches for organic pollutants: A critical perspective. *Environment International, 37*(8), 1362–1375. DOI: 10.1016/j.envint.2011.06.003 PMID: 21722961

Mekonnen, B. A., Aragaw, T. A., & Genet, M. B. (2024). Bioremediation of petroleum hydrocarbon contaminated soil: A review on principles, degradation mechanisms, and advancements. *Frontiers in Environmental Science, 12*, 1354422. DOI: 10.3389/fenvs.2024.1354422

Mendes, M., Cassoni, A. C., Alves, S., Moreira, P., Castro, P. M., & Pintado, M. E. (2023). Assessment of drying conditions for storage of a yeast-based decolorization solution for application in textile industrial wastewater treatment plants. *Fungal Biology, 127*(7–8), 1111–1117. DOI: 10.1016/j.funbio.2023.06.002 PMID: 37495301

Mendes, M., Moreira, I. S., Moreira, P., Pintado, M., & Castro, P. M. L. (2023). Bioaugmentation of Aerobic Granular Sludge with Dye-Decolorizing Yeast for Textile Industrial Wastewater. *Processes (Basel, Switzerland), 11*(6), 1–16. DOI: 10.3390/pr11061654

Meng, S., Li, Z., Zhang, P., Contreras, F., Ji, Y., & Schwaneberg, U. (2023). Deep learning guided enzyme engineering of Thermobifida fusca cutinase for increased PET depolymerization. *Chinese Journal of Catalysis, 50*, 229–238. DOI: 10.1016/S1872-2067(23)64470-5

Michaelis, C., & Grohmann, E. (2023). Horizontal Gene Transfer of Antibiotic Resistance Genes in Biofilms. *Antibiotics (Basel, Switzerland), 12*(2), 328. DOI: 10.3390/antibiotics12020328 PMID: 36830238

Minhas, W. R., Saleemi, H., Raheem, M. F., Khalid, S. S., Aslam, H., & Ahmad, W. (2024). CRISPR-Engineered Microbes for Enhanced biodegradation of recalcitrant pollutants a genomic approach to environmental remediation. *Frontiers in Health Informatics*, •••, 8281–8304. DOI: 10.63682/fhi787

Ministry of Ecology and Environment of China. (2020). Biosafety management regulations [Press release/Regulation summary]. Retrieved from http://english.mee.gov.cn

Ministry of Environment. Forest and Climate Change (MoEFCC). (2016). The Harzadous and other wastes (Management and Transboundary Movement) Rules, 2016. Government of India. Retrieved from https://upload.indiacode.nic.in/showfile?actid=AC_RJ_83-1096-00001_00001_1563872109827&filename=harzadous_and_other_wastes_%28management_and_transboundary_movement%29_rules%2C_2016.pdf&type=rule

Mishra, K., Siwal, S. S., Nayaka, S. C., Guan, Z., & Thakur, V. K. (2023). Waste-to-chemicals: Green solutions for bioeconomy markets. *The Science of the Total Environment*, *887*, 164006. DOI: 10.1016/j.scitotenv.2023.164006 PMID: 37172858

Mishra, S. (2025). Warming Seas: The Impact of Climate Change on Ocean Life and Ecosystems. In *Navigating Climate Change: Impacts on Biodiversity and Ecosystem Resilience* (pp. 467–501). Springer Nature Singapore. DOI: 10.1007/978-981-95-0409-1_24

Mishra, S., & Maiti, A. (2018). The efficacy of bacterial species to decolourise reactive azo, anthroquinone and triphenylmethane dyes from wastewater: A review. *Environmental Science and Pollution Research International*, *25*(9), 8286–8314. DOI: 10.1007/s11356-018-1273-2 PMID: 29383646

Mishra, T., Tiwari, P. B., Kanchan, S., & Kesheri, M. (2025). Advances in Microbial Bioremediation for Effective Wastewater Treatment. *Water (Basel)*, *17*(22), 3196. DOI: 10.3390/w17223196

Mitikie, B. B., & Elsaigh, W. A. (2025). Bio-based bacterial concrete: A sustainable alternative in construction. *Innovative Infrastructure Solutions*, *10*(6), 234. DOI: 10.1007/s41062-025-02058-y

Mkilima, T., & Baimukasheva, S. (2025). Performance of a multi-chamber microbial fuel cell with biochar anode for industrial wastewater treatment and energy recovery. *Journal of Ecological Engineering*, *26*(7), 367–380. DOI: 10.12911/22998993/203696

Mlost, J., Dawli, R., Liu, X., Costa, A. R., & Pollak Dorocic, I. (2025). Evaluation of unsupervised learning algorithms for the classification of behavior from pose estimation data. *Patterns (New York, N.Y.)*, *6*(5), 101237. DOI: 10.1016/j.patter.2025.101237 PMID: 40486967

Moe-Behrens, G. H. G., Davis, R., & Haynes, K. A. (2013). Preparing synthetic biology for the world. *Frontiers in Microbiology*, *4*, 5. DOI: 10.3389/fmicb.2013.00005 PMID: 23355834

Mohamed, B. A., & Samer, M. (2023). *Biobutanol production from agricultural wastes. Valorization of Wastes for Sustainable Development*. Elsevier.

Mohanty, S. S., & Kumar, A. (2025). Microbial decolorization of anthraquinone dyes: Batch and continuous treatment systems - a mini-review. *Frontiers in Environmental Engineering*, *4*(June), 1–21. DOI: 10.3389/fenve.2025.1553712

Mohapatra, B., & Phale, P. S. (2021). Microbial degradation of naphthalene and substituted naphthalenes: Metabolic diversity and genomic insight for bioremediation. *Frontiers in Bioengineering and Biotechnology*, *9*, 602445. DOI: 10.3389/fbioe.2021.602445 PMID: 33791281

Mohd Noor, M. (2021). Environmental impacts of oil industry: An overview of the impacts and source management EEEE Environmental Impacts of Oil Industry: An overview of the impacts and source management. *Environmental Management*, *4*, 1–8.

Moino, B. P., Costa, C. S., da Silva, M. G., & Vieira, M. G. (2017). Removal of nickel ions on residue of alginate extraction from Sargassum filipendula seaweed in packed bed. *Canadian Journal of Chemical Engineering*, *95*(11), 2120–2128. DOI: 10.1002/cjce.22859

Monteiro, C. M., Castro, P. M., & Malcata, F. X. (2012). Metal uptake by microalgae: Underlying mechanisms and practical applications. *Biotechnology Progress*, *28*(2), 299–311. DOI: 10.1002/btpr.1504 PMID: 22228490

Morales, N., Val del Río, A., Vázquez-Padín, J. R., Gutiérrez, R., Fernández-González, R., Icaran, P., Rogalla, F., Campos, J. L., Méndez, R., & Mosquera-Corral, A. (2015). Influence of dissolved oxygen concentration on the start-up of the anammox-based process: ELAN®. *Water Science and Technology*, *72*(4), 520–527. DOI: 10.2166/wst.2015.233 PMID: 26247749

Moreira, I. S., Murgolo, S., Mascolo, G., & Castro, P. M. L. (2022). Biodegradation and Metabolic Pathway of 17β-Estradiol by *Rhodococcus* sp. ED55. *International Journal of Molecular Sciences*, *23*(11), 6181. DOI: 10.3390/ijms23116181 PMID: 35682859

Morrissey, J., Barberi, G., Strain, B., Facco, P., & Kontoravdi, C. (2025). NEXT-FBA: A hybrid stoichiometric/data-driven approach to improve intracellular flux predictions. *Metabolic Engineering*, *91*, 130–144. DOI: 10.1016/j.ymben.2025.03.010 PMID: 40118205

Moura de Sousa, J., Lourenço, M., & Gordo, I. (2023). Horizontal gene transfer among host-associated microbes. *Cell Host & Microbe*, *31*(4), 513–527. DOI: 10.1016/j.chom.2023.03.017 PMID: 37054673

Mousavi, S. A., & Khodadoost, F. (2019). Effects of detergents on natural ecosystems and wastewater treatment processes: A review. *Environmental Science and Pollution Research International, 26*(26), 26439–26448. DOI: 10.1007/s11356-019-05802-x PMID: 31352596

Moyo, A. (2024). Controlling acid and metalliferous drainage generated by legacy mine wastes in Tasmania using industrial wastes (Doctoral dissertation, University of Tasmania).

Moyo, S., Makhanya, B. P., & Zwane, P. E. (2022). Use of bacterial isolates in the treatment of textile dye wastewater: A review. *Heliyon, 8*(6), e09632. DOI: 10.1016/j.heliyon.2022.e09632 PMID: 35677403

Mozumder, M. S. I., Picioreanu, C., Van Loosdrecht, M. C. M., & Volcke, E. I. P. (2014). Effect of heterotrophic growth on autotrophic nitrogen removal in a granular sludge reactor. *Environmental Technology, 35*(8), 1027–1037. DOI: 10.1080/09593330.2013.859711 PMID: 24645487

Mrozik, A., & Piotrowska-Seget, Z. (2010). Bioaugmentation as a strategy for cleaning up of soils contaminated with aromatic compounds. *Microbiological Research, 165*(5), 363–375. DOI: 10.1016/j.micres.2009.08.001 PMID: 19735995

Mudhoo, A., & Kumar, S. (2013). Effects of heavy metals as stress factors on anaerobic digestion processes and biogas production from biomass. *International Journal of Environmental Science and Technology, 10*(6), 1383–1398. DOI: 10.1007/s13762-012-0167-y

Mukhtar, Z. (2023). Environmental pollution and regulatory and non-regulatory environmental responsibility (reviewing Pakistan environmental protection act). *American Journal of Industrial and Business Management, 13*(6), 443–456. DOI: 10.4236/ajibm.2023.136028

Müller, B., & Schackmann, M. (2023). *Coatings formulation*. Lack in Vincentz GmbH and Company KG.

Murphy, N., & Krimsky, S. (2003). Implicit precaution, scientific inference, and indirect evidence: The basis for the US Environmental Protection Agency's regulation of genetically modified crops. *New Genetics and Society, 22*(2), 159–175. DOI: 10.1080/1463677032000102602

Musah, B. I. (2024). Effects of heavy metals and metalloids on plant-animal interaction and biodiversity of terrestrial ecosystems - An overview. *Environmental Monitoring and Assessment, 197*(1), 12. DOI: 10.1007/s10661-024-13490-5 PMID: 39623084

Mussatto, S. I. (2014). Brewer's spent grain: A valuable feedstock for industrial applications. *Journal of the Science of Food and Agriculture, 94*(7), 1264–1275. DOI: 10.1002/jsfa.6486 PMID: 24254316

Muter, O. (2023). Current trends in bioaugmentation tools for bioremediation: A critical review of advances and knowledge gaps. *Microorganisms, 11*(3), 710. DOI: 10.3390/microorganisms11030710 PMID: 36985282

Nadu, T.Sanmathi K R. (2023). Biodegradation on the Removal of Dyes from Textile Effluent. *International Journal of Advanced Research in Science. Tongxin Jishu, 4–8*, 4–8. Advance online publication. DOI: 10.48175/IJARSCT-13102

Nagarajan, K., Ni, C., & Lu, T. (2022). Agent-Based Modeling of Microbial Communities. *ACS Synthetic Biology, 11*(11), 3564–3574. DOI: 10.1021/acssynbio.2c00411 PMID: 36315012

Naik, S., & Eswari, J. S. (2022). Electrical waste management: Recent advances challenges and future outlook. *Total Environment Research Themes, 1*, 100002. DOI: 10.1016/j.totert.2022.100002

Nair, S., Manu, B., & Azhoni, A. (2021). Sustainable treatment of paint industry wastewater: Current techniques and challenges. *Journal of Environmental Management, 296*, 113105. DOI: 10.1016/j.jenvman.2021.113105 PMID: 34216906

Naizabayeva, L., Nurzhanov, C. A., Satymbekov, M. N., & Elle, V. Z. (2024). Information system for remediation and cleanup of contaminated soil with machine learning. *Procedia Computer Science, 231*, 145–150. DOI: 10.1016/j.procs.2023.12.186

Nanda, M., Kumar, V., & Sharma, D. K. (2019). Multimetal tolerance mechanisms in bacteria: The resistance strategies acquired by bacteria that can be exploited to 'clean-up'heavy metal contaminants from water. *Aquatic Toxicology, 212*, 1–10. DOI: 10.1016/j.aquatox.2019.04.011 PMID: 31022608

Nasr, M., Eid, C., Habchi, R., Miele, P., & Bechelany, M. (2018). Recent progress on titanium dioxide nanomaterials for photocatalytic applications. *ChemSusChem, 11*(18), 3023–3047. DOI: 10.1002/cssc.201800874 PMID: 29984904

Natarajan, A., & Bhatt, A. S. (2020). Microbes and microbiomes in 2020 and beyond. *Nature Communications, 11*(1), 4988. DOI: 10.1038/s41467-020-18850-6 PMID: 33020496

Nayak, N. P. (2023). Microorganisms and their application in mining and allied industries. *Materials Today: Proceedings, 72*, 2886–2891. DOI: 10.1016/j.matpr.2022.07.392

Nazlı, S. B., & Karatay, S. E. (2024). Evaluation of Food Wastes in Chlorella vulgaris Cultivation for Remazol Brilliant Blue R Biosorption. *Aquatic Sciences and Engineering*, *39*(1), 1–7. DOI: 10.26650/ASE20241305101

Negri, E., Fumagalli, L., & Macchi, M. (2017). A review of the roles of digital twin in CPS-based production systems. *Procedia Manufacturing*, *11*, 939–948. DOI: 10.1016/j.promfg.2017.07.198

Neifar, M., Chouchane, H., Mahjoubi, M., Jaouani, A., & Cherif, A. (2020). *Pseudomonas extremorientalis* BU118 : A new salt-tolerant laccase-secreting bacterium with biotechnological potential in textile azo dye decolourization. *Biotech*, *6*(1), 1–9. DOI: 10.1007/s13205-016-0425-7 PMID: 35822774

Nematollahi, M. J., Rastegari Mehr, M., Shakeri, A., Amjadian, K., Ebrahimi, P., & Pirouei, M. (2022). Polycyclic aromatic hydrocarbons (PAHs) in soils around oil refineries; Case study: Erbil and Kirkuk refineries, Iraq-A modelling approach. *Journal of Environmental Health Science & Engineering*, *20*(2), 699–716. DOI: 10.1007/s40201-022-00809-y PMID: 36406609

Nevin, K. P., Woodard, T. L., Franks, A. E., Summers, Z. M., & Lovley, D. R. (2010). Microbial electrosynthesis: Feeding microbes electricity to convert carbon dioxide and water to multicarbon extracellular organic compounds. *mBio*, *1*(2), e00103–e00110. DOI: 10.1128/mBio.00103-10 PMID: 20714445

Newby, D., & Pepper, I. (2002). Dispersal of plasmid pJP4 in unsaturated and saturated 2, 4-dichlorophenoxyacetic acid contaminated soil. *FEMS Microbiology Ecology*, *39*(2), 157–164. DOI: 10.1111/j.1574-6941.2002.tb00917.x PMID: 19709195

Newhart, K. B., Holloway, R. W., Hering, A. S., & Cath, T. Y. (2019). Data-driven performance analyses of wastewater treatment plants: A review. *Water Research*, *157*, 498–513. DOI: 10.1016/j.watres.2019.03.030 PMID: 30981980

Ngatia, L., Grace, J. M., III, & Moriasi, D. (2019). Nitrogen and phosphorus eutrophication in marine environments. *Monitoring of marine pollution*, pp.77-93.

Nguyen, L. N., Vu, M. T., Johir, M. A. H., Pernice, M., Ngo, H. H., Zdarta, J., Jesionowski, T., Nghiem, L. D., & Hai, F. I. (2020). Promotion of direct interspecies electron transfer and potential impact of conductive materials in anaerobic digestion and its downstream processing: A critical review. *Bioresource Technology*, *316*, 123925. DOI: 10.1016/j.biortech.2020.123925 PMID: 34467893

Nicula, N. O., Lungulescu, E. M., Rîmbu, G. A., Marinescu, V., Corbu, V. M., & Csutak, O. (2023). Bioremediation of Wastewater Using Yeast Strains: An Assessment of Contaminant Removal Efficiency. *International Journal of Environmental Research and Public Health, 20*(6), 4795. Advance online publication. DOI: 10.3390/ijerph20064795 PMID: 36981703

Niederdorfer, R., Fragner, L., Yuan, L., Hausherr, D., Wei, J., Magyar, P., Joss, A., Lehmann, M. F., Ju, F., & Bürgmann, H. (2021). Distinct growth stages controlled by the interplay of deterministic and stochastic processes in functional anammox biofilms. *Water Research, 200*, 117225. DOI: 10.1016/j.watres.2021.117225 PMID: 34052477

Nielsen, A. A. K., Der, B. S., Shin, J., Vaidyanathan, P., Paralanov, V., Strychalski, E. A., Ross, D., Densmore, D., & Voigt, C. A. (2016). Genetic circuit design automation. *Science, 352*(6281), aac7341. Advance online publication. DOI: 10.1126/science.aac7341 PMID: 27034378

Nielsen, J., & Keasling, J. D. (2016). Engineering cellular metabolism. *Cell, 164*(6), 1185–1197. DOI: 10.1016/j.cell.2016.02.004 PMID: 26967285

Nieto, E. E., Macchi, M., Valacco, M. P., Festa, S., Morelli, I. S., & Coppotelli, B. M. (2023). Metaproteomic and gene expression analysis of interspecies interactions in a PAH-degrading synthetic microbial consortium constructed with the key microbes of a natural consortium. *Biodegradation, 34*(2), 181–197. DOI: 10.1007/s10532-022-10012-3 PMID: 36596914

Nikel, P. I., & De Lorenzo, V. (2018). *Pseudomonas putida* as a functional chassis for industrial biocatalysis: From native biochemistry to trans-metabolism. *Metabolic Engineering, 50*, 142–155. DOI: 10.1016/j.ymben.2018.05.005 PMID: 29758287

Nikel, P. I., & de Lorenzo, V. (2021). Metabolic engineering for large-scale environmental bioremediation. *Metabolic Engineering, 13*, 859–890. DOI: 10.1002/9783527823468.ch22

Nikolopoulou, M., & Kalogerakis, N. (2009). Biostimulation strategies for fresh and chronically polluted marine environments with petroleum hydrocarbons. *Journal of Chemical Technology & Biotechnology: International Research in Process. Environmental & Clean Technology, 84*, 802–807.

Nikolopoulou, M., Pasadakis, N., & Kalogerakis, N. (2013). Evaluation of autochthonous bioaugmentation and biostimulation during microcosm-simulated oil spills. *Marine Pollution Bulletin, 72*(1), 165–173. DOI: 10.1016/j.marpolbul.2013.04.007 PMID: 23660443

Ni, L., Wang, Y., Lin, X., Yan, Y., Zhang, Y., & Wang, W. (2018). Enhancement of the adaptability of anammox granules to zinc shock by appropriate organic carbon treatment. *Bioresource Technology, 268*, 496–503. DOI: 10.1016/j.biortech.2018.08.035 PMID: 30114669

Nirmala, A., Soundariya, S., & Praveena, A. R. (2021). *On the Phytotoxicity of Waste-Water from Textile Industry on Selected Crops Seed Germination and its Treatment Using Bacteria and Zinc Oxide Nanoparticles. 14*(December), 1564–1570.

Ni, S. Q., & Zhang, J. (2013). Anaerobic ammonium oxidation: From laboratory to full-scale application. *BioMed Research International, 2013*, 1–10. Advance online publication. DOI: 10.1155/2013/469360 PMID: 23956985

Nishikawa, E., da Silva, M. G. C., & Vieira, M. G. A. (2018). Cadmium biosorption by alginate extraction waste and process overview in life cycle assessment context. *Journal of Cleaner Production, 178*, 166–175. DOI: 10.1016/j.jclepro.2018.01.025

Nnaji, N. D., Anyanwu, C. U., Miri, T., & Onyeaka, H. (2024). Mechanisms of heavy metal tolerance in bacteria: A review. *Sustainability (Basel), 16*(24), 11124. DOI: 10.3390/su162411124

Nnaji, N. D., Onyeaka, H., Miri, T., & Ugwa, C. (2023). Bioaccumulation for heavy metal removal: A review. *SN Applied Sciences, 5*(5), 125. DOI: 10.1007/s42452-023-05351-6

Nombamba, A., Fadiji, A. E., & Babalola, O. O. (2024). Exploring the role of endophytic fungi in the amelioration of drought stress in plants. *Plant Protection Science, 60*(3), 213–228. DOI: 10.17221/25/2023-PPS

Nordin, A. R. R., Navarro, A. R., Reyes, J. C., Maragathavalli, S., Kristanti, R. A., Wulandari, R., & Bunrith, S. (2025). Microbial Bioremediation of Petroleum Contaminated Soil: A Sustainable Approach. *Topical Aquatic and Soil Pollution, 5*(1), 71–87. DOI: 10.53623/tasp.v5i1.683

Noruzman, A. H., & Apandi, N. M. (2020). Paint waste management in industry. *Journal of Advanced Research in Business and Management Studies, 20*(1), 28–33. DOI: 10.37934/arbms.20.1.2833

Novak, V., Andeer, P. F., King, E., Calabria, J., Fitzpatrick, C., Kelm, J. M., & Northen, T. R. (2025). Breaking the reproducibility barrier with standardized protocols for plant–microbiome research. *PLoS Biology, 23*(9), e3003358. DOI: 10.1371/journal.pbio.3003358 PMID: 40920748

Novello, G., Bona, E., Nasuelli, M., Massa, N., Sudiro, C., Campana, D. C., Gorrasi, S., Hochart, M. L., Altissimo, A., Vuolo, F., & Gamalero, E. (2024). The Impact of Nitrogen-Fixing Bacteria-Based Biostimulant Alone or in Combination with Commercial Inoculum on Tomato Native Rhizosphere Microbiota and Production: An Open-Field Trial. *Biology (Basel)*, *13*(6), 400. DOI: 10.3390/biology13060400 PMID: 38927280

Nozari, M., Samaei, M. R., Dehghani, M., & Ebrahimi, A. A. (2018). Bioremediation of alkane hydrocarbons using bacterial consortium from soil. *Health Scope*, *7*(In Press), 12524. DOI: 10.5812/jhealthscope.12524

Nunes, P. S., Lacerda-Junior, G. V., Mascarin, G. M., Guimarães, R. A., Medeiros, F. H., Arthurs, S., & Bettiol, W. (2024). Microbial consortia of biological products: Do they have a future? *Biological Control*, *188*, 105439. DOI: 10.1016/j.biocontrol.2024.105439

Nzila, A. (2018). Biodegradation of high-molecular-weight polycyclic aromatic hydrocarbons under anaerobic conditions: Overview of studies, proposed pathways and future perspectives. *Environmental Pollution*, *239*, 788–802. DOI: 10.1016/j.envpol.2018.04.074 PMID: 29751337

Nzila, A., Razzak, S. A., & Zhu, J. (2016). Bioaugmentation: An emerging strategy of industrial wastewater treatment for reuse and discharge. *International Journal of Environmental Research and Public Health*, *13*(9), 846. DOI: 10.3390/ijerph13090846 PMID: 27571089

Obidi, O. F., Kade, E. A., & Fadipe, O. T. (2022a). A first report on the identification of a novel archaea, *Methanospirillum lacunae* from spoilt paints in Lagos, Nigeria, using a metagenomics approach. *Scientific African*, *15*, 2468–2276.

Obidi, O. F., Olushina, O. A., Peter, V., Weinan, C., Matthew, C., Bright, O., Adetunji, A., Damilare, O., Kehinde, A., Malak, A., Idowu, S., Favour, O., & Obaloluwa, S. (2022b). Vibrational spectroscopy and non-parametric analysis for biodegradation evaluation: The acrylic-emulsion paint scenario. *Vibrational Spectroscopy*, *123*, 103427. DOI: 10.1016/j.vibspec.2022.103427

Odelade, K. A., & Babalola, O. O. (2019). Bacteria, fungi and archaea domains in rhizospheric soil and their effects in enhancing agricultural productivity. *International Journal of Environmental Research and Public Health*, *16*(20), 3873. DOI: 10.3390/ijerph16203873 PMID: 31614851

Ogunseitan, O. (2005). *Microbial Diversity: Form and Function in Prokaryotes*. John Wiley & Sons.

Ogunseitan, O. A. (2023). Planetary boundaries and safe operating space for novel entities. In Hester, R., & Harrison, R. (Eds.), *Environmental impacts of modern agriculture* (pp. 1–24). Royal Society of Chemistry.

Ogunseitan, O. A. (2026). Bacterial Diversity, Introduction to. In Wolf, J. B., & Russo, C. A. M. (Eds.), *Encyclopedia of Evolutionary Biology* (Vol. 5, pp. 42–47). Elsevier, Academic Press., DOI: 10.1016/B978-0-443-15750-9.00073-2

Ohanwemadu, O. S., & Nwakanma, C. (2025). Ecological Risk Evaluation of Heavy Metal Accumulation in Aba River Bottom Sediments: Implications for Soap and Detergent Industry Wastewater Management. *Journal of Agricultural Science and Technology A*, *15*(1). Advance online publication. DOI: 10.17265/2161-6256/2025.01.002

Ohimain, E. I. (2025). Microbial Biotechnology for Food Waste Reduction and Upcycling. Babalola O. O. et al. (eds.), Ecofriendly Frontiers: Harnessing Microbial Applications for Food Security. https://doi.org/. Springer Nature SwitzerlandDOI: 10.1007/978-3-031-98700-7_12

Oj, A.-S., Meiczinger, M., Al-Juboori, R., Grmasha, R., Andredaki, M., Somogyi, V., Idowu, I., Stenger-Kovács, C., Jakab, M. & Lengyel, E. 2023. The synergistic effect of degradation and adsorption for emerging pollutants removal from water and real wastewater by using immobilized activated carbon from pomegranate peels.

Ojonimi, T. I., Chanda, T. P., & Ameh, E. G. (2021). Acid mine drainage (AMD) contamination in coal mines and the need for extensive prediction and remediation: A review. *Journal of Degraded and Mining Lands Management*, *9*(1), 3129–3136. DOI: 10.15243/jdmlm.2021.091.3129

Ojuederie, O. B., & Babalola, O. O. (2017). Microbial and plant-assisted bioremediation of heavy metal polluted environments: A review. *International Journal of Environmental Research and Public Health*, *14*(12), 1504. DOI: 10.3390/ijerph14121504 PMID: 29207531

Okafor, C. P., Udemang, N. L., Chikere, C. B., Akaranta, O., & Ntushelo, K. (2021). Indigenous microbial strains as bioresource for remediation of chronically polluted Niger Delta soils. *Scientific African*, *11*, e00682. DOI: 10.1016/j.sciaf.2020.e00682

Okechukwu, V. U., & Onwukeme, V. (2025). Health Risk Assessment of Selected Heavy Metals in Surface Water and Sediments from Aba River in Abia state, Nigeria. *Chemical Research and Technology*, *2*(3), 115–127.

Okoye, A. U., Selvarajan, R., Chikere, C. B., Okpokwasili, G. C., & Mearns, K. (2024). Characterization and identification of long-chain hydrocarbon-degrading bacterial communities in long-term chronically polluted soil in Ogoniland: An integrated approach using culture-dependent and independent methods. *Environmental Science and Pollution Research International, 31*(21), 30867–30885. DOI: 10.1007/s11356-024-33326-6 PMID: 38622422

Olawade, D. B., Oladipo, P., Ajisafe, O., Egbon, E., Fapohunda, O., & Kade, A. (2025). Genetically engineered microorganisms: A promising frontier for PFAS bioremediation. [REMOVED HYPERLINK FIELD]. *Process Safety and Environmental Protection, 202*, 107699. DOI: 10.1016/j.psep.2025.107699

Olawuyi, D. S., & Tubodenyefa, Z. (2018). *Review of the environmental guidelines and standards for the petroleum industry in Nigeria (EGASPIN)*. OGEES Institute.

Oldiges, M., Eikmanns, B. J., & Blombach, B. (2014). Application of metabolic engineering for the biotechnological production of L-valine. *Applied Microbiology and Biotechnology, 98*(13), 5859–5870. DOI: 10.1007/s00253-014-5782-8 PMID: 24816722

Oliveira, N. M., Wheeler, J. H., Deroy, C., Booth, S. C., Walsh, E. J., Durham, W. M., & Foster, K. R. (2022). Suicidal chemotaxis in bacteria. *Nature Communications, 13*(1), 7608. DOI: 10.1038/s41467-022-35311-4 PMID: 36494355

Olukanni, O. D., Osuntoki, A. A., Kalyani, D. C., Gbenle, G. O., & Govindwar, S. P. (2010). Decolorization and biodegradation of Reactive Blue 13 by *Proteus mirabilis* LAG. *Journal of Hazardous Materials, 184*(1–3), 290–298. DOI: 10.1016/j.jhazmat.2010.08.035 PMID: 20832936

Oluwagbemi, O., & Awe, O. (2018). A comparative computational genomics of Ebola Virus Disease strains: In-silico Insight for Ebola control, Informatics in Medicine Unlocked, Volume 12, 2018; Pages 106-119, ISSN 2352-9148; https://doi.org/DOI: 10.1016/j.imu.2018.07.004

Oluwagbemi, O. O., Oladipo, E. K., Kolawole, O. M., Oloke, J. K., Adelusi, T. I., Irewolede, B. A., Dairo, E. O., Ayeni, A. E., Kolapo, K. T., Akindiya, O. E., Oluwasegun, J. A., Oluwadara, B. F., & Fatumo, S. (2022). Bioinformatics, Computational Informatics, and Modeling Approaches to the Design of mRNA COVID-19 Vaccine Candidates. *Computation (Basel, Switzerland), 10*(7), 117. DOI: 10.3390/computation10070117

Omotuyi, O. Y. (2023). A Critical Assessment of the Regulatory Framework for Oil and Gas Decommissioning in Nigeria. *Journal of Sustainable Developmental Law & Policy, 14*(1), 140–163. DOI: 10.4314/jsdlp.v14i1.7s

Organisation for Economic Co-operation and Development (OECD). (2020). The Mutual Acceptance of Data (MAD) system-harmonised test methods and data quality requirements. Retrieved from https://www.oecd.org/en/topics/sub-issues/testing-of-chemicals/mutual-acceptance-of-data-system.html

Oro, C. E. D., Saorin Puton, B. M., Venquiaruto, L. D., Dallago, R. M., & Tres, M. V. (2024). Effective microbial strategies to remediate contaminated agricultural soils and conserve functions. *Agronomy (Basel), 14*(11), 2637. DOI: 10.3390/agronomy14112637

Ossai, I. C., Ahmed, A., Hassan, A., & Hamid, F. S. (2020). Remediation of soil and water contaminated with petroleum hydrocarbon: A review. *Environmental Technology & Innovation, 17*, 100526. DOI: 10.1016/j.eti.2019.100526

Oyewola, D. O., Dada, E. G., Emebo, O., & Oluwagbemi, O. O. (2022b). Using Deep 1D Convolutional Grated Recurrent Unit Neural Network to Optimize Quantum Molecular Properties and Predict Intramolecular Coupling Constants of Molecules of Potential Health Medications and Other Generic Molecules. *Applied Sciences (Basel, Switzerland), 12*(14), 7228. DOI: 10.3390/app12147228

Oyewola, D. O., Dada, E. G., Omotehinwa, T. O., Emebo, O., & Oluwagbemi, O. O. (2022a). Application of Deep Learning Techniques and Bayesian Optimization with Tree Parzen Estimator in the Classification of Supply Chain Pricing Datasets of Health Medications. *Applied Sciences (Basel, Switzerland), 12*(19), 10166. DOI: 10.3390/app121910166

Pacheco, A. R., Moel, M., & Segrè, D. (2019). Costless Metabolic Secretions as Drivers of Interspecies Interactions in Microbial Ecosystems. *Nature Communications, 10*(1), 1–12. DOI: 10.1038/s41467-018-07946-9 PMID: 30626871

Palermo, A. (2022). Metabolomics- and systems-biology-guided discovery of metabolite lead compounds and druggable targets. *Drug Discovery Today, 28*(2), 1–10, 103460. DOI: 10.1016/j.drudis.2022.103460 PMID: 36427778

Palladino, F., Marcelino, P. R. F., Schlogl, A. E., José, Á. H. M., Rodrigues, R. C. L. B., Fabrino, D. L., Santos, I. J. B., & Rosa, C. A. (2024). Bioreactors: Applications and innovations for a sustainable and healthy future—a critical review. *Applied Sciences (Basel, Switzerland), 14*(20), 9346. DOI: 10.3390/app14209346

Panasia, G., Oetermann, S., Steinbüchel, A., & Philipp, B. (2019). Sulfate ester detergent degradation in *Pseudomonas aeruginosa* is subject to both positive and negative regulation. *Applied and Environmental Microbiology, 85*(23), e01352–e19. DOI: 10.1128/AEM.01352-19 PMID: 31540990

Pandey, J., Chauhan, A., & Jain, R. K. (2009). Integrative approaches for assessing the ecological sustainability of *in situ* bioremediation. *FEMS Microbiology Reviews*, *33*(2), 324–375. DOI: 10.1111/j.1574-6976.2008.00133.x PMID: 19178567

Pandey, K., Singh, B., Pandey, A. K., Badruddin, I. J., Pandey, S., Mishra, V. K., & Jain, P. A. (2017). Application of Microbial Enzymes in Industrial Waste Water Treatment. *International Journal of Current Microbiology and Applied Sciences*, *6*(8), 1243–1254. DOI: 10.20546/ijcmas.2017.608.151

Pandey, P., & Kiran, U. V. (2020). Degradation of paints and its microbial effect on health and environment. *Journal of Critical Reviews*, *7*(19), 4879–4884.

Panhwar, A., Sattar Jatoi, A., Ali Mazari, S., Kandhro, A., Rashid, U., & Qaisar, S. (2024). Water resources contamination and health hazards by textile industry effluent and glance at treatment techniques: A review. *Waste Management Bulletin*, *1*(4), 158–163. DOI: 10.1016/j.wmb.2023.09.002

Papenfort, K., & Bassler, B. L. (2016). Quorum sensing signal-response system in Gram-negative bacteria. *Nature Reviews. Microbiology*, *14*(9), 576–588. DOI: 10.1038/nrmicro.2016.89 PMID: 27510864

Parajuli, A., Grönroos, M., Siter, N., Puhakka, R., Vari, H. K., Roslund, M. I., Jumpponen, A., Nurminen, N., Laitinen, O. H., Hyöty, H., Rajaniemi, J., & Sinkkonen, A. (2018). Urbanization reduces transfer of diverse environmental microbiota indoors. *Frontiers in Microbiology*, *9*, 84. DOI: 10.3389/fmicb.2018.00084 PMID: 29467728

Parde, D., Ghosh, R., Rajpurohit, P., Bhaduri, S., & Behera, M. (2024). Nutrient Retrieval Techniques in Wastewater Treatment. In Ghangrekar, M. M., Yadav, S., & Yadava, R. N. (Eds.), *Biological and Hybrid Wastewater Treatment Technology. Earth and Environmental Sciences Library*. Springer., DOI: 10.1007/978-3-031-63046-0_7

Paritosh, K., Kushwaha, S. K., Yadav, M., Pareek, N., Chawade, A., & Vivekanand, V. (2017). Food waste to energy: An overview of sustainable approaches for food waste management and nutrient recycling. *BioMed Research International*, *2370927*, 1–19. Advance online publication. DOI: 10.1155/2017/2370927 PMID: 28293629

Park, H., Seo, S. I., Lim, J. H., Song, J., Seo, J. H., & Kim, P. I. (2022). Screening of Carbofuran-Degrading Bacteria *Chryseobacterium* sp. BSC2-3 and Unveiling the Change in Metabolome during Carbofuran Degradation. *Metabolites*, *12*(3), 1–16. DOI: 10.3390/metabo12030219 PMID: 35323662

Park, S. Y., Yang, D., Ha, S. H., & Lee, S. Y. (2018). Metabolic engineering of microorganisms for the production of natural compounds. *Advanced Biosystems*, *2*(1), 1700190. DOI: 10.1002/adbi.201700190

Parmar, K. S., & Patel, K. M. (2025). Biosorption and bioremediation of heavy metal ions from wastewater using algae: A comprehensive review. *World Journal of Microbiology & Biotechnology*, *41*(7), 262. DOI: 10.1007/s11274-025-04424-5 PMID: 40640643

Parshetti, G. K., Telke, A. A., Kalyani, D. C., & Govindwar, S. P. (2010). Decolorization and detoxification of sulfonated azo dye methyl orange by *Kocuria rosea* MTCC 1532. *Journal of Hazardous Materials*, *176*(1-3), 503–509. DOI: 10.1016/j.jhazmat.2009.11.058 PMID: 19969416

Passi, A., Tibocha-Bonilla, J. D., Kumar, M., Tec-Campos, D., Zengler, K., & Zuniga, C. (2021). Genome-Scale Metabolic Modeling Enables In-Depth Understanding of Big Data. *Metabolites*, *12*(1), 14. DOI: 10.3390/metabo12010014 PMID: 35050136

Patel, A. B., Manvar, T., Jain, K. R., Desai, C., & Madamwar, D. (2021). Metagenomic insights into bacterial communities' structures in polycyclic aromatic hydrocarbons degrading consortia. *Journal of Environmental Chemical Engineering*, *9*(6), 106578. DOI: 10.1016/j.jece.2021.106578

Patel, A. I., Vyas, D. R., Vyas, A. J., Patel, A. B., Dudhrejiya, A. V., & Chotaliya, U. J. (2023). Proteomics: Technology, challenges and advances. *Asian Journal of Pharmaceutical Research*, *13*(4), 249–255. DOI: 10.52711/2231-5691.2023.00046

Patel, J. G., Nirmal Kumar, J. I., Kumar, R. N., & Khan, S. R. (2015). Enhancement of pyrene degradation efficacy of Synechocystis sp., by construction of an artificial microalgal-bacterial consortium. *Cogent Chemistry*, *1*(1), 1064193. DOI: 10.1080/23312009.2015.1064193

PatSnap. (2025, April 29). *What is a microbial consortium and how does it work in synthetic biology?* Synapse. https://synapse.patsnap.com/article/what-is-a-microbial-consortium-and-how-does-it-work-in-synthetic-biology

Pearce, C. I., Lloyd, J. R., & Guthrie, J. T. (2003). The removal of colour from textile wastewater using whole bacterial cells: A review. *Dyes and Pigments*, *58*(3), 179–196. DOI: 10.1016/S0143-7208(03)00064-0

Peaudecerf, F. J., Bunbury, F., Bhardwaj, V., Bees, M. A., Smith, A. G., Goldstein, R. E., & Croze, O. A. (2017). *Microbial mutualism at a distance: The role of geometry in diffusive exchanges*. ArXiv., DOI: 10.1103/PhysRevE.97.022411

Peng, Q., Quan, L., Zheng, H., Li, J., & Xie, G. (2025). Analyzing the contribution of top-down and bottom-up methods to the construction of synthetic microbial communities in Jiuyao. *Food Microbiology*, *129*, 104759. DOI: 10.1016/j.fm.2025.104759 PMID: 40086988

Perfumo, A., Banat, I. M., Marchant, R., & Vezzulli, L. (2007). Thermally enhanced approaches for bioremediation of hydrocarbon-contaminated soils. *Chemosphere*, *66*(1), 179–184. DOI: 10.1016/j.chemosphere.2006.05.006 PMID: 16782171

Periyasamy, A. P. (2025). A review of bioremediation of textile dye containing wastewater. *Cleaner Water*, *4*(June), 100092. DOI: 10.1016/j.clwat.2025.100092

Pfaff, G. (2022). The world of inorganic pigments. *Chemtexts*, *8*(3), 15. DOI: 10.1007/s40828-022-00166-1

Phale, P. S., Mohapatra, B., Malhotra, H., & Shah, B. A. (2022). Eco-physiological portrait of a novel *Pseudomonas* sp. CSV86: An ideal host/candidate for metabolic engineering and bioremediation. *Environmental Microbiology*, *24*(6), 2797–2816. DOI: 10.1111/1462-2920.15694 PMID: 34347343

Philp, J., & Winickoff, D. E. (2018). *Realising the circular bioeconomy. OECD Science, Technology and Industry Policy Papers No. 60*. OECD Publishing.

Pickering, L., Castro-Gutierrez, V., Holden, B., Haley, J., Jarvis, P., Campo, P., & Hassard, F. (2024). How bioaugmentation for pesticide removal influences the microbial community in biologically active sand filters. *Chemosphere*, *363*, 142956. DOI: 10.1016/j.chemosphere.2024.142956 PMID: 39074664

Pilalis, E., Zisis, D., Andrinopoulou, C., Karamanidou, T., Antonara, M., Stavropoulos, T. G., & Chatziioannou, A. (2025). Genome-wide functional annotation of variants: A systematic review of state-of-the-art tools, techniques and resources. *Frontiers in Pharmacology*, *16*, 1474026. DOI: 10.3389/fphar.2025.1474026 PMID: 40098614

Pinu, F. R., Beale, D. J., Paten, A. M., Kouremenos, K., Swarup, S., Schirra, H. J., & Wishart, D. (2019). Systems Biology and Multi-Omics Integration: Viewpoints from the Metabolomics Research Community. *Metabolites*, *9*(4), 76. DOI: 10.3390/metabo9040076 PMID: 31003499

Pisera, A. O., & Liu, C. C. (2025). The role of plasmid copy number and mutation rate in evolutionary outcomes. *Nature Ecology and Evolution*, *9*(9), 1694–1704. DOI: 10.1038/s41559-025-02792-7 PMID: 40659873

Pointing, S. (2001). Feasibility of bioremediation by white-rot fungi. *Applied Microbiology and Biotechnology*, *57*(1-2), 20–33. DOI: 10.1007/s002530100745 PMID: 11693920

Pollmann, K., Kutschke, S., Matys, S., Raff, J., Hlawacek, G., & Lederer, F. L. (2018). Bio-recycling of metals: Recycling of technical products using biological applications. *Biotechnology Advances*, *36*(4), 1048–1062. DOI: 10.1016/j.biotechadv.2018.03.006 PMID: 29555455

Potts, L. D., Douglas, A., Perez Calderon, L. J., Anderson, J. A., Witte, U., Prosser, J. I., & Gubry-Rangin, C. (2022). Chronic environmental perturbation influences microbial community assembly patterns. *Environmental Science & Technology*, *56*(4), 2300–2311. DOI: 10.1021/acs.est.1c05106 PMID: 35103467

Poulos, C. J., Bayer, A., Beaupre, L., Clare, L., Poulos, R. G., Wang, R. H., & McGilton, K. S. (2017). A comprehensive approach to reablement in dementia. *Alzheimer's & Dementia: Translational Research & Clinical Interventions*, *3*(3), 450–458. DOI: 10.1016/j.trci.2017.06.005 PMID: 29067351

Poursat, B. A., van Spanning, R. J., de Voogt, P., & Parsons, J. R. (2019). Implications of microbial adaptation for the assessment of environmental persistence of chemicals. *Critical Reviews in Environmental Science and Technology*, *49*(23), 2220–2255. DOI: 10.1080/10643389.2019.1607687

Prasad, M. N. V. (2023). Minerogenic dust and human health. *Medical Geology: En route to One Health*, pp.65-82.

Prempeh, K. B. (2024). The role of economic growth, financial development, globalization, renewable energy and industrialization in reducing environmental degradation in the economic community of West African States. *Cogent Economics & Finance*, *12*(1), 2308675. DOI: 10.1080/23322039.2024.2308675

Prévoteau, A., Carvajal-Arroyo, J. M., Ganigué, R., & Rabaey, K. (2020). Microbial electrosynthesis from CO_2: Forever a promise? *Current Opinion in Biotechnology*, *62*, 48–57. DOI: 10.1016/j.copbio.2019.08.014 PMID: 31593911

Prince, J. F., van den Bosch, M. A., Nijsen, J. F., Smits, M. L., van den Hoven, A. F., Nikolakopoulos, S., Wessels, F. J., Bruijnen, R. C. G., Braat, M. N. G. J. A., Zonnenberg, B. A., & Lam, M. G. (2018). Efficacy of radioembolization with 166Ho-microspheres in salvage patients with liver metastases: A phase 2 study. *Journal of Nuclear Medicine*, *59*(4), 582–588. DOI: 10.2967/jnumed.117.197194 PMID: 28916623

Przymus, P., Rykaczewski, K., Truu, J., Kolev, M., Naskinova, I., Gruca, A., Sampri, A., Frohme, M., & Nechyporenko, A. (2025). Deep learning in microbiome analysis: A comprehensive review of neural network models. *Frontiers in Microbiology*, *15*, 1516667. DOI: 10.3389/fmicb.2024.1516667 PMID: 39911715

Pundir, A., Thakur, M. S., Prakash, S., Kumari, N., Sharma, N., Parameswari, E., He, Z., Nam, S., Thakur, M., Puri, S., Puranik, S., Kumar, S., Madhu, , & Kumar, M. (2024). Fungi as versatile biocatalytic tool for treatment of textile wastewater effluents. *Environmental Sciences Europe*, *36*(1), 185. Advance online publication. DOI: 10.1186/s12302-024-01007-3

Punzi, M. (2015). *Treatment of textile wastewater by combining biological processes and advanced oxidation* [Lund University]. https://portal.research.lu.se/portal/files/6076976/5367871.pdf

Purohit, H. J., Tikariha, H., & Kalia, V. C. (2018). Current scenario on application of computational tools in biological systems. In Purohit, H., Kalia, V., & More, R. (Eds.), *Soft Computing for Biological Systems* (pp. 1–12). Springer. DOI: 10.1007/978-981-10-7455-4_1

Pushkar, B., Sevak, P., Parab, S., & Nilkanth, N. (2021). Chromium pollution and its bioremediation mechanisms in bacteria: A review. *Journal of Environmental Management*, *287*, 112279. DOI: 10.1016/j.jenvman.2021.112279 PMID: 33706095

Puteri, M. N., Gew, L. T., Ong, H. C., & Ming, L. C. (2025). Technologies to eliminate microplastic from water: Current approaches and future prospects. *Environment International*, *199*, 109397. DOI: 10.1016/j.envint.2025.109397 PMID: 40279687

Qattan, S. Y. (2025). Harnessing bacterial consortia for effective bioremediation: Targeted removal of heavy metals, hydrocarbons, and persistent pollutants. *Environmental Sciences Europe*, *37*(1), 85. DOI: 10.1186/s12302-025-01103-y

Qi, L. S., Larson, M. H., Gilbert, L. A., Doudna, J. A., Weissman, J. S., Arkin, A. P., & Lim, W. A. (2013). Repurposing CRISPR as an RNA-guided platform for sequence-specific control of gene expression. *Cell*, *152*(5), 1173–1183. DOI: 10.1016/j.cell.2013.02.022 PMID: 23452860

Qiu, W., Zheng, M., Sun, J., Tian, Y., Fang, M., Zheng, Y., Zhang, T., & Zheng, C. (2019). Photolysis of enrofloxacin, pefloxacin and sulfaquinoxaline in aqueous solution by UV/H2O2, UV/Fe(II), and UV/H2O2/Fe(II) and the toxicity of the final reaction solutions on zebrafish embryos. *The Science of the Total Environment*, *651*, 1457–1468. DOI: 10.1016/j.scitotenv.2018.09.315 PMID: 30360275

Qi, X., Ma, Y., Chang, H., Li, B., Ding, M., & Yuan, Y. (2021). Evaluation of PET Degradation Using Artificial Microbial Consortia. *Frontiers in Microbiology*, *12*, 778828. DOI: 10.3389/fmicb.2021.778828 PMID: 35003008

Quek, E., Ting, Y.-P., & Tan, H. M. (2006). Rhodococcus sp. F92 immobilized on polyurethane foam shows ability to degrade various petroleum products. *Bioresource Technology, 97*(1), 32–38. DOI: 10.1016/j.biortech.2005.02.031 PMID: 16154500

Qu, Y., Shi, S., Ma, F., & Yan, B. (2010). Bioresource Technology Decolorization of Reactive Dark Blue K-R by the synergism of fungus and bacterium using response surface methodology. *Bioresource Technology, 101*(21), 8016–8023. DOI: 10.1016/j.biortech.2010.05.025 PMID: 20566285

Radojević, I. D., & Jakovljević, V. D. (2025). Bacterial Potential for Bioremediation of Surfactants and Heavy Metals: Current Knowledge and Trends in Wastewater Treatment Processes. *Separations, 12*(11), 308. DOI: 10.3390/separations12110308

Rafeeq, H., Afsheen, N., Rafique, S., Arshad, A., Intisar, M., Hussain, A., Bilal, M., & Iqbal, H. M. (2023). Genetically engineered microorganisms for environmental remediation. *Chemosphere, 310*, 136751. DOI: 10.1016/j.chemosphere.2022.136751 PMID: 36209847

Rajendran, S., Kalairaj, A., & Senthilvelan, T. J. B. C. (2025). A comprehensive review on enzymatic decolorization of various azo dyes using laccase for the abatement of industrial pollution. *Biomass Conversion and Biorefinery, 15*(9), 13079–13101. DOI: 10.1007/s13399-024-06104-0

Rajkumar, R., & Kurinjimalar, C. (2021). *Microbes and plant mineral nutrition. Microbiological activity for soil and plant health management.* Springer.

Raju, C. A., Anitha, J., Kalyani, R. M., Satyanandam, K., & Jagadeesh, P. (2021). Sorption of cobalt using marine macro seaweed graciliariacorticatared algae powder. *Materials Today: Proceedings, 44*, 1816–1827. DOI: 10.1016/j.matpr.2020.12.009

Raklami, A., Meddich, A., Oufdou, K., & Baslam, M. (2022). Plants Microorganisms-based bioremediation for heavy metal cleanup: Recent developments, phytoremediation techniques, regulation mechanisms, and molecular responses. *International Journal of Molecular Sciences, 23*(9), 5031. DOI: 10.3390/ijms23095031 PMID: 35563429

Ram, R. M., Debnath, A., Negi, S. & Singh, H. 2022. Use of microbial consortia for broad spectrum protection of plant pathogens: Regulatory hurdles, present status and future prospects. *Biopesticides*, 319-335.

Ramamurthy, K., Priya, P. S., Murugan, R., & Arockiaraj, J. (2024). Hues of risk: Investigating genotoxicity and environmental impacts of azo textile dyes. *Environmental Science and Pollution Research International, 31*(23), 33190–33211. DOI: 10.1007/s11356-024-33444-1 PMID: 38676865

Rambabu, K., Banat, F., Pham, Q. M., Ho, S. H., Ren, N. Q., & Show, P. L. (2020). Biological remediation of acid mine drainage: Review of past trends and current outlook. *Environmental Science and Ecotechnology*, *2*, 100024. DOI: 10.1016/j.ese.2020.100024 PMID: 36160925

Rane, A., & Joshi, S. J. (2021). Biodecolorization and Biodegradation of Dyes: A Review. *The Open Biotechnology Journal*, *15*(1), 97–108. DOI: 10.2174/1874070702115010097

Ranganathan, S., Suthers, P. F., & Maranas, C. D. (2009). Flux balance analysis of biological systems. *Briefings in Bioinformatics*, *10*(4), 435–449. DOI: 10.1093/bib/bbp011 PMID: 19287049

Rani, A., Snyder, S. W., Kim, H., Lei, Z., & Pan, S. Y. (2022). Pathways to a net-zero- carbon water sector through energy-extracting wastewater technologies. *NPJ Clean Water*, *5*(1), 49. DOI: 10.1038/s41545-022-00197-8

Rather, M. A., Gupta, K., & Mandal, M. (2021). Microbial biofilm: formation, architecture, antibiotic resistance, and control strategies. *Brazilian journal of microbiology: [publication of the Brazilian Society for Microbiology]*, *52*(4), 1701–1718. https://doi.org/DOI: 10.1007/s42770-021-00624-x

Rathour, R. K., Sharma, D., Ullah, S., Mahmoud, E.-H. M., Sharma, N., Kumar, P., Bhatt, A. K., Ahmad, I., & Bhatia, R. K. (2024). Bacterial–microalgal consortia for bioremediation of textile industry wastewater and resource recovery for circular economy. *Biotechnology for the Environment*, *1*(1), 1–21. DOI: 10.1186/s44314-024-00005-2

Rehan, Z. A., & Usman, A. (2023). Polymeric paints and coatings. In *Advanced Functional Polymers: Synthesis to Applications* (pp. 49–76). Springer Nature Singapore. DOI: 10.1007/978-981-99-0787-8_4

Reichl, C., & Schatz, M. (2022). *World Mining Data 2022* (Vol. 37). Austrian Federal Ministry of Agriculture, Regions and Tourism.

Reiman, D., Layden, B. T., & Dai, Y. (2021). MiMeNet: Exploring microbiome–metabolome relationships using neural networks. *PLoS Computational Biology*, *17*(5), e1009021. DOI: 10.1371/journal.pcbi.1009021 PMID: 33999922

Renda, B. A., Hammerling, M. J., & Barrick, J. E. (2014). Engineering reduced evolutionary potential for synthetic biology. *Molecular BioSystems*, *10*(7), 1668–1678. DOI: 10.1039/C3MB70606K PMID: 24556867

Rendón-Castrillón, L., Ramírez-Carmona, M., Ocampo-López, C., & Gómez-Arroyave, L. (2023). Bioleaching techniques for sustainable recovery of metals from solid matrices. *Sustainability (Basel), 15*(13), 10222. DOI: 10.3390/su151310222

Renganathan, P., Gaysina, L. A., Gutiérrez, C. G., Rueda Puente, E. O., & Sainz-Hernández, J. C. (2025). Harnessing Engineered Microbial Consortia for Xenobiotic Bioremediation: Integrating Multi-Omics and AI for Next-Generation Wastewater Treatment. *Journal of Xenobiotics, 15*(4), 133. DOI: 10.3390/jox15040133 PMID: 40863340

Rengasamy, D., Mase, J. M., Kumar, A., Rothwell, B., Torres, M. T., Alexander, M. R., Winkler, D. A., & Figueredo, G. P. (2022). Feature importance in machine learning models: A fuzzy information fusion approach. *Neurocomputing, 511*(1), 163–174. DOI: 10.1016/j.neucom.2022.09.053

Rezaei, Z., Amoozegar, M. A., & Moghimi, H. (2025). Innovative approaches in bioremediation: The role of halophilic microorganisms in mitigating hydrocarbons, toxic metals, and microplastics in hypersaline environments. *Microbial Cell Factories, 24*(1), 184. DOI: 10.1186/s12934-025-02817-7 PMID: 40804729

Rezaei, Z., & Moghimi, H. (2024). Fungal-bacterial consortia: A promising strategy for the removal of petroleum hydrocarbons. *Ecotoxicology and Environmental Safety, 280*, 116543. DOI: 10.1016/j.ecoenv.2024.116543 PMID: 38833981

Ribicic, D., Mcfarlin, K. M., Netzer, R., Brakstad, O. G., Winkler, A., Throne-Holst, M., & Størseth, T. R. (2018). Oil type and temperature dependent biodegradation dynamics-Combining chemical and microbial community data through multivariate analysis. *BMC Microbiology, 18*(1), 83. DOI: 10.1186/s12866-018-1221-9 PMID: 30086723

Richmond, J. Y., & McKinney, R. W. (2009). *Biosafety in microbiological and biomedical laboratories*. US Government Printing Office.

Rios Miguel, A. B., Jetten, M. S. M., & Welte, C. U. (2020). The role of mobile genetic elements in organic micropollutant degradation during biological wastewater treatment. *Water Research X, 9*, 100065. DOI: 10.1016/j.wroa.2020.100065 PMID: 32984801

Riza, M., Ehsan, M. N., & Hoque, S. (2021). *Portrayal of Textile Based Pollutants and its Impact on Soil, Plants and Fisheries. 20.*

Rizwana, M., Darshan, M., & Nilesh, D. (2014). Phytoremediation of Textile Waste Water Using Potential Wetland Plant: Eco Sustainable Approach. *International Journal of Interdisciplinary and Multidisciplinary Studies, 1*(4), 130–138. http://www.ijims.com

Roberts, C., Edwards, S., Vague, M., León-Zayas, R., Scheffer, H., Chan, G., Swartz, N. A., & Mellies, J. L. (2020). Environmental consortium containing pseudomonas and bacillus species synergistically degrades polyethylene terephthalate plastic. *MSphere, 5*(6), 1–20. DOI: 10.1128/mSphere.01151-20 PMID: 33361127

Robinson, S. L. (2022). Artificial intelligence for microbial biotechnology: Beyond the hype. *Microbial Biotechnology, 15*(1), 65–69. DOI: 10.1111/1751-7915.13943 PMID: 34606686

Rodrigues, J. L., Maltseva, O. V., Tsoi, T. V., Helton, R. R., Quensen, J. F., Fukuda, M., & Tiedje, J. M. (2001). Development of a Rhodococcus recombinant strain for degradation of products from anaerobic dechlorination of PCBs. *Environmental Science & Technology, 35*(4), 663–668. DOI: 10.1021/es001308t PMID: 11349275

Rodriguez-couto, S. (2013). Treatment of Textile Wastewater by White-rot Fungi: Still a Far Away Reality? *Textiles and Light Industrial Science and Technology, 2*(3), 113–119.

Roell, G. W., Zha, J., Carr, R. R., Koffas, M. A., Fong, S. S., & Tang, Y. J. (2019). Engineering microbial consortia by division of labour. *Microbial Cell Factories, 18*(35), 1–11. DOI: 10.1186/s12934-019-1083-3 PMID: 30736778

Rollakanti, C. R., & Srinivasu, K. (2022). Self-healing Concrete Based on Bacteria for Sustainable Infrastructure: A State-of-the-Art Review. In: Kumar, P.G., Subramaniam, K.V.L., Santhakumar, S.M., Satyam D., N. (eds) Recent Advances in Civil Engineering. Lecture Notes in Civil Engineering, vol 233. Springer, Singapore. DOI: 10.1007/978-981-19-0189-8_15

Romantschuk, M., Lahti-Leikas, K., Kontro, M., Galitskaya, P., Talvenmäki, H., Simpanen, S., Allen, J. A., & Sinkkonen, A. (2023). Bioremediation of contaminated soil and groundwater by in situ biostimulation. *Frontiers in Microbiology, 14*, 1258148. DOI: 10.3389/fmicb.2023.1258148 PMID: 38029190

Romantschuk, M., Sarand, I., Petänen, T., Peltola, R., Jonsson-Vihanne, M., Koivula, T., Yrjälä, K., & Haahtela, K. (2000). Means to improve the effect of in situ bioremediation of contaminated soil: An overview of novel approaches. *Environmental Pollution, 107*(2), 179–185. DOI: 10.1016/S0269-7491(99)00136-0 PMID: 15092994

Romero, M., Gallego, D., Blaz, J., Lechuga, A., Martínez, J. F., Barajas, H. R., Hayano-Kanashiro, C., Peimbert, M., Cruz-Ortega, R., & Molina-Freaner, F. E. (2019). Rhizosphere metagenomics of mine tailings colonizing plants: Assembling and selecting synthetic bacterial communities to enhance in situ bioremediation. *bioRxiv.*, 664805. DOI: 10.1101/664805

Rosli, N. A., Aziz, H. A., Pueh, L. L. L., Othman, I. B., Zawawi, M. H., & Hung, Y. T. (2023). Management of various sources of hazardous waste. In *Waste Treatment in the Biotechnology, Agricultural and Food Industries* (Vol. 2, pp. 19–64). Springer International Publishing.

Roslund, M. I., Rantala, S., Oikarinen, S., Puhakka, R., Hui, N., Parajuli, A., Laitinen, O. H., Hyöty, H., Rantalainen, A.-L., Sinkkonen, A., Grönroos, M., Hui, N., Kalvo, R., Nurminen, N., Oikarinen, S., Parajuli, A., Puhakka, R., Rantala, S., Roslund, M. I., & Sinkkonen, A. (2019). Endocrine disruption and commensal bacteria alteration associated with gaseous and soil PAH contamination among daycare children. *Environment International, 130*, 104894. DOI: 10.1016/j.envint.2019.06.004 PMID: 31220749

Rotaru, A. E., Shrestha, P. M., Liu, F., Markovaite, B., Chen, S., Nevin, K. P., & Lovley, D. R. (2014). Direct interspecies electron transfer between *Geobacter metallireducens* and *Methanosarcina barkeri. Applied and Environmental Microbiology, 80*(15), 4599–4605. DOI: 10.1128/AEM.00895-14 PMID: 24837373

Ruan, Z., Chen, K., Cao, W., Meng, L., Yang, B., Xu, M., Xing, Y., Li, P., Freilich, S., Chen, C., Gao, Y., Jiang, J., & Xu, X. (2024). Engineering natural microbiomes toward enhanced bioremediation by microbiome modeling. *Nature Communications, 15*(1), 4694. DOI: 10.1038/s41467-024-49098-z PMID: 38824157

Ruaud, A., Sancaktar, C., Bagatella, M., Ratzke, C., & Martius, G. (2024, July 8). *Modelling Microbial Communities with Graph Neural Networks.* PMLR. https://proceedings.mlr.press/v235/ruaud24a.html

Ruffino, B., Campo, G., Idris, S. S., Salihoğlu, G., & Zanetti, M. (2023). Automotive paint sludge: A review of pretreatments and recovery options. *Resources, 12*(4), 45. DOI: 10.3390/resources12040045

Saeed, H., Padmesh, S., Singh, A., Nandy, A., Singh, S. P., & Deshwal, R. K. (2024). Impact of veterinary pharmaceuticals on environment and their mitigation through microbial bioremediation. *Frontiers in Microbiology, 15*, 1396116. DOI: 10.3389/fmicb.2024.1396116 PMID: 39040911

Saha, L., Tiwari, J., Bauddh, K., & Ma, Y. (2021). Recent developments in microbe–plant-based bioremediation for tackling heavy metal-polluted soils. *Frontiers in Microbiology, 12*, 731723. DOI: 10.3389/fmicb.2021.731723 PMID: 35002995

Saini, S., Tewari, S., Dwivedi, J., & Sharma, V. (2023). Biofilm-mediated wastewater treatment: A comprehensive review. *Materials Advances, 4*(6), 1415–1443. DOI: 10.1039/D2MA00945E

Salazar-García, L. M., Damas-Ramos, L. C., Trejo-Alarcón, L. M., Rago, D., Ahonen, L., Cruz-Morales, P., Ponce-Noyola, P., & Licona-Cassani, C. (2025). CRISPR-driven enhanced hydrocarbon emulsification in an environmental *Pseudomonas aeruginosa* strain. *Microbial Cell Factories, 24*(1), 151. DOI: 10.1186/s12934-025-02769-y PMID: 40605043

Salihoglu, G., & Salihoglu, N. K. (2016). A review on paint sludge from automotive industries: Generation, characteristics and management. *Journal of Environmental Management, 169*, 223–235. DOI: 10.1016/j.jenvman.2015.12.039 PMID: 26773426

Salis, H. M., Mirsky, E. A., & Voigt, C. A. (2009). Automated design of synthetic ribosome binding sites to control protein expression. *Nature Biotechnology, 27*(10), 946–950. DOI: 10.1038/nbt.1568 PMID: 19801975

Salleh, A. B., Ghazali, F. M., Rahman, R., & Basri, M. (2003). Bioremediation of petroleum hydrocarbon pollution. *Indian Journal of Biotechnology, 2*, 411–425.

Samin, G., Pavlova, M., Arif, M. I., Postema, C. P., Damborsky, J., & Janssen, D. B. (2014). A *Pseudomonas putida* strain genetically engineered for 1, 2, 3-trichloropropane bioremediation. *Applied and Environmental Microbiology, 80*(17), 5467–5476. DOI: 10.1128/AEM.01620-14 PMID: 24973068

Samuchiwal, S., Gola, D., & Malik, A. (2021). Decolourization of textile effluent using native microbial consortium enriched from textile industry effluent. *Journal of Hazardous Materials, 402*(September 2020), 123835. DOI: 10.1016/j.jhazmat.2020.123835

Sandbhor, S., Apte, S., Puranik, S., Asati, M., & Rai, S. (2025). Technique for improving concrete characteristics with bacillus: Insights on strength and durability. *MethodsX, 14*, 103151. DOI: 10.1016/j.mex.2024.103151 PMID: 39866201

Sanghvi, G., Thanki, A., Pandey, S., & Singh, N. K. (2020). Bioremediation of Pollutants. Elsevier, Amsterdam, The Netherlands: *Engineered bacteria for bioremediation*, pp. 359–374. DOI: 10.1016/B978-0-12-819025-8.00017-X

Sani, Z. M., Dalhatu, A. S., Adam, B. S., Mohammed, K., Muhammad, Y. Y., & Ibrahim, S. (2021). Bioremediation of Some Reactive Dyes Commonly used in Fabric Re-dyeing by *Chlorella vulgaris*. *Asian Journal of Environment & Ecology, 15*(December 2019), 10–19. DOI: 10.9734/ajee/2021/v15i430233

Santala, S., Efimova, E., & Santala, V. (2018). Dynamic decoupling of biomass and wax ester biosynthesis in Acinetobacter baylyi by an autonomously regulated switch. *Metabolic Engineering Communications, 7*, e00078. DOI: 10.1016/j.mec.2018.e00078 PMID: 30271720

Santos, B., Araújo, J., Carvalho, B., Cotrim, C., Bernardino, R., Freitas, F., & Encarnação, T. (2025). Bioremediation of Synthetic Wastewater with Contaminants of Emerging Concern by Nannochloropsis sp. and Lipid Production: A Circular Approach. *Bioengineering (Basel, Switzerland), 12*(3), 246. DOI: 10.3390/bioengineering12030246 PMID: 40150710

Saranya, K., Sundaramanickam, A., Shekhar, S., & Swaminathan, S. (2019). Biosorption of mercury by Bacillus thuringiensis (CASKS3) isolated from mangrove sediments of southeast coast India. *Indian Journal of Geo-Marine Sciences, 48*, 143–150.

Saratale, R. G., Saratale, G. D., Chang, J. S., & Govindwar, S. P. (2010). Decolorization and biodegradation of reactive dyes and dye wastewater by a developed bacterial consortium. *Biodegradation, 21*(6), 999–1015. DOI: 10.1007/s10532-010-9360-1 PMID: 20407917

Saratale, R. G., Saratale, G. D., Chang, J. S., & Govindwar, S. P. (2011). Bacterial decolorization and degradation of azo dyes: A review. *Journal of the Taiwan Institute of Chemical Engineers, 42*(1), 138–157. DOI: 10.1016/j.jtice.2010.06.006

Sarkar, J., Kazy, S. K., Gupta, A., Dutta, A., Mohapatra, B., Roy, A., Bera, P., Mitra, A., & Sar, P. (2016). Biostimulation of indigenous microbial community for bioremediation of petroleum refinery sludge. *Frontiers in Microbiology, 7*, 1407. DOI: 10.3389/fmicb.2016.01407 PMID: 27708623

Sarker, M. A. B., Imtiaz, M. H., Holsen, T. M., & Baki, A. B. (2024). Real-time detection of microplastics using an ai camera. *Sensors (Basel), 24*(13), 4394. DOI: 10.3390/s24134394 PMID: 39001173

Satam, H., Joshi, K., Mangrolia, U., Waghoo, S., Zaidi, G., Rawool, S., Thakare, R. P., Banday, S., Mishra, A. K., Das, G., & Malonia, S. K. (2023). Next-Generation Sequencing Technology: Current Trends and Advancements. *Biology (Basel), 12*(7), 997. DOI: 10.3390/biology12070997 PMID: 37508427

Sathe, S. M., Bhowmick, G. D., Dubey, B. K., & Ghangrekar, M. M. (2020). Surfactant removal from wastewater using photo-cathode microbial fuel cell and laterite-based hybrid treatment system. *Bioprocess and Biosystems Engineering*, *43*(11), 2075–2084. DOI: 10.1007/s00449-020-02396-4 PMID: 32596770

Satpathy, R. (2021). Application of Dehalogenase Enzymes in Bioremediation of Halogenated Pollutants: A Short Review. In Malik, J. (Ed.), *Handbook of Research on Microbial Remediation and Microbial Biotechnology for Sustainable Soil* (pp. 373–385)., DOI: 10.4018/978-1-7998-7062-3.ch014

Sayed, K., Baloo, L., & Sharma, N. K. (2021). Bioremediation of total petroleum hydrocarbons (TPH) by bioaugmentation and biostimulation in water with floating oil spill containment booms as bioreactor basin. *International Journal of Environmental Research and Public Health*, *18*(5), 2226. DOI: 10.3390/ijerph18052226 PMID: 33668225

Scheuerl, T., Hopkins, M., Nowell, R. W., Rivett, D. W., Barraclough, T. G., & Bell, T. (2020). Bacterial adaptation is constrained in complex communities. *Nature Communications*, *11*(1), 754. DOI: 10.1038/s41467-020-14570-z PMID: 32029713

Scholz, D. A. Y. M. (2019). Textile dye wastewater characteristics and constituents of synthetic effluents: a critical review. In *International Journal of Environmental Science and Technology* (Vol. 16, Number 2). Springer Berlin Heidelberg. DOI: 10.1007/s13762-018-2130-z

Schroven, K., Aertsen, A. & Lavigne, R. (2021). Bacteriophages as drivers of bacterial virulence and their potential for biotechnological exploitation, *FEMS Microbiology Reviews*, 45(1), 041, https://doi.org/DOI: 10.1093/femsre/fuaa041

Schwalm, N. D.III, Mojadedi, W., Gerlach, E. S., Benyamin, M., Perisin, M. A., & Akingbade, K. L. (2019). Developing a Microbial Consortium for Enhanced Metabolite Production from Simulated Food Waste. *Fermentation (Basel, Switzerland)*, *5*(4), 98–105. DOI: 10.3390/fermentation5040098

Schwuger, M. J., & Smulders, E. J. (2020). Inorganic builders. In *Detergency* (pp. 371–439). CRC Press. DOI: 10.1201/9781003065494-6

Scott, S. R., & Hasty, J. (2016). Quorum sensing communication modules for microbial consortia. *ACS Synthetic Biology*, *5*(9), 969–977. DOI: 10.1021/acssynbio.5b00286 PMID: 27172092

Scow, K. M., & Hicks, K. A. (2005). Natural attenuation and enhanced bioremediation of organic contaminants in groundwater. *Current Opinion in Biotechnology*, *16*(3), 246–253. DOI: 10.1016/j.copbio.2005.03.009 PMID: 15961025

Sebesta, J., Xiong, W., Guarnieri, M. T., & Yu, J. (2022). Biocontainment of genetically engineered algae. *Frontiers in Plant Science*, *13*, 839446. DOI: 10.3389/fpls.2022.839446 PMID: 35310623

Secretariat of the Convention on Biological Diversity. (2000). *Cartagena Protocol on Biosafety to the Convention on Biological Diversity*. Secretariat of the Convention on Biological Diversity.

Selle, K., & Barrangou, R. (2015). Harnessing CRISPR–Cas systems for bacterial genome editing. *Trends in Microbiology*, *23*(4), 225–232. DOI: 10.1016/j.tim.2015.01.008 PMID: 25698413

Selvasembian, R., Azelee, N. I. B. W., Saravanan, R., Ponnusami, V., & Mishra, A. K. (2023). *Valorization of Wastes for Sustainable Development: Waste to Wealth*. Elsevier.

Senthilkumar, R., Prasad, D. R., Govindarajan, L., Saravanakumar, K., & Prasad, B. N. (2019). Green alga-mediated treatment process for removal of zinc from synthetic solution and industrial effluent. *Environmental Technology*, *40*(10), 1262–1270. DOI: 10.1080/09593330.2017.1420696 PMID: 29284361

Seressiotis, A., & Bailey, J. E. (1986). MPS: An algorithm and data base for metabolic pathway synthesis. *Biotechnology Letters*, *8*(12), 837–842. DOI: 10.1007/BF01078641

Shade, A., Dunn, R. R., Blowes, S. A., Keil, P., Bohannan, B. J., Herrmann, M., Küsel, K., Lennon, J. T., Sanders, N. J., Storch, D., & Chase, J. (2018). Macroecology to unite all life, large and small. *Trends in Ecology & Evolution*, *33*(10), 731–744. DOI: 10.1016/j.tree.2018.08.005 PMID: 30209011

Shahidi-Rizi, M. S. S., Emtiazi, G., Sepahi, A. A., Maal, K. B., & Hosseini, F. (2017). Synergistic effect of nanoparticles and *Thalassospira* sp. isolated from Bahregan area in biodegradation of oil sludge. *Petroleum Science and Technology*, *35*(9), 831–838. DOI: 10.1080/10916466.2016.1278386

Shajil, U. K., Sujith, A., & Nair, B. G. (2025). Comparison of human intelligence with artificial intelligence. In *Artificial Intelligence and Biological Sciences* (pp. 68–92). CRC Press. DOI: 10.1201/9781003492726-4

Shanmukha, N. T., Vinayaka, M., & Lokeshappa, B. (2024). Biodiversity loss due to mining activities. In *Impact of societal development and infrastructure on biodiversity decline* (pp. 166–191). IGI Global.

Sharghi, E. A., & Davarpanah, L. (2022). Optimization of chemical coagulation–flocculation process of detergent manufacturing plant wastewater treatment for full scale applications: A case study. *Desalination and Water Treatment, 262*, 38–53. DOI: 10.5004/dwt.2022.28494

Sharma, I. (2020). Bioremediation Techniques for Polluted Environment: Concept, Advantages, Limitations, and Prospects. In Trace Metals in the Environment-New Approaches and Recent Advances. *IntechOpen*. DOI: DOI: 10.5772/intechopen.90453

Sharma, R., & Malaviya, P. (2022). Constructed wetlands for textile wastewater remediation: A review on concept, pollutant removal mechanisms, and integrated technologies for efficiency enhancement. *Chemosphere, 290*(December 2021), 133358. DOI: 10.1016/j.chemosphere.2021.133358

Sharma, A., Singh, S. B., Sharma, R., Chaudhary, P., Pandey, A. K., Ansari, R., Vasudevan, V., Arora, A., Singh, S., Saha, S., & Nain, L. (2016). Enhanced biodegradation of PAHs by microbial consortium with different amendment and their fate in in-situ condition. *Journal of Environmental Management, 181*, 728–736. DOI: 10.1016/j.jenvman.2016.08.024 PMID: 27558829

Sharma, B., & Shukla, P. (2022). Futuristic avenues of metabolic engineering techniques in bioremediation. *Biotechnology and Applied Biochemistry, 69*(1), 51–60. DOI: 10.1002/bab.2080 PMID: 33242354

Sharma, I. (2020). Bioremediation techniques for polluted environment: concept, advantages, limitations, and prospects. In *Trace metals in the environment-new approaches and recent advances*. IntechOpen.

Sharma, M., Sharma, S., Mazumder, S., & Negi, R. K. (2022). Application of "OMICs" Approaches in Bioremediation. In Tripathi, M., & Singh, D. S. (Eds.), *Bioremediation: Challenges and Advancements* (Vol. 69, pp. 191–223). Bentham Science Publisher. DOI: 10.2174/9789815036039122010010

Sharma, P. K., Sharma, V., Sharma, S., Bhatia, G., Singh, K., & Sharma, R. (2019). Comparative metatranscriptome analysis revealed broad response of microbial communities in two soil types, agriculture versus organic soil. *Journal of Genetic Engineering and Biotechnology, 17*(1), 6. DOI: 10.1186/s43141-019-0006-3 PMID: 31659568

Sharma, P., Gaur, V. K., Sirohi, R., Varjani, S., Hyoun Kim, S., & Wong, J. W. C. (2021). Sustainable processing of food waste for production of bio-based products for circular bioeconomy. *Bioresource Technology, 325*, 124684. DOI: 10.1016/j.biortech.2021.124684 PMID: 33493748

Shayanthan, A., Ordoñez, P. C., & Oresnik, I. J. (2022). The role of Synthetic Microbial Communities (SYNCOM) in sustainable agriculture. *Frontiers in Agronomy*, *4*, 896307. Advance online publication. DOI: 10.3389/fagro.2022.896307

Sheng, J., Amankwah-Amoah, J., & Wang, X. (2019). Technology in the 21st century: New challenges and opportunities. *Technological Forecasting and Social Change*, *143*, 321–335. DOI: 10.1016/j.techfore.2018.06.009

Shen, L., Zhao, Q., Wu, X., Li, X., Li, Q., & Wang, Y. (2016). Interspecies electron transfer in syntrophic methanogenic consortia: From cultures to bioreactors. *Renewable & Sustainable Energy Reviews*, *54*, 1358–1367. DOI: 10.1016/j.rser.2015.10.102

Sheoran, K., Siwal, S. S., Kapoor, D., Singh, N., Saini, A. K., Alsanie, W. F., & Thakur, V. K. (2022). Air pollutants removal using biofiltration technique: A challenge at the frontiers of sustainable environment. *ACS Engineering Au*, *2*(5), 378–396. DOI: 10.1021/acsengineeringau.2c00020 PMID: 36281334

Shi, L., Liu, Z., Yang, L., & Fan, W. (2022). Effect of plant waste addition as exogenous nutrients on microbial remediation of petroleum-contaminated soil. *Annals of Microbiology*, *72*(1), 22. DOI: 10.1186/s13213-022-01679-3

Shirani Faradonbeh, R., Shah, M. I., Bahadori, M., & Jang, H. (2025). Toward Sustainable Mining: Exploring Alternative Mineral Resources and Innovative Extraction Techniques. *Mining (Basel)*, *5*(4), 66. DOI: 10.3390/mining5040066

Shi, Y., Zhang, L., Peterson, C. B., Do, A., & Jenq, R. R. (2022). Performance determinants of unsupervised clustering methods for microbiome data. *Microbiome*, *10*(1), 25. DOI: 10.1186/s40168-021-01199-3 PMID: 35120564

Shmakov, S., Smargon, A., Scott, D., Cox, D., Pyzocha, N., Yan, W., Abudayyeh, O. O., Gootenberg, J. S., Makarova, K. S., Wolf, Y. I., Severinov, K., Zhang, F., & Koonin, E. V. (2017). Diversity and evolution of class 2 CRISPR–Cas systems. *Nature Reviews. Microbiology*, *15*(3), 169–182. DOI: 10.1038/nrmicro.2016.184 PMID: 28111461

Shoukat, R., Khan, S. J., & Jamal, Y. (2019). Hybrid anaerobic-aerobic biological treatment for real textile wastewater. *Journal of Water Process Engineering*, *29*(July 2018), 100804. DOI: 10.1016/j.jwpe.2019.100804

Shrestha, H. K., Appidi, M. R., Villalobos Solis, M. I., Wang, J., Carper, D. L., Burdick, L., Pelletier, D. A., Doktycz, M. J., Hettich, R. L., & Abraham, P. E. (2021). Metaproteomics reveals insights into microbial structure, interactions, and dynamic regulation in defined communities as they respond to environmental disturbance. *BMC Microbiology*, *21*(308), 1–17. DOI: 10.1186/s12866-021-02370-4 PMID: 34749649

Shu, W.-S., & Huang, L.-N. (2022). Microbial diversity in extreme environments. *Nature Reviews. Microbiology, 20*(4), 219–235. DOI: 10.1038/s41579-021-00648-y PMID: 34754082

Shu, X., He, J., Zhou, Z., Xia, L., Hu, Y., Zhang, Y., Zhang, Y., Luo, Y., Chu, H., Liu, W., Yuan, S., Gao, X., & Wang, C. (2022). Organic amendments enhance soil microbial diversity, microbial functionality and crop yields: A meta-analysis. *The Science of the Total Environment, 829*, 154627. DOI: 10.1016/j.scitotenv.2022.154627 PMID: 35306065

Siles, J. A., & Margesin, R. (2018). Insights into microbial communities mediating the bioremediation of hydrocarbon-contaminated soil from an Alpine former military site. *Applied Microbiology and Biotechnology, 102*(10), 4409–4421. DOI: 10.1007/s00253-018-8932-6 PMID: 29594357

Silva, E., Fialho, A. M., Sá-Correia, I., Burns, R. G., & Shaw, L. J. (2004). Combined bioaugmentation and biostimulation to cleanup soil contaminated with high concentrations of atrazine. *Environmental Science & Technology, 38*(2), 632–637. DOI: 10.1021/es0300822 PMID: 14750741

Silva-Rocha, R., de Jong, H., Tamames, J., & de Lorenzo, V. (2011). The logic layout of the TOL network of *Pseudomonas putida* pWW0 plasmid stems from a metabolic amplifier motif (MAM) that optimizes biodegradation of m-xylene. *BMC Systems Biology, 5*(1), 191. DOI: 10.1186/1752-0509-5-191 PMID: 22078029

Sinduja, M., Sathya, V., Maheswari, M., Kalpana, P., Dhevagi, P., Dinesh, G. K., & Chitdeshwari, T. (2023). Chemical transformation and bioavailability of chromium in the contaminated soil amended with bioamendments. *Bioremediation Journal, 27*(3), 229–250. DOI: 10.1080/10889868.2022.2049677

Singh Arora, D., & Kumar Sharma, R. (2010). Ligninolytic fungal laccases and their biotechnological applications. *Applied Biochemistry and Biotechnology, 160*(6), 1760–1788. DOI: 10.1007/s12010-009-8676-y PMID: 19513857

Singh, R., Kumar, M., Mittal, A., & Mehta, P. K. (2016). Microbial enzymes: industrial progress in 21st century. *3 Biotech, 6*(2), 1–15. DOI: 10.1007/s13205-016-0485-8

Singh, S. N. (2015). Microbial degradation of synthetic dyes in wastewaters. *Environmental Science and Engineering (Subseries: Environmental Science)*, (9783319109411), iii–iv. DOI: 10.1007/978-3-319-10942-8

Singh, S., & Ramakrishna, W. (2021). Application of CRISPR–Cas9 in plant–plant growth-promoting rhizobacteria interactions for next Green Revolution. *3 Biotech, 11*(12), 492. https://doi.org/DOI: 10.1007/s13205-021-03041-x

Singh, A. K., Bilal, M., Iqbal, H. M., & Raj, A. (2021). Trends in predictive biodegradation for sustainable mitigation of environmental pollutants: Recent progress and future outlook. *The Science of the Total Environment*, *770*, 144561. DOI: 10.1016/j.scitotenv.2020.144561 PMID: 33736422

Singh, A., Sharma, A., Verma, R. K., Chopade, R. L., Pandit, P. P., Nagar, V., & Sankhla, M. S. (2022). Heavy metal contamination of water and their toxic effect on living organisms. In *The toxicity of environmental pollutants*. IntechOpen. DOI: 10.5772/intechopen.105075

Singh, A., Xue, A., Tai, J., Mbadugha, F., Obi, P., Mascarenhas, R., Tyagi, A., Siena, A., & Chen, Y. G. (2024). A scalable and cost-efficient rRNA depletion approach to enrich RNAs for molecular biology investigations. *RNA (New York, N.Y.)*, *30*(6), 728–738. DOI: 10.1261/rna.079761.123 PMID: 38485192

Singh, B. K., & Walker, A. (2006). Microbial degradation of organophosphorus compounds. *FEMS Microbiology Reviews*, *30*(3), 428–471. DOI: 10.1111/j.1574-6976.2006.00018.x PMID: 16594965

Singh, D. K. (2008). Biodegradation and bioremediation of pesticide in soil: Concept, method and recent developments. *Indian Journal of Microbiology*, *48*(1), 35–40. DOI: 10.1007/s12088-008-0004-7 PMID: 23100698

Singh, D. P., Prabha, R., Gupta, V. K., & Verma, M. K. (2018). Metatranscriptome Analysis Deciphers Multifunctional Genes and Enzymes Linked With the Degradation of Aromatic Compounds and Pesticides in the Wheat Rhizosphere. *Frontiers in Microbiology*, *9*(1331), 1–15. DOI: 10.3389/fmicb.2018.01331 PMID: 30034370

Singh, J. S. (2015). Microbes: The chief ecological engineers in reinstating equilibrium in degraded ecosystems. *Agriculture, Ecosystems & Environment*, *203*, 80–82. DOI: 10.1016/j.agee.2015.01.026

Singh, J. S., Abhilash, P., Singh, H., Singh, R. P., & Singh, D. (2011). Genetically engineered bacteria: An emerging tool for environmental remediation and future research perspectives. *Gene*, *480*(1-2), 1–9. DOI: 10.1016/j.gene.2011.03.001 PMID: 21402131

Singh, J., & Singh, S. (2019). Biochar applications in bioremediation. In Prasad, M., Favas, M., & Maiti, S. (Eds.), *Bio-geotechnologies for mine site rehabilitation* (pp. 417–433). Elsevier., DOI: 10.1016/B978-0-12-812986-9.00023-7

Singh, S., Sukla, L. B., & Goyal, S. K. (2020). Mine waste and circular economy. *Materials Today: Proceedings*, *30*, 332–339. DOI: 10.1016/j.matpr.2020.01.616

Skaar, I., Asare, N. Y. O., Klein, J., Okoli, A. S., Ruus, A., Gjøen, T., & Ytrehus, B. (2016). *Health and environmental risk evaluation of microorganisms used in bioremediation. Scientific Opinion of the Panel on Microbial Ecology of the Norwegian Scientific Committee for Food Safety. (VKM Report 2016: 42)*. Norwegian Scientific Committee for Food Safety.

Slate, A. J., Whitehead, K. A., Brownson, D. A. C., & Banks, C. E. (2019). Microbial fuel cells: An overview of current technology. *Renewable & Sustainable Energy Reviews, 101*, 60–81. DOI: 10.1016/j.rser.2018.09.044

Smith, K. R., Bruce, N., Balakrishnan, K., Adair-Rohani, H., Balmes, J., Chafe, Z., Dherani, M., Hosgood, H. D., Mehta, S., Pope, D., & Rehfuess, E. (2014). Millions dead: How do we know and what does it mean? Methods used in the comparative risk assessment of household air pollution. *Annual Review of Public Health, 35*(1), 185–206. DOI: 10.1146/annurev-publhealth-032013-182356 PMID: 24641558

Smith, K., Fearnley, C. J., Dixon, D., Bird, D. K., & Kelman, I. (2023). *Environmental Hazards: Assessing Risk and Reducing Disaster* (7th ed.). Routledge., DOI: 10.4324/9781351261647

Smolke, C. D. (2009). Building outside of the box: iGEM and the BioBricks Foundation. *Nature Biotechnology, 27*(12), 1099–1102. DOI: 10.1038/nbt1209-1099 PMID: 20010584

Soberón-Chávez, G. (2023). Some insights on traditional and novel approaches in microbial biotechnology that contribute to the United Nations Sustainable Development Goals. *Microbial Biotechnology, 16*(11), 2015–2018. DOI: 10.1111/1751-7915.14318 PMID: 37452713

Solanki, K., Subramanian, S., & Basu, S. (2013). Microbial fuel cells for azo dye treatment with electricity generation: A review. *Bioresource Technology, 131*, 564–571. DOI: 10.1016/j.biortech.2012.12.063 PMID: 23403060

Song, H. G., & Bartha, R. (1990). Effects of jet fuel spills on the microbial community of soil. *Applied and Environmental Microbiology, 56*(3), 646–651. DOI: 10.1128/aem.56.3.646-651.1990 PMID: 16348138

Song, Y., Yuan, Q., Chen, S., Zeng, Y., Zhao, H., & Yang, Y. (2024). Accurately predicting enzyme functions through geometric graph learning on ESMFold-predicted structures. *Nature Communications, 15*(1), 8180. DOI: 10.1038/s41467-024-52533-w PMID: 39294165

Son, H.-I., Weiss, A., & You, L. (2021). Design patterns for engineering genetic stability. *Current Opinion in Biomedical Engineering*, *19*, 100297. DOI: 10.1016/j.cobme.2021.100297 PMID: 34308010

Sonune, N. A., & Garode, A. M. (2018). Isolation and Characterization of Efficient Bacterial Isolates for Treatment of Municipal Wastewater. Int. J. Curr. Microbiol. App. Sci, 4(7), 577-584.

Sonune, N. (2021). *Microbes: a potential tool for bioremediation. Rhizobiont in bioremediation of hazardous waste*. Springer.

Sørensen, S. J., Bailey, M., Hansen, L. H., Kroer, N., & Wuertz, S. (2005). Studying plasmid horizontal transfer in situ: A critical review. *Nature Reviews. Microbiology*, *3*(9), 700–710. DOI: 10.1038/nrmicro1232 PMID: 16138098

Springer, A. (1993). Impact of present and future regulations on bioremediation. *Bioremediation Journal*, *1*(1), 1–12. DOI: 10.1007/BF01575586

Srinivasan, S., Bankole, P. O., & Sadasivam, S. K. (2022). Biodecolorization and degradation of textile azo dyes using *Lysinibacillus sphaericus* MTCC 9523. *Frontiers in Environmental Science*, *10*(October), 1–13. DOI: 10.3389/fenvs.2022.990855

Srinivasan, S., Jnana, A., & Murali, T. S. (2024). Modeling Microbial Community Networks: Methods and Tools for Studying Microbial Interactions. *Microbial Ecology*, *87*(1), 56. DOI: 10.1007/s00248-024-02370-7 PMID: 38587642

Stahel, W. R. (2016). The circular economy. *Nature*, *531*(7595), 435–438. DOI: 10.1038/531435a PMID: 27008952

Stams, A. J. M., & Plugge, C. M. (2021). Electron transfer in syntrophic communities. *Nature Reviews. Microbiology*, *19*(7), 457–469. DOI: 10.1038/nrmicro2166 PMID: 19609258

Stankiewicz, A. I., & Moulijn, J. A. (2000). Process intensification: Transforming chemical engineering. *Chemical Engineering Progress*, *96*(1), 22–34.

Stávková, J., & Maroušek, J. (2021). Novel sorbent shows promising financial results on P recovery from sludge water. *Chemosphere*, *276*, 130097. DOI: 10.1016/j.chemosphere.2021.130097 PMID: 33711796

Stockinger, P., Niederhauser, C., Farnaud, S., & Buller, R. (2025). Computational analysis reveals temperature-induced stabilization of FAST-PETase. *Computational and Structural Biotechnology Journal*, *27*, 969–977. DOI: 10.1016/j.csbj.2025.03.006 PMID: 40151525

Strong, L. C., Mctavish, H., Sadowsky, M. J., & Wackett, L. P. (2000). Field-scale remediation of atrazine-contaminated soil using recombinant Escherichia coli expressing atrazine chlorohydrolase. *Environmental Microbiology*, *2*(1), 91–98. DOI: 10.1046/j.1462-2920.2000.00079.x PMID: 11243266

Strong, P. J., Xie, S., & Clarke, W. P. (2015). Methane as a resource: Can the methanotrophs add value? *Environmental Science & Technology*, *49*(7), 4001–4018. DOI: 10.1021/es504242n PMID: 25723373

Suarez, C., Persson, F., & Hermansson, M. (2015). Predation of nitritation–anammox biofilms used for nitrogen removal from wastewater. *FEMS Microbiology Ecology*, *91*(11), 124. DOI: 10.1093/femsec/fiv124 PMID: 26472578

Sudhakar, P., Santosh, R., Asthalakshmi, B., Kumaresan, G., & Velraj, R. J. R. E. (2021). Performance augmentation of solar photovoltaic panel through PCM integrated natural water circulation cooling technique. *Renewable Energy*, *172*, 1433–1448. DOI: 10.1016/j.renene.2020.11.138

Sun, S., Wang, Y., He, B., Chen, J., Leng, F., & Luo, W. (2024). Comparative transcriptomics revealed the mechanism of Stenotrophomonas rhizophila JC1 response and biosorption to Pb2+. *Environmental Geochemistry and Health*, *46*(7), 1–18. DOI: 10.1007/s10653-024-02019-x PMID: 38849682

Szarka, A., Viktoryová, N., Horváth, G., Szalay, Z., Šimo, F., & Hrouzková, S. (2022). GC–MS methods for the evaluation of the performance of electrochemical water treatment for the degradation of pollutants from paint industry effluents. *Monatshefte für Chemie*, *153*(2), 161–169. DOI: 10.1007/s00706-022-02890-3

Tabacchioni, S., Passato, S., Ambrosino, P., Huang, L., Caldara, M., Cantale, C., Hett, J., Del Fiore, A., Fiore, A., Schlüter, A., Sczyrba, A., Maestri, E., Marmiroli, N., Neuhoff, D., Nesme, J., Sørensen, S. J., Aprea, G., Nobili, C., Presenti, O., & Bevivino, A. (2021). Identification of beneficial microbial consortia and bioactive compounds with potential as plant biostimulants for a sustainable agriculture. *Microorganisms*, *9*(2), 426. DOI: 10.3390/microorganisms9020426 PMID: 33669534

Takada, N., Kobayashi, T., & Oya, M. (2025). Interfacial Activity and Acute Toxicity of Surfactants to Aquatic Animals. *Journal of Water and Environment Technology*, *23*(3), 133–144. DOI: 10.2965/jwet.24-058

Talmage, S. S. (2020). *Environmental and human safety of major surfactants: alcohol ethoxylates and alkylphenol ethoxylates*. CRC Press. DOI: 10.1201/9780138757199

Tanasupawat, S., Takehana, T., Yoshida, S., Hiraga, K., & Oda, K. (2016). *Ideonella sakaiensis* sp. nov., isolated from a microbial consortium that degrades poly(ethylene terephthalate). *International Journal of Systematic and Evolutionary Microbiology, 66*(8), 2813–2818. DOI: 10.1099/ijsem.0.001058 PMID: 27045688

Tan, L., He, M., Song, L., Fu, X., & Shi, S. (2016). Bioresource Technology Aerobic decolorization, degradation and detoxification of azo dyes by a newly isolated salt-tolerant yeast *Scheffersomyces spartinae* TLHS-SF1. *Bioresource Technology, 203*, 287–294. DOI: 10.1016/j.biortech.2015.12.058 PMID: 26744802

Tan, L., Li, H., Ning, S., & Xu, B. (2014). Bioresource Technology Aerobic decolorization and degradation of azo dyes by suspended growing cells and immobilized cells of a newly isolated yeast *Magnusiomyces ingens* LH-F1. *Bioresource Technology, 158*, 321–328. DOI: 10.1016/j.biortech.2014.02.063 PMID: 24632410

Tao, S., Chen, H., Li, N., Wang, T., & Liang, W. (2022). The spread of antibiotic resistance genes *in vivo* model. *The Canadian Journal of Infectious Diseases & Medical Microbiology, 2022*(1), 3348695. DOI: 10.1155/2022/3348695 PMID: 35898691

Tariq, A., Guo, S., Farhat, F., & Shen, X. (2025). Engineering Synthetic Microbial Communities: Diversity and applications in soil for plant resilience. *Agronomy (Basel), 15*(3), 513. DOI: 10.3390/agronomy15030513

Tawfiq, R., Niu, K., Hoehndorf, R., & Kulmanov, M. (2024). DeepGOMeta for functional insights into microbial communities using deep learning-based protein function prediction. *Scientific Reports, 14*(1), 31813. DOI: 10.1038/s41598-024-82956-w PMID: 39738309

Techtmann, S. M., & Hazen, T. C. (2016). Metagenomics applications in environmental monitoring and bioremediation. *Journal of Industrial Microbiology & Biotechnology, 43*(10), 1345–1354. DOI: 10.1007/s10295-016-1809-8 PMID: 27558781

Thompson, I. P., Van Der Gast, C. J., Ciric, L., & Singer, A. C. (2005). Bioaugmentation for bioremediation: The challenge of strain selection. *Environmental Microbiology, 7*(7), 909–915. DOI: 10.1111/j.1462-2920.2005.00804.x PMID: 15946288

Tigini, V., Giansanti, P., Mangiavillano, A., Pannocchia, A., & Varese, G. C. (2011). Evaluation of toxicity, genotoxicity and environmental risk of simulated textile and tannery wastewaters with a battery of biotests. *Ecotoxicology and Environmental Safety, 74*(4), 866–873. DOI: 10.1016/j.ecoenv.2010.12.001 PMID: 21176963

Tikilili, P. V., & Chirwa, E. (2016). Searching for indigenous anaerobic ammonium oxidizing (anammox) bacteria in South African habitats: Pretoria region. *Biotechnology, Biotechnological Equipment*, *30*(6), 1097–1105. DOI: 10.1080/13102818.2016.1225515

Toczyłowska-Mamińska, R., Mamiński, M. Ł., & Kwasowski, W. (2025). Microbial Fuel Cell Technology as a New Strategy for Sustainable Management of Soil-Based Ecosystems. *Energies*, *18*(4), 970. DOI: 10.3390/en18040970

Toju, H., Abe, M. S., Ishii, C., Hori, Y., Fujita, H., & Fukuda, S. (2020). Scoring Species for Synthetic Community Design: Network Analyses of Functional Core Microbiomes. *Frontiers in Microbiology*, *11*(543143), 11–17. DOI: 10.3389/fmicb.2020.01361 PMID: 32676061

Tomaszewski, M., Cema, G., & Ziembińska-Buczyńska, A. (2017). Influence of temperature and pH on the anammox process: A review and meta-analysis. *Chemosphere*, *182*, 203–214. DOI: 10.1016/j.chemosphere.2017.05.003 PMID: 28499181

Tosi, S., Sfeir, C., Carnesecchi, E., & Chauzat, M. P. (2022). Lethal, sublethal, and combined effects of pesticides on bees: A meta-analysis and new risk assessment tools. *The Science of the Total Environment*, *844*, 156857. DOI: 10.1016/j.scitotenv.2022.156857 PMID: 35760183

Tournier, V., Topham, C. M., Gilles, A., David, B., Folgoas, C., Moya-Leclair, E., Kamionka, E., Desrousseaux, M. L., Texier, H., Gavalda, S., Cot, M., Guémard, E., Dalibey, M., Nomme, J., Cioci, G., Barbe, S., Chateau, M., André, I., Duquesne, S., & Marty, A. (2020). An engineered PET depolymerase to break down and recycle plastic bottles. *Nature*, *580*(7802), 216–219. DOI: 10.1038/s41586-020-2149-4 PMID: 32269349

Toussaint, P. A., Leiser, F., Thiebes, S., Schlesner, M., Brors, B., & Sunyaev, A. (2024). Explainable artificial intelligence for omics data: A systematic mapping study. *Briefings in Bioinformatics*, *25*(1), bbad453. DOI: 10.1093/bib/bbad453 PMID: 38113073

Towell, M. G., Paton, G. I., & Semple, K. T. (2011). The biodegradation of cable oil components: Impact of oil concentration, nutrient addition and bioaugmentation. *Environmental Pollution*, *159*(12), 3777–3783. DOI: 10.1016/j.envpol.2011.06.043 PMID: 21872976

Tripathi, M., Pandey, R., & Kumar, S. (2018). Biodecolorization of Orange II dye by native *Bacillus sp* and *Staphylococcus sp* in simulated medium. *Journal of Pharmacognosy and Phytochemistry*, •••, 1366–1368.

Tripathi, M., & Singh, D. N. (Eds.). (2022). *Bioremediation: challenges and advancements*. Bentham Science Publishers. DOI: 10.2174/97898150360391220101

Tripathi, M., Singh, D. N., Prasad, N., & Gaur, R. (2021). Advanced Bioremediation Strategies for Mitigation of Chromium and Organics Pollution in Tannery. In Kumar, V., Prasad, R., & Kumar, M. (Eds.), *Rhizobiont in Bioremediation of Hazardous Waste* (pp. 195–215). Springer., DOI: 10.1007/978-981-16-0602-1_10

Trump, B. D., Cummings, C. L., Klasa, K., Galaitsi, S. E., & Linkov, I. (2020). A decision analytic model to guide early-stage government regulatory action: Applications for synthetic biology oversight. *Regulation & Governance, 14*(2), 251–272. DOI: 10.1111/rego.12245

Tudi, M., Daniel Ruan, H., Wang, L., Lyu, J., Sadler, R., Connell, D., Chu, C., & Phung, D. T. (2021). Agriculture development, pesticide application and its impact on the environment. *International Journal of Environmental Research and Public Health, 18*(3), 1112. DOI: 10.3390/ijerph18031112 PMID: 33513796

Turco, F., Garavaglia, M., Houdt, R. V., Hill, P., Rawson, F. J., & Kovacs, K. (2022). Synthetic Biology Toolbox, Including a Single-Plasmid CRISPR-Cas9 System to Biologically Engineer the Electrogenic, Metal-Resistant Bacterium Cupriavidus metallidurans CH34. *ACS Synthetic Biology, 11*(11), 3617–3628. DOI: 10.1021/acssynbio.2c00130 PMID: 36278822

Tyagi, B., & Kumar, N. (2021). Bioremediation: Principles and applications in environmental management. In Bioremediation for Environmental Sustainability (pp. 3-28). Elsevier.

Tyagi, M., Da Fonseca, M. M. R., & De Carvalho, C. C. (2011). Bioaugmentation and biostimulation strategies to improve the effectiveness of bioremediation processes. *Biodegradation, 22*(2), 231–241. DOI: 10.1007/s10532-010-9394-4 PMID: 20680666

Tyagi, S., Kapoor, R. T., Singh, R., & Shah, M. P. (2025). Insights on microbial enzymes mediated biodegradation of azo dyes: A sustainable strategy for environment clean up. *Bioremediation Journal, 0*(0), 1–43. DOI: 10.1080/10889868.2025.2498695

U.S. Environmental Protection Agency (EPA). (2022). Overview of Biotechnology under the Toxic Substances Control Act (TSCA): Role of the office of pollution Prevention and Toxics (OPPT). U.S. EPA, office of Chemical Safety and Pollution Prevention. Retrieved from: https//www.epa.gov/regulation-biotechnology-under-tsca-and-fifra/overview-biotechnology-under-tsca

Ubani, O., Atagana, H. I., Thantsha, M. S. & Adeleke, R. 2016. Characterisation of oil degrading bacteria from tailored compost containing crude oil sludge.

Uddin, M. N., Tafsirojjaman, T., Shanmugasundaram, N., Praveenkumar, S., & Li, L. (2023). Smart self-healing bacterial concrete for sustainable goal. *Innovative Infrastructure Solutions*, *8*(1), 46. DOI: 10.1007/s41062-022-01020-6

Uddin, M., Islam, M. K., & Dev, S. (2024). Investigation of the performance of the combined moving bed bioreactor-membrane bioreactor (MBBR-MBR) for textile wastewater treatment. *Heliyon*, *10*(10), e31358. DOI: 10.1016/j.heliyon.2024.e31358 PMID: 38813235

Unal, Y. E., Gundogdu, H., Bilgin, O. N., Yokus, S., Turk, O. K., Yazici Guvenc, S., Can-Güven, E., & Varank, G. (2024). Effective treatment of textile industry wastewater by combined ultraviolet assisted and Fe2+ catalyzed percarbonate oxidation. *Journal of Dispersion Science and Technology*, *45*(8), 1585–1598. DOI: 10.1080/01932691.2023.2222823

United Nations Environment Programme (UNEP). (2019). Environmental Governance Update-December 2019. Retrieved from https://www.unep.org/resources/report/environmental-governance-update-december-2019

United Nations Environment Programme (UNEP). (2024). Why dirty air costs us trillions every year. Retrieved from https;//www.unep.org

United States Congress. (1976). Toxic Substances Control Act (TSCA), 15 U.S.C. § 2601–2629. U.S. Government Publishing Office. https://www.govinfo.gov/content/pkg/USCODE-2011-title15/pdf/USCODE-2011-title15-chap53.pdf

Upadhyay, S. K., Rani, N., Kumar, V., Mythili, R., & Jain, D. (2023). A review on simultaneous heavy metal removal and organo-contaminants degradation by potential microbes: Current findings and future outlook. *Microbiological Research*, *273*, 127419. DOI: 10.1016/j.micres.2023.127419 PMID: 37276759

Urbanek, A. K., Rymowicz, W., & Mirończuk, A. M. (2018). Degradation of plastics and plastic-degrading bacteria in cold marine habitats. *Applied Microbiology and Biotechnology*, *102*(18), 7669–7678. DOI: 10.1007/s00253-018-9195-y PMID: 29992436

Vaksmaa, A., Guerrero-Cruz, S., Ghosh, P., Zeghal, E., Hernando-Morales, V., & Niemann, H. (2023). Role of fungi in bioremediation of emerging pollutants. *Frontiers in Marine Science*, *10*, 1070905. DOI: 10.3389/fmars.2023.1070905

Valls, M., & De Lorenzo, V. (2002). Exploiting the genetic and biochemical capacities of bacteria for the remediation of heavy metal pollution. *FEMS Microbiology Reviews*, *26*(4), 327–338. DOI: 10.1016/S0168-6445(02)00114-6 PMID: 12413663

Van Niftrik, L., & Jetten, M. S. M. (2012). Anaerobic ammonium-oxidizing bacteria: Microorganisms with exceptional properties. *Microbiology and Molecular Biology Reviews : MMBR, 76*(3), 585–596. DOI: 10.1128/MMBR.05025-11 PMID: 22933561

Van Stempvoort, D., & Biggar, K. (2008). Potential for bioremediation of petroleum hydrocarbons in groundwater under cold climate conditions: A review. *Cold Regions Science and Technology, 53*(1), 16–41. DOI: 10.1016/j.coldregions.2007.06.009

Vanwonterghem, I., Jensen, P. D., Dennis, P. G., Hugenholtz, P., Rabaey, K., & Tyson, G. W. (2014). Deterministic processes guide long-term synchronised population dynamics in replicate anaerobic digesters. *The ISME Journal, 8*(10), 2015–2028. DOI: 10.1038/ismej.2014.50 PMID: 24739627

Varjani, S. J. (2017). Microbial degradation of petroleum hydrocarbons. *Bioresource Technology, 223*, 277–286. DOI: 10.1016/j.biortech.2016.10.037 PMID: 27789112

Varjani, S., & Upasani, V. N. (2019). Influence of abiotic factors, natural attenuation, bioaugmentation and nutrient supplementation on bioremediation of petroleum crude contaminated agricultural soil. *Journal of Environmental Management, 245*, 358–366. DOI: 10.1016/j.jenvman.2019.05.070 PMID: 31158688

Varner, P. M., Allemann, M. N., Michener, J. K., & Gunsch, C. K. (2022). The effect of bacterial growth strategies on plasmid transfer and naphthalene degradation for bioremediation. *Environmental Technology & Innovation, 28*, 102910. DOI: 10.1016/j.eti.2022.102910 PMID: 37091576

Vera, M. L., Traid, H. D., Dwojak, A. N., Rosenberger, M. R., & Schvezov, C. E. (2023). Advances in nanostructured TiO2 coatings for industrial applications. In *Industrial Applications of Nanoparticles* (pp. 228–255). CRC Press. DOI: 10.1201/9781003183525-14

Verdezoto-Prado, J., Chicaiza-Ortiz, C., Mejía-Pérez, A. B., Freire-Torres, C., Viteri-Yánez, M., Deng, L., Barba-Ostria, C., & Guamán, L. P. (2025). Advances in environmental biotechnology with CRISPR/Cas9: Bibliometric review and cutting-edge applications. *Discover Applied Sciences, 7*(3), 167. DOI: 10.1007/s42452-025-06609-x

Verlicchi, P., Al Aukidy, M., & Zambello, E. (2012). Occurrence of pharmaceutical compounds in urban wastewater: Removal, mass load and environmental risk after a secondary treatment—A review. *The Science of the Total Environment, 429*, 123–155. DOI: 10.1016/j.scitotenv.2012.04.028 PMID: 22583809

Vidanage, P. W. (2024). State of Art Technologies for Nutrient Removal from Wastewater. In Yadav, S., Ghangrekar, M. M., & Yadava, R. N. (Eds.), *Emerging Technologies in Biological and Hybrid Wastewater Treatment: Lessons from Developed to Enhancing Practices in Developing Countries*. Springer., DOI: 10.1007/978-3-031-74515-7_7

Viegas, C., & Gonçalves, M. (2021). Integrated treatment of pig production wastewaters using pre-treatment with biomass ash and bioremediation by microalgae. In *International Conference on Water Energy Food and Sustainability* (pp. 245-257). Cham: Springer International Publishing. DOI: 10.1007/978-3-030-75315-3_29

Vijayaraghavan, G., & Shanthakumar, S. (2020). Removal of crystal violet dye in textile effluent by coagulation using algal alginate from brown algae *Sargassum sp. Desalination and Water Treatment, 196*(September 2019), 402–408. DOI: 10.5004/dwt.2020.25569

Viktoryová, N., Szarka, A., & Hrouzková, S. (2022). Recent developments and emerging trends in paint industry wastewater treatment methods. *Applied Sciences (Basel, Switzerland), 12*(20), 10678. DOI: 10.3390/app122010678

Villaverde, J., Rubio-Bellido, M., Merchán, F., & Morillo, E. (2017). Bioremediation of diuron contaminated soils by a novel degrading microbial consortium. *Journal of Environmental Management, 188*, 379–386. DOI: 10.1016/j.jenvman.2016.12.020 PMID: 28011373

Villiger, L., Joung, J., Koblan, L., Weissman, J., Abudayyeh, O. O., & Gootenberg, J. S. (2024). CRISPR technologies for genome, epigenome and transcriptome editing. *Nature Reviews. Molecular Cell Biology, 25*(6), 464–487. DOI: 10.1038/s41580-023-00697-6 PMID: 38308006

Visvanathan, P., Kirubakaran, D., Selvam, K., Manimegalai, P., Rajkumar, M., & Vasantharaj, K. (2024). Recent strategies for natural bioremediation of emerging pollutants: Development of a green and sustainable environment. *ACS Symposium Series, 2*, 1–19. DOI: 10.1021/bk-2024-1476.ch001

Voit, E. O. (2022). Perspective: Systems biology beyond biology. *Frontiers in Systems Biology, 2*, 987135. DOI: 10.3389/fsysb.2022.987135

Wackett, L. P. (2022). The future of *Microbial Biotechnology. Microbial Biotechnology, 15*(1), 79–80. DOI: 10.1111/1751-7915.13920 PMID: 34612578

Waghmode, T. R., Kurade, M. B., & Govindwar, S. P. (2011). International Biodeterioration & Biodegradation Time dependent degradation of mixture of structurally different azo and non azo dyes by using *Galactomyces geotrichum* MTCC 1360. *International Biodeterioration & Biodegradation, 65*(3), 479–486. DOI: 10.1016/j.ibiod.2011.01.010

Wagner, M., & Loy, A. (2002). Bacterial community composition and function in sewage treatment systems. *Current Opinion in Biotechnology, 13*(3), 218–227. DOI: 10.1016/S0958-1669(02)00315-4 PMID: 12180096

Wanapaisan, P., Laothamteep, N., Vejarano, F., Chakraborty, J., Shintani, M., Muangchinda, C., Morita, T., Suzuki-Minakuchi, C., Inoue, K., Nojiri, H., & Pinyakong, O. (2018). Synergistic degradation of pyrene by five culturable bacteria in a mangrove sediment-derived bacterial consortium. *Journal of Hazardous Materials, 342*, 561–570. DOI: 10.1016/j.jhazmat.2017.08.062 PMID: 28886568

Wang, J. Y., & Doudna, J. A. (2023). CRISPR technology: A decade of genome editing is only the beginning. *science, 379*(6629), eadd8643.

Wang, X., Jiang, J., & Gao, W. (2022). *Reviewing textile wastewater produced by industries: characteristics, environmental impacts, and treatment strategies. 85*(7), 2076–2096. DOI: 10.2166/wst.2022.088

Wang, C., Sitong, L., Xu, X., & Zhao, C. (2017). Potential coupling effects of ammonia-oxidizing and anaerobic ammonium-oxidizing bacteria on completely autotrophic nitrogen removal over nitrite biofilm formation induced by the second messenger cyclic diguanylate. *Applied Microbiology and Biotechnology, 101*(9), 101. DOI: 10.1007/s00253-016-7981-y PMID: 28078398

Wang, F., Xiang, L., Sze-Yin Leung, K., Elsner, M., Zhang, Y., Guo, Y., Pan, B., Sun, H., An, T., Ying, G., Brooks, B. W., Hou, D., Helbling, D. E., Sun, J., Qiu, H., Vogel, T. M., Zhang, W., Gao, Y., Simpson, M. J., & Tiedje, J. M. (2024a). Emerging contaminants: A One Health perspective. *Innovation (Cambridge (Mass.)), 5*(4), 100612. DOI: 10.1016/j.xinn.2024.100612 PMID: 38756954

Wang, F., & Zhang, W. (2019). Synthetic biology: Recent progress, biosafety and biosecurity concerns, and possible solutions. *Journal of Biosafety and Biosecurity, 1*(1), 22–30. DOI: 10.1016/j.jobb.2018.12.003

Wang, H., Liu, X., Liu, S., Yu, Y., Lin, J., Lin, J., Pang, X., & Zhao, J. (2012). Development of a markerless gene replacement system for *Acidithiobacillus ferrooxidans* and construction of a pfkB mutant. *Applied and Environmental Microbiology, 78*(6), 1826–1835. DOI: 10.1128/AEM.07230-11 PMID: 22210219

Wang, H., & Ren, Z. J. (2014). Bio electrochemical metal recovery from wastewater: A review. *Water Research, 66*, 219–232. DOI: 10.1016/j.watres.2014.08.013 PMID: 25216302

Wang, S., Zhang, D., Wang, T., Wang, X., & Yuan, L. (2018). Preliminary performance and biofilm characteristics of Anammox process in an improved FBR with mixture of honeycomb-like and non-woven carriers. *Desalination and Water Treatment, 116*, 67–74. DOI: 10.5004/dwt.2018.22273

Wang, S., Zhan, Y., Jiang, X., & Lai, Y. (2024). Engineering microbial consortia as living materials: Advances and prospectives. *ACS Synthetic Biology, 13*(9), 2653–2666. DOI: 10.1021/acssynbio.4c00313 PMID: 39174016

Wang, T., Wang, X., Yuan, L., Luo, Z., & Indira, H. K. (2019). Start-up and operational performance of Anammox process in an anaerobic baffled biofilm reactor (ABBR) at a moderate temperature. *Bioresource Technology, 279*, 1–9. DOI: 10.1016/j.biortech.2019.01.114 PMID: 30710814

Wang, X.-W., Wang, T., & Liu, Y.-Y. (2024b). Artificial intelligence for microbiology and microbiome research. *Comprehensive Reviews in Microbiology, 18*(4), 601–629.

Wang, Y., Jiang, Q., Zhou, C., Chen, B., Zhao, W., Song, J., Fang, R., Chen, J., & Xiao, M. (2014). In-situ remediation of contaminated farmland by horizontal transfer of degradative plasmids among rhizosphere bacteria. *Soil Use and Management, 30*(2), 303–309. DOI: 10.1111/sum.12105

Wang, Y., Wang, H., Wang, X., Xiao, Y., Zhou, Y., Su, X., Cai, J., & Sun, F. (2020). Resuscitation, isolation and immobilization of bacterial species for efficient textile wastewater treatment: A critical review and update. *The Science of the Total Environment, 730*, 139034. DOI: 10.1016/j.scitotenv.2020.139034 PMID: 32416505

Wang, Z. (2022). *Geological and hydrgeological characteristics of acid mine drainage from an abandoned coal mine: A case study of Shandi coal mine in Niangziguan spring catchment.*

Wang, Z., Wang, S., He, Q., Yang, X., Zhao, B., Zhang, H., & Deng, Y. (2025). Ecological design of high-performance synthetic microbial communities: From theoretical foundations to functional optimization. *ISME Communications, 5*(1), ycaf133. Advance online publication. DOI: 10.1093/ismeco/ycaf133 PMID: 40860568

Wang, Z., Zheng, M., Duan, H., Yuan, Z., & Hu, S. (2022). A 20-year journey of partial nitritation and anammox (PN/A): From sidestream toward mainstream. *Environmental Science & Technology, 56*(12), 7522–7531. DOI: 10.1021/acs.est.1c06107 PMID: 35657148

Wani, A. K., Akhtar, N., Mir, T. U. G., Chopra, C., Singh, R., Hong, J. C., & Kadam, U. S. (2024). CRISPR/Cas12a-based biosensors for environmental monitoring and diagnostics. *Environmental Technology & Innovation*, *34*, 103625. DOI: 10.1016/j.eti.2024.103625

Wan, X., Volpetti, F., Petrova, E., French, C., Maerkl, S. J., & Wang, B. (2019). Cascaded amplifying circuits enable ultrasensitive cellular sensors for toxic metals. *Nature Chemical Biology*, *15*(5), 540–548. DOI: 10.1038/s41589-019-0244-3 PMID: 30911179

Weis, J. S. (2024). *Marine pollution*. Oxford University Press. DOI: 10.1093/wentk/9780197753804.001.0001

Weissman, J. L., Hou, S., & Fuhrman, J. A. (2021). Estimating maximal microbial growth rates from cultures, metagenomes, and single cells via codon usage patterns. *Proceedings of the National Academy of Sciences of the United States of America*, *118*(12), e2016810118. DOI: 10.1073/pnas.2016810118 PMID: 33723043

Wei, Y., Fu, J., Wu, J., Jia, X., Zhou, Y., Li, C., & Chen, F. (2018). Bioinformatics analysis and characterization of highly efficient polyvinyl alcohol (PVA)-degrading enzymes from the novel PVA degrader Stenotrophomonas rhizophila QL-P4. *Applied and Environmental Microbiology*, *84*(1), e01898–e17. DOI: 10.1128/AEM.01898-17 PMID: 29079625

Wesseler, J., Kleter, G., Meulenbroek, M., & Purnhagen, K. P. (2023). EU regulation of genetically modified microorganisms in light of new policy developments: Possible implications for EU bioeconomy investments. *Applied Economic Perspectives and Policy*, *45*(2), 839–859. DOI: 10.1002/aepp.13259

Whiteley, M., Diggle, S. P., & Greenberg, E. P. (2017). Progress in and promise of bacterial quorum sensing research. *Nature*, *551*(7680), 313–320. DOI: 10.1038/nature24624 PMID: 29144467

WHO. (2024). *Compendium of WHO and other UN guidance on health and environment: version with International Classification of Health Intervention (ICHI) codes*. World Health Organization.

Wierckx, N., Prieto, M. A., Pomposiello, P., de Lorenzo, V., O'Connor, K., & Blank, L. M. (2015). Plastic waste as a novel substrate for industrial biotechnology. *Microbial Biotechnology*, *8*(6), 900–903. DOI: 10.1111/1751-7915.12312 PMID: 26482561

Wieszczycka, K. (2018). Wastes generated by mineral extraction industries. *Physical Sciences Reviews*, *3*(6). Advance online publication. DOI: 10.1515/psr-2018-0026

Wongkiew, S., Hu, Z., Chandran, K., Lee, J. W., & Khanal, S. K. (2017). Nitrogen transformations in aquaponic systems: A review. *Aquacultural Engineering*, *76*, 9–19. DOI: 10.1016/j.aquaeng.2017.01.004

Wood, C. M., & Eom, J. (2021). The Osmo respiratory compromise in the fish gill. *Comparative Biochemistry and Physiology. Part A, Molecular & Integrative Physiology*, *254*, 110895. DOI: 10.1016/j.cbpa.2021.110895 PMID: 33429056

Wood, S., Biessy, L., Latchford, J. L., von Ammon, U., Laroche, O., Zaiko, A., & Pochon, X. (2019). Release and degradation of environmental DNA and RNA in a marine system. *The Science of the Total Environment*, *704*, 135314. DOI: 10.1016/j.scitotenv.2019.135314 PMID: 31780169

World Bank. (2025). Accelerating access to clean air for a livable planet (English). Washington, D. C., World Bank Group. https://documents.worldbank.org/curated/en/099032625132535486

World Health Organization (WHO). (2017). Internal Code of Conduct on Pesticide Mangement; Guidelines for the registration of microbial, botanical and semiochemical pest control agents for plant protection and public health uses. Retrieved from https://www.who.int/publications/i/item/WHO-HTM-NTD-WHOPES-2017.05

World Health Organization. (2020). *Laboratory biosafety manual* (4th ed.). WHO Press. https://www.who.int/publications/i/item/9789240011311

World Water Assessment Programme (WWAP)/UN-Water (2018). *The United Nations World Water Development Report 2018: Nature-based Solutions for Water*. Paris: UNESCO. Available at: https://www.unwater.org/publications/world-water-development-report-2018 (Accessed 7 Aug. 2019).

Wozniak, C. A., McClung, G., Gagliardi, J., & Matthews, K. (2012). Regulation of genetically engineered microorganisms under FIFRA, FFDCA and TSCA. In C. A. Wozniak, G. McClung, J. Gagliardi, and K. Matthews (Eds.), Regulation of agricultural biotechnology: The United States and Canada (pp. 57–94). Springer. DOI: 10.1007/978-94-007-2156-2_4

Wright, O., Stan, G. B., & Ellis, T. (2013). Building-in biosafety for synthetic biology. *Microbiology (Reading, England)*, *159*(Pt 7), 1221–1235. DOI: 10.1099/mic.0.066308-0 PMID: 23519158

Wu, K., Su, D., Liu, J., Saha, R., & Wang, J. P. (2019). Magnetic nanoparticles in nanomedicine: A review of recent advances. *Nanotechnology*, *30*(50), 502003. DOI: 10.1088/1361-6528/ab4241 PMID: 31491782

Xie, B. B., Li, M., Anantharaman, K., & Ravin, N. V. (2021). Editorial: the uncultured microorganisms: novel technologies and applications. *Frontiers in Microbiology, 12*, 756287. DOI: 10.3389/fmicb.2021.756287 PMID: 34899641

Xu, C., & Yu, H. (2021). Insights into constructing a stable and efficient microbial consortium. *Chinese Journal of Chemical Engineering, 30*, 112–120. 112-120.doi .org/10.1016/j.cjche.2020.12.012. DOI: 10.1016/j.cjche.2020.12.012

Xue, F., Yu, X., Shang, Y., Peng, C., Zhang, L., Xu, Q., & Li, A. (2020). Heterologous overexpression of a novel halohydrin dehalogenase from *Pseudomonas pohangensis* and modification of its enantioselectivity by semi-rational protein engineering. *International Journal of Biological Macromolecules, 146*, 80–88. DOI: 10.1016/j.ijbiomac.2019.12.203 PMID: 31891700

Xu, H., Chang, J., Wang, H., Liu, Y., Zhang, X., Liang, P., & Huang, X. (2019). Enhancing direct interspecies electron transfer in syntrophic-methanogenic associations with (semi) conductive iron oxides: Effects and mechanisms. *The Science of the Total Environment, 695*, 133876. DOI: 10.1016/j.scitotenv.2019.133876 PMID: 31756846

Xu, M., Guo, J., Zeng, G., Zhong, X., & Sun, G. (2006). Decolorization of anthraquinone dye by *Shewanella decolorationis* S12. *Applied Microbiology and Biotechnology, 71*(2), 246–251. DOI: 10.1007/s00253-005-0144-1 PMID: 16160829

Xu, Y., & Zhao, F. (2018). Single-cell metagenomics: Challenges and applications. *Protein & Cell, 9*(5), 501–510. DOI: 10.1007/s13238-018-0544-5 PMID: 29696589

Yadav, A. K., Banerjee, S. K., Das, B., & Chaudhary, K. (2022). Editorial: Systems Biology and Omics Approaches for Understanding Complex Disease Biology. *Frontiers in Genetics, 13*, 896818. DOI: 10.3389/fgene.2022.896818 PMID: 35495146

Yadav, A., & Yadav, K. (2025). Circular biorefineries for rural India: Turning rice straw and bagasse into biofuels. *Academia Green Energy, 2*(4). Advance online publication. DOI: 10.20935/AcadEnergy7949

Yadav, D., & Dutta, J. (2024). Advancing environmental sustainability: Recent trends and developments in treatment methods for paint industry wastewater. *Journal of Water Process Engineering, 61*, 105290. DOI: 10.1016/j.jwpe.2024.105290

Yalçın, S., & Özyürek, M. (2018). Biosorption potential of two brown seaweeds in the removal of chromium. *Water Science and Technology, 78*(12), 2564–2576. DOI: 10.2166/wst.2019.007 PMID: 30767921

Yamagishi, J. F., Saito, N., & Kaneko, K. (2020). Advantage of leakage of essential metabolites for cells. *Physical Review Letters*, *124*(4), 048101. DOI: 10.1103/PhysRevLett.124.048101 PMID: 32058757

Yang, G. F., & Jin, R. C. (2012). The joint inhibitory effects of phenol, copper (II), oxytetracycline (OTC) and sulfide on Anammox activity. *Bioresource Technology*, *126*, 187–192. DOI: 10.1016/j.biortech.2012.09.023 PMID: 23073107

Yang, J., Kim, B., Kim, G. Y., Jung, G. Y., & Seo, S. W. (2019). Synthetic biology for evolutionary engineering: From perturbation of genotype to acquisition of desired phenotype. *Biotechnology for Biofuels*, *12*(1), 113. DOI: 10.1186/s13068-019-1460-5 PMID: 31086565

Yang, Q., Li, C., Li, H., Li, Y., & Yu, N. (2009). Degradation of synthetic reactive azo dyes and treatment of textile wastewater by a fungi consortium reactor. *Biochemical Engineering Journal*, *43*(3), 225–230. DOI: 10.1016/j.bej.2008.10.002

Yang, X., López-Grimau, V., Vilaseca, M., & Crespi, M. (2020). Treatment of Textile Wastewater by CAS, MBR, and MBBR: A Comparative Study from Technical, Economic, and Environmental Perspectives. *Water (Basel)*, *12*(5), 1306. DOI: 10.3390/w12051306

Yaseen, Z. M., Melini Wan Mohtar, W. H., Homod, R. Z., Alawi, O. A., Abba, S. I., Oudah, A. Y., Togun, H., Goliatt, L., Ul Hassan Kazmi, S. S., & Tao, H. (2024). Heavy metals prediction in coastal marine sediments using hybridized machine learning models with metaheuristic optimization algorithm. *Chemosphere*, *352*, 141329. DOI: 10.1016/j.chemosphere.2024.141329 PMID: 38296204

Ye, X., Peng, T., Feng, J., Yang, Q., Pratush, A., Xiong, G., & Hu, Z. (2019). A novel dehydrogenase 17β-HSDx from Rhodococcus sp. P14 with potential application in bioremediation of steroids contaminated environment. *Journal of Hazardous Materials*, *362*, 170–177. DOI: 10.1016/j.jhazmat.2018.09.023 PMID: 30236938

Yin, X., Li, W., Zhang, M., Du, R., & Chen, C. (2025). Advances in anammox process for municipal wastewater treatment plants: A review focusing on applied study and full-scale implementation cases. *Critical Reviews in Environmental Science and Technology*, *55*(13), 1001–1024. DOI: 10.1080/10643389.2025.2483522

Yoshida, S., Hiraga, K., Takehana, T., Taniguchi, I., Yamaji, H., Maeda, Y., Toyohara, K., Miyamoto, K., Kimura, Y., & Oda, K. (2016). A bacterium that degrades and assimilates poly (ethylene terephthalate). *Science*, *351*(6278), 1196–1199. DOI: 10.1126/science.aad6359 PMID: 26965627

Yuan, G., Deng, S., Czajka, J. J., Dai, Z., Hofstad, B. A., Kim, J., & Pomraning, K. R. (2024). CRISPR-Cas9/Cas12a systems for efficient genome editing and large genomic fragment deletions in Aspergillus niger. *Frontiers in Bioengineering and Biotechnology, 12*, 1452496. DOI: 10.3389/fbioe.2024.1452496 PMID: 39479294

Yuan, K., & Nault, R. (2025). Application of a metabolic network-based graph neural network for the identification of toxicant-induced perturbations. *Toxicological Sciences : An Official Journal of the Society of Toxicology, 206*(1), 19–29. DOI: 10.1093/toxsci/kfaf065 PMID: 40445920

Yuan, Y., Guo, S., Li, F., Wu, B., Yang, X., & Li, X. (2016). Coupling electrokinetics with microbial biodegradation enhances the removal of cycloparaffinic hydrocarbons in soils. *Journal of Hazardous Materials, 320*, 591–601. DOI: 10.1016/j.jhazmat.2016.07.043 PMID: 27501882

Yu, L., Gong, S., & Liang, J. (2024). Assistance of aminated straw on enhanced biodegradation of phenol through bacteria and potential prediction of machine learning for bioremediation. *Water Science and Technology, 90*(7), 2174–2185. DOI: 10.2166/wst.2024.323

Zafar, H., Ishaq, S., Peleato, N., & Roberts, D. (2022). Meta-analysis of operational performance and response metrics of microbial fuel cells (MFCs) fed with complex food waste. *Journal of Environmental Management, 315*, 115152. DOI: 10.1016/j.jenvman.2022.115152 PMID: 35525044

Zawierucha, I., & Malina, G. (2006). Bioaugmentation as a method of biodegradation enhancement in oil hydrocarbons contaminated soil. *Ecohydrology & Hydrobiology, 6*(1-4), 163–169. DOI: 10.1016/S1642-3593(06)70138-0

Zawierucha, I., & Malina, G. (2011). *Bioremediation of contaminated soils: effects of bioaugmentation and biostimulation on enhancing biodegradation of oil hydrocarbons. Bioaugmentation, biostimulation and biocontrol.* Springer.

Zawierucha, I., & Malina, G. (2014). Application of Zeolite to Trace Metal Removal from Groundwater in Permeable Reactive Barrier Systems (PRB). *Ochrona Środowiska, 36*, 39–44.

Zelezniak, A., Andrejev, S., Ponomarova, O., Mende, D. R., Bork, P., & Patil, K. R. (2015). Metabolic dependencies drive species co-occurrence in diverse microbial communities. *Proceedings of the National Academy of Sciences of the United States of America, 112*(20), 6449–6454. DOI: 10.1073/pnas.1421834112 PMID: 25941371

Zhang, H., & Wang, X. (2016). Modular co-culture engineering, a new approach for metabolic engineering. *Metabolic Engineering*, *37*, 114–121. DOI: 10.1016/j.ymben.2016.05.007 PMID: 27242132

Zhang, J., Zhu, T., Meng, T., Zhang, Y., Yang, J., Yang, W., Müller, C., & Cai, Z. (2013). Agricultural land use affects nitrate production and conservation in humid subtropical soils in China. *Soil Biology & Biochemistry*, *62*, 107–114. DOI: 10.1016/j.soilbio.2013.03.006

Zhang, K., Wang, S., Guo, P., & Guo, S. (2021). Characteristics of organic carbon metabolism and bioremediation of petroleum-contaminated soil by a mesophilic aerobic biopile system. *Chemosphere*, *264*, 128521. DOI: 10.1016/j.chemosphere.2020.128521 PMID: 33039688

Zhang, L., Chen, F., Zeng, Z., Xu, M., Sun, F., Yang, L., Bi, X., Lin, Y., Gao, Y., Hao, H., Yi, W., Li, M., & Xie, Y. (2021). Advances in Metagenomics and Its Application in Environmental Microorganisms. *Frontiers in Microbiology*, *12*, 766364. DOI: 10.3389/fmicb.2021.766364 PMID: 34975791

Zhang, L., Chen, L., Diao, J., Song, X., Shi, M., & Zhang, W. (2020). Construction and analysis of an artificial consortium based on the fast-growing cyanobacterium Synechococcus elongatus UTEX 2973 to produce the platform chemical 3-hydroxypropionic acid from CO2. *Biotechnology for Biofuels*, *13*(1), 1–12. DOI: 10.1186/s13068-020-01720-0 PMID: 32391082

Zhang, L., Liu, H., & Chen, Q. (2021). Recent development of the ANAMMOX process in China: Challenges and opportunities. *Processes (Basel, Switzerland)*, *9*(7), 1207. DOI: 10.3390/pr9071207

Zhang, M., Tao, S., & Wang, X. (2020). Interactions between organic pollutants and carbon nanomaterials and the associated impact on microbial availability and degradation in soil: A review. *Environmental Science. Nano*, *7*(9), 2486–2508. DOI: 10.1039/D0EN00515K

Zhang, T., & Zhang, H. (2022). Microbial consortia are needed to degrade soil pollutants. *Microorganisms*, *10*(2), 261. DOI: 10.3390/microorganisms10020261 PMID: 35208716

Zhang, Y., Cao, B., Yin, H., Meng, L., Jin, W., Wang, F., & Al-Tabbaa, A. (2022). Application of zeolites in permeable reactive barriers (PRBs) for in-situ groundwater remediation: A critical review. *Chemosphere*, *308*, 136290. DOI: 10.1016/j.chemosphere.2022.136290 PMID: 36058373

Zhang, Y., Geng, Y., Li, S., Shi, T., Ma, X., Hua, R., & Fang, L. (2023). Efficient knocking out of the organophosphorus insecticides degradation gene opdB in *Cupriavidus nantongensis* X1T via CRISPR/Cas9 with red system. *International Journal of Molecular Sciences*, *24*(6), 6003. DOI: 10.3390/ijms24066003 PMID: 36983076

Zhao, S., Su, X., Xu, C., Gao, X., & Lu, S. (2025). Microbial adaptation and genetic modifications for enhanced remediation in low-permeability soils. *The Science of the Total Environment*, *958*, 177916. DOI: 10.1016/j.scitotenv.2024.177916 PMID: 39647202

Zhili, N., Lilin, Z., Ziyang, L., Bin, K., Gefu, Z., Jianbo, J., & Tao, W. (2023). Iron-modified biochar boosts anaerobic digestion of sulfamethoxazole pharmaceutical wastewater: Performance and microbial mechanism. *Journal of Hazardous Materials*, *452*, 131314. DOI: 10.1016/j.jhazmat.2023.131314 PMID: 37030222

Zhou, Y., & Zeeshan Ul Haq, M. (2024). Engineering of Synthetic Microbial Consortia for Sustainable Management of Wastewater and Polyethylene Terephthalate: A Comprehensive Review. *International Journal of Molecular Sciences*, *26*(23), 11623. DOI: 10.3390/ijms262311623 PMID: 41373776

Zhuang, M., Sanganyado, E., Zhang, X., Xu, L., Zhu, J., Liu, W., & Song, H. (2020). Azo dye degrading bacteria tolerant to extreme conditions inhabit nearshore ecosystems: Optimization and degradation pathways. *Journal of Environmental Management, 261*(October 2019), 110222. DOI: 10.1016/j.jenvman.2020.110222

Ziesack, M., Gibson, T., Oliver, J. K., Shumaker, A. M., Hsu, B. B., Riglar, D. T., Giessen, T. W., DiBenedetto, N. V., Bry, L., Way, J. C., Silver, P. A., & Gerber, G. K. (2019). Engineered interspecies amino acid cross-feeding increases population evenness in a synthetic bacterial consortium. *mSystems*, *4*(4), 10–1128. DOI: 10.1128/mSystems.00352-19 PMID: 31409662

Zimmermann, W., & Dombrowski, A. (2020). Enzymatic degradation and recycling of poly(ethylene terephthalate): A critical review. In Kabasci, S. (Ed.), *Bio-based Plastics: Materials and Applications* (pp. 227–242). John Wiley & Sons., DOI: 10.1002/9781119592525.ch10

Zuo, D., Han, Y., Gao, X., Ma, G., Xu, Z., Bi, Y., Abbaspour, K. C., & Yang, H. (2022). Identification of priority management areas for non-point source pollution based on critical source areas in an agricultural watershed of Northeast China. *Environmental Research*, *214*(2), 113892. DOI: 10.1016/j.envres.2022.113892 PMID: 35863450

Zupanic, A., Bernstein, H. C., & Heiland, I. (2020). Systems biology: Current status and challenges. *Cellular and Molecular Life Sciences : CMLS*, *77*(3), 379–380. DOI: 10.1007/s00018-019-03410-z PMID: 31932855

About the Contributors

Olubukola Oluranti Babalola (Pr.Sci.Nat, MASSAF, FTWAS, FAS, FAAS, MAAAS, FISC, FRSB, FAAM) is the Vice President of The World Academy of Sciences (TWAS, Africa region), the Foreign Secretary of the Nigerian Academy of Sciences, quondam Vice President of the Organization for Women in Science for the Developing World (OWSD), former Head of Department and the Director of Research at the North-West University. She is a member of the Think Tank for The Forum for Agricultural Research in Africa (FARA) and an Honorary Visiting Professor at Imperial College London, contributing to global research agendas, policy dialogue, and interdisciplinary collaboration. She is an AAAS-TWAS Science diplomacy alumnus, impacting the world after the eye-opening awareness of policymaking, diplomacy, and diplomatic engagement. She is taking action using the gender lens and forging collaborations with diplomats. She holds advanced academic qualifications culminating in a professorship in Plant Health Management and an MBA in General Management, reflecting a rare integration of scientific depth and strategic leadership. Her leadership vision centers on excellence, inclusivity, and translating science into societal benefit. With an h-index of over 90, Prof Babalola ranks among the most influential scientists in agricultural and environmental microbiology. Her research has fundamentally advanced the understanding of plant-microbe interactions, microbial ecology, biofertilizers, biocontrol, and climate-resilient agriculture, shaping sustainable food systems, particularly in Africa and the Global South. She is a distinguished Principal Investigator and mentor, and has supervised 39 PhD and 29 MSc graduates, built a vibrant, high-impact research ecosystem and trained the next generation of scientists. Her scholarly output also includes books and book chapters, underscoring her commitment to knowledge dissemination and capacity building. Prof Babalola is a prolific author with over 450 publications. She serves on the editorial boards of Elsevier, Springer, SAGE, and the American Society for Microbiology (ASM), and is an International Advisor to several others. Her efforts have received many awards, including being a finalist in

GenderInSite 2020, a place in Clarivate's Highly Cited Research Top 1% for the field in 2022 and 2023, and, notably, in 2024, the South African Women in Science Awards (SAWiSA), celebrating her as a trailblazing woman in STEM. In 2025, she rose to an NRF B-rating, and became an internationally acclaimed scientist. Prof Babalola is #1 in Africa for Soil Science and Plant Nutrition. Through her enduring professional trajectory, she continues to redefine scientific leadership, leaving a lasting, transformative imprint on the global STEM landscape.

Olukayode Amund is a reputable Microbiologist specializing in Petroleum and Environmental Microbiology. He earned his first degree in Biology graduating in First Class (Hons.) from the University of Lagos in 1977 and a Doctorate degree in Biotechnology from Cranfield University, UK, in 1982. He began his academic career as Graduate Assistant at the University of Lagos in 1978 and rose through the ranks to become a Professor in 1998. He held several leadership positions as Head, Department of Botany and Microbiology, Dean, Faculty of Science, Dean of Student Affairs, Director of the Distance Learning Institute and Member of the Governing Council at the University of Lagos. He was Foundation Vice-Chancellor of Wesley University of Science and Technology, Ondo and later served as the third substantive Vice-Chancellor of Elizade University, Ilara-Mokin, Ondo State, Nigeria. He is a Fellow of the Nigerian Academy of Science (FAS), African Academy of Sciences (FAAS), Institute of Policy Managemnt Development (FIPMD), Royal Society of Biology (FRSB) and African Scientific Institute (FASI). He has published over 150 scientific articles in learned journals, contributed book chapters and presented numerous articles at conferences, while 38 publications were contributed to the database of National Center for Biotechnology Information (NCBI). Emeritus Prof Amund is affiliated with the Department of Microbiology, University of Lagos from where he officially retired in 2023. He secured several research grants from International and local agencies and provided consultancy services to industry. He has received numerous awards for his contributions to science, industry and the society. He is a co-Editor to the book Ecofriendly Frontiers:Harnessing Microbial Applications for Food Security published by Springer in 2025. He is an Alumnus of the Galilee International Management Institute, Nahalal, Galilee, Israel.

Olayinka Osuolale embarked on his academic journey with a PhD in Microbiology from the University of Fort Hare in South Africa, further enriching his expertise through postgraduate studies at Loughborough University in the United Kingdom. His research delves deeply into the complex intersections of infectious diseases and environmental health. Dr. Osuolale's expertise spans critical areas such as Leishmaniasis and Dengue virus, antimicrobial resistance, healthcare-associated waterborne infections, microbial water quality monitoring, and the broader field of environmental microbiology and public health. Since 2016, Dr. Osuolale has been a dedicated member of Elizade University's faculty, progressing from Lecturer I to Reader by 2025. Throughout his tenure, he has actively fostered collaborations with both local and international research institutions, championing interdisciplinary approaches to tackle microbial and environmental challenges. Dr. Osuolale's scholarly work has garnered significant recognition, with over 8,661 citations, an h-index of 20, and an i10-index of 22, according to Google Scholar.

***

Rosemary Agbeko is a microbiologist and emerging researcher with a strong interest in environmental microbiology, molecular biology and antibiotic resistance. Rosemary is passionate about advancing sustainable environmental research and contributing to global efforts aimed at mitigating antimicrobial resistance.

Jerome Bidu is a Lecturer in the Department of Mechanical and Industrial Engineering at the University of Dar es Salaam. He earned his PhD in Chemical Engineering from KU Leuven, Belgium, in 2023. Prior to that, he completed a master's degree in Textile Trade and Technology Management, specializing in Technical Textiles, at Niederrhein University of Applied Sciences, Germany, in 2013. He also holds a BSc in Chemical and Process Engineering from the University of Dar es Salaam, awarded in 2010. Dr. Jerome has a strong track record in developing sustainable solutions for industrial and commercial applications, supported by extensive academic research and hands-on consultancy in textiles, chemical engineering, and process engineering. His research has focused on various aspects, including natural and synthetic fibers, as well as sustainable methods for industrial and domestic wastewater treatment. He has also explored the use of biodegradable materials in textiles and packaging, contributing to sustainability within the circular economy. Currently, his research is centered on the valorization of pre-and post-consumer waste to promote circularity. As a textile expert, chemical and process engineer, and academic, Dr. Jerome has published extensively in areas such as textile functionalization, wastewater treatment, biomaterials, and nanotechnology. He served as a co-principal investigator in the project "Food Packaging Materials from Agricultural Biomass." Dr. Jerome has also participated in several specialized training programs, including an advanced course in sustainable packaging at the Indian Institute of Packaging (IIP) in Delhi, India, from January 3 to 13, 2025, and a training program on Food Security and Climate Change in Dar es Salaam from July 19 to 31, 2015.

Chioma Chikere, a proudly Nigerian and internationally acclaimed scholar is a Professor of Environmental Microbiology and Biotechnology at the University of Port Harcourt (UniPort), Nigeria, where she also served as Head, Department of Microbiology, Faculty of Science from May 2024 to May 2026. Her leadership extends across borders as a Professor Extraordinarius at the University of South Africa (Unisa) and a visiting professor at the World Bank Africa Centre of Excellence in Oilfield Chemicals Research (ACE-CEFOR). With over 80 peer-reviewed publications, more than 80 conference presentations, and a robust portfolio of global collaborations, including the Erasmus+ AMIGO mobility agreement, Prof. Chikere is a powerhouse of innovation and impact. Her award-winning project under the Elsevier Foundation's 2017 Chemistry for Climate Action Challenge restored biodiversity in a heavily oil-polluted farmland in Nigeria's Niger Delta. Read more about her project here: https://elsevierfoundation.org/gscc-5th-anniversary-interviews-with-past-winners-series-1-dr-chioma-blaise-chikere/ Prof. Chikere is a passionate advocate for the African Union Agenda 2063, women's and girls' empowerment, and the United Nations Sustainable Development Goals (UN SDGs), making her a quintessential role model for other women for the advancement of diversity, equity, inclusivity and belonging (DEIB) in higher education leadership and research.

Kawther Ibrahem is a Lecturer in the Department of Soil and Water Resources at the University of Kirkuk, Iraq. She obtained her Ph.D. in Microbiology from Mustansiriyah University, Baghdad, in 2016. Her research focuses on microbiology, probiotics, nanobiotechnology, and the biosynthesis of metal nanoparticles using beneficial bacteria. Dr. Ibrahem has published several scientific articles on the biosynthesis, characterization, and antimicrobial applications of titanium dioxide and other nanoparticles. Her work contributes to the development of eco-friendly nanomaterials for biomedical and environmental applications.

Funso Kutu is a practicing agronomist with over 32 years of research and teaching experience focused on agronomy of grain legumes, particularly, cowpea.

Obinna Nwinyi is currently a rated established researcher in Nigeria. He is currently a Professor of Microbiology and Head of Department, Biological Sciences, Covenant University, Nigeria. In year 2021, He won the 2nd National prize for the Polymer Institute of Nigeria / Raw Materials Research and Development Council (RMRDC) Institutional award in Nigeria; In 2015 he won the Professor Oyewale Tomori National Prize for Young Scientist in Microbiology in Nigeria. In 2010, he was awarded an Honorary Citizen of the great State of Nebraska awarded by the Governor of Nebraska, United States of America. In addition, he won the First

Place Position/Certificate of Achievement by the Director, Office of Academic Exchange Program, Bureau of Educational and Cultural Affairs, United States Department of State. Before these, he won and completed the competitive Global Stature Postdoctoral Fellowship of the University of Johannesburg, South Africa. He successfully conducted Fulbright Foundation-funded PhD research at the School of Public and Environmental Affairs (SPEA) at Indiana University, Bloomington, United States of America. He has received many other certifications from trainings within Nigeria and other part of the world. He received his MSc degree in Microbiology from the University of Lagos, Nigeria and his B.Sc degree in Microbiology from Nnamdi Azikiwe University, Awka, Nigeria. He has served as Ag. Dean, School of Postgraduate Studies, Covenant University; Associate Dean, School of Postgraduate Studies and Deputy-Director Research and Development Covenant University, Ota. He has served as internal and external examiner/assessor of postgraduate dissertations and Thesis, and Assessor for professorial appointment in Nigeria and other parts of the world. Prof Obinna has 19 years of teaching experience in teaching Microbiology in different University settings. He has widely conducted and gained research experience in different universities in Africa and America. He has also won individual and prestigious collaborative international research grants. He has published over 100 articles in many prestigious national and international journals. Has authored over two hundred and sixty (260) new genomic sequences of Bacteria and Fungi under the United States National Centre for Biotechnology Information. He has been actively involved in research through self-initiation and collaboration with other researchers from Nigerian Universities and foreign Universities. Obinna's research interests include Environmental Microbiology-Bioremediation and Biodegradation, Petroleum Microbiology, Molecular Biology and Bioinformatics and sustainable environment (Soil, water/wastewater treatment, pollution reduction), Biodegradation and Bioremediation of Organic pollutants, and public health research. He is also involved with teaching and research supervision at both undergraduate and postgraduate levels.

Oladele Ogunseitan, PhD, MPH, is a Distinguished Professor and University of California Presidential Chair at the University of California, Irvine. He serves as Professor of Population Health and Disease Prevention in the UCI Program in Public Health, where he was founding chair of the Department of Population Health & Disease Prevention and led the establishment of UCI's public health program. publichealth.uci.eduBorn in Nigeria, Ogunseitan earned his PhD in microbiology from the University of Tennessee, Knoxville, where he discovered transducing bacteriophage in natural environments. He also holds an MPH in environmental health sciences from UC Berkeley. His research focuses on the environmental and human health effects of industrial development, with expertise in toxic environmental pollution, microbial diversity, ecology, antimicrobial resistance, and e-waste impacts. Ogunseitan is an elected Fellow of the American Association for the Advancement of Science in both Medical Sciences and Societal Impacts of Science and Technology, the American Academy of Microbiology, Collegium Ramazzini, and the British Royal Society of Arts, Manufactures and Commerce. He was a Jefferson Science Fellow with the U.S. Department of State and received a meritorious honor award for contributions to the third UN Environment Assembly. Beyond research, he co-directs the World Institute for Sustainable Development of Materials and directs education initiatives for the UC Global Health Institute. He leads workforce training for USAID's One Health Workforce project across 17 countries and co-chairs Apple Inc.'s Green Chemistry Advisory Board. Ogunseitan is author of Microbial Diversity and editor of Green Health, with publications in Science, Nature, and The Lancet Global Health.

Elijah Ohimain is Professor of Bioenergy, Agricultural and Environmental Microbiology at Niger Delta University, Wilberforce Island, Bayelsa State, Nigeria. He previously served as Head of the Department of Biological Sciences at the university. He holds a PhD in Environmental Microbiology from the University of Benin, Nigeria, and a Postgraduate Diploma in Sustainable Development from Staffordshire University, UK. His research focuses on bioenergy, veterinary and agricultural microbiology, environmental science, and petroleum microbiology, with particular emphasis on the Niger Delta region. His work addresses critical issues such as biofuel development, cassava-based ethanol production, methanol contamination in fermented beverages, and the environmental impacts of cassava mill effluents and crude oil pollution. He has published extensively on microbial ecology, environmental biotechnology, and public health microbiology, with over 250 publications to his name. He is an editorial board member for over 10 academic journals and serves as a reviewer for more than 100 journal titles. His research on topics like small-scale ethanol production from cassava and methanol contamination in indigenous fermented alcoholic beverages has contributed to policy discussions on renewable energy and food safety in Nigeria. Prof. Ohimain is recognized for his contributions to environmental and agricultural microbiology in Nigeria and remains active in mentoring postgraduate students and advancing sustainable biotechnology solutions for the Niger Delta

Olubenga Oluwagbemi has over 18 years of teaching experience teaching Computer Science and Information Technology courses in different university settings. He has won some prestigious individual and collaborative international research grants. He has published in many prestigious international journals and has been the first author and or corresponding author in many of his scientific research publications. He has over 10 years research experience at various universities in the United Kingdom, USA, South Africa, and Nigeria. Dr. Oluwagbemi's research interests include Artificial Intelligence (AI) in Health; Robotics; Computational modelling in health; Applied Health Informatics; AI in biomedical data science; Modelling the impact of climate change effects on the proliferation of infectious and non-infectious diseases; Inclusive; Bioinformatics (Computational Genomics); Informatics in medicine; Computer-Aided Diagnosis; Applied AI/Data Science; Applied Machine learning. Dr. Oluwagbemi has a strong publication record, extensive international research experience, and has received several prestigious grants.

Index

A

B

C

E

G

H

I

M

N

O

P

R

S

T

W

X

www.ingramcontent.com/pod-product-compliance
Lightning Source LLC
LaVergne TN
LVHW062335300926
849161LV00007B/1698